Problem Books in Mathematics

Series Editor

Peter Winkler
Department of Mathematics
Dartmouth College
Hanover, NH
USA

More information about this series at http://www.springer.com/series/714

Dorin Andrica • Ovidiu Bagdasar

Recurrent Sequences

Key Results, Applications, and Problems

Dorin Andrica
Department of Mathematics
"Babeş-Bolyai" University
Cluj-Napoca, Romania

Ovidiu Bagdasar
College of Engineering and Technology
University of Derby
Derby, UK

ISSN 0941-3502 ISSN 2197-8506 (electronic)
Problem Books in Mathematics
ISBN 978-3-030-51501-0 ISBN 978-3-030-51502-7 (eBook)
https://doi.org/10.1007/978-3-030-51502-7

Mathematics Subject Classification: 05A18, 11B39, 11B83, 11D45, 11N37, 11N56, 11N64, 11Y55, 11Y70, 40A05, 40A10

This Springer imprint is published by the registered company Springer Nature Switzerland AG
The registered company address is: Gewerbestrasse 11, 6330 Cham, Switzerland

“Your days increase with each Tomorrow” **(“Cu mâine zilele-ţi adaogi”)**

by Mihai Eminescu

Your days increase with each Tomorrow,
Your life grows less with Yesterday,
In front of you there lies, however,
For all of the eternity: Today.

These famous verses of the national poet of the Romanians capture the close connections between poetry and mathematics. Indeed, if D_n denotes the number of days in someone’s life, then the equation of life described by the poet can be encoded in the following well-known recurrence:

$$D_{n+1} - D_{n-1} = D_n,$$

which coincides with the recurrence satisfied by the Fibonacci numbers.

Preface

Overview and Goals

This book presents the state-of-the-art results concerning recurrent sequences and their practical applications in algebra, number theory, geometry of the complex plane, discrete mathematics, or combinatorics. The purpose of this book is to familiarize more readers with recent developments in this area and to encourage further research.

The content of the book is recent and reflects current research in the field of recurrent sequences. Some new approaches promoted by this book are visual representations of recurrences in the complex plane and the use of multiple methods for deriving novel identities or number sequences.

The first part of the book is dedicated to fundamental results and key examples of recurrences and their properties. We also present the geometry of linear recurrences in the complex plane in some detail. Here, some recent research has led to unexpected developments within combinatorics, number theory, integer sequences, and random number generation.

The second part of the book presents 123 olympiad training problems with full solutions and some appendices. These relate to linear recurrences of first, second, and higher orders, classical sequences, homographic recurrences, systems of recurrences, complex recurrent sequences, and recurrent sequences in combinatorics.

Audience

This book will be useful for researchers and scholars who are interested in recent advances in the field of recurrences, postgraduate students in college or university and their instructors, or advanced high school students.

Students training for mathematics competitions and their coaches will find numerous worked examples and problems with detailed solutions. Many of these problems are original, while others are selected from international olympiads or from various specialist journals.

Organization and Features

The book is organized in eight chapters and an appendix. The first six chapters are dedicated to theoretical results, examples, and applications, while the last two contain olympiad problems accompanied by full solutions.

Chapter 1 presents the fundamental aspects concerning recurrence relations, such as explicit and implicit forms, order, systems of recurrent sequences, or existence and uniqueness of the solution. The chapter also presents some recurrence sequences arising in mathematical modeling, algebra, combinatorics, geometry, analysis, or iterative numerical methods.

Chapter 2 is dedicated to first- and second-order linear recurrences with general coefficients, various classical sequences and polynomials (including Fibonacci, Lucas, Pell, or Lucas–Pell), and homographic recurrences defined by linear fractional transformations in the complex plane.

Chapter 3 presents the arithmetic properties and trigonometric formulae for classical recurrent sequences, extending some results for Fibonacci and Lucas numbers.

Chapter 4 is dedicated to ordinary and exponential generating functions for classical functions and polynomials and presents numerous new results. It also presents some new useful versions of Cauchy's integral formula.

Chapter 5 explores the dynamics of second-order linear recurrences in the complex plane, referred to as Horadam sequences. We first formulate periodicity conditions, which are then used to investigate the geometric structure and the number of periodic Horadam patterns. An atlas of non-periodic Horadam patterns is also presented, followed by an application to pseudo-random number generators. We conclude with some examples of periodic nonhomogeneous Horadam sequences.

Chapter 6 presents the dynamics of complex linear recurrent sequences of higher order and investigates the periodicity, geometric structure, and enumeration of the periodic patterns. Some results concerning systems of linear recurrence sequences are also analyzed, together with applications to Diophantine equations. An atlas of complex linear recurrent patterns (both periodic and non-periodic) for third-order recurrences is also discussed, along with connections to finite differences.

Chapter 7 contains 123 olympiad training problems involving recurrent sequences, solved in detail in Chapter 8. Many problems are original and concern linear recurrence sequences of first, second, and higher orders, some classical sequences, homographic sequences, systems of sequences, complex recurrence sequences, and recursions in combinatorics.

Prerequisites

Our intention was to make the book as self-contained as possible, so we have included many definitions together with examples. Chapters 1, 2, and 5 only require good understanding of college algebra, complex numbers, analysis, and basic combinatorics. For Chapters 3, 4, and 6, the prerequisites include number theory, linear algebra, and complex analysis.

How to Use the Book

The book presents recurrent sequences in connection with a wide range of topics and is illustrated with numerous examples and diagrams.

The introductory chapter and some of the properties of first- and second-order linear recurrences (which include classical number sequences) are elementary and can be understood by high school students.

More advanced topics such as homographic recurrences, generating functions, higher-order recurrent sequences, or systems of recurrent sequences require concepts usually covered in college or undergraduate courses.

We have also included many original and recent results on arithmetic and trigonometric properties of classical sequences, new applications of Cauchy's integral formula in the study of polynomial coefficients, or the complex recurrent patterns applied in pseudo-random number generation, which expose the reader to the state of the art in the field.

The collection of problems with full solutions illustrates the wide range of topics where recurrent sequences can be found and represents an ideal material for preparing students for Olympiads.

The book also includes 177 references and an index which will help the readers to further investigate key notions and concepts.

Acknowledgements

We would like to thank Dr George Cătălin Ţurcaş and Remus Mihăilescu, for checking the document carefully and providing feedback.

We are indebted to the anonymous referees for their constructive remarks and suggestions, which helped us to improve the quality of the manuscript.

Special thanks to our families, for their continuous and discrete support.

Cluj-Napoca, Romania
Derby, UK
May 2020

Dorin Andrica
Ovidiu Bagdasar

Contents

Chapter 1
Introduction to Recurrence Relations

In this chapter we present fundamental concepts and motivating examples of recurrent sequences, and show connections of recurrence relations to mathematical modeling, algebra, combinatorics, and analysis. There are numerous sources presenting the classical theory (see, e.g., [41, 62, 128]).

Let X be an arbitrary set. A function $f : \mathbb{N} \to X$ defines a *sequence* $(x_n)_{n\geq 0}$ of elements of X, where $x_n = f(n)$, $n = 0, 1, \ldots$. The set of all sequences with elements in X is denoted by $X^{\mathbb{N}}$, while X^n denotes the Cartesian product of n copies of X. In practice, X will be chosen as $\mathbb{C}$, the Euclidean space $\mathbb{R}^m$, the algebra $M_r(A)$ of the $r \times r$ matrices with entries in a ring A, etc.

The set $X^{\mathbb{N}}$ has numerous important subsets. For instance, when $X = \mathbb{R}$, the set of real numbers $\mathbb{R}^{\mathbb{N}}$ includes sequences which are bounded, monotonous, convergent, positive, nonzero, periodic, etc.

When $a \in X$ is fixed, in *implicit form*, a recurrence relation is defined by

$$F_n(x_n, x_{n-1}, \ldots, x_0) = a, \quad n = 1, 2, \ldots, \tag{1.1}$$

where $F_n : X^{n+1} \to X$ is a function of $n+1$ variables, $n = 0, 1, \ldots$. Notice that, in general, the implicit form of a recurrence relation does not define uniquely the sequence $(x_n)_{n\geq 0}$.

The *explicit form* of a recurrence relation is

$$x_n = f_n(x_{n-1}, \ldots, x_0), \quad n = 1, 2, \ldots, \tag{1.2}$$

where $f_n : X^n \to X$ is a function for all $n \geq 1$.

The relations (1.2) give the rule to construct the term x_n of the sequence $(x_n)_{n\geq 0}$ from the first term x_0. We have $x_1 = f_1(x_0)$, $x_2 = f_2(x_1, x_0)$, $\ldots$, i.e., the relation (1.2) is a functional type relation.

D. Andrica, O. Bagdasar, *Recurrent Sequences*, Problem Books in Mathematics,
https://doi.org/10.1007/978-3-030-51502-7_1

1.1 Recursive Sequences of Order k

Let k be a positive integer and X. A kth order recurrence relation is written

- implicitly, for $a \in X$ and $F_n : X^{k+1} \to X$, with $n = k, k+1, \ldots,$ as

$$F_n(x_n, x_{n-1}, \ldots, x_{n-k}) = a; \tag{1.3}$$

- explicitly, for $f_n : X^k \to X, n = k, k+1, \ldots,$ as

$$x_n = f_n(x_{n-1}, \ldots, x_{n-k}). \tag{1.4}$$

Sometimes, for $n = 0, 1, \ldots,$ these are also written in the forms

$$F_n(x_{n+k}, x_{n+k-1}, \ldots, x_n) = a, \tag{1.5}$$

$$x_{n+k} = f_n(x_{n+k-1}, \ldots, x_n). \tag{1.6}$$

If $f_n \in \mathbb{C}[u_1, \ldots, u_k]$ is a polynomial of degree 1 in k variables, defined by $f_n(u_1, \ldots, u_k) = a_n^1 u_1 + \cdots + a_n^k u_k + b_n$, where the sequences of complex numbers $(a_n^j)_{n\geq 0}$, $j = 1, \ldots, k$ and $(b_n)_{n\geq 0}$ are given, then the recurrence relation (1.6) is called *linear* and it is written as

$$x_{n+k} = a_n^1 x_{n+k-1} + \cdots + a_n^k x_n + b_n, \quad n = 0, 1, \ldots. \tag{1.7}$$

In particular, when sequences $(a_n^j)_{n\geq 0}$ are constant, i.e., $a_n^j = a_j$, $j = 1, \ldots, k$, and $b_n = 0$, the recurrence relation is called a *homogeneous*, kth order linear recurrence relation with constant coefficients, and can be written as

$$x_{n+k} = a_1 x_{n+k-1} + \cdots + a_k x_n, \quad n = 0, 1, \ldots. \tag{1.8}$$

The *general solution* of the recurrence relations (1.3) or (1.4) consists of the set of all sequences $(x_n)_{n\geq 0}$ satisfying the recurrence relation.

Moreover, if the sequence $(x_n)_{n\geq 0}$ satisfying the recurrence equations (1.3) or (1.4) also satisfies the initial conditions

$$x_0 = \alpha_0, \; x_1 = \alpha_1, \; \ldots, x_{k-1} = \alpha_{k-1},$$

with $\alpha_0, \ldots, \alpha_{k-1}$ fixed complex numbers, then the sequence $(x_n)_{n\geq 0}$ is called a *particular solution* of the recurrence relation.

A *first-order* linear nonhomogeneous recurrence sequence is defined by

$$x_{n+1} = a_n x_n + b_n, \quad n = 0, 1, \ldots, \tag{1.9}$$

while the homogeneous sequence has the form $x_{n+1} = a_n x_n$, $n = 0, 1, \ldots.$

In the following chapters we shall discuss in more detail about linear recursive sequences of first, second, and arbitrary order, as well as about some other important nonlinear sequences and their applications.

1.2 Recurrent Sequences Defined by a Sequence of Functions

Let x_0 be a point in X and consider $f_n : X \to X$, $n = 1, 2, \ldots$, to be a sequence of functions. The following recurrence sequence can be defined as

$$x_{n+1} = f_{n+1}(x_n), \quad n = 0, 1, \ldots. \tag{1.10}$$

When the sequence of functions $(f_n)_{n\geq 1}$ is constant, i.e., $f_n = f$, $n = 1, 2, \ldots$, this reduces to a sequence defined by a function

$$x_{n+1} = f(x_n), \quad n = 0, 1, \ldots. \tag{1.11}$$

Denoting the iterations of f by $f^n = f$, $f^2 = f \circ f, \ldots, f^n = f^{n-1} \circ f, \ldots$, the sequence $(x_n)_{n\geq 0}$ can be represented by $x_n = f^n(x_0)$, referred to as the *sequence of successive approximations* associated with f and x_0.

An important problem with many applications is to determine the classes of functions f, for which the sequence associated with f and x_0 is convergent. Recall here the famous Banach Fixed Point Theorem, when (X, d) is a complete metric space and $f : X \to X$ is a contraction, where for every starting point $x_0 \in X$, the sequence of successive approximations converges.

In general, if the function f is not continuous, then the associated sequence $(x_n)_{n\geq 0}$ does not converge.

1.3 Systems of Recurrent Sequences

The following example is a special case of the recurrence relation defined by a function. Let m be a positive integer, X an arbitrary set, $x_0 = (x_0^1, \ldots, x_0^m)$ a point of X^m and $f_n : X^m \to X^m$, $n = 1, 2, \ldots$, a sequence of functions. An important first-order recurrence relation is defined by

$$x_{n+1} = f_{n+1}(x_n), \quad n = 0, 1, \ldots. \tag{1.12}$$

This is equivalent to the following system

$$\begin{cases} x_{n+1}^1 = f_{n+1}^1\left(x_n^1, \ldots, x_n^m\right) \\ \ldots\ldots\ldots\ldots\ldots\ldots\ldots\ldots \\ x_{n+1}^m = f_{n+1}^m\left(x_n^1, \ldots, x_n^m\right), \end{cases}$$

where $f_n = \left(f_n^1, \ldots, f_n^m\right)$, $n = 0, 1, \ldots$. When the sequence of functions $(f_n)_{n\geq 1}$ is constant, i.e., $f_n = f$, $n = 1, 2, \ldots$, where $f : X^m \to X^m$, $f = \left(f^1, \ldots, f^m\right)$ is a function, the above system of equations becomes

$$\begin{cases} x_{n+1}^1 = f^1\left(x_n^1, \ldots, x_n^m\right) \\ \ldots\ldots\ldots\ldots\ldots\ldots\ldots\ldots \\ x_{n+1}^m = f^m\left(x_n^1, \ldots, x_n^m\right), \end{cases} \quad n = 0, 1, \ldots. \tag{1.13}$$

An important example is given by the system of linear recurrent sequences

$$\begin{cases} x_{n+1}^1 = a_{11}x_n^1 + \cdots + a_{1m}x_n^m \\ \ldots\ldots\ldots\ldots\ldots\ldots\ldots\ldots \\ x_{n+1}^m = a_{m1}x_n^1 + \cdots + a_{mm}x_n^m, \end{cases} \quad n = 0, 1, \ldots, \tag{1.14}$$

where $a_{ij} \in \mathbb{C}$ for $i, j = 1, \ldots, m$, and $x_0 = (x_0^1, \ldots, x_0^m) \in \mathbb{C}^m$. Introducing the matrices $A = (a_{ij})_{1\leq i,j\leq m}$ and $X_n = (x_n^1, \ldots, x_n^m)^t$, the transpose of $(x_n^1, \ldots, x_n^m)$, then the relation (1.14) can be written as $X_{n+1} = AX_n$, $n = 0, 1, \ldots$, and we have $X_n = A^n X_0$, $n = 0, 1, \ldots$. Hence, finding the solution of the recurrence system (1.14) is equivalent to determining the powers A^n of the matrix A.

1.4 Existence and Uniqueness of the Solution

The existence and the uniqueness of solution to the recurrence relation (1.2) are important problems in various mathematical processes. We present here a few general aspects.

Theorem 1.1 *Let $\alpha \in X$ and let $(f_n)_{n\geq 1}$ be a sequence of functions $f_n : X^n \to X$. There is a unique sequence $(x_n)_{n\geq 0}$ satisfying the recurrence relation* (1.2) *and the initial condition $x_0 = \alpha$.*

Proof Clearly, $x_1 = f_1(x_0) = f_1(\alpha)$ and assume that the terms $x_1, x_2, \ldots, x_m$ are defined. Then, from we have $x_{m+1} = f_{m+1}(x_m, \ldots, x_0)$, and the conclusion follows by the strong form of Mathematical Induction. □

Theorem 1.2 *Let $\alpha_0, \alpha_1, \ldots, \alpha_{k-1} \in X$ and let $(f_n)_{n\geq k}$ be a sequence of functions $f_n : X^k \to X$. There is a unique sequence $(x_n)_{n\geq 0}$ satisfying the recurrence relation* (1.4) *and the initial conditions $x_0 = \alpha_0$, $x_1 = \alpha_1$, $\ldots$, $x_{k-1} = \alpha_{k-1}$.*

Proof Clearly, we have $x_k = f_k(x_{k-1}, \ldots, x_0) = f_k(\alpha_{k-1}, \ldots, \alpha_0)$. Assume that the terms $x_{m-1}, \ldots, x_{m-k}$ are defined. Then, from the relation (1.4) one can deduce that $x_m = f_m(x_{m-1}, \ldots, x_{m-k})$, i.e., the term x_m is uniquely defined. The conclusion follows by the step k form of Mathematical Induction. □

A special case is when the sequence of functions $(f_n)_{n\geq k}$ is constant, i.e., $f_n = f$ for all $n \geq k$, where $f : X^k \to X$. We obtain the following result.

Corollary 1.1 *Let $\alpha_0, \alpha_1, \ldots, \alpha_{k-1}, \in X$ and $f : X^k \to X$ be a function. There is a unique sequence $(x_n)_{n\geq 0}$ satisfying*

$$x_n = f(x_{n-1}, \ldots, x_{n-k}), \quad n = k, k+1, \ldots, \tag{1.15}$$

and the initial conditions $x_0 = \alpha_0$, $x_1 = \alpha_1$, ..., $x_{k-1} = \alpha_{k-1}$.

Another special situation is when the function $f_n : X^k \to X$ only depends on the variable u_k, $n = k, \ldots$, i.e., $f(u_1, \ldots, u_k) = g_n(u_k)$, $n = k, \ldots$, where $(g_n)_{n\geq k}$ is a sequence of functions $g_n : X \to X$. In this respect we have the following.

Corollary 1.2 *Let $\alpha_0, \alpha_1, \ldots, \alpha_{k-1}, \in X$ and $(g_n)_{n\geq k}$ be a sequence of functions $g_n : X \to X$. There is a unique sequence $(x_n)_{n\geq 0}$ satisfying*

$$x_n = g(x_{n-k}), \quad n = k, k+1, \ldots, \tag{1.16}$$

and the initial conditions $x_0 = \alpha_0$, $x_1 = \alpha_1$, ..., $x_{k-1} = \alpha_{k-1}$.

When the sequence $(g_n)_{n\geq k}$ is constant, i.e., $g_n = g$ for every $n \geq k$, where $g : X \to X$, we obtain a k-order recurrence relation generated by the function g and the initial values $\alpha_0, \alpha_1, \ldots, \alpha_{k-1}$. For $k = 1$ we recover the first-order recurrence relation seen in Section 1.2.

Example 1.1 Consider α and r arbitrary complex numbers and the sequence $(x_n)_{n\geq 0}$ defined by $x_0 = \alpha$ and

$$x_n = x_{n-1} + r, \quad n = 1, 2, \ldots.$$

According to Corollary 1.2 for $k = 1$ and the function $g : \mathbb{C} \to \mathbb{C}$, $g(u) = u + r$, there is a unique such sequence. This is called *arithmetic sequence* and it is easy to prove the formula $x_n = \alpha + rn$, $n = 0, 1, \ldots$.

Similarly, for β, q, a *geometric sequence* $(y_n)_{n\geq 0}$ is defined by $y_0 = \beta$ and

$$y_{n+1} = q y_{n-1}, \quad n = 1, 2, \ldots.$$

Applying again Corollary 1.2 for $k = 1$ and the function $g : \mathbb{C} \to \mathbb{C}$, $g(u) = qz$, the uniqueness of the sequence $(y_n)_{n\geq 0}$ follows. The explicit formula for the term y_n is $y_n = \beta q^n$, $n = 0, 1, \ldots$.

Example 1.2 Consider the recurrence relation

$$x_{n+2} = e^{x_{n+1}} + x_n, \quad n = 0, 1, \ldots,$$

and $x_0 = 0$, $x_1 = 1$. From Corollary 1.1 with $k = 2$ and the function $f : \mathbb{R}^2 \to \mathbb{R}$, $f(u, v) = e^u + v$, it follows the uniqueness of the sequence $(x_n)_{n\geq 0}$.

Remark 1.1 If the recurrence relations are not of functional type, then, in general, we don't have the uniqueness of the sequence.

Example 1.3 Let $(x_n)_{n\geq 0}$ be a sequence with $x_0 = 0$ and $x_{n+1}^2 = x_n^2 + 1$, for $n = 0, 1, \ldots$. One can easily check that the sequences $(\sqrt{n})_{n\geq 0}$ and $(-\sqrt{n})_{n\geq 0}$ both satisfy the above conditions.

1.5 Recurrent Sequences Arising in Practical Problems

In this section, let us look at some specific examples of recurrence sequences in mathematics and science. The list of applications is extensive, and we suggest the interested reader to consult the monographs of Koshy [101, 102] and Vorobiev [164], or the papers of Newell [130] and Vogel [163].

1.5.1 Applications in Mathematical Modeling

1. Fibonacci numbers In his book Liber Abaci (the Book of Calculations), the Italian mathematician Leonardo Pisano (also called Fibonacci) proposed the following theoretical problem concerning a population of rabbits.

"*Having a population of rabbits, one male and one female, in a field. How many rabbits will they produce after one year?*"

The following assumptions were made

1. No rabbits die due to natural causes or predation
2. Reaching maturity in a month, each mature female reproduces monthly
3. A female always gives birth to a pair of rabbits (one male and one female)

Start of month	Rabbits at start of month	Pairs of rabbits at end of month
1	1 pair (the original one)	1
2	1	2
3	2	3 (1 mature pair breeds)
4	3	5 (2 mature pairs breed)
5	5	8 (3 mature pairs breed)
6	8	13 (5 mature pairs breed)
7	13	21 (8 mature pairs breed)
8	21	34 (13 mature pairs breed)
9	34	55 (21 mature pairs breed)
10	55	89 (34 mature pairs breed)
11	89	144 (55 mature pairs breed)
12	144	233 (89 mature pairs breed)

Denote by F_n the number of pairs of rabbits at the start of month n. During this month, the F_{n-1} pairs existent at the start of month $n-1$ will give birth to new pairs, while the rest $F_n - F_{n-1}$ just become mature. At the start of month $n+1$, we will then have $F_{n+1} = F_n + F_{n-1}$, which recovers Fibonacci numbers. This is a second-order homogeneous recurrence relation, as (1.8).

2. The logistic model Based on research on population growth in the 1830s, Verhulst introduced the logistic model described by the equation

$$\frac{dN}{dt} = rN - \alpha N^2, \tag{1.17}$$

where $N(t)$ is the population size at time t, r is the intrinsic growth rate, and α is the density-dependent crowding effect (also known as intraspecific competition). The quantity $\frac{\alpha}{r}$ is sometimes denoted by K and can be related to the capacity of the environment [161].

The logistic map is a discrete version of this model, popularized by the 1976 paper of May [116], described by the nonlinear recurrence relation

$$x_{n+1} = rx_n\left(1 - x_n\right), \tag{1.18}$$

where x_n is a positive number, representing the ratio of existing population to the maximum possible population, and the parameter r represents the reproduction rate. This recurrence relation is defined by a function and can be obtained by setting $f(x) = rx(1-x)$ in (1.11).

3. Recurrent sequences in chemistry Recurrent sequences were used to model changes in physicochemical constants of organic compounds (A) in homologous series, as shown in some works by Zenkevich (see, e.g., [174]). The author is using the simple recurrence equations

$$A(n+k) = aA(n) + b, \quad k = 1, 2, \ldots,$$

where (n) is the number of carbon atoms in the molecule and a and b are coefficients are computed or fitted from practical experiments. For $k = 1$, the solution given by $A(x) = ka^x + b(a^x - 1)/(a-1)$ represents a good approximation for the solubility of organic compounds in water, while for $k = 2$, the resulting equation represents an accurate model for the change in the melting point.

Another interesting application of second-order recurrences was proposed by Challacombe et al. [51], in relation to the efficient calculation of the Cartesian multipole interaction tensor. Using the Hermite polynomials satisfying

$$H_{n+1}(x) = 2xH_n(x) - 2nH_{n-1}(x), \quad H_0(x) = 1, \quad H_1(x) = 2x,$$

the CPU calculation time was lowered to $O(n^4)$, from the $O(n^6)$ required by classical methods available in 1995. Since then, the approach inspired other similar applications in other areas of Chemistry and Physics.

1.5.2 Algebra

1. The Cayley–Hamilton theorem for 2 × 2 matrices Consider the square matrix $A = \begin{pmatrix} a & b \\ c & d \end{pmatrix}$. Direct computations show that A satisfies

$$A^2 - (\mathrm{Tr} A)\, A + (\det A)\, I_2 = O_2,$$

where $\mathrm{Tr} A = a + d$ denotes the trace, and $\det A = ad - bc$ the determinant of A. This is called the characteristic equation of the matrix A.

By induction, one can find two sequences $(x_n)_{n\geq 0}$ and $(y_n)_{n\geq 0}$ of complex numbers such that the following formula holds: $A^n = x_n A + y_n I_2, \quad n = 0, 1, \ldots$. We have $x_0 = 0$ and $y_0 = 1$ from $A^0 = I_2$, and $x_1 = 1$ and $y_1 = 0$. From the characteristic equation, we have $x_2 = \mathrm{Tr} A$ and $y_2 = -\det A$, hence

$$\begin{aligned} x_{n+1} A + y_{n+1} I_2 &= A^{n+1} = AA^n = A\,(x_n A + y_n I_2) = x_n A^2 + y_n A \\ &= x_n (x_2 A + y_2 I_2) + y_n A = (x_2 x_n + y_n)\, A + y_2 x_n I_2, \end{aligned}$$

hence, the sequences $(x_n)_{n\geq 0}$ and $(y_n)_{n\geq 0}$ satisfy the recurrence relation

$$x_{n+1} = x_2 x_n + y_n, \quad y_{n+1} = y_2 x_n, \quad n = 0, 1, \ldots.$$

One can notice that this is in fact a system of two linear recurrence equations.

2. Solutions of a Pell equation Let $(a_n)_{n\geq 0}$ and $(b_n)_{n\geq 0}$ be the sequence of integers defined by the relation $\left(2 + \sqrt{3}\right)^n = a_n + b_n\sqrt{3}$, $n = 0, 1, \ldots$. We have $a_0 = 1$, $b_0 = 0$ and $a_1 = 2$, $b_1 = 1$. A simple inductive argument shows that a_n and b_n are positive integers for $n \geq 1$. Observe that

$$\begin{aligned} a_{n+1} + b_{n+1}\sqrt{3} &= \left(2 + \sqrt{3}\right)^{n+1} = \left(2 + \sqrt{3}\right)\left(2 + \sqrt{3}\right)^n \\ &= \left(2 + \sqrt{3}\right)\left(a_n + b_n\sqrt{3}\right) = 2a_n + 3b_n + (a_n + 2b_n)\sqrt{3}, \end{aligned}$$

hence we obtain the recurrence relations

$$a_{n+1} = 2a_n + 3b_n, \quad b_{n+1} = a_n + 2b_n, \quad n = 0, 1, \ldots.$$

By the binomial expansion we get $\left(2-\sqrt{3}\right)^n = a_n - b_n\sqrt{3}$, with $n = 0, 1, \ldots$. Multiplying by the initial relation we get

$$1 = \left(2+\sqrt{3}\right)^n \left(2-\sqrt{3}\right)^n = \left(a_n + b_n\sqrt{3}\right)\left(a_n - b_n\sqrt{3}\right) = a_n^2 - 3b_n^2,$$

hence (a_n, b_n), $n = 0, 1, \ldots$, are solutions to Pell's equation $x^2 - 3y^2 = 1$.

1.5.3 Combinatorics

1. The number of subsets Let $X = \{x_1, \ldots, x_n\}$ be a set with n elements, where n is a fixed positive integer, and let us denote by α_n the number of subsets of X. Clearly, we have $\alpha_1 = 2$, because the only subsets are $\emptyset$ and $\{x_1\}$. If $Y = X \cup \{x_{n+1}\}$, where $x_{n+1} \neq x_j$, $j = 1, \ldots, n$, then the subsets of Y are all the subsets of X and the subsets of the form $A \cup \{x_{n+1}\}$, with A being any subset of X. This leads to the recurrence formula $\alpha_{n+1} = \alpha_n + \alpha_n = 2\alpha_n$, which gives $\alpha_n = 2^n$, $n = 0, 1, \ldots$, which is a geometric sequence.

2. Binomial coefficients Let $0 \le k \le n$ be two natural numbers. The binomial coefficient represents the coefficient of the x^k term in the polynomial expansion of the binomial power $(1+x)^n$, and have the expression

$$\binom{n}{k} = \frac{n!}{k!(n-k)!}. \tag{1.19}$$

For example, the power of x^3 in $(1+x)^5$ is $\binom{5}{3} = \frac{5!}{3!2!} = 10$, where

$$\begin{aligned}(1+x)^5 &= \binom{5}{0}x^0 + \binom{5}{1}x^1 + \binom{5}{2}x^2 + \binom{5}{3}x^3 + \binom{5}{4}x^4 + \binom{5}{5}x^5 \\ &= 1 + 5x + 10x^2 + 10x^3 + 5x^4 + x^5.\end{aligned}$$

Binomial coefficients can be arranged in Pascal's triangle (Table 1.1), where each number is obtained by summing the numbers immediately to its left and right in the previous row. This can be written as

$$\binom{n}{k} = \binom{n-1}{k} + \binom{n-1}{k-1}, \quad 0 \le k \le n-1.$$

The sum of the elements on the nth row is

$$\sum_{k=0}^{n} \binom{n}{k} = (1+1)^n = 2^n.$$

Table 1.1 Pascal's triangle

$n=0$							1						
$n=1$						1		1					
$n=2$					1		2		1				
$n=3$				1		3		3		1			
$n=4$			1		4		6		4		1		
$n=5$		1		5		10		10		5		1	
$n=6$	1		6		15		20		15		6		1

Since the number of subsets with $0 \leq k \leq n$ elements selected from a set with n elements is $\binom{n}{k}$, this is another proof of the previous example.

3. The Stirling numbers of the first kind For a positive integer n, let us consider the polynomial of degree n

$$[x]_n = x(x-1)\cdots(x-n+1).$$

Expanding the product we get the algebraic expression of $[x]_n$, i.e., we have

$$[x]_n = \sum_{k=0}^{n} s(n,k)x^k,$$

where the integers $s(n,k)$, $k = 0, 1, \ldots, n$, are called the Stirling numbers of the first kind. Clearly, we have $s(n,0) = 0$, $s(n, n-1) = (-1)^{n-1}(n-1)!$, and $s(n,n) = 1$. A convenient way to compute these numbers is to observe that from $[x]_{n+1} = [x]_n\,(x-n)$, we obtain

$$\sum_{k=0}^{n+1} s(n+1,k)x^k = (x-n)\sum_{j=0}^{n} s(n,j)x^j = \sum_{j=0}^{n} s(n,j)x^{j+1} - \sum_{j=0}^{n} ns(n,j)x^j$$

$$= -ns(n,0) + \sum_{k=1}^{n}\Big[s(n,k-1)x^k - ns(n,k)\Big]x^k + s(n,n)x^{n+1}.$$

Identifying the coefficients of x^k for $k = 1, \ldots, n$, it follows that

$$s(n+1,k) = s(n,k-1) - ns(n,k), \quad k = 1, \ldots, n, \tag{1.20}$$

with $s(n+1,0) = -ns(n,0) = 0$ and $s(n+1,n+1) = s(n,n) = 1$. By this recurrence relation, the Stirling numbers of the first kind can be computed successively, some terms being given in Table 1.2.

Table 1.2 Stirling numbers of the second kind

	k									
n	0	1	2	3	4	5	6	7	8	9
0	1									
1	0	1								
2	0	1	1							
3	0	2	3	1						
4	0	6	11	6	1					
5	0	24	50	35	10	1				
6	0	120	274	225	85	15	1			
7	0	720	1764	1624	735	175	21	1		
8	0	5040	13068	13132	6769	1960	322	28	1	

1.5.4 Geometry

1. Lines dividing a plane Let d_n be the maximum number of regions defined by n lines in a plane, $n = 1, 2, \dots$. It is clear that $d_1 = 2$ and $d_2 = 4$.

One may easily notice that the number of regions defined by n lines is maximum if and only if any two lines are not parallel and any three lines are not concurrent. Moreover, in this situation, the number d_n does not depend on the configuration of the lines.

Suppose that we have $n+1$ lines $l_1, l_2, \dots, l_{n+1}$ in a configuration as above. The line l_{n+1} intersects the lines $l_1, l_2, \dots, l_n$ in n points and these points cut the line l_{n+1} into $n+1$ parts ($n-1$ segments and 2 half-lines). Therefore, the line l_{n+1} cuts exactly $n+1$ of the existing regions, each of them in two regions. This leads to the recurrence relation

$$d_{n+1} = d_n + n + 1, \quad n = 1, 2, \dots.$$

This is a first-order, linear, nonhomogeneous recurrence relation, obtained by setting $a_n = 1$ and $b_n = n+1$ (1.9), having the solution $d_n = \frac{n(n+1)}{2} + 1$.

2. Catalan numbers Let P be a convex polygon in $\mathbb{R}^2$. We shall derive the number of ways to divide a labeled convex $(n+2)$-gon into triangles, denoted by C_n. It is known that $C_1 = 1$, $C_2 = 2$, and $C_3 = 5$.

We shall find a recurrence relation for C_{n+1} as a function of $C_0, C_1, \dots, C_n$. Consider a convex $(n+3)$-polygon having the vertices $v_1, v_2, \dots, v_{n+3}$, denoted by $P_{v_1,v_2,\dots,v_{n+3}}$. The side v_1v_{n+3} belongs to a certain triangle Δ, whose third vertex is denoted by v_{k+2} $(0 \le k \le n)$, as seen in Figure 1.1.

The polygon is divided into the convex $(k+2)$-polygon $P_{v_1,v_2,\dots,v_{k+2}}$, the triangle Δ, and the convex $(n-k+2)$-polygon $P_{v_{k+2},v_{k+3},\dots,v_{n+3}}$. By the induction hypothesis, there are C_k ways of dividing $P_{v_1,v_2,\dots,v_{k+2}}$ into triangles, and C_{n-k} ways of dividing $P_{v_{k+2},v_{k+3},\dots,v_{n+3}}$ into triangles, which gives

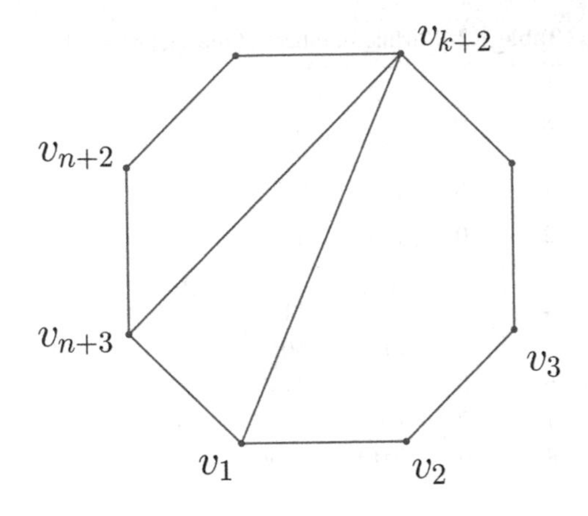

Fig. 1.1 v_{k+2} the third vertex of the triangle Δ with the side $v_1 v_{n+3}$

$$C_{n+1} = \sum_{k=0}^{n} C_k C_{n-k}, \quad C_0 = 1.$$

Catalan numbers are a classical example of recurrent sequence given in explicit form, which depends on all the previous terms. It can be obtained from formula (1.2), by setting $f_n(x_n, \ldots, x_0) = \sum_{k=0}^{n} x_k x_{n-k}$.

3. Fractals and the Mandelbrot set Let $c \in \mathbb{C}$ and let us define the recursion

$$z_{n+1} = z_n^2 + c, \quad z_1 = 0. \tag{1.21}$$

The Mandelbrot set consists of the numbers c, for which the terms of the recursive sequence are bounded. Noticeably, z_n is a polynomial in c, whose terms are given by the Catalan numbers. The Mandelbrot set is represented in Figure 1.2, where a point c is colored black if it belongs to the set, and white otherwise.

1.5.5 Analysis

1. Trigonometric limits For a positive integer n, we denote by L_n the limit

$$\lim_{x\to 0} \frac{1 - \cos x \cos 2x \cdots \cos nx}{x^2}.$$

Observe that

$$L_{n+1} = \lim_{x\to 0} \frac{1 - \cos x \cdots \cos nx \cos(n+1)x}{x^2}$$

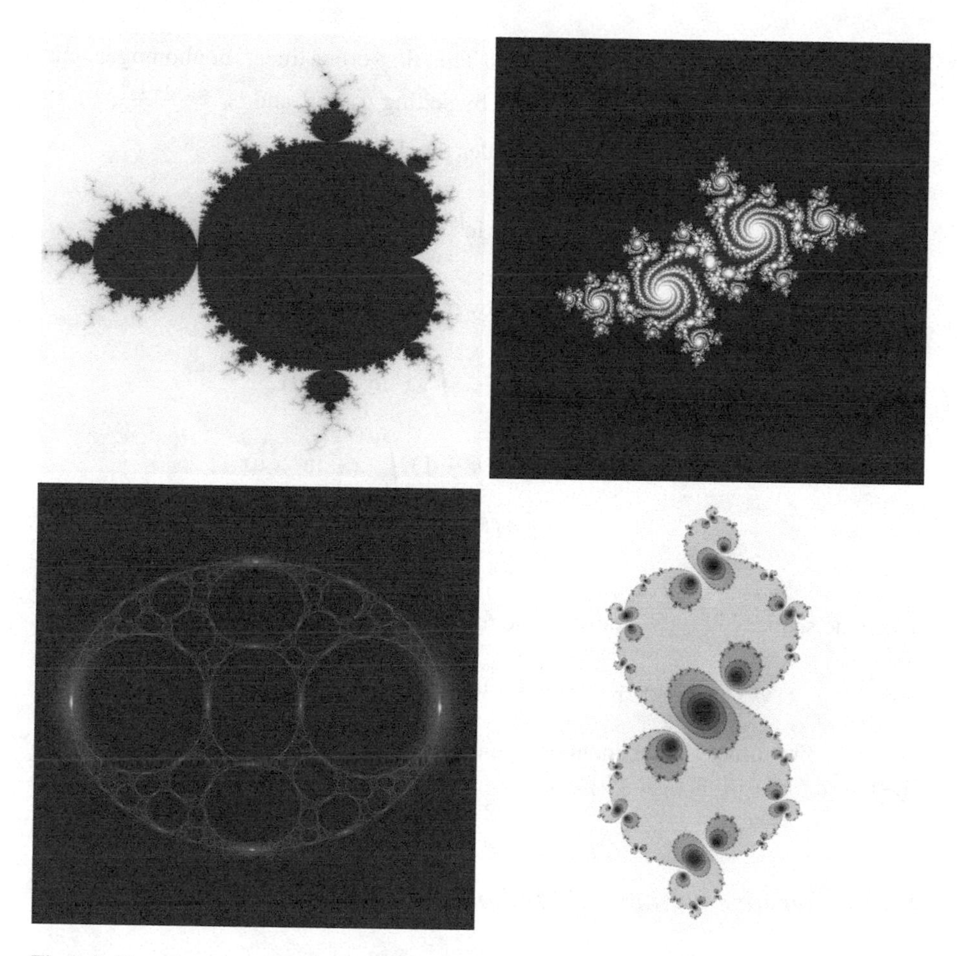

Fig. 1.2 Fractals generated by the Matlab routines of Pawar [137]: fractal6.m (top-left, Mandelbrot set); juliaset3.m (top-right, Julia 12 set); otherjulia1.m (bottom-left); otherjulia2.m (bottom-right)

$$= \lim_{x\to 0}\left[\frac{1-\cos(n+1)x}{x^2} + \frac{\cos(n+1)x\,(1-\cos x\cdots\cos nx)}{x^2}\right]$$
$$= \lim_{x\to 0}\frac{1-\cos(n+1)x}{x^2} + L_n.$$

Because

$$\lim_{x\to 0}\frac{1-\cos(n+1)x}{x^2} = \lim_{x\to 0}\frac{2\sin^2\frac{n+1}{2}x}{x^2}$$
$$= \frac{(n+1)^2}{2}\lim_{x\to 0}\left(\frac{\sin\frac{n+1}{2}x}{\frac{n+1}{2}x}\right)^2 = \frac{(n+1)^2}{2},$$

we get the equation $L_{n+1} = L_n + \frac{(n+1)^2}{2}$. This first-order linear, nonhomogeneous recurrence relation is obtained from (1.9) by setting $a_n = 1$ and $b_n = \frac{(n+1)^2}{2}$.

2. Integrals For the integer $n \geq 0$, consider the defined integral

$$I_n = \int_1^e \ln^n x \, dx.$$

Clearly, we have $I_0 = e - 1$. Integrating by parts, we obtain

$$\begin{aligned} I_{n+1} &= \int_1^e \ln^{n+1} x \, dx = \int_1^e x' \ln^{n+1} x \, dx \\ &= x \ln^{n+1} x \,|_1^e - (n+1) \int_1^e x \frac{1}{x} \ln^n x \, dx \\ &= e - (n+1) \int_1^e \ln^n x \, dx, \end{aligned}$$

hence the sequence $(I_n)_{n \geq 0}$ satisfies the recurrence relation

$$I_{n+1} = e - (n+1) I_n, \quad n = 0, 1, \ldots.$$

This is a first-order, linear, nonhomogeneous recurrence equation obtained from (1.9) for $a_n = -(n+1)$ and $b_n = e$.

1.5.6 Iterative Numerical Methods

Many iterative numerical methods which approximate the exact solutions to polynomial or differential equations lead naturally to recurrence sequences.

1. Approximating solutions of equations Let $h : \mathbb{R} \to \mathbb{R}$ be a continuous real function. We present some numerical methods for solving the equation $h(x) = 0$.

Bisection method Consider an interval $[a_0, b_0]$ over which the continuous function h changes sign, i.e., $h(a_0) \cdot h(b_0) < 0$. Testing the value of the function at the midpoint c_0 of the segment $[a_0, b_0]$, one can construct a new interval $[a_1, b_1]$ containing a point x^*, for which $h(x^*) = 0$. The sequence $c_0, c_1, \ldots,$ produced in this way is an approximation for the solution x^*, which after n iterations satisfies $|c_n - x^*| \leq (b - a)/2^{n+1}$.

Newton–Raphson method Provided that function h is also differentiable, one can build a recursive sequence of iterations starting from an initial point x_0, involving the function's values and the first derivative, through the formula

$$x_{n+1} = x_n - \frac{h(x_n)}{h'(x_n)}, \qquad n = 0, 1, \ldots .$$

The sequence $(x_n)_{n\geq 0}$ may converge to a solution of $h(x) = 0$. This recurrence relation is defined by the function $f(x) = x - \frac{h'(x)}{h(x)}$, as in (1.11).

Secant method The solution is estimated by building a recursive sequence of iterations, starting from two points x_0 and x_1, by the formula

$$x_{n+2} = \frac{x_n h(x_{n+1}) - x_{n+1} h(x_n)}{h(x_{n+1}) - h(x_n)}, \qquad n = 0, 1, \ldots .$$

This is a second-order, nonlinear recurrence relation, defined explicitly by using a bivariate function $f_n(x, y) = f(x, y) = \frac{yh(x) - xh(y)}{h(x) - h(x)}$ in (1.6), $n = 0, 1, \ldots$.

2. Finite differences and the harmonic oscillator The method of finite differences is a classical approach for solving differential equations. The process involves the transformation of the initial equation into a recurrence relation, which can be solved to produce an approximate solution.

Many finite difference schemes exist for derivatives of first, second, or higher order, with different degrees of accuracy and complexity. Such schemes are obtained by truncating Taylor series of a real function $x(t)$ (assumed to be infinitely differentiable) at a real point t, and various choices of a in

$$x(t) + \frac{x'(t)}{1!}(a - t) + \frac{x''(t)}{2!}(a - t)^2 + \frac{x'''(t)}{3!}(a - t)^3 + \cdots .$$

To illustrate this approach, we consider the case of the driven harmonic oscillator with damping [146, Chapter 15]. When such a system of mass m is displaced from the equilibrium state by an externally applied force $F(t)$ in the direction of x, it is subject to a restoring force proportional to the displacement x, and a friction force proportional to the velocity v.

Explicitly, the motion of the oscillator is described by the equation

$$F(t) - kx - cx'(t) = mx''(t), \tag{1.22}$$

where m, k, and c are positive constants. Some special cases present interest.

The *simple harmonic oscillator* obtained from (1.22) for $c = 0$ and $F(t) \equiv 0$, and its motion is given by the equation $mx''(t) = -kx$, having the solution

$$x(t) = A\cos(\omega t + \varphi).$$

The constants $\omega = \sqrt{\frac{k}{m}}$, A, and φ depend on the initial position and velocity.

The *damped harmonic oscillator* is obtained from (1.22) for $F(t) \equiv 0$, and its motion is described by the equation $mx''(t) + cx'(t) + kx = 0$. Solutions can be obtained by rewriting the equation in the form

$$x''(t) + 2\zeta\omega_0 x'(t) + \omega_0^2 x(t) = 0, \tag{1.23}$$

where $\omega = \sqrt{\frac{k}{m}}$ is the undamped angular frequency, and $\zeta = \frac{c}{2\sqrt{mk}}$ is the damping ratio. We illustrate the finite difference method for solving (1.23).

Let n be an integer, with x_0, y_0, and $t > 0$ real numbers. Let $x(t)$ be a real function satisfying the equation (1.23), and $x(0) = x_0$ and $x'(0) = y_0$.

Consider a partition of the time interval $[0, t]$, by the points $t_0, t_1, \ldots, t_n$ defined by $t_k = k\Delta t, k = 0, \ldots, n$, with $t_n = n\Delta t = t$. The exact values of the function $x(t)$ evaluated at these points will be approximated by a sequence of points $x_k \simeq x(t_k)$, $k = 0, \ldots, n$, generated by a numerical method.

Writing the equation (1.23) at $t = t_k$, $k = 0, \ldots, n$, one obtains

$$x''(t_k) + 2\zeta\omega_0 x'(t_k) + \omega_0^2 x(t_k) = 0. \tag{1.24}$$

The derivatives $x'(t_k)$ and $x''(t_k)$, $k = 0, \ldots, n$, are replaced by finite difference formulae involving neighboring nodes (see, e.g., [52, Chapter 4])

$$x'(t_k) \simeq \frac{x_{k+1} - x_k}{\Delta t}, \tag{1.25}$$

$$x'(t_k) \simeq \frac{x_{k+1} - x_{k-1}}{2\Delta t}, \tag{1.26}$$

$$x''(t_k) \simeq \frac{x_{k+1} - 2x_k + x_{k-1}}{(\Delta t)^2}. \tag{1.27}$$

The first formula is the forward difference approximation for x', while the following two are the central difference approximations for x' and x'', respectively. Substituting into (1.24), one obtains

$$\frac{x_{k+1} - 2x_k + x_{k-1}}{(\Delta t)^2} + 2\zeta\omega_0 \frac{x_{k+1} - x_k}{\Delta t} + \omega_0^2 x_k = 0, \quad k = 1, \ldots, n-1.$$

This is a second-order linear recurrence relation of the form

$$x_{k+1} + a x_k + b x_{k-1} = 0, \quad k = 1, \ldots, n-1, \tag{1.28}$$

where the terms a and b are the constants

$$a = \frac{(\omega_0 \Delta t)^2 - 2\zeta\omega_0\Delta t - 2}{1 + 2\zeta\omega_0\Delta t}, \quad b = \frac{1}{1 + 2\zeta\omega_0\Delta t}.$$

The terms of the recurrence sequence (1.28) can be obtained explicitly once we have the starting values x_0 and x_1. Clearly, x_0 is already known, while x_1 can be estimated from the condition $x'(0) = y_0$.

Applying the forward difference approximation for $x'(t)$ at $t = 0$ we have

$$y_0 = x'(0) \simeq \frac{x_1 - x_0}{\Delta t},$$

from where we obtain $x_1 = x_0 + (\Delta t)\, y_0$.

The points $x_2, x_3, \ldots, x_n$ can now be computed recursively by (1.28).

The error of the above-mentioned schemes is $O(\Delta t)$ for the forward differences method and $O((\Delta t)^2)$ for the central differences methods. Therefore, accuracy is then improved when using central differences to approximate the derivatives $x'(t_k)$ and the initial condition $x'(0)$.

Chapter 2
Basic Recurrent Sequences

In this chapter we present basic results and examples of first-order and second-order linear recurrent sequences with arbitrary coefficients, some classical sequences, polynomials, and linear fractional transformations.

In Section 2.1 we discuss general formulae for first-order linear recurrent sequences and illustrative examples. Section 2.2 is dedicated to second-order linear recurrent sequences and their properties. We present formulae for the general term, examples, including Fibonacci, Lucas, Pell, and Lucas–Pell numbers, and associated polynomials and generalizations. In Section 2.3 we analyze homographic recurrent sequences in the complex plane, presenting exact formulae, geometric patterns, and periodicity conditions. The results motivate the study of periodic Horadam sequences and their generalizations in Chapters 5 and 6.

2.1 First-Order Linear Recurrent Sequences

Often, we have to find the terms x_n, where $(x_n)_{n\geq 0}$ is a sequence defined by

$$x_{n+1} = ax_n + b, \quad n = 0, 1, 2, \ldots, \quad x_0 = \alpha, \tag{2.1}$$

where α, a, and b are given complex numbers.

One can identify the following special cases:

1. If $a = 0$, then $(x_n)_{n\geq 0}$ is the constant sequence $x_n = b$, $n = 1, 2, \ldots$.
2. If $a = 1$, $b = 0$, then $(x_n)_{n\geq 0}$ is a geometric sequence and

$$x_n = a\alpha^{n-1}, \quad n = 1, 2, \ldots.$$

D. Andrica, O. Bagdasar, *Recurrent Sequences*, Problem Books in Mathematics,
https://doi.org/10.1007/978-3-030-51502-7_2

3. If $a = 1$, $b \neq 0$, then $(x_n)_{n\geq 0}$ is an arithmetic sequence and

$$x_n = \alpha + nb.$$

In the general case, one has $a \neq 1$, $a \neq 0$, and $b \neq 0$. To find the general formula, one may first identify a real number x such that

$$x_{n+1} + x = a(x_n + x), \quad n = 0, 1, \ldots.$$

For $n = 0$, we get $x_1 + x = a(x_0 + x)$ and $x_1 = ax_0 + b$, hence $x = \dfrac{b}{a-1}$.

For the second step, we introduce the sequence $(y_n)_{n\geq 0}$, $y_n = x_n + x$. We obtain $y_{n+1} = ay_n$, $n = 0, 1, \ldots$, hence $(y_n)_{n\geq 0}$ is a geometric sequence.

From here it follows that $y_n = a^n y_0$, $n = 0, 1, 2, \ldots$, so

$$x_n = a^n \left(\alpha + \frac{b}{a-1}\right) - \frac{b}{a-1}, \quad n = 0, 1, 2, \ldots. \tag{2.2}$$

Example 2.1 Consider the sequence $(x_n)_{n\geq 0}$ defined by $x_0 = 0$ and

$$x_{n+1} = -\frac{1}{2}x_n + 1, \quad n = 0, 1, 2, \ldots.$$

Find a formula for x_n.

Solution Applying formula (2.2) for $\alpha = 0$, $a = -\frac{1}{2}$, $b = 1$, we obtain

$$\begin{aligned} x_n &= \left(-\frac{1}{2}\right)^n \left(\frac{1}{-\frac{1}{2} - 1}\right) + \frac{1}{\frac{1}{2} + 1} \\ &= \left(-\frac{1}{2}\right)^n \left(-\frac{2}{3}\right) + \frac{2}{3} \\ &= \frac{(-1)^{n+1}}{3 \cdot 2^{n-1}} + \frac{2}{3}, \quad n = 0, 1, 2, \ldots. \end{aligned}$$

Example 2.2 Let $(x_n)_{n\geq 0}$ be a sequence defined by $x_0 = \alpha$ and

$$x_{n+1} = ax_n + 1, \quad n = 0, 1, \ldots.$$

Find all complex numbers a, for the sequence $(x_n)_{n\geq 0}$ is convergent.

Solution Applying formula (2.2) for $a \neq 1$ and $b = 1$, we get

$$x_n = a^n \left(\alpha + \frac{1}{a-1}\right) - \frac{1}{a-1}, \quad n = 0, 1, 2, \ldots.$$

In this case $(x_n)_{n\geq 0}$ is convergent if and only if $|\ a\ | \in (-1, 1)$, and we have $\lim_{n\to\infty} x_n = -\frac{1}{a-1}$ since $\lim_{n\to\infty} a^n = 0$. If $a = 1$, then $(x_n)_{n\geq 0}$ is the arithmetic sequence given by $x_n = \alpha + n$, $n = 0, 1, 2, \ldots$. Clearly, one can easily deduce that $\lim_{n\to\infty} x_n = +\infty$.

General Context Let $(a_n)_{n\geq 0}$ and $(b_n)_{n\geq 0}$ be two given sequences of complex numbers, and let $(x_n)_{n\geq 0}$ be the sequence defined by

$$x_{n+1} = a_n x_n + b_n, \quad n = 0, 1, 2, \ldots, \tag{2.3}$$

where $x_0 = \alpha$ is a complex number. We have the following result.

Theorem 2.1 *If $a_n \neq 0$ for every $n \neq 0$, then the following formula holds:*

$$x_n = a_0 \cdots a_{n-1} \left(\alpha + \sum_{k=0}^{n-1} \frac{b_k}{a_0 \cdots a_k} \right), \quad n = 0, 1, 2, \ldots. \tag{2.4}$$

Proof We are looking for a solution to (2.3), of the form $x_n = a_0 \cdots a_{n-1} y_n$, where $y_0 = \alpha$. We can determine the sequence $(y_n)_{n\geq 0}$ from the recursive relation (2.3). Indeed, since we have

$$a_0 \cdots a_n y_{n+1} = a_0 \cdots a_n y_n + b_n,$$

it follows that

$$y_{n+1} - y_n = \frac{b_n}{a_0 \cdots a_n}, \quad n = 0, 1, 2, \ldots.$$

From the above relation we obtain

$$y_n = y_0 + \frac{b_0}{a_0} + \frac{b_1}{a_0 a_1} + \frac{b_2}{a_0 a_1 a_2} + \cdots + \frac{b_{n-1}}{a_0 \cdots a_{n-1}},$$

therefore

$$x_n = a_0 \cdots a_{n-1} \left(\alpha + \frac{b_0}{a_0} + \frac{b_1}{a_0 a_1} + \frac{b_2}{a_0 a_1 a_2} + \cdots + \frac{b_{n-1}}{a_0 \cdots a_{n-1}} \right).$$

□

Remark 2.1

1° If the sequence $(a_n)_{n\geq 0}$ is constant, i.e., $a_n = a$, $n = 0, 1, \ldots$, then the formula (2.4) becomes

$$x_n = a^n \left(\alpha + \sum_{k=0}^{n-1} \frac{b_k}{a^k} \right), \quad n = 0, 1, 2, \ldots. \tag{2.5}$$

2° When the sequence $(b_n)_{n \geq 0}$ is constant, $b_n = b$, $n = 0, 1, \ldots,$ then formula (2.4) becomes

$$x_n = a_0 \cdots a_{n-1} \left(\alpha + b \sum_{k=0}^{n-1} \frac{1}{a_0 \cdots a_k} \right), \quad n = 0, 1, 2, \ldots. \tag{2.6}$$

3° When $a_n = a$, and $b_n = b$, $n = 0, 1, \ldots,$ we recover formula (2.2).

Example 2.3 Let $(x_n)_{n \geq 0}$ be the sequence defined by $x_0 = \alpha$ and

$$x_{n+1} = x_n + n, \quad n = 0, 1, 2, \ldots.$$

Find the value of x_{2019}.

Solution By (2.5) for $a_n = 1$ and $b_n = n$, $n = 0, 1, \ldots,$ we obtain

$$x_n = \alpha + \sum_{k=0}^{n-1} k = \alpha + \frac{(n-1)n}{2},$$

hence $x_{2019} = \alpha + \frac{2018 \cdot 2019}{2}$.

Example 2.4 Let $(x_n)_{n \geq 0}$ be the sequence defined by

$$x_0 = \alpha \text{ and } x_{n+1} = \frac{1}{2} x_n + 3, \quad n = 0, 1, 2, \ldots.$$

Find the formula of x_n.

Solution One has to apply formula (2.2) for $a = \frac{1}{2}$, $b = 3$ to obtain

$$x_n = \left(\frac{1}{2}\right)^n \left(\alpha + \frac{3}{\frac{1}{2} - 1} \right) - \frac{3}{\frac{1}{2} - 1} = \left(\frac{1}{2}\right)^n (\alpha - 6) + 6, \quad n = 0, 1, \ldots.$$

Example 2.5 Prove that the sequence $(x_n)_{n \geq 0}$, where

$$x_0 = 1 \text{ and } x_{n+1} = -\frac{1}{2} x_n + (-1)^n, \quad n = 0, 1, 2, \ldots,$$

is divergent.

Solution This is not a recursion with constant coefficients. The general term can be computed explicitly using Theorem 2.1, or Remark 2.1 1° for $a = -\frac{1}{2}$ and

$b_n = (-1)^n$. It follows that for $n = 1, 2, \ldots$, we have

$$x_n = \left(-\frac{1}{2}\right)^n x_0 + \frac{\left(-\frac{1}{2}\right)^n - (-1)^n}{\frac{1}{2}} = \left(-\frac{1}{2}\right)^n x_0 + 2\left[\left(-\frac{1}{2}\right)^n - (-1)^n\right].$$

Clearly,

$$x_{2n} = \frac{1}{2^{2n}} x_0 + 2\frac{1}{2^{2n}} - 2 \to -2, \text{ when } n \to +\infty.$$

At the same time, we have

$$x_{2n+1} = \frac{1}{2^{2n+1}} x_0 - 2\frac{1}{2^{2n+1}} + 2 \to 2, \text{ when } n \to +\infty.$$

This proves that $(x_n)_{n\geq 0}$ is divergent.

Example 2.6 Consider the sequence $(x_n)_{n\geq 0}$ defined by $x_0 = 1$ and

$$x_{n+1} = ax_n + b^n, \quad n = 0, 1, 2, \ldots.$$

Find all real numbers a, b such that the sequence $(x_n)_{n\geq 0}$ is convergent.

Solution Indeed, from the recurrence relation we obtain successively

$$a^{n-1}x_1 - a^n x_0 = a^{n-1}$$
$$a^{n-2}x_2 - a^{n-1}x_1 = a^{n-2}b$$
$$\cdots\cdots\cdots$$
$$ax_{n-1} - a^2 x_{n-2} = ab^{n-2}$$
$$x_n - ax_{n-1} = b^{n-1}.$$

Adding these relations, for $n = 0, 1, \ldots$, one obtains

$$x_n = a^n x_0 + \left(a^{n-1} + a^{n-2}b + \cdots + ab^{n-2} + b^{n-1}\right) = a^n + \frac{a^n - b^n}{a - b}.$$

The formula can also be obtained by Remark 2.1 1° for $a_n = a$ and $b_n = b^n$.

The sequence $(x_n)_{n\geq 0}$ is convergent if and only if

1. $a \in (-1, 1)$, $b \in (-1, 1)$ and in this situation $\lim\limits_{n\to\infty} x_n = 0$.
2. $a = 1$, $b \in (-1, 1)$ and here $\lim\limits_{n\to\infty} x_n = 1 + \frac{1}{1-b}$.
3. $a \in (-1, 1)$, $b = 1$ where $\lim\limits_{n\to\infty} x_n = \frac{1}{1-a}$.

The Stolz–Cesàro Theorem is often useful in applications [129, pp. 85–88].

Theorem 2.2 (Stolz–Cesàro) *Let $(x_n)_{n\geq 1}$ and $(y_n)_{n\geq 1}$ be two sequences of real numbers. Assume that the sequence $(y_n)_{n\geq 1}$ is strictly monotone and divergent (i.e., increasing and approaching $+\infty$, or decreasing and approaching $-\infty$). If the limit $\lim_{n\to\infty} \frac{x_{n+1}-x_n}{y_{n+1}-y_n} = l$ exists, then $\lim_{n\to\infty} \frac{x_n}{y_n} = l$.*

The proof of the following result illustrates this fact.

Theorem 2.3 *Let $(a_n)_{n\geq 0}$ and $(b_n)_{n\geq 0}$ be two sequences of complex numbers satisfying the properties:*

1° $|a_n| \in (0, 1)$, $n = 0, 1, \ldots$, *and* $\lim\limits_{n\to\infty} |a_n| = a \in [0, 1)$;

2° $\lim\limits_{n\to\infty} b_n = 0$.

The sequence $(x_n)_{n\geq 0}$ defined by the recurrence equation $x_{n+1} = a_n x_n + b_n$, $n = 0, 1, 2, \ldots$, is convergent and its limit is 0.

Proof Using formula (2.4) we have

$$
\begin{aligned}
|x_n| =& |a_0 \cdots a_{n-1} \left(\alpha + \sum_{k=0}^{n-1} \frac{b_k}{a_0 \cdots a_k} \right) | \\
\leq& |a_0| \cdots |a_{n-1}| \left(|\alpha| + \sum_{k=0}^{n-1} \frac{|b_k|}{|a_0| \cdots |a_k|} \right).
\end{aligned} \tag{2.7}
$$

Consider $u_n = |a_0| \cdots |a_n|$, $n = 0, 1, \ldots$. Because $\frac{u_{n+1}}{u_n} = |a_{n+1}| < 1$, $n = 0, 1, \ldots$, it follows that the sequence $(u_n)_{n\geq 0}$ is decreasing and bounded, hence it is convergent. Let us denote by $u = \lim\limits_{n\to\infty} u_n$. Clearly, $u \in [0, 1)$. If $u \neq 0$, then from $\frac{u_{n+1}}{u_n} = |a_{n+1}| < 1$ for $n \to \infty$, it follows $1 = a$, which is not possible in the condition $a \in [0, 1)$. Therefore, $u = 0$.

The inequality (2.7) is equivalent to

$$
|x_n| \leq u_{n-1} \left(|\alpha| + \sum_{k=0}^{n-1} \frac{|b_k|}{u_k} \right),
$$

and

$$
\lim_{n\to\infty} u_{n-1} \left(|\alpha| + \sum_{k=0}^{n-1} \frac{|b_k|}{u_k} \right) = \lim_{n\to\infty} \frac{|\alpha| + \sum_{k=0}^{n-1} \frac{|b_k|}{u_k}}{\frac{1}{u_{n-1}}}.
$$

The conditions of Theorem 2.2 are satisfied, hence

$$\lim_{n\to\infty}\frac{|\alpha|+\sum_{k=0}^{n-1}\frac{|b_k|}{u_k}}{\frac{1}{u_{n-1}}}=\lim_{n\to\infty}\frac{\frac{|b_n|}{u_n}}{\frac{1}{u_n}-\frac{1}{u_{n-1}}}=\lim_{n\to\infty}\frac{|b_n|}{1-\frac{u_n}{u_{n-1}}}=\lim_{n\to\infty}\frac{|b_n|}{1-|a_n|}$$

$$=\frac{1}{1-a}\lim_{n\to\infty}|b_n|=0.$$

Therefore, we have $\lim_{n\to\infty}|x_n|=0$, and the conclusion follows. □

Corollary 2.1 *Let $0<\beta<\alpha$ be two real numbers. If $(x_n)_{n\geq 0}$ is a sequence of complex numbers, then the following statements are equivalent*

1° $\lim_{n\to\infty}|x_n|=0$;
2° $\lim_{n\to\infty}(\alpha x_{n+1}+\beta x_n)=0$.

Proof The implication 1° $\Longrightarrow$ 2° is obvious.

To prove the implication 2° $\Longrightarrow$ 1°, let us consider the sequence $(y_n)_{n\geq 0}$, where $y_n=\alpha x_{n+1}+\beta x_n$. We have

$$x_{n+1}=-\frac{\beta}{\alpha}x_n+\frac{1}{\alpha}y_n,\quad n=0,1,\ldots,$$

hence we are in the hypothesis of Theorem 2.3, therefore $\lim_{n\to\infty}|x_n|=0$. □

Example 2.7 Consider the sequence $(x_n)_{n\geq 0}$ defined by $x_0=\alpha$ and the formula $x_{n+1}=\frac{1}{2^{n+1}}x_n+\frac{n}{3^n}$, $n=0,1,\ldots$. Find when the sequence is convergent.

Solution We have $a_n=\frac{1}{2^{n+1}}$ and $b_n=\frac{n}{3^n}$, $n=0,1,\ldots$. Clearly, $a_n\in(0,1)$, $n=0,1,\ldots$, and $\lim_{n\to\infty}a_n=0$. Moreover, $\lim_{n\to\infty}b_n=0$, hence we are in a position to apply Theorem 2.3. It follows that $\lim_{n\to\infty}x_n=0$.

Example 2.8 Consider the sequence $(y_n)_{n\geq 0}$ defined by $y_0=1$ and the equation $y_{n+1}=\frac{1}{5^{n+1}}y_n+(-1)^n$, $n=0,1,\ldots$. Decide if the sequence is convergent.

Solution Note that $a_n=\frac{1}{5^{n+1}}$ and $b_n=(-1)^n$, $n=0,1,\ldots$. Clearly, $a_n\in(0,1)$, but the sequence $(b_n)_{n\geq 0}$ is not convergent, so we cannot apply Theorem 2.3. On the other hand, using the formula (2.4) we obtain

$$y_n=\frac{1}{5^{\frac{n(n+1)}{2}}}\left(1+\sum_{k=0}^{n-1}(-1)^k5^{k+1}\right)=\frac{1}{5^{\frac{n(n+1)}{2}}}\left(1+5\frac{(-5)^n-1}{-6}\right)$$

$$=\frac{1}{5^{\frac{n(n+1)}{2}}}\left[1-\frac{5}{6}\left((-5)^n-1\right)\right],$$

and clearly, we have $\lim_{n\to\infty}y_n=0$, i.e., the sequence is convergent.

This example shows that the hypotheses 1° and 2° in Theorem 2.3 are only sufficient to have the convergence to 0.

2.2 Second-Order Linear Recurrent Sequences

In Section 1.5, we have discussed some applications of second-order recurrent sequences. Probably the most famous number sequence of all is the one named after Fibonacci, which has numerous generalizations in integers, real, and complex numbers (i.e., the Horadam sequences discussed detail in Chapter 5). In this section we present basic results regarding second-order linear recurrent sequences, to familiarize the reader with the fundamental ideas and formulae used later on throughout this book.

2.2.1 *Homogeneous Recurrent Sequences*

Let $(x_n)_{n\geq 0}$ be the sequence defined by $x_0 = \alpha_0, x_1 = \alpha_1$, and

$$x_{n+2} = ax_{n+1} + bx_n, \quad n = 0, 1, 2, \ldots, \tag{2.8}$$

where α_0, α_1, a, b are given real (or complex) numbers.

A formula for the general term of sequence $(x_n)_{n\geq 0}$ satisfying the recurrence relation (2.8) can be given in terms of α_0, α_1, a and b.

In the search for solutions of the equation (2.8), one may try expressions of the form $x_n = t^n$. By substitution in the original equation

$$t^n \left(t^2 - at - b\right) = 0.$$

Clearly, a trivial solution is obtained for $t = 0$, so we shall assume for now that $t \neq 0$. The *characteristic equation* of sequence $(x_n)_{n\geq 0}$ is defined by

$$t^2 - at - b = 0. \tag{2.9}$$

One can distinguish two cases, depending on whether the two roots t_1, t_2 of the characteristic equation (2.9) are distinct or equal.

Case 1. Distinct roots If the roots t_1, t_2 of (2.9) are distinct (real or complex), then the sequences $(t_1^n)_{n\geq 0}$ and $(t_2^n)_{n\geq 0}$ are both solutions of (2.8).

The general formula is given by the linear combination

$$x_n = c_1 t_1^n + c_2 t_2^n, \quad n = 0, 1, 2, \ldots, \tag{2.10}$$

where the coefficients c_1 and c_2 are determined by the system of linear equations

$$\begin{cases} c_1 + c_2 = \alpha_0 \\ c_1 t_1 + c_2 t_2 = \alpha_1. \end{cases} \tag{2.11}$$

Case 2. Equal roots If $t_1 = t_2$, then the solution of (2.8) is given by

$$x_n = (c_1 + c_2 n)\, t_1^n, \quad n = 0, 1, 2, \ldots, \tag{2.12}$$

where the constant coefficients c_1 and c_2 are determined by the system

$$\begin{cases} c_1 = \alpha_0 \\ (c_1 + c_2)\, t_2 = \alpha_1. \end{cases}$$

The relations (2.10) and (2.12) are often called *Binet-type formulae*. Initially formulated by Binet in 1843 in the context of the Fibonacci sequence, numerous Binet-type formulae have been obtained for generalizations of the Fibonacci numbers [92], including the case of recurrent sequences of higher order. We shall discuss such examples in Chapter 6.

Remark 2.2 When the initial conditions are not specified, the recurrence relation (2.8) is satisfied by a family of sequences. The structure of this family is that of a vector space of dimension 2, for which the sequences $(t_1^n)_{n\geq 0}$ and $(t_2^n)_{n\geq 0}$ form a basis when $t_1 \neq t_2$. When $t_1 = t_2$, a basis for this vector space is given by the sequences $(t_1^n)_{n\geq 0}$ and $(nt_1^n)_{n\geq 0}$. This superposition principle has general applicability, and will be encountered again in Chapter 6, when we solve linear recurrent sequences of higher order.

Negative indices The recurrence relation (2.8) can be extended for negative indices and a link can be established between x_{-n} and x_n. We focus on the nondegenerate case, when the roots t_1 and t_2 of (2.9) are distinct.

Indeed, considering $n \geq 0$, one may use (2.10) to write the expressions corresponding to x_{-n}, x_{-n-1} and x_{-n-2}. Indeed, we have

$$x_{-n} - ax_{-n-1} - bx_{-n-2} = c_1 t_1^{-n-2}(t_1^2 - at_1 - b) + c_2 t_2^{-n-2}(t_2^2 - at_2 - b) = 0.$$

Moreover,

$$\begin{aligned} x_{-n} &= \frac{c_1}{t_1^n} + \frac{c_2}{t_2^n} = \frac{c_1 t_2^n + c_2 t_1^n}{(t_1 t_2)^n} = \frac{(-1)^n}{b^n}\left[-x_n + (c_1 + c_2)(t_1^n + t_2^n)\right] \\ &= \frac{(-1)^n}{b^n}\left[-x_n + \alpha_0(t_1^n + t_2^n)\right], \end{aligned} \tag{2.13}$$

where we have used the relations $c_1 + c_2 = \alpha_0$ and $t_1 t_2 = -b$.

Matrix form The linear recurrence relation of second order (2.8) can be expressed in matrix form as follows:

$$\begin{pmatrix} x_{n+2} \\ x_{n+1} \end{pmatrix} = \begin{pmatrix} a & b \\ 1 & 0 \end{pmatrix} \begin{pmatrix} x_{n+1} \\ x_n \end{pmatrix}. \tag{2.14}$$

Considering the notations

$$X_n = \begin{pmatrix} x_{n+1} \\ x_n \end{pmatrix}, \quad A = \begin{pmatrix} a & b \\ 1 & 0 \end{pmatrix},$$

the relation (2.14) can be written as $X_{n+1} = AX_n$, and we have

$$X_n = A^n X_0. \tag{2.15}$$

The sequence $(x_n)_{n\geq 0}$ can be determined if we can compute the powers of matrix A, which can be done using the eigenvalues of the matrix A, i.e., the roots t_1, t_2 of the characteristic equation (2.9).

Indeed, for distinct eigenvalues, there is a matrix P such that

$$A = P \begin{pmatrix} t_1 & 0 \\ 0 & t_2 \end{pmatrix} P^{-1},$$

hence by (2.15) we obtain the relation

$$\begin{pmatrix} x_{n+1} \\ x_n \end{pmatrix} = X_n = A^n X_0 = P \begin{pmatrix} t_1 & 0 \\ 0 & t_2 \end{pmatrix}^n P^{-1} X_0 = P \begin{pmatrix} t_1^n & 0 \\ 0 & t_2^n \end{pmatrix} P^{-1} X_0,$$

which clearly produces the Binet-type formula (2.10).

As shown in [17, 18], second-order linear recurrent sequences can be written as first-order nonlinear recurrent sequences.

Theorem 2.4

1° *The following formula holds:*

$$\begin{pmatrix} x_{n-1} & x_n \\ x_n & x_{n+1} \end{pmatrix} = \begin{pmatrix} 0 & 1 \\ b & a \end{pmatrix}^{n-1} \begin{pmatrix} x_0 & x_1 \\ x_1 & x_2 \end{pmatrix}, \quad n = 1, 2, \ldots. \tag{2.16}$$

2° *The recurrence sequence* (2.8) *satisfies the identity below for* $n = 1, 2, \ldots$

$$x_n^2 - ax_nx_{n-1} - bx_{n-1}^2 = (-1)^{n-1}b^{n-1}(\alpha_1^2 - a\alpha_0\alpha_1 - b\alpha_0^2). \tag{2.17}$$

Proof

1° Note that the following matrix relation holds for $n \geq 2$

$$\begin{pmatrix} x_{n-1} & x_n \\ x_n & x_{n+1} \end{pmatrix} = \begin{pmatrix} 0 & 1 \\ b & a \end{pmatrix} \begin{pmatrix} x_{n-2} & x_{n-1} \\ x_{n-1} & x_n \end{pmatrix}, \tag{2.18}$$

hence, we can write

$$\begin{pmatrix} x_{n-1} & x_n \\ x_n & x_{n+1} \end{pmatrix} = \begin{pmatrix} 0 & 1 \\ b & a \end{pmatrix}^{n-1} \begin{pmatrix} x_0 & x_1 \\ x_1 & x_2 \end{pmatrix}.$$

2° Taking determinants in both sides, we obtain the relation

$$x_{n-1}x_{n+1} - x_n^2 = (-b)^{n-1}\left(x_0x_2 - x_1^2\right),$$

which by the recurrence formula (2.8) gives

$$x_{n-1}(ax_n + bx_{n-1}) - x_n^2 = (-b)^{n-1}\left[x_0(ax_1 + bx_0) - x_1^2\right],$$

which represents the relation (2.17).

□

Remark 2.3 Using formula (2.16) we can show that

$$\begin{pmatrix} x_{n+1} & x_n \\ x_n & x_{n-1} \end{pmatrix} = \begin{pmatrix} a & b \\ 1 & 0 \end{pmatrix}^{n-1} \begin{pmatrix} x_2 & x_1 \\ x_1 & x_0 \end{pmatrix}. \tag{2.19}$$

Indeed, we have

$$\begin{pmatrix} 0 & 1 \\ b & a \end{pmatrix} = \begin{pmatrix} 0 & 1 \\ 1 & 0 \end{pmatrix} \begin{pmatrix} a & b \\ 1 & 0 \end{pmatrix} \begin{pmatrix} 0 & 1 \\ 1 & 0 \end{pmatrix} = \begin{pmatrix} 0 & 1 \\ 1 & 0 \end{pmatrix} \begin{pmatrix} a & b \\ 1 & 0 \end{pmatrix} \begin{pmatrix} 0 & 1 \\ 1 & 0 \end{pmatrix}^{-1},$$

and

$$\begin{pmatrix} 0 & 1 \\ b & a \end{pmatrix}^k = \begin{pmatrix} 0 & 1 \\ 1 & 0 \end{pmatrix} \begin{pmatrix} a & b \\ 1 & 0 \end{pmatrix}^k \begin{pmatrix} 0 & 1 \\ 1 & 0 \end{pmatrix}^{-1}.$$

Therefore, it follows that

$$\begin{aligned} \begin{pmatrix} x_{n+1} & x_n \\ x_n & x_{n-1} \end{pmatrix} &= \begin{pmatrix} 0 & 1 \\ 1 & 0 \end{pmatrix} \begin{pmatrix} x_{n-1} & x_n \\ x_n & x_{n+1} \end{pmatrix} \begin{pmatrix} 0 & 1 \\ 1 & 0 \end{pmatrix} \\ &= \begin{pmatrix} 0 & 1 \\ 1 & 0 \end{pmatrix} \begin{pmatrix} 0 & 1 \\ b & a \end{pmatrix}^{n-1} \begin{pmatrix} x_0 & x_1 \\ x_1 & x_2 \end{pmatrix} \begin{pmatrix} 0 & 1 \\ 1 & 0 \end{pmatrix} \\ &= \begin{pmatrix} 0 & 1 \\ 1 & 0 \end{pmatrix} \begin{pmatrix} 0 & 1 \\ 1 & 0 \end{pmatrix} \begin{pmatrix} a & b \\ 1 & 0 \end{pmatrix}^{n-1} \begin{pmatrix} 0 & 1 \\ 1 & 0 \end{pmatrix} \begin{pmatrix} x_0 & x_1 \\ x_1 & x_2 \end{pmatrix} \begin{pmatrix} 0 & 1 \\ 1 & 0 \end{pmatrix} \\ &= \begin{pmatrix} a & b \\ 1 & 0 \end{pmatrix}^{n-1} \begin{pmatrix} x_2 & x_1 \\ x_1 & x_0 \end{pmatrix}. \end{aligned}$$

Remark 2.4 For a, b, c, d real numbers, the following identity holds:

$$\begin{pmatrix} 0 & 1 \\ 1 & 0 \end{pmatrix} \begin{pmatrix} a & b \\ c & d \end{pmatrix} \begin{pmatrix} 0 & 1 \\ 1 & 0 \end{pmatrix} = \begin{pmatrix} d & c \\ b & a \end{pmatrix}. \tag{2.20}$$

Note that the matrix $\begin{pmatrix} 0 & 1 \\ 1 & 0 \end{pmatrix}$ is equal to its inverse, while the effect of the matrix multiplications in (2.20) is the interchange of the rows and the columns.

Example 2.9 The sequence $(x_n)_{n\geq 1}$ is defined by $x_1 = 0$ and

$$x_{n+1} = 5x_n + \sqrt{24x_n^2 + 1}, \quad n = 1, 2, \ldots.$$

Prove that all x_n are positive integers.

Solution It is clear that $x_1 < x_2 < \ldots$. The recursive relation is equivalent to

$$x_{n+1}^2 - 10x_n x_{n+1} + x_n^2 - 1 = 0, \quad n = 1, 2, \ldots.$$

Replacing n by $n - 1$ we get

$$x_n^2 - 10x_n x_{n-1} + x_{n-1}^2 - 1 = 0, \quad n = 2, 3, \ldots.$$

It follows that x_{n+1} and x_{n-1} are the roots of the quadratic equation

$$t^2 - 10x_n t + x_n^2 - 1 = 0,$$

hence

$$x_{n+1} + x_{n-1} = 10x_n.$$

We obtain

$$x_{n+1} = 10x_n - x_{n-1}, \quad n = 2, 3, \ldots, \quad x_1 = 0, \quad x_2 = 1.$$

An inductive argument shows that x_n is a positive integer for any n.

Example 2.10 Determine x_n if the sequence $(x_n)_{n\geq 1}$ is defined by $x_1 = 1$ and the formula $x_{n+1} = 3x_n + \lfloor x_n\sqrt{5} \rfloor$ for all $n = 1, 2, \ldots$, where $\lfloor x \rfloor$ denotes the greatest integer that does not exceed x.

Solution We have $x_{n+1} = \lfloor (3 + \sqrt{5})x_n \rfloor$ for all $n \geq 1$, and we deduce that

$$x_{n+2} = \lfloor (3 + \sqrt{5})x_{n+1} \rfloor = 6x_{n+1} + \lfloor -(3 - \sqrt{5}) \lfloor (3 + \sqrt{5})x_n \rfloor \rfloor$$

$$= 6x_{n+1} + \left\lfloor -4\frac{\lfloor (3+\sqrt{5})x_n \rfloor}{(3+\sqrt{5})} \right\rfloor = 6x_{n+1} - 4x_n,$$

with $x_1 = 1$ and $x_2 = 5$. We know that

$$x_n = c_1 t_1^n + c_2 t_2^n, \quad n = 1, 2, \ldots,$$

where t_1, t_2 are the roots of the quadratic equation $t^2 - 6t + 4 = 0$, that is $t_1 = 3+\sqrt{5}$ and $t_2 = 3 - \sqrt{5}$. Solving the system $c_1 t_1 + c_2 t_2 = 1$, $c_1 t_1^2 + c_2 t_2^2 = 5$, we obtain $c_1 = \frac{1+\sqrt{5}}{8\sqrt{5}}$ and $c_2 = \frac{-1+\sqrt{5}}{8\sqrt{5}}$, hence the formula for x_n is

$$x_n = \frac{1}{8\sqrt{5}} \left[(1+\sqrt{5})(3+\sqrt{5})^n + (-1+\sqrt{5})(3-\sqrt{5})^n \right].$$

One may now check by induction that $x_n = 2^{n-2} F_{2n+1}$, $n = 1, 2, \ldots$, where F_n denotes the nth Fibonacci number (clearly, $x_2 = 5 = 2^0 F_5$, $x_3 = 26 = 2^1 F_7$).

Example 2.11 Determine all functions $f : [0, \infty) \to \mathbb{R}$ such that $f(0) = 0$ and

$$f(x) = 1 + 5f\left(\left\lfloor \frac{x}{2} \right\rfloor\right) - 6f\left(\left\lfloor \frac{x}{4} \right\rfloor\right), \quad x > 0.$$

Solution Let $x \geq 0$. If $x \subset (0, 2)$, then $f(x) = 1 + 5f(0) - 6f(0) = 1$. If $x \in [2, 4)$, then $f(x) = 1 + 5f(1) - 6f(0) = 6 = a_1$. If $x \in [4, 8)$, then $\left\lfloor \frac{x}{2} \right\rfloor \in [2, 4)$ and $\left\lfloor \frac{x}{4} \right\rfloor \in [1, 2)$, hence $f(x) = 1 + 5f(2) - 6f(1) = 1 + 5 \cdot 6 - 6 \cdot 1 = 25 = a_2$. We shall now proceed by induction. Assume for $n \geq 1$, that the function f is constant on $[2^n, 2^{n+1})$, having the value a_n, and constant on $[2^{n+1}, 2^{n+2})$, with the value a_{n+1}. If $x \in [2^{n+2}, 2^{n+3})$, then we have $\left\lfloor \frac{x}{2} \right\rfloor \in [2^{n+1}, 2^{n+2})$ and $\left\lfloor \frac{x}{4} \right\rfloor \in [2^n, 2^{n+1})$, hence

$$f(x) = 1 + 5a_{n+1} - 6a_n.$$

We deduce that f is constant on the interval $[2^{n+2}, 2^{n+3})$, taking the particular value $a_{n+2} = 1 + 5a_{n+1} - 6a_n$, which can be rewritten as

$$a_{n+2} - \frac{1}{2} = 5\left(a_{n+1} - \frac{1}{2}\right) - 6\left(a_n - \frac{1}{2}\right).$$

Since the roots of the characteristic equation $t^2 - 5t + 6 = 0$ are 2 and 3, there exist two constants c_1 and c_2 such that

$$a_n = \frac{1}{2} + c_1 2^n + c_2 3^n, \quad n \geq 1.$$

From $a_1 = 6$ and $a_2 = 25$ we get $c_1 = -4$ and $c_2 = \frac{9}{2}$, and finally

$$f(x) = \begin{cases} 0 & \text{if } x = 0, \\ 1 & \text{if } x \in (0, 2) \\ -2^{n+2} + \frac{3^{n+2}+1}{2} & \text{if } x \in [2^n, 2^{n+1}) \text{ and } n \geq 1. \end{cases}$$

Example 2.12 Find all functions $f : \mathbb{R} \to \mathbb{R}$ such that

$$2f(x) = f(x + y) + f(x + 2y),$$

for all $x \in \mathbb{R}$ and for all $y \geq 0$.

Solution Without loss of generality, we may assume that $f(0) = 0$. Let $y > 0$ and n a positive integer. For $x = ny$ we get

$$2f(ny) = f((n + 1)y) + f((n + 2)y). \tag{2.21}$$

The sequence $a_n = f(ny)$ satisfies the second-order linear recursive relation

$$a_{n+2} = -a_{n+1} + 2a_n, \tag{2.22}$$

with $a_0 = f(0) = 0$, $a_1 = f(y)$. The characteristic equation is $t^2 - t - 2 = 0$, with the roots $t_1 = 1$ and $t_2 = -2$. It follows that $a_n = c_1 + c_2(-2)^n$, where the constants c_1, c_2 satisfy $c_1 + c_2 = 0$ and $c_1 - 2c_2 = f(y)$. We get

$$f(ny) = a_n = \frac{1 - (-2)^n}{3} \cdot f(y). \tag{2.23}$$

In particular, we have $f(4y) = -5f(y)$, while from (2.23) get $f(2y) = -f(y)$, hence $f(4y) = -f(2y) = f(y)$, for any $y \geq 0$. This confirms that $f(y) = 0$ for any $y \geq 0$. For any $x \in \mathbb{R}$, we have

$$2f(x) = f(x + |x|) + f(x + 2|x|) = 0,$$

hence $f(x) = 0$. This proves that all the desired functions are constant.

Example 2.13 Consider a nonzero real number a such that

$$\{a\} + \left\{\frac{1}{a}\right\} = 1,$$

where $\{x\}$ is the fractional part of x. Prove that for every integer $n > 0$

$$\{a^n\} + \left\{\frac{1}{a^n}\right\} = 1.$$

Solution We have

$$a+\frac{1}{a}=\lfloor a\rfloor+\left\lfloor\frac{1}{a}\right\rfloor+\{a\}+\left\{\frac{1}{a}\right\}=\lfloor a\rfloor+\left\lfloor\frac{1}{a}\right\rfloor+1,$$

is an integer, which we denote by k. Letting $S_n=a^n+\frac{1}{a^n}$, $n=0,1,\ldots$. Since a and $\frac{1}{a}$ are the roots of the quadratic equation $x^2-kx+1=0$, it follows that

$$x_{n+2}=kx_{n+1}-x_n,\quad n=0,1,2,\ldots,$$

where $S_0=2$ and $S_1=k$. By induction with step 2, we show that $S_n\in\mathbb{Z}$ for $n=2,3,\ldots$. It follows that

$$\{a^n\}+\left\{\frac{1}{a^n}\right\}=a+\frac{1}{a}-\lfloor a\rfloor=\left\lfloor\frac{1}{a}\right\rfloor=x_n-\lfloor a\rfloor+\left\lfloor\frac{1}{a}\right\rfloor\in\mathbb{Z},$$

hence $\{a^n\}+\left\{\frac{1}{a^n}\right\}=m\in\mathbb{Z}$, from where we deduce that $m\in\{0,1\}$. If $m=0$, then a^n and $\frac{1}{a^n}$ are integers, which is not possible. This shows that $m=1$, which ends the proof.

Example 2.14 (IMO Shortlist 2013) Let $n>0$ be an integer and let $a_1,a_2,\ldots,a_{n-1}$ be arbitrary real numbers. Define the sequences $u_0,u_1,\ldots,u_n$ and $v_0,v_1,\ldots,v_n$ inductively by $u_0=u_1=v_0=v_1$ and the recursive relations $u_{k+1}=u_k+a_ku_{k-1}$, $v_{k+1}=v_k+a_{n-k}v_{k-1}$, $k=1,\ldots,n-1$. Prove that $u_n=v_n$.

Solution For $k=1,2,\ldots,n-1$, let $x_{k+1}=u_{k+1}-u_k$ and $y_{k+1}=v_{k+1}-v_k$, and define the matrix $\begin{pmatrix}1+a_k & -a_k\\ a_k & -a_k\end{pmatrix}$. The following relations hold:

$$\begin{pmatrix}u_{k+1}\\ x_{k+1}\end{pmatrix}=A_k\begin{pmatrix}u_k\\ x_k\end{pmatrix}\text{ and }\left(v_{k+1};\ y_{k+1}\right)=\left(v_k;\ y_k\right)A_{n-k}.$$

From these relations we deduce that

$$\begin{pmatrix}u_n\\ x_n\end{pmatrix}=A_{n-1}A_{n-2}\cdots A_2A_1\begin{pmatrix}u_1\\ x_1\end{pmatrix}=A_{n-1}A_{n-2}\cdots A_2A_1\begin{pmatrix}1\\ 0\end{pmatrix}$$

$$\left(v_n;\ y_n\right)=\left(v_1;\ y_1\right)A_{n-1}A_{n-2}\cdots A_2A_1=(1;\ 0)\,A_{n-1}A_{n-2}\cdots A_2A_1.$$

We deduce that

$$(u_n)=(1;\ 0)\begin{pmatrix}u_n\\ x_n\end{pmatrix}=(1;\ 0)\,A_{n-1}\cdots A_1\begin{pmatrix}1\\ 0\end{pmatrix}=\left(v_n;\ y_n\right)\begin{pmatrix}1\\ 0\end{pmatrix}=(v_n).$$

This implies $u_n=v_n$.

The problem can also be related to the Fibonacci sequence. For example, when $a_1 = \cdots = a_{n-1}$, one has $u_k = v_k = F_{k+1}$, the $(k+1)$th Fibonacci number.

Also, the problem is equivalent to

$$\frac{u_{k+1}}{u_k} = 1 + \frac{a_k}{1 + \frac{a_{k-1}}{1 + \cdots + \frac{a_2}{1+a_1}}} \text{ and } \frac{v_{k+1}}{v_k} = 1 + \frac{a_{n-k}}{1 + \frac{a_{n-k+1}}{1 + \cdots + \frac{a_{n-2}}{1+a_{n-1}}}},$$

that is, the fractions $\frac{u_n}{u_{n-1}}$ and $\frac{v_n}{v_{n-1}}$ have the same numerator.

2.2.2 *Nonhomogeneous Recurrent Sequences*

Let $(a_n)_{n\geq 0}$ be a sequence of real (or complex) numbers. We want to determine the sequence $(x_n)_{n\geq 0}$ defined by the second-order recurrence relation

$$x_{n+2} = ax_{n+1} + bx_n + a_n, \quad n = 0, 1, \ldots, \tag{2.24}$$

with $x_0 = \alpha_0$, $x_1 = \alpha_1$, where α_0, α_1, a, b are real (or complex) numbers. By Theorem 1.2, (2.24), this special case of (1.7) has a unique solution. If $b = 0$, then the recurrence relation (2.24) is a special case of the first-order recurrence equation (2.3) seen in Section 2.1. We shall assume that $b \neq 0$.

A sequence $(v_n)_{n\geq 0}$ satisfying (2.24) is called a *particular solution* to the recurrence relation (2.24). It can be easily shown that if $(u_n)_{n\geq 0}$ satisfies the homogeneous second-order recurrence equation

$$u_{n+2} = au_{n+1} + bu_n, \quad n = 0, 1, \ldots,$$

with $u_0 = \alpha_0 - v_0$ and $u_1 = \alpha_1 - v_1$, then the sequence $(x_n)_{n\geq 0}$ given by

$$x_n = u_n + v_n, \quad n = 0, 1, \ldots, \tag{2.25}$$

is a general solution of the recurrence relation (2.24). For $n \geq 0$, we have

$$\begin{aligned} ax_{n+1} + bx_n + a_n &= a\,(u_{n+1} + v_{n+1}) + b\,(u_n + v_n) + a_n \\ &= (au_{n+1} + bu_n) + (av_{n+1} + bv_n + a_n) \\ &= u_{n+2} + v_{n+2} = x_{n+2}, \end{aligned}$$

and $x_0 = u_0 + v_0 = \alpha_0 - v_0 + v_0 = \alpha_0$ and $x_1 = u_1 + v_1 = \alpha_1 - v_1 + v_1 = \alpha_1$.

Let t_1 and t_2 be the roots of (2.9). We have the following cases.

Case 1. Distinct roots If t_1, t_2 are distinct, then

$$x_n = c_1 t_1^n + c_2 t_2^n + v_n, \quad n = 0, 1, \ldots, \tag{2.26}$$

where the coefficients c_1 and c_2 are determined by the system of linear equations (2.11), corresponding to the initial values x_0, x_1, v_0, v_1.

Case 2. Equal roots If $t_1 = t_2$, then

$$x_n = (c_1 + c_2 n)\, t_1^n + v_n, \quad n = 0, 1, \ldots, \tag{2.27}$$

where the coefficients c_1 and c_2 are determined by the system

$$\begin{cases} c_1 = \alpha_0 - v_0 \\ (c_1 + c_2)\, t_1 = \alpha_1 - v_1. \end{cases}$$

In concrete applications, the particular solution $(v_n)_{n\geq 0}$ depends on the algebraic form of the sequence $(a_n)_{n\geq 0}$. The following result clarifies the form of v_n when a_n is a product of a polynomial and an exponential of n.

Theorem 2.5 *Assume that $a_n = P(n)\beta^n$, $n = 0, 1, \ldots$, where $\beta \neq 0$ and P is a polynomial of degree s. Then, there is a polynomial Q of degree $s + \theta$ such that $v_n = Q(n)\beta^n$, $n = 0, 1, \ldots$, where*

$$\theta = \begin{cases} 0 & \text{if } \beta \neq t_1 \text{ and } \beta \neq t_2 \\ 1 & \text{if } \beta = t_1 \text{ and } \beta \neq t_2, \text{ or } \beta = t_2 \text{ and } \beta \neq t_1 \\ 2 & \text{if } \beta = t_1 = t_2. \end{cases}$$

Proof If $v_n = Q(n)\beta^n$, $n = 0, 1, \ldots$, satisfies (2.24), then we obtain

$$Q(n+2)\beta^{n+2} = aQ(n+1)\beta^{n+1} + bQ(n)\beta^n + P(n)\beta^n, \quad n = 0, 1, \ldots,$$

hence

$$Q(n+2)\beta^2 - aQ(n+1)\beta - bQ(n) = P(n), \quad n = 0, 1, \ldots. \tag{2.28}$$

Let q be the leading coefficient of Q. Clearly, one can write

$$q\left[\beta^2(n+2)^{s+\theta} - a\beta(n+1)^{s+\theta} - bn^{s+\theta}\right] = c_s n^s, \quad n = 0, 1, \ldots, \tag{2.29}$$

where $P(n) = c_s n^s + c_{s-1} n^{s-1} + \cdots + c_1 n + c_0$.

The relation (2.29) is equivalent to

$$q\Bigg[\left(\beta^2 - a\beta - b\right) n^{s+\theta} + (s+\theta)\beta(2\beta - a)n^{s+\theta-1} + \frac{(s+\theta)(s+\theta-1)}{2}(4\beta - a)n^{s+\theta-2} + \cdots\Bigg] = c_s n^s, \quad n = 0, 1, \ldots.$$

If $\beta^2 - a\beta - b \neq 0$, then we choose $\theta = 0$ and obtain $q = \frac{c_s}{\beta^2 - a\beta - b}$.
If $\beta^2 - a\beta - b = 0$ and $\beta \neq \frac{a}{2}$, then $\theta = 1$ and obtain $q = \frac{c_s}{(s+1)\beta(2\beta - a)}$.
If $\beta^2 - a\beta - b = 0$ and $\beta = \frac{a}{2}$, then $\theta = 2$ and obtain $q = \frac{2c_s}{(s+2)(s+1)\beta(4\beta - a)}$.
After we have determined the leading coefficient of Q, we proceed in a similar fashion to obtain the next coefficient. □

A special situation is the case when $\beta = 1$, where a_n is a polynomial of degree s in n. We have the following consequence.

Corollary 2.2 *Given the nonhomogeneous linear recurrence relation* (2.24), *where* $a_n = P(n)$, $n = 0, 1, \ldots,$ *and* P *is a polynomial of degree* s.

1° *If* $a + b \neq 1$, *then* (2.24) *has a particular solution of the form* $v_n = Q(n)$, *where* Q *is a polynomial of degree* s.
2° *If* $a + b = 1$, *then* (2.24) *can be reduced to the first-order recurrence equation*

$$y_{n+2} = (a - 1)y_{n+1} + a_n, \quad n = 0, 1, \ldots,$$

where $(y_n)_{n \geq 1}$ *is given by* $y_n = x_n - x_{n-1}$, $n = 1, 2, \ldots$.

Example 2.15 Determine the sequences $(x_n)_{n \geq 0}$ defined by

$$x_{n+2} = 6x_{n+1} - 9x_n + 8n^2 - 24n, \quad n = 0, 1, \ldots,$$

with $x_0 = 5$ and $x_1 = 5$.

Solution We are in the first case of Corollary 2.2, where $P(n) = 8n^2 - 24n$. Put $v_n = A_2n^2 + A_1n + A_0$ into the recurrence relation and obtain

$$A_2(n+2)^2 + A_1(n+2) + A_0 = 6\left[A_2(n+1)^2 + A_1(n+1) + A_0\right] - 9\left(A_2n^2 + A_1n + A_0\right) + 8n^2 - 24n.$$

Collecting the coefficients of n^2, n, and the constant, we have

$$(4A_2 - 8)\,n^2 + (-8A_2 + 4A_1 + 24)\,n + (-2A_2 - 4A_1 + 4A_0) = 0.$$

It follows that $4A_2 - 8 = 0$, $-8A_2 + 4A_1 + 24 = 0$, and $-2A_2 - 4A_1 + 4A_0 = 0$. This has the solution $A_2 = 2$, $A_1 = -2$, and $A_0 = -1$, hence $v_n = 2n^2 - 2n - 1$ is a particular solution. It follows that the general solution of the recurrence relation is given by formula (2.27) as

$$x_n = (c_1 + c_2n)3^n + 2n^2 - 2n - 1, \quad n = 0, 1, \ldots.$$

Using the initial conditions $x_0 = 5$ and $x_1 = 5$, we have $c_1 = 8$ and $c_2 = -\frac{14}{3}$. The sequence is finally obtained as

$$x_n = \left(8 - \frac{14}{3}n\right)3^n + 2n^2 - 2n - 1$$
$$= 8 \cdot 3^n - 14n3^{n-1} + 2n^2 - 2n - 1, \quad n = 0, 1, \ldots.$$

Example 2.16 Find the sequences $(x_n)_{n \geq 0}$ satisfying the recurrence relation

$$x_{n+2} = 3x_{n+1} - 2x_n + 2^n, \quad n = 0, 1, \ldots.$$

Solution We have $a_n = 2^n$, and 2 is a simple root of the equation $t^2 - 3t + 2 = 0$. Applying Theorem 2.5, a particular solution to this relation is of the form $v_n = Q(n)2^n$, where Q is a first degree polynomial, therefore we have $v_n = (An + b)2^n$, $n = 0, 1, \ldots$. Replacing in the recurrence relation we obtain $A = \frac{1}{2}$ and B arbitrary, hence we can choose $v_n = n2^{n-1}$, $n = 0, 1, \ldots$. The general solution is given by formula (2.26) as

$$x_n = c_1 + c_2 2^n + n2^{n-1}, \quad n = 0, 1, \ldots,$$

where c_1 and c_2 are constants.

Example 2.17 Consider the sequences $(x_n)_{n \geq 0}$ satisfying the relation

$$x_{n+2} = -4x_{n+1} - 4x_n + (-1)^n, \quad n = 0, 1, \ldots.$$

Solution We have $a_n = (-1)^n$, and -1 is different from the double root -2 of the equation $t^2 + 4t + 4 = 0$. Using Theorem 2.5, it follows that a particular solution is of the form $v_n = A(-1)^n$, $n = 0, 1, \ldots$. Replacing in the recurrence relation, we obtain $A = 1$. The general solution is given by (2.27)

$$x_n = (c_1 + c_2 n)(-2)^n + (-1)^n, \quad n = 0, 1, \ldots,$$

where the coefficients c_1 and c_2 are arbitrary constants.

Example 2.18 Let $(x_n)_{n \geq 0}$ be the sequence defined by the recursive relation

$$x_{n+2} = ax_{n+1} + bx_n + c, \quad n = 0, 1, \ldots,$$

where $a > 0$, $b > 0$, and $c > 0$ are real numbers. Prove that if $a + 2b + c < 1$, then $(x_n)_{n \geq 0}$ is convergent.

Solution Indeed, from the condition $a + 2b + c < 1$, it follows that $a + b < 1$, hence 1 is not a root of the equation $t^2 - at - b = 0$. According to Theorem 2.5, a particular solution is of the form $v_n = \alpha$, $n = 0, 1, \ldots$, where α is a real number. Replacing in the recurrence relation, we get $\alpha = \frac{c}{1-a-b}$. The general solution to this recurrence relation is then given by formula (2.26)

$$x_n = c_1 t_1^n + c_2 t_2^n + \frac{c}{1-a-b}, \quad n = 0, 1, \ldots,$$

where t_1, t_2 are the roots of $t^2 - at - b = 0$. The discriminant of this equation is $\Delta = a^2 + 4b > 0$, hence t_1, t_2 are real numbers. On the other hand, considering the quadratic function $h(t) = t^2 - at - b$, we have $h(1) = 1 - a - b > 0$, $h(-1) = 1 + a - b > 0$, and $h(0) = -b < 0$, therefore $t_1, t_2 \in (-1, 1)$. It follows that $\lim_{n\to\infty} x_n = \frac{c}{1-a-b}$.

Some examples of nonhomogeneous second-order recurrence sequences in the complex plane are further discussed in Section 5.8.

2.2.3 *Fibonacci, Lucas, Pell, and Pell–Lucas Numbers*

The Fibonacci numbers $(F_n)_{n\geq 0}$ satisfy the recurrence relation $F_{n+2} = F_{n+1} + F_n$, and start with the terms

$$0, 1, 1, 2, 3, 5, 8, 13, 21, 34, 55, 89, \ldots \text{ (to A000045 in the OEIS [157]).}$$

They are ubiquitous in nature (flower patterns or bee crawling) and play roles in chemistry, music, poetry, or... stock exchange. Numerous identities, extensions, and abstractions exist [143, 147, 175], as well as dedicated monographs [101], and even journals, like Fibonacci Quarterly.

They have applications in Graph Theory (Fibonacci graphs) [97], data structures (Fibonacci heap) [66], Combinatorics [115], or search algorithms (Fibonacci search) [91]. Numerous links with Chebyshev polynomials and generating functions are known (see [43, 45, 46], or [118, 119, 121–123]).

Fermat's spiral, a geometric pattern related to the golden section, provided optimal solutions for the layout of mirrors in a concentrated solar power plant [131]. Fibonacci numbers were related to other optimality problems [100].

Lucas numbers $(L_n)_{n\geq 0}$ are another famous sequence, satisfying the recurrence relation $L_{n+2} = L_{n+1} + L_n$, and starting with

$$2,\ 1,\ 3,\ 4,\ 7,\ 11,\ 18,\ 29,\ 47,\ 76,\ 123,\ \ldots \text{ (to A000032 in OEIS [157]),}$$

which is closely related to the Fibonacci numbers, and satisfying the same recurrence equation, but starting with different values.

Pell numbers $(P_n)_{n\geq 0}$ are another classical second-order recurrent sequence, satisfying $P_{n+2} = 2P_{n+1} + P_n$, and starting with

$$0, 1, 2, 5, 12, 29, 70, 169, 408, 985, 2378, 5741, 13860, \ldots \text{ (to A000129 in OEIS [157]).}$$

This sequence has numerous applications, related to the approximation of $\sqrt{2}$ by rationals, numerous identities, and Diophantine equations.

The integer solutions to the equation $x^2 - 2y^2 = 4(-1)^n$ are (Q_n, P_n), where Q_n is the nth Pell–Lucas number. The sequences $(P_n)_{n\geq 0}$ and $(Q_n)_{n\geq 0}$ satisfy the same recurrence equation, the only difference being in the initial conditions, i.e., we have $P_0 = 0$ and $P_1 = 1$, whereas $Q_0 = 2$ and $Q_1 = 2$.

The Pell–Lucas numbers are companion to Pell numbers, starting with

2, 2, 6, 14, 34, 82, 198, 478, 1154, 2786, 6726, 16238, ... (to A002203 in OEIS [157]).

Recall that a Binet-type formula gives the general term of a recurrent sequence for general second-order linear recurrent sequences, as in (2.10) and (2.12). Such formulae are common for Fibonacci, Lucas, Pell, and Pell–Lucas numbers.

Theorem 2.6

1° *(Fibonacci numbers) Let F_n be given by $F_0 = 0$, $F_1 = 1$ and*

$$F_{n+2} = F_{n+1} + F_n, \quad n = 0, 1, 2, \ldots.$$

The formula of the general term is given by

$$F_n = \frac{1}{\sqrt{5}}\left[\left(\frac{1+\sqrt{5}}{2}\right)^n - \left(\frac{1-\sqrt{5}}{2}\right)^n\right]. \tag{2.30}$$

2° *(Lucas numbers) Let L_n be given by $L_0 = 2$, $L_1 = 1$ and*

$$L_{n+2} = L_{n+1} + L_n, \quad n = 0, 1, 2, \ldots.$$

The formula of the general term is given by

$$L_n = \left(\frac{1+\sqrt{5}}{2}\right)^n + \left(\frac{1-\sqrt{5}}{2}\right)^n. \tag{2.31}$$

3° *(Pell numbers) Let P_n be given by $P_0 = 0$, $P_1 = 1$ and*

$$P_{n+2} = 2P_{n+1} + P_n, \quad n = 0, 1, 2, \ldots.$$

The formula of the general term is given by

$$P_n = \frac{1}{2\sqrt{2}}\left[\left(1+\sqrt{2}\right)^n - \left(1-\sqrt{2}\right)^n\right]. \tag{2.32}$$

4° *(Pell–Lucas numbers) Let Q_n be given by $Q_0 = 2,\ Q_1 = 2$ and*

$$Q_{n+2} = 2Q_{n+1} + Q_n, \quad n = 0, 1, 2, \ldots.$$

The formula of the general term is given by

$$Q_n = \left(1+\sqrt{2}\right)^n + \left(1-\sqrt{2}\right)^n. \tag{2.33}$$

Proof

1° The associated characteristic equation is

$$t^2 - t - 1 = 0,$$

having the distinct roots $t_1 = \dfrac{1+\sqrt{5}}{2}$ and $t_2 = \dfrac{1-\sqrt{5}}{2}$. Solving the system

$$\begin{cases} c_1 + c_2 = 0 \\ c_1 t_1 + c_2 t_2 = 1, \end{cases}$$

we get $c_1 = \dfrac{1}{\sqrt{5}},\ c_2 = -\dfrac{1}{\sqrt{5}}$ and the desired formula follows.

2° The proof is similar to that used at 1° for Fibonacci numbers, but we have different starting points.

3° The associated characteristic equation is

$$t^2 - 2t - 1 = 0,$$

having the distinct roots $t_1 = 1 + \sqrt{2}$ and $t_2 = 1 - \sqrt{2}$, while from the initial conditions $P_0 = 0,\ P_1 = 1$ one obtains the formula.

4° The proof follows the same ideas as for Pell numbers, but here we have the starting points $Q_0 = 2,\ Q_1 = 2$.

□

These formulae can be extended naturally to negative integers, where they yield the identities $F_{-n} = (-1)^{n-1}F_n,\ L_{-n} = (-1)^n L_n,\ P_{-n} = (-1)^{n-1}P_n$, and $Q_{-n} = (-1)^n Q_n$, valid for all n.

Using the expression (2.16), we can obtain matrix forms for the terms of the Fibonacci, Lucas, Pell, and Pell–Lucas sequences.

Theorem 2.7 *The following formulae hold for all integers $n \geq 1$:*

$$\begin{pmatrix} F_{n-1} & F_n \\ F_n & F_{n+1} \end{pmatrix} = \begin{pmatrix} 0 & 1 \\ 1 & 1 \end{pmatrix}^{n-1} \begin{pmatrix} 0 & 1 \\ 1 & 1 \end{pmatrix} = \begin{pmatrix} 0 & 1 \\ 1 & 1 \end{pmatrix}^n, \tag{2.34}$$

$$\begin{pmatrix} L_{n-1} & L_n \\ L_n & L_{n+1} \end{pmatrix} = \begin{pmatrix} 0 & 1 \\ 1 & 1 \end{pmatrix}^{n-1} \begin{pmatrix} 2 & 1 \\ 1 & 3 \end{pmatrix}, \tag{2.35}$$

$$\begin{pmatrix} P_{n-1} & P_n \\ P_n & P_{n+1} \end{pmatrix} = \begin{pmatrix} 0 & 1 \\ 1 & 2 \end{pmatrix}^{n-1} \begin{pmatrix} 0 & 1 \\ 1 & 2 \end{pmatrix} = \begin{pmatrix} 0 & 1 \\ 1 & 2 \end{pmatrix}^{n}, \tag{2.36}$$

$$\begin{pmatrix} Q_{n-1} & Q_n \\ Q_n & Q_{n+1} \end{pmatrix} = \begin{pmatrix} 0 & 1 \\ 1 & 2 \end{pmatrix}^{n-1} \begin{pmatrix} 2 & 2 \\ 2 & 6 \end{pmatrix}. \tag{2.37}$$

Remark 2.5 Using formula (2.19) it follows that for $n \geq 1$ we have

$$\begin{pmatrix} F_{n+1} & F_n \\ F_n & F_{n-1} \end{pmatrix} = \begin{pmatrix} 1 & 1 \\ 1 & 0 \end{pmatrix}^{n};$$

$$\begin{pmatrix} L_{n+1} & L_n \\ L_n & L_{n-1} \end{pmatrix} = \begin{pmatrix} 1 & 1 \\ 1 & 0 \end{pmatrix}^{n-1} \begin{pmatrix} 3 & 1 \\ 1 & 2 \end{pmatrix};$$

$$\begin{pmatrix} P_{n+1} & P_n \\ P_n & P_{n-1} \end{pmatrix} = \begin{pmatrix} 2 & 1 \\ 1 & 0 \end{pmatrix}^{n};$$

$$\begin{pmatrix} Q_{n+1} & Q_n \\ Q_n & Q_{n-1} \end{pmatrix} = \begin{pmatrix} 2 & 1 \\ 1 & 0 \end{pmatrix}^{n-1} \begin{pmatrix} 6 & 2 \\ 2 & 2 \end{pmatrix}.$$

Taking determinants for the identities in Theorem 2.7 or Remark 2.5, one obtains the Cassini identities.

Theorem 2.8 *The following formulae hold for $n \geq 1$:*

$$F_n^2 - F_{n-1}F_{n+1} = (-1)^{n-1} \tag{2.38}$$

$$L_n^2 - L_{n-1}L_{n+1} = 5(-1)^n \tag{2.39}$$

$$P_n^2 - P_{n-1}P_{n+1} = (-1)^{n-1} \tag{2.40}$$

$$Q_n^2 - Q_{n-1}Q_{n+1} = 8(-1)^n. \tag{2.41}$$

Some history and related generalizations of these results are discussed by Melham in [120], or by Fairgrieve and Gould in [63]. For example, the relation (2.38) was apparently proved by Simson in 1753.

We give below an application of the classical Cassini identity.

Example 2.19 Compute $\sum_{n=1}^{\infty} \frac{(-1)^{n+1}}{F_n F_{n+1}}$, where $(F_n)_{n\geq 0}$ is the Fibonacci sequence.

Solution Indeed, by the Cassini relation

$$F_n^2 = F_{n-1}F_{n+1} + (-1)^{n+1},$$

the sum of the first n terms is given by

$$S_n = \sum_{k=1}^{n} \frac{(-1)^{k+1}}{F_k F_{k+1}} = 1 - \sum_{k=2}^{n} \frac{F_{k-1}F_{k+1} - F_k^2}{F_k F_{k+1}}$$
$$= 1 - \sum_{k=2}^{n} \left(\frac{F_{k-1}}{F_k} - \frac{F_k}{F_{k+1}} \right) = \frac{F_n}{F_{n+1}}.$$

Taking limits, the sum of the series is given by

$$\sum_{n=1}^{\infty} \frac{(-1)^{n+1}}{F_n F_{n+1}} = \lim_{n\to\infty} S_n = \frac{2}{1+\sqrt{5}} = \frac{\sqrt{5}-1}{2}.$$

One may consult [3, 85, 104] for more results on reciprocal sums.

In what follows we are also going to discuss some further generalizations. Using Theorem 2.4 for Fibonacci, Lucas, Pell, and Pell–Lucas numbers, or directly from Theorem 2.8, one can obtain some other classical results.

Theorem 2.9 (Cassini-type identities) *The following formulae hold for $n \geq 1$:*

$$F_n^2 - F_n F_{n-1} - F_{n-1}^2 = (-1)^{n-1} \tag{2.42}$$
$$L_n^2 - L_n L_{n-1} - L_{n-1}^2 = 5(-1)^n \tag{2.43}$$
$$P_n^2 - 2P_n P_{n-1} - P_{n-1}^2 = (-1)^{n-1} \tag{2.44}$$
$$Q_n^2 - Q_n Q_{n-1} - Q_{n-1}^2 = 8(-1)^n. \tag{2.45}$$

Solving the quadratic equations for F_n, L_n, P_n, and Q_n in Theorem 2.9, the following first-order nonlinear recurrent sequences can be obtained.

Corollary 2.3 *The following relations hold for all integers $n \geq 1$:*

$$F_n = \frac{1}{2}\left(F_{n-1} + \sqrt{5F_{n-1}^2 + 4(-1)^{n-1}} \right) \tag{2.46}$$
$$L_n = \frac{1}{2}\left(L_{n-1} + \sqrt{5L_{n-1}^2 + 20(-1)^n} \right) \tag{2.47}$$
$$P_n = P_{n-1} + \sqrt{2P_{n-1}^2 + (-1)^{n-1}} \tag{2.48}$$
$$Q_n = Q_{n-1} + \sqrt{2Q_{n-1}^2 + 8(-1)^n}. \tag{2.49}$$

A formula involving products of three Fibonacci terms is given by Melham in [120], for Fibonacci numbers and for arbitrary second-order linear recurrent sequences. We present the general result below.

Theorem 2.10 *Consider the sequence* $(x_n)_{n\geq 0}$ *defined by the recurrence equation*

$$x_{n+2} = ax_{n+1} + bx_n, \quad n = 0, 1, \ldots,$$

with $x_0 = \alpha_0$, $x_1 = \alpha_1$, *where* α_0, α_1, a, b *are given real (or complex) numbers. The following identity holds:*

$$x_{n+1}x_{n+2}x_{n+6} - x_{n+3}^3 = D(-b)^{n+1}(a^3x_{n+2} - b^2x_{n+1}), \tag{2.50}$$

where $D = a\alpha_0\alpha_1 + b\alpha_0^2 - \alpha_1^2$.

Proof The proof uses the identities

$$x_{n+1}x_{n+3} = x_{n+2}^2 + D(-b)^{n+1},$$

$$x_{n+3} = ax_{n+2} + bx_{n+1},$$

$$x_{n+6} = (a^4 + 3a^2b + b^2)x_{n+2} + (a^3b + 2ab^2)x_{n+1}.$$

Other matrix-based proofs are referenced in [120]. □

The following Melham-type identities are obtained for Fibonacci, Lucas, Pell, and Pell–Lucas numbers. The first of these was given in [120].

Theorem 2.11 *The following formulae hold for* $n \geq 1$*:*

$$F_{n+1}F_{n+2}F_{n+6} - F_{n+3}^3 = (-1)^n F_n \tag{2.51}$$

$$L_{n+1}L_{n+2}L_{n+6} - L_{n+3}^3 = 3(-1)^{n+1}L_n \tag{2.52}$$

$$P_{n+1}P_{n+2}P_{n+6} - P_{n+3}^3 = (-1)^n (8P_{n+2} - P_{n+1}) \tag{2.53}$$

$$Q_{n+1}Q_{n+2}Q_{n+6} - Q_{n+3}^3 = 8(-1)^{n+1} (8Q_{n+2} - Q_{n+1}). \tag{2.54}$$

Proof Simple computations confirm the values of D in (2.50) in each case. For Fibonacci and Lucas numbers, $a = b = 1$, hence $a^3x_{n+2} - b^2x_{n+1}$ simplifies to x_n. This is no longer the case for Pell and Pell–Lucas numbers. □

We derive results for products of four terms of second-order linear recurrent sequences with arbitrary coefficients, adapting notations from [139].

Theorem 2.12 *Consider the sequence* $(x_n)_{n\geq 0}$ *defined by the recurrence relation*

$$x_{n+2} = ax_{n+1} + bx_n, \quad n = 0, 1, \ldots,$$

with $x_0 = \alpha_0$, $x_1 = \alpha_1$, *where* α_0, α_1, a, b *are given real (or complex) numbers.*

The following identity holds:

$$x_{n-2}x_{n-1}x_{n+1}x_{n+2} = x_n^4 + D(-b)^{n-2}\left(a^2 - b\right)x_n^2 - D^2a^2b^{2n-3}, \tag{2.55}$$

where $D = a\alpha_0\alpha_1 + b\alpha_0^2 - \alpha_1^2$.

Proof The proof uses the identities

$$x_{n+1}x_{n-1} = x_n^2 + D(-b)^{n-1}, \quad x_{n+2}x_{n-2} = x_n^2 + D(-b)^{n-2}a,$$

which can be derived by representing x_n, x_{n+1}, x_{n+2} as functions of a, b, x_{n-1}, and x_{n-2} based on the recurrence equation, or recovered by setting $r = 1$ and $r = 2$ in Catalan's identity

$$x_{n+r}x_{n-r} = x_n^2 + D(-b)^{n-r}y_r^2, \quad r = 0, 1, \ldots, n,$$

where $(y_n)_{n\geq 0}$ is the sequence satisfying the same recurrence equation, but with initial values $y_0 = 0$, $y_1 = 1$.

Notice that the term corresponding to x_n^2 vanishes when $a^2 = b$. □

Many extensions are also known (see, e.g., [84]). In particular, for Fibonacci, Lucas, Pell, and Pell–Lucas numbers, we get the following result.

Theorem 2.13 *The following formulae hold for* $n \geq 1$*:*

$$F_n^4 - F_{n-2}F_{n-1}F_{n+1}F_{n+2} = 1 \tag{2.56}$$

$$L_n^4 - L_{n-2}L_{n-1}L_{n+1}L_{n+2} = 9 \tag{2.57}$$

$$P_n^4 - P_{n-2}P_{n-1}P_{n+1}P_{n+2} = 3(-1)^n P_n^2 + 4 \tag{2.58}$$

$$Q_n^4 - Q_{n-2}Q_{n-1}Q_{n+1}Q_{n+2} = 24(-1)^{n-1}Q_n^2 + 256. \tag{2.59}$$

Proof Simple computations confirm the values of D in (2.55) in each case. For Fibonacci and Lucas numbers, $a = b = 1$, hence the term corresponding to $b - a^2$ vanishes, which is not the case for Pell and Pell–Lucas numbers. □

Remark 2.6 The formula for Fibonacci numbers was given by E. Gelin and proved by E. Cesàro in 1880. We give an alternative proof. Define

$$A_n = F_{n+2}^4 - F_nF_{n+1}F_{n+3}F_{n+4}, \quad n \geq 0.$$

One can easily check that

$$\begin{aligned}
A_n &= (F_{n+1} + F_n)^4 - F_nF_{n+1}(2F_{n+1} + F_n)(2F_{n+1} + 3F_n + F_{n-1}) \\
&= F_{n+1}^4 + F_n^4 - 2F_{n+1}^2F_nF_{n-1} - (F_{n-1} + F_n)F_n^2(F_n + 3F_{n-1}) \\
&= F_{n+1}^4 - F_{n-1}F_n(F_n^2 + 3F_nF_{n+1} + F_{n+1}^2)
\end{aligned}$$

$$= F_{n+1}^4 - F_{n-1}F_n(F_n + F_{n+1})(F_n + 2F_{n+1})$$
$$= F_{n+1}^4 - F_{n-1}F_nF_{n+2}F_{n+3} = A_{n-1}.$$

It follows that $A_n = A_0 = 1$ for all $n \geq 0$. □

Using the same methodology, one can also prove another elegant formula between Pell and Pell–Lucas numbers [102, p. 156]

$$Q_n^4 - P_{n-1}P_{n+1}(2P_n)^2 = 1.$$

Expressions involving products of more Fibonacci factors exist, but these are less elegant when more than 4 terms are involved, as shown in [63]

$$F_{n-1}^3 F_{n+1}F_{n+2} - F_n^5 = -F_n + (-1)^n F_{n-1}F_{n+1}F_{n+2}$$
$$F_{n-3}F_{n-2}F_{n-1}F_{n+1}F_{n+2}F_{n+3} - F_n^6 = (-1)^n \left[4F_n^4 - (-1)^n F_n^2 - 4\right].$$

Theorem 2.14

1° *(Golden ratio) The sequence of ratios of Fibonacci numbers*

$$\frac{F_2}{F_1}, \frac{F_3}{F_2}, \frac{F_4}{F_3}, \ldots, \frac{F_{n+1}}{F_n}, \ldots,$$

has the limit $\phi = \frac{1+\sqrt{5}}{2} \approx 1.61803\,39887\cdots$, *called the* golden ratio.

2° *The sequence of ratios of Lucas numbers*

$$\frac{L_2}{L_1}, \frac{L_3}{L_2}, \frac{L_4}{L_3}, \ldots, \frac{L_{n+1}}{L_n}, \ldots,$$

also has the limit $\phi = \frac{1+\sqrt{5}}{2}$.

3° *The sequence of ratios of Pell numbers*

$$\frac{P_2}{P_1}, \frac{P_3}{P_2}, \frac{P_4}{P_3}, \ldots, \frac{P_{n+1}}{P_n}, \ldots,$$

has the limit $1 + \sqrt{2}$, *referred to as the* silver ratio.

4° *The sequence of ratios of Pell–Lucas numbers*

$$\frac{Q_2}{Q_1}, \frac{Q_3}{Q_2}, \frac{Q_4}{Q_3}, \ldots, \frac{Q_{n+1}}{Q_n}, \ldots,$$

has the limit $1 + \sqrt{2}$.

Proof

1° From the Binet formula (2.30), we have

$$F_n = \frac{1}{\sqrt{5}}\left[\phi^n - (-1)^n\phi^{-n}\right].$$

One can check that for $n \geq 0$

$$\frac{F_{n+1}}{F_n} = \frac{\phi^{n+1} - (-1)^{n+1}\phi^{-n-1}}{\phi^n - (-1)^n\phi^{-n}}.$$

Clearly, as $\phi > 1$ the limit of this expression is ϕ.

2° From the Binet-type formula (2.31), we have

$$L_n = \phi^n + (-1)^n\phi^{-n},$$

hence

$$\frac{L_{n+1}}{L_n} = \frac{\phi^{n+1} + (-1)^{n+1}\phi^{-n-1}}{\phi^n + (-1)^n\phi^{-n}}.$$

and the solution follows.

3° Similarly, from the Binet-type formula (2.32), we have

$$P_n = \frac{1}{2\sqrt{2}}\left[\alpha^n - (-1)^n\alpha^{-n}\right],$$

where $\alpha = 1 + \sqrt{2}$. Therefore,

$$\frac{P_{n+1}}{P_n} = \frac{\alpha^{n+1} - (-1)^{n+1}\alpha^{-n-1}}{\alpha^n - (-1)^n\alpha^{-n}},$$

hence the limit is $\alpha = 1 + \sqrt{2}$.

4° By the Binet-type formula (2.33) we have

$$Q_n = \alpha^n + (-1)^n\alpha^{-n},$$

where $\alpha = 1 + \sqrt{2}$. We obtain

$$\frac{Q_{n+1}}{Q_n} = \frac{\alpha^{n+1} + (-1)^{n+1}\alpha^{-n-1}}{\alpha^n + (-1)^n\alpha^{-n}},$$

and the conclusion follows.

□

Remark 2.7 The golden ratio is linked to the proportions between parts of the human body, musical notes, and various patterns in nature. Many of these occurrences are presented in the monograph of Koshy [101].

The formula below for Fibonacci numbers was proved by Lucas in 1876.

Theorem 2.15 *The Fibonacci number F_n can be written as*

$$F_n = \sum_{k=0}^{\lfloor \frac{n-1}{2} \rfloor} \binom{n-k-1}{k}, \qquad n \geq 0.$$

Proof Let $x_0 = 0$ and

$$x_n = \sum_{k=0}^{\lfloor \frac{n-1}{2} \rfloor} \binom{n-k-1}{k}, \qquad n \geq 1.$$

Note that $k > \lfloor \frac{n-1}{2} \rfloor$ is equivalent to $k > n-k-1$. Since $\binom{m}{p} = 0$ whenever $p > m$, one may also write

$$x_n = \sum_{k=0}^{n-1} \binom{n-k-1}{k}, \qquad n \geq 1.$$

Clearly, $x_0 = 0$ and $x_1 = \binom{0}{0} = 0$ It suffices to show that the sequence x_n satisfies the Fibonacci recurrence relation. Also, notice that

$$\begin{aligned}
x_{n+1} + x_n &= \sum_{k=0}^{n} \binom{n-k}{k} + \sum_{k=0}^{n-1} \binom{n-k-1}{k} \\
&= \binom{n}{0} + \sum_{k=1}^{n} \binom{n-k}{k} + \sum_{k=1}^{n} \binom{n-k}{k-1} \\
&= \binom{n}{0} + \sum_{k=1}^{n} \left[\binom{n-k}{k} + \binom{n-k}{k-1} \right] \\
&= \binom{n}{0} + \sum_{k=1}^{n} \binom{n-k+1}{k} \\
&= \binom{n+1}{0} + \sum_{k=1}^{n} \binom{n-k+1}{k} + \binom{0}{n+1}
\end{aligned}$$

$$= \sum_{k=0}^{n+1} \binom{(n+2)-k-1}{k} = x_{n+2}.$$

This ends the proof. □

2.2.4 *The Polynomials $U_n(x, y)$ and $V_n(x, y)$*

Inspired by the Binet formulae for Fibonacci, Lucas, Pell, and Pell–Lucas sequences, one can define the polynomials $U_n, V_n \in \mathbb{Z}[x, y]$, $n = 0, 1, \ldots$

$$U_n(x, y) = \frac{x^n - y^n}{x - y}, \quad V_n(x, y) = x^n + y^n. \tag{2.60}$$

The polynomials U_n and V_n are symmetric and satisfy $\deg U_n = n-1$, $n = 1, 2, \ldots$, and $\deg V_n = n$, $n = 0, 1, \ldots$. By (2.60) we can extend the sequence $U_n(x, y)$ for $n = 0$, by $U_0(x, y) = 0$.

On the other hand, the sequences $(U_n(x, y))_{n\geq 0}$ and $(V_n(x, y))_{n\geq 0}$ satisfy the same recursive relation of order 2, with different initial values

$$U_{n+2} = (x + y)U_{n+1} - xyU_n, \quad U_0 = 0,\ U_1 = 1, \tag{2.61}$$

$$V_{n+2} = (x + y)V_{n+1} - xyV_n, \quad V_0 = 2,\ V_1 = x + y. \tag{2.62}$$

The Fibonacci, Lucas, Pell, and Pell–Lucas numbers are obtained as particular instances of U_n and V_n for special pairs of real numbers (x, y) as follows:

$$F_n = U_n\left(\frac{1+\sqrt{5}}{2}, \frac{1-\sqrt{5}}{2}\right), \quad n = 0, 1, \ldots,$$

$$L_n = V_n\left(\frac{1+\sqrt{5}}{2}, \frac{1-\sqrt{5}}{2}\right), \quad n = 0, 1, \ldots,$$

$$P_n = U_n\left(1+\sqrt{2}, 1-\sqrt{2}\right), \quad n = 0, 1, \ldots,$$

$$Q_n = V_n\left(1+\sqrt{2}, 1-\sqrt{2}\right), \quad n = 0, 1, \ldots.$$

The polynomials U_n and V_n have similar algebraic properties as the Fibonacci, Lucas, Pell, and Pell–Lucas numbers. For instance, we have the following matrix forms valid for $n = 1, 2, \ldots$

$$\begin{pmatrix} U_{n+1} & -xyU_n \\ U_n & -xyU_{n-1} \end{pmatrix} = \begin{pmatrix} x+y & -xy \\ 1 & 0 \end{pmatrix}^n, \tag{2.63}$$

$$\begin{pmatrix} V_{n+1} & -xyV_n \\ V_n & -xyV_{n-1} \end{pmatrix} = \begin{pmatrix} x+y & -xy \\ 1 & 0 \end{pmatrix}^n \begin{pmatrix} x+y & -2xy \\ 2 & -(x+y) \end{pmatrix}, \tag{2.64}$$

which contain all the matrix forms in Remark 2.5 obtained for the particular values $(x, y) = \left(\frac{1+\sqrt{5}}{2}, \frac{1-\sqrt{5}}{2}\right)$ and $(x, y) = \left(1+\sqrt{2}, 1-\sqrt{2}\right)$, respectively.

From (2.63) we can write

$$\begin{pmatrix} U_{m+n} & -xyU_{m+n-1} \\ U_{m+n-1} & -xyU_{m+n-2} \end{pmatrix} = \begin{pmatrix} x+y & -xy \\ 1 & 0 \end{pmatrix}^{m-1} \begin{pmatrix} x+y & -xy \\ 1 & 0 \end{pmatrix}^n$$
$$= \begin{pmatrix} U_m & -xyU_{m-1} \\ U_{m-1} & -xyU_{m-2} \end{pmatrix} \begin{pmatrix} U_{n+1} & -xyU_n \\ U_n & -xyU_{n-1} \end{pmatrix},$$

hence we obtain the following relation:

$$U_{m+n} = U_m U_{n+1} - xyU_{m-1}U_n. \tag{2.65}$$

Similarly, from (2.64) we have

$$\begin{pmatrix} V_{m+n} & -xyV_{m+n-1} \\ V_{m+n-1} & -xyV_{m+n-2} \end{pmatrix} = \begin{pmatrix} x+y & -xy \\ 1 & 0 \end{pmatrix}^{n-1} \begin{pmatrix} x+y & -xy \\ 1 & 0 \end{pmatrix}^m \begin{pmatrix} x+y & -2xy \\ 2 & -x-y \end{pmatrix}$$
$$= \begin{pmatrix} U_n & -xyU_{n-1} \\ U_{n-1} & -xyU_{n-2} \end{pmatrix} \begin{pmatrix} V_{m+1} & -xyV_m \\ V_m & -xyV_{m-1} \end{pmatrix},$$

and we derive the following identity:

$$V_{m+n} = V_{m+1}U_n - xyV_mU_{n-1}. \tag{2.66}$$

We mention here some important polynomials which can be recovered directly as particular instances of the polynomials U_n and V_n.

Fibonacci polynomials $f_n \in \mathbb{Z}[x]$, $n = 0, 1, \ldots$

$$f_n(x) = U_n\left(\frac{x+\sqrt{x^2+4}}{2}, \frac{x-\sqrt{x^2+4}}{2}\right)$$
$$= \frac{1}{\sqrt{x^2+4}}\left[\left(\frac{x+\sqrt{x^2+4}}{2}\right)^n - \left(\frac{x-\sqrt{x^2+4}}{2}\right)^n\right].$$

These satisfy the recurrence equation

$$f_{n+2} = xf_{n+1} + f_n, \quad f_0 = 0,\ f_1 = 1. \tag{2.67}$$

The first few Fibonacci polynomials are

$$\begin{aligned} f_0(x) &= 0 \\ f_1(x) &= 1 \\ f_2(x) &= x \\ f_3(x) &= x^2 + 1 \\ f_4(x) &= x^3 + 2x \\ f_5(x) &= x^4 + 3x^2 + 1. \end{aligned}$$

Lucas polynomials $l_n \in \mathbb{Z}[x], n = 0, 1, \dots$

$$\begin{aligned} l_n(x) &= V_n\left(\frac{x+\sqrt{x^2+4}}{2}, \frac{x-\sqrt{x^2+4}}{2}\right) \\ &= \left(\frac{x+\sqrt{x^2+4}}{2}\right)^n + \left(\frac{x-\sqrt{x^2+4}}{2}\right)^n. \end{aligned}$$

These satisfy the recurrence relation

$$l_{n+2} = x l_{n+1} + l_n, \quad l_0 = 2,\ l_1 = x. \tag{2.68}$$

The first few Lucas polynomials are

$$\begin{aligned} l_0(x) &= 2 \\ l_1(x) &= x \\ l_2(x) &= x^2 + 2 \\ l_3(x) &= x^3 + 3x \\ l_4(x) &= x^4 + 4x^2 + 2 \\ l_5(x) &= x^5 + 5x^3 + 5x. \end{aligned}$$

Pell polynomials $p_n \in \mathbb{Z}[x], n = 0, 1, \dots$

$$\begin{aligned} p_n(x) &= U_n\left(x+\sqrt{x^2+1}, x-\sqrt{x^2+1}\right) \\ &= \frac{1}{2\sqrt{x^2+1}}\left[\left(x+\sqrt{x^2+1}\right)^n - \left(x-\sqrt{x^2+1}\right)^n\right]. \end{aligned}$$

These satisfy the recurrence equation

$$p_{n+2} = 2xp_{n+1} + p_n, \quad p_0 = 0,\ p_1 = 1. \tag{2.69}$$

The first few Pell polynomials are

$$\begin{aligned}
p_0(x) &= 0\\
p_1(x) &= 1\\
p_2(x) &= 2x\\
p_3(x) &= 4x^2 + 1\\
p_4(x) &= 8x^3 + 4x\\
p_5(x) &= 16x^4 + 12x^2 + 1.
\end{aligned}$$

Pell–Lucas polynomials $q_n \in \mathbb{Z}[x], n = 0, 1, \dots$

$$\begin{aligned}
q_n(x) &= V_n\left(x + \sqrt{x^2+1}, x - \sqrt{x^2+1}\right)\\
&= \left(x + \sqrt{x^2+1}\right)^n + \left(x - \sqrt{x^2+1}\right)^n.
\end{aligned}$$

These satisfy the recurrence relation

$$q_{n+2} = 2xq_{n+1} + q_n, \quad q_0 = 2,\ q_1 = 2x. \tag{2.70}$$

The first few Pell–Lucas polynomials are

$$\begin{aligned}
q_0(x) &= 2\\
q_1(x) &= 2x\\
q_2(x) &= 4x^2 + 2\\
q_3(x) &= 8x^3 + 6x\\
q_4(x) &= 16x^4 + 16x^2 + 2\\
q_5(x) &= 32x^5 + 40x^3 + 10x.
\end{aligned}$$

Chebyshev polynomials of the first kind $T_n \in \mathbb{Z}[x], n = 0, 1, \dots$

$$\begin{aligned}
T_n(x) &= \frac{1}{2}V_n\left(x + \sqrt{x^2-1}, x - \sqrt{x^2-1}\right)\\
&= \frac{1}{2}\left[\left(x + \sqrt{x^2-1}\right)^n + \left(x - \sqrt{x^2-1}\right)^n\right].
\end{aligned}$$

These satisfy the recurrence relation

$$T_{n+2} = 2xT_{n+1} - T_n, \quad T_0 = 1,\ T_1 = x. \tag{2.71}$$

The first few Chebyshev polynomials of the first kind are

$$\begin{aligned}
T_0(x) &= 1\\
T_1(x) &= x\\
T_2(x) &= 2\,x^2 - 1\\
T_3(x) &= 4\,x^3 - 3\,x\\
T_4(x) &= 8\,x^4 - 8\,x^2 + 1\\
T_5(x) &= 16\,x^5 - 20\,x^3 + 5\,x.
\end{aligned}$$

Chebyshev polynomials of the second kind $u_{n-1} \in \mathbb{Z}[x], n = 0, 1, \ldots$

$$\begin{aligned}
u_{n-1}(x) &= U_n\left(x + \sqrt{x^2-1}, x - \sqrt{x^2-1}\right)\\
&= \frac{1}{\sqrt{x^2-1}}\left[\left(x + \sqrt{x^2-1}\right)^n - \left(x - \sqrt{x^2-1}\right)^n\right].
\end{aligned}$$

These satisfy the recurrence equation

$$u_{n+2} = 2xu_{n+1} - u_n, \quad u_0 = 0,\ u_1 = 1. \tag{2.72}$$

The first few Chebyshev polynomials of the second kind are

$$\begin{aligned}
u_0(x) &= 0\\
u_1(x) &= 1\\
u_2(x) &= 2\,x\\
u_3(x) &= 4\,x^2 - 1\\
u_4(x) &= 8\,x^3 - 4\,x\\
u_5(x) &= 16\,x^4 - 12\,x^2 + 1.
\end{aligned}$$

Hoggatt–Bicknell–King polynomials of Fibonacci kind $g_n \in \mathbb{Z}[x], n = 0, 1, \ldots$

$$g_n(x) = U_n\left(\frac{x + \sqrt{x^2-4}}{2}, \frac{x - \sqrt{x^2-4}}{2}\right)$$

$$= \frac{1}{\sqrt{x^2-4}}\left[\left(\frac{x+\sqrt{x^2-4}}{2}\right)^n - \left(\frac{x-\sqrt{x^2-4}}{2}\right)^n\right].$$

These satisfy the recurrence relation

$$g_{n+2} = xg_{n+1} - g_n, \quad g_0 = 0,\ g_1 = 1. \tag{2.73}$$

The first few Hoggatt–Bicknell–King polynomial of Fibonacci kind are

$$\begin{aligned} g_0(x) &= 0 \\ g_1(x) &= 1 \\ g_2(x) &= x \\ g_3(x) &= x^2 - 1 \\ g_4(x) &= x^3 - 2x \\ g_5(x) &= x^4 - 3x^2 + 1. \end{aligned}$$

Hoggatt–Bicknell–King polynomials of Lucas kind

$$\begin{aligned} h_n(x) &= V_n\left(\frac{x+\sqrt{x^2-4}}{2}, \frac{x-\sqrt{x^2-4}}{2}\right) \\ &= \left(\frac{x+\sqrt{x^2-4}}{2}\right)^n + \left(\frac{x-\sqrt{x^2-4}}{2}\right)^n, \quad n = 0, 1, \ldots. \end{aligned}$$

These satisfy the recurrence relation

$$h_{n+2} = xh_{n+1} - h_n, \quad h_0 = 2,\ h_1 = x. \tag{2.74}$$

The first few Hoggatt–Bicknell–King polynomial of Lucas kind are

$$\begin{aligned} h_0(x) &= 2 \\ h_1(x) &= x \\ h_2(x) &= x^2 - 2 \\ h_3(x) &= x^3 - 3x \\ h_4(x) &= x^4 - 4x^2 + 2 \\ h_5(x) &= x^5 - 5x^3 + 5x. \end{aligned}$$

Jacobsthal polynomials $J_n \in \mathbb{Z}[x], n = 0, 1, \ldots$

$$J_n(x) = U_n\left(\frac{1+\sqrt{8x+1}}{2}, \frac{1-\sqrt{8x+1}}{2}\right)$$
$$= \frac{1}{\sqrt{8x+1}}\left[\left(\frac{1+\sqrt{8x+1}}{2}\right)^n - \left(\frac{1-\sqrt{8x+1}}{2}\right)^n\right].$$

These satisfy the recurrence equation

$$J_{n+2} = J_{n+1} + 2xJ_n, \quad J_0 = 0,\ J_1 = 1. \tag{2.75}$$

The first few Jacobsthal polynomials are

$$J_0(x) = 0$$
$$J_1(x) = 1$$
$$J_2(x) = 1$$
$$J_3(x) = 2x + 1$$
$$J_4(x) = 4x + 1$$
$$J_5(x) = 4x^2 + 6x + 1.$$

Morgan–Voyce polynomials $B_n \in \mathbb{Z}[x], n = 0, 1, \ldots$

$$B_n(x) = U_{n+1}\left(\frac{x+2+\sqrt{x^2+4x}}{2}, \frac{x+2-\sqrt{x^2+4x}}{2}\right)$$
$$= \frac{1}{\sqrt{x^2+4x}}\left[\left(\frac{x+2+\sqrt{x^2+4x}}{2}\right)^{n+1} - \left(\frac{x+2-\sqrt{x^2+4x}}{2}\right)^{n+1}\right].$$

These satisfy the recurrence relation

$$B_{n+2} = (x+2)B_{n+1} - B_n, \quad B_0 = 1,\ B_1 = x + 2, \tag{2.76}$$

and are used in the study of electrical ladder networks of resistors. Clearly, we have $B_{n-1}(x) = g_n(x+2), n = 1, 2, \ldots$.

The first few Morgan–Voyce polynomials are

$$B_0(x) = 1$$
$$B_1(x) = x + 2$$

$$B_2(x) = x^2 + 4x + 3$$

$$B_3(x) = x^3 + 6x^2 + 10x + 4$$

$$B_4(x) = x^4 + 8x^3 + 21x^2 + 20x + 5$$

$$B_5(x) = x^5 + 10x^4 + 36x^3 + 56x^2 + 35x + 6.$$

Brahmagupta polynomials These polynomials $x_n, y_n \in \mathbb{Z}[x]$, $n = 0, 1, \ldots$, depend on the integer parameter $t > 0$ as

$$x_n(x, y) = \frac{1}{2} V_n\left(x + y\sqrt{t}, x - y\sqrt{t}\right) = \frac{1}{2}\left[\left(x + y\sqrt{t}\right)^n + \left(x - y\sqrt{t}\right)^n\right],$$

$$y_n(x, y) = yU_n\left(x + y\sqrt{t}, x - y\sqrt{t}\right) = \frac{1}{2\sqrt{t}}\left[\left(x + y\sqrt{t}\right)^n - \left(x - y\sqrt{t}\right)^n\right].$$

Here it is assumed that t is square free. These polynomials arise naturally from the so called multiplication principle, considered for the first time by the Indian astronomer and mathematician Brahmagupta, and generate solutions of the general Pell equation $x^2 - ty^2 = \pm N$, with N a positive integer.

These satisfy the recurrence relation

$$x_{n+2} = 2x \cdot x_{n+1} - (x^2 - y^2 t)x_n, \quad x_0 = 1,\ x_1 = x, \tag{2.77}$$

$$y_{n+2} = 2x \cdot y_{n+1} - (x^2 - y^2 t)y_n, \quad y_0 = 1,\ y_1 = y. \tag{2.78}$$

The first few Brahmagupta polynomials are

$$x_0(x, y) = 1, \qquad y_0(x, y) = 0$$

$$x_1(x, y) = x, \qquad y_1(x, y) = y$$

$$x_2(x, y) = ty^2 + x^2, \qquad y_2(x, y) = 2xy$$

$$x_3(x, y) = 3txy^2 + x^3, \qquad y_3(x, y) = y\left(ty^2 + 3x^2\right)$$

$$x_4(x, y) = t^2y^4 + 6tx^2y^2 + x^4, \qquad y_4(x, y) = 4xy\left(ty^2 + x^2\right)$$

$$x_5(x, y) = 5t^2xy^4 + 10tx^3y^2 + x^5, \qquad y_5(x, y) = y\left(t^2y^4 + 10tx^2y^2 + 5x^4\right).$$

The following arithmetic properties involving the polynomials U_n and V_n are useful as they can be used to prove some divisibility properties of the Fibonacci, Lucas, Pell, and Pell–Lucas numbers.

Lemma 2.1 *Assume that a and b are nonzero real numbers such that $a + b$ and ab are integers. The following properties hold:*

1° *$U_n(a,b)$ and $V_n(a,b)$ are integers for all $n = 0, 1, 2, \ldots$;*
2° *If k is a positive integer, then $U_m(a,b) \mid U_{km}(a,b)$;*
3° *If k is an odd positive integer, then $V_m(a,b) \mid V_{km}(a,b)$.*

Proof

1° The results follow by using induction with step 2, using the recursive relations (2.61) and (2.62). Alternatively, one can use the fundamental theorem of symmetric polynomials.
2° By the definition, we have

$$\begin{aligned}U_{km}(a,b) &= \frac{a^{km} - b^{km}}{a-b} = \frac{(a^m)^k - (b^m)k}{a-b} \\ &= U_m(a,b)\left(a^{(k-1)m} + a^{(k-2)m}b^m + \cdots + a^m b^{(k-2)m} + b^{(k-1)m}\right) \\ &= U_m(a,b)\left[V_{(k-1)m}(a,b) + a^m b^m V_{(k-3)m}(a,b) + \cdots\right],\end{aligned}$$

and the conclusion follows since $V_s(a,b)$ is an integer, $s = 0, 1, \ldots$.
3° Let $k = 2s+1$, for some positive integer s. By definition, we have

$$\begin{aligned}V_{km}(a,b) &= V_{(2s+1)m}(a,b) = (a^m)^{2s+1} + (b^m)2s+1 \\ &= V_m(a,b)\left(a^{2sm} - a^{(2s-1)m}b^m + \cdots - a^m b^{(2s-1)m} + b^{2sm}\right) \\ &= V_m(a,b)\left[V_{2sm}(a,b) - a^m b^m V_{(2s-2)m}(a,b) + \cdots\right],\end{aligned}$$

and we obtain the result, since $V_s(a,b)$ is an integer, $s = 0, 1, \ldots$.

□

Theorem 2.16 *Let a and b be nonzero real numbers such that $a+b$ and ab are integers with $\gcd(a+b, ab) = 1$. Then, the following relation holds:*

$$\gcd(U_m(a,b), U_n(a,b)) = U_{\gcd(m,n)}(a,b). \tag{2.79}$$

Proof Using Lemma 2.1 2°, it follows that $U_{\gcd(m,n)}(a,b)$ divides $U_m(a,b)$ and $U_n(a,b)$, hence also divides $\gcd(U_m(a,b), U_n(a,b))$.

Conversely, we will now show that $\gcd(U_m(a,b), U_n(a,b))$ also divides $U_{\gcd(m,n)}(a,b)$. If $\gcd(U_m, U_n) = 1$, then we are done. We now assume that $\gcd(U_m, U_n) > 1$. Let u and v be integers such that $um + vn = \gcd(m,n)$. It follows that $M^{\gcd(m,n)} = (M^m)^u (M^n)^v$, where the matrix

$$M = \begin{pmatrix} a+b & -ab \\ 1 & 0 \end{pmatrix},$$

is diagonal modulo $U_{\gcd(m,n)}$. Since M^m and M^n are also diagonal modulo $U_{\gcd(m,n)}$, so is $M^{\gcd(m,n)}$. Therefore, $\gcd(U_m, U_n)$ also divides $U_{\gcd(m,n)}$. □

Theorem 2.17 *Let a and b be nonzero real numbers such that $a+b$ and ab are integers with $\gcd(a+b, ab) = 1$. Then for every odd positive integers m and n, the following relation holds:*

$$\gcd(V_m(a,b), V_n(a,b)) = V_{\gcd(m,n)}(a,b). \tag{2.80}$$

Proof By Lemma 2.1 3°, we just have to prove the case when m and n are relatively prime. Therefore, suppose that $\gcd(m,n) = 1$ and let us prove that

$$\gcd(V_m(a,b), V_n(a,b)) = V_1(a,b) = a+b. \tag{2.81}$$

By Lemma 2.1 3°, we know that $\frac{V_m(a,b)}{a+b}$ and $\frac{V_n(a,b)}{a+b}$ are both integers, so by (2.81) we have to show that $\frac{V_m(a,b)}{a+b}$ and $\frac{V_n(a,b)}{a+b}$ are in fact relatively prime.

We will first show that if p is a prime dividing $\frac{V_m(a,b)}{a+b}$ and $\frac{V_n(a,b)}{a+b}$, then p also divides $a+b$, which will further allow us to reach a contradiction.

Without loss of generality, we can assume that $m < n$. It follows that

$$p \mid \left(a^{n-m} + b^{n-m}\right) \frac{V_m(a,b)}{a+b} - \frac{V_n(a,b)}{a+b},$$

hence

$$p \mid \frac{a^m b^{n-m} + a^{n-m} b^m}{a+b}.$$

If $n > 2m$, then we have $p \mid a^m b^m (a^{n-2m} + b^{n-2m})$.

If $p \mid ab$, from the relation $V_m = (a+b)V_{m-1} - abV_{m-2}$, we get $p \mid V_{m-1}$, because $p \mid V_m$ from the hypothesis. From $p \mid V_{m-1}$ and the recurrence equation, if follows that $p \mid V_{m-1} + abV_{m-3} = (a+b)V_{m-2}$. Because $\gcd(a+b, ab) = 1$, we get $p \mid V_{m-2}$. Similarly, we establish that $p \mid V_{m-4}$, $p \mid V_{m-6}, \ldots, p \mid V_1 = a+b$, which is a contradiction.

Hence, p does not divide ab, so we must have $p \mid a^{n-2m} + b^{n-2m} = V_{n-2m}$. We have reduced the problem from the pair (m,n) with $\gcd(m,n) = 1$ to the pair $(n-2m, m)$ with $\gcd(n-2m, m) = 1$. Continuing this procedure we get, after a finite number of steps, $p \mid V_1 = a+b$, and we are done.

From $p \mid \frac{V_m(a,b)}{a+b}$ and $p \mid a+b$, using the relations

$$\frac{V_m(a,b)}{a+b} = V_{m-1}(a,b) - ab\frac{V_{m-2}(a,b)}{a+b}$$

$$V_{m-1}(a,b) = (a+b)V_{m-2}(a,b) - abV_{m-3}(a,b),$$

and

$$p \mid \frac{V_m(a,b)}{a+b} - (a+b)V_{m-2}(a,b),$$

it follows that

$$p \mid ab\left[\frac{V_{m-2}(a,b)}{a+b} + V_{m-3}(a,b)\right].$$

Since $\gcd(a+b, ab) = 1$, we obtain

$$p \mid \frac{V_{m-2}(a,b)}{a+b} + V_{m-3}(a,b),$$

hence it follows that

$$p \mid 2V_{m-3}(a,b) - ab\frac{V_{m-4}(a,b)}{a+b}.$$

Recursively, for every positive integer s we have

$$p \mid sV_{m-(2s-1)}(a,b) - ab\frac{V_{m-2s}(a,b)}{a+b}.$$

Denoting $m = 2k+1$, we get $p \mid kV_2(a,b) - ab$, hence $p \mid k(a+b)^2 - mab$. Because $p \mid a+b$ and $\gcd(a+b, ab) = 1$, we get $p \mid m$.

Similarly, starting from $p \mid \frac{V_n(a,b)}{a+b}$ and $p \mid a+b$, we obtain $p \mid n$.

This is a contradiction, since we have $\gcd(m,n) = 1$. □

Remark 2.8 A different proof in the spirit of the proof of Theorem 2.16 is given by Jeffery and Pereira in [88].

2.2.5 *Properties of Fibonacci, Lucas, Pell, and Lucas–Pell Numbers*

Here we present some key properties of the Fibonacci, Lucas, Pell, and Lucas–Pell numbers, concerning primarily certain recurrent formulae and divisibility relations involving the sequence terms.

Theorem 2.18 (properties of Fibonacci numbers)

1° *For any positive integers m, n, we have the identity*

$$F_{m+n} = F_{m-1}F_n + F_mF_{n+1}. \tag{2.82}$$

2° *If $m \mid n$, then $F_m \mid F_n$.*

3° *For any* $m, n \geq 0$, $\gcd(F_m, F_n) = F_{\gcd(m,n)}$.
4° *If* $\gcd(m, n) = 1$, *then* $F_m F_n$ *divides* F_{mn}.
5° *(Catalan's identity) For an integer* $0 \leq r \leq n$, *the following identity holds:*

$$F_n^2 - F_{n-r}F_{n+r} = (-1)^{n-r}F_r^2.$$

6° *(Vajda's identity) For positive integers* i, j, *the following identity holds:*

$$F_{n+i}F_{n+j} - F_nF_{n+i+j} = (-1)^n F_i F_j.$$

Proof

1° Using the identity (2.65) and the relation $F_s = U_s\left(\frac{1+\sqrt{5}}{2}, \frac{1-\sqrt{5}}{2}\right)$, $s = 0, 1, \ldots,$ the desired formula follows.
2° It follows by Lemma 2.1 2°, and $F_s = U_s\left(\frac{1+\sqrt{5}}{2}, \frac{1-\sqrt{5}}{2}\right)$, $s = 0, 1, \ldots.$
3° **First Proof** We use Theorem 2.16 for $(a, b) = \left(\frac{1+\sqrt{5}}{2}, \frac{1-\sqrt{5}}{2}\right)$. Observe that $a + b = 1$ and $ab = -1$, and $\gcd(a + b, ab) = 1$, hence (2.79) holds, and the conclusion follows.

Second Proof Suppose that $n > m$ and let $d = \gcd(m, n)$. By applying Euclid's Algorithm, we get

$$n = mq_1 + r_1$$
$$m = r_1q_2 + r_2$$
$$r_1 = r_2q_3 + r_3$$
$$\ldots$$
$$r_{i-1} = r_iq_{i+1},$$

and so $d = r_i$. By the identity (2.82) we have

$$\gcd(F_m, F_n) = \gcd(F_m, F_{mq_1+r_1}) = \gcd(F_m, F_{mq_1-1}F_{r_1} + F_{mq_1+r_1}F_{r_1+1})$$
$$= \gcd(F_m, F_{mq_1-1}F_{r_1}) = \gcd(F_m, F_{r_1}),$$

where we have used that $\gcd(F_m, F_{mq_1-1}) = 1$.
By applying the procedure repeatedly, we arrive at

$$\gcd(F_m, F_n) = \gcd(F_{r_1}, F_m) = \gcd(F_{r_2}, F_{r_1}) \cdots = \gcd(F_{r_{i-1}}, F_{r_i}) = F_{r_i} = d.$$

4° The property follows from 2° and 3°, by observing that

$$\gcd(F_m, F_m) = F_{\gcd(m,m)} = F_1 = 1.$$

5° Using the notations at point 1°, one obtains

$$F_n^2 - F_{n-r}F_{n+r} = \frac{1}{5}\left[(\alpha^n - \beta^n)^2 - (\alpha^{n-r} - \beta^{n-r})(\alpha^{n+r} - \beta^{n+r})\right]$$
$$= \frac{1}{5}\left[\alpha^{n+r}\beta^{n-r} - 2\alpha^n\beta^n + \alpha^{n-r}\beta^{n+r}\right]$$
$$= \frac{1}{5}(\alpha\beta)^{n-r}\left(\alpha^r - \beta^r\right)^2 = (-1)^{n-r}F_r^2,$$

where we have used the Binet-type formula and the identity $\alpha\beta = -1$.

6° Similar to point 5°, we obtain successively

$$F_{n+i}F_{n+j} - F_nF_{n+i+j}$$
$$= \frac{1}{5}\left[(\alpha^{n+i} - \beta^{n+i})(\alpha^{n+j} - \beta^{n+j}) - (\alpha^n - \beta^n)(\alpha^{n+i+j} - \beta^{n+i+j})\right]$$
$$= \frac{1}{5}\left[-\alpha^{n+i}\beta^{n+j} - \alpha^{n+j}\beta^{n+i} + \alpha^n\beta^{n+i+j} + \alpha^{n+i+j}\beta^n\right]$$
$$= \frac{1}{5}(\alpha\beta)^n\left(\alpha^i - \beta^i\right)\left(\alpha^j - \beta^j\right) = (-1)^nF_iF_j.$$

□

Theorem 2.19 (properties of Lucas numbers)

1° *For any positive integers m, n, we have the identity*

$$L_{m+n} = L_{m+1}F_n + L_mF_{n-1}.$$

2° *If k is a positive integer, then $F_{n+k} + (-1)^kF_{n-k} = L_kF_n$.*

3° $F_{2n} = L_nF_n$.

4° *If $k, n \in \mathbb{N}$ with k odd, then $L_n \mid L_{kn}$.*

5° *If $k, n \in \mathbb{N}$ with k even, then $L_nF_n \mid F_{kn}$.*

6° *If $d = \gcd(m, n)$ and m/d, n/d are both odd, then $\gcd(L_m, L_n) = L_{\gcd(m,n)}$.*

Proof Denoting $\alpha = \dfrac{1+\sqrt{5}}{2}$, $\beta = \dfrac{1-\sqrt{5}}{2}$, we have $\alpha + \beta = 1$ and $\alpha\beta = -1$.

1° Using the identity (2.66) and the relation $L_k = V_k\left(\frac{1+\sqrt{5}}{2}, \frac{1-\sqrt{5}}{2}\right)$, valid for $k = 0, 1, \ldots$, the desired formula follows.

2° This relation follows from the exact formulae. Since $\alpha\beta = -1$, we have

$$L_kF_n = \frac{1}{\sqrt{5}}\left(\alpha^k + \beta^k\right)(\alpha^n - \beta^n) = \frac{1}{\sqrt{5}}\left(\alpha^{n+k} - \beta^{n+k} + \alpha^n\beta^k - \alpha^k\beta^n\right)$$
$$= F_{n+k} + \frac{(-1)^k}{\sqrt{5}}\left(\alpha^{n-k} - \beta^{n-k}\right) = F_{n+k} + (-1)^kF_{n-k}.$$

3° In the previous relation we take $k = n$.

4° Follows from Lemma 2.1 3° and $L_s = V_s\left(\frac{1+\sqrt{5}}{2}, \frac{1-\sqrt{5}}{2}\right)$, $s = 0, 1, \ldots$.

5° Denoting $k = 2l$ one obtains by Binet's formula

$$\frac{F_{kn}}{L_n} = \frac{1}{\sqrt{5}}\frac{\alpha^{nk} - \beta^{kn}}{\alpha^n + \beta^n} = \frac{1}{\sqrt{5}}\frac{(\alpha^{2n})^l - (\beta^{2n})^l}{\alpha^n + \beta^n}$$

$$= \frac{1}{\sqrt{5}}\frac{\alpha^{2n} - \beta^{2n}}{\alpha^n + \beta^n}\left(\alpha^{2n(l-1)} + \alpha^{2n(l-2)}\beta^{2n} + \cdots + \alpha^{2n}\beta^{2n(l-2)} + \beta^{2n(l-1)}\right)$$

$$= F_n\left(L_{2n(l-1)} + L_{2n(l-3)} + \cdots\right),$$

where we have used the relation $\alpha\beta = -1$.

6° **First Proof** We apply Theorem 2.17 for $(a, b) = \left(\frac{1+\sqrt{5}}{2}, \frac{1-\sqrt{5}}{2}\right)$ and the relation holds since $L_k = V_k\left(\frac{1+\sqrt{5}}{2}, \frac{1-\sqrt{5}}{2}\right)$, for $k = 0, 1, \ldots$.

Second Proof Note that we have

$$\begin{pmatrix} L_{k+1} & L_k \\ L_k & L_{k-1} \end{pmatrix} = \begin{pmatrix} 1 & 1 \\ 1 & 0 \end{pmatrix}^{k-1} \begin{pmatrix} 3 & 1 \\ 1 & 2 \end{pmatrix} = \begin{pmatrix} 1 & 1 \\ 1 & 0 \end{pmatrix}^{k} \begin{pmatrix} 1 & 2 \\ 2 & -1 \end{pmatrix} = L^k A,$$

where

$$L = \begin{pmatrix} 1 & 1 \\ 1 & 0 \end{pmatrix} \text{ and } A = \begin{pmatrix} 1 & 2 \\ 2 & -1 \end{pmatrix}.$$

The matrices L and A satisfy $A = 2L - I_2$, hence A and L commute. Also, let us observe that $A^2 = 5I_2$.

From the property 4° above, we obtain that $\gcd(L_m, L_n)$ divides L_d. We now show that $\gcd(L_m, L_n)$ divides L_d. If $\gcd(L_m, L_n) = 1$, then we are done. Suppose that $\gcd(L_m, L_n) > 1$. Since no element of the Lucas sequence is divisible by 5 and $\det A = -5$, we deduce that A must be invertible modulo L_n, for any positive integer n. Clearly, there exist integers a and b such that $am + bn = d$. The matrices $L^m A$ and $L^n A$ are both diagonal modulo $\gcd(L_m, L_n)$ and so is the matrix $(L^m A)^a (L^n A)^b$, which is equal to a power of 5 times $L^d A$. Therefore, $\gcd(L_m, L_n)$ divides $L_{\gcd(m,n)}$, and the desired conclusion follows.

□

Items 4°–6° were stated by Jaffery and Pereira [88].

Remark 2.9 The relation $\gcd(L_m, L_n) = L_{\gcd(m,n)}$ is not true in general. For instance, considering $m = 5$ and $n = 10$ we have $L_5 = 11$, $L_{10} = 123$ and $\gcd(L_5, L_{10}) = 1$, $L_{\gcd(5,10)} = L_5 = 11$, hence $\gcd(L_5, L_{10}) \neq L_{\gcd(5,10)}$.

Here we have $5 \mid 10$ but $L_5 \nmid L_{10}$, hence the implication $m \mid n \Rightarrow L_m \mid L_n$ does not hold in general.

More related theorems and results can be found in [101, 102].

Theorem 2.20 (properties of Pell numbers)

1° *For any positive integers m, n, we have the identity*

$$P_{m+n} = P_m P_{n+1} + P_{m-1} P_n.$$

2° *We have $P_m \mid P_n$ if and only if $m \mid n$;*
3° *For any $m, n \geq 0$,* $\gcd(P_m, P_n) = P_{\gcd(m,n)}$.
4° *The following identity holds:*

$$\sum_{i=0}^{4n+1} P_i = (P_{2n} + P_{2n+1})^2 = \left(\sum_{r=0}^{n} 2^r \binom{2n+1}{2r}\right)^2,$$

i.e., the sum of the Pell numbers up to P_{4n+1} is always a square [144].
5° *Pell numbers can be used to form Pythagorean triples in which a and b are one unit apart, corresponding to right triangles that are nearly isosceles*

$$\left(2P_n P_{n+1}, P_{n+1}^2 - P_n^2, P_{n+1}^2 + P_n^2 = P_{2n+1}\right).$$

Proof

1° Using the identity (2.65) and the relation $P_k = U_k\left(1+\sqrt{2}, 1-\sqrt{2}\right)$, $k = 0, 1, \ldots$, the desired formula follows.
2° The implication "⇐" follows by induction. More precisely, we show that $P_m \mid P_{km}$, for every integer $k \geq 1$. Clearly, this holds for $k = 1$ and we assume that it is true up to some fixed value $k > 2$. By part 1° we have

$$P_{(k+1)m} = P_{km+m} = P_{km} P_{m+1} + P_{km-1} P_m,$$

therefore $P_m \mid P_{(k+1)m}$.
For the implication "⇒" assume that $P_m \mid P_n$ and let $n = km + r$, where $0 \leq r < m$. Using again formula 1° we have

$$P_n = P_{km+r} = P_{km} P_{r+1} + P_{km-1} P_r.$$

Because $P_m \mid P_n$ and $P_m \mid P_{km}$, it follows that $P_m \mid P_{km-1} P_r$. From formula (2.44) we obtain $\gcd(P_{km}, P_{km-1}) = 1$, hence it is necessary to have $P_m \mid P_r$, which is not possible unless $r = 0$.

3° **First Proof** We use Theorem 2.16 for $(a, b) = \left(1+\sqrt{2}, 1-\sqrt{2}\right)$. Observe that $a+b=2$ and $ab=-1$, and $\gcd(a+b, ab)=1$, hence (2.79) holds, and our conclusion follows.

Second Proof Alternatively, a proof of this property may follow the steps of the proof of Theorem 2.18 for Fibonacci numbers.

4° Using the Binet-type formula we have

$$\sum_{i=0}^{m} P_i = \frac{1}{2\sqrt{2}}\left[\frac{(1+\sqrt{2})^{m+1}-1}{1+\sqrt{2}-1} - \frac{(1-\sqrt{2})^{m+1}-1}{1-\sqrt{2}-1}\right]$$
$$= \frac{1}{4}\left[(1+\sqrt{2})^{m+1}+(1-\sqrt{2})^{m+1}-2\right].$$

On the other hand, we obtain

$$P_{2n}+P_{2n+1} = \frac{1}{2\sqrt{2}}\left[(1+\sqrt{2})^{2n}-(1-\sqrt{2})^{2n}+(1+\sqrt{2})^{2n+1}-(1-\sqrt{2})^{2n+1}\right]$$
$$= \frac{1}{2\sqrt{2}}\left[(2+\sqrt{2})(1+\sqrt{2})^{2n}-(2-\sqrt{2})(1-\sqrt{2})^{2n}\right]$$
$$= \frac{1}{2}\left[(1+\sqrt{2})^{2n+1}-(1-\sqrt{2})^{2n+1}\right], \tag{2.83}$$

hence

$$(P_{2n}+P_{2n+1})^2 = \frac{1}{4} = \left[(1+\sqrt{2})^{4n+2}+(1-\sqrt{2})^{4n+2}-2\right],$$

which shows that indeed we have

$$\sum_{i=0}^{4n+1} P_i = (P_{2n}+P_{2n+1})^2.$$

The other identity follows by the binomial expansion of $P_{2n}+P_{2n+1}$ in (2.83).

5° Substituting $m=n+1$ in 1° one obtains the identity

$$P_{2n+1} = P_{n+1}^2 + P_n^2,$$

i.e., $\left(2P_nP_{n+1}, P_{n+1}^2-P_n^2, P_{n+1}^2+P_n^2=P_{2n+1}\right)$ is a Pythagorean triple.

□

Theorem 2.21 (properties of Pell–Lucas numbers)

1° *For any positive integers m, n, we have the identity*

$$Q_{m+n} = P_{n-1}Q_m + P_n Q_{m+1}.$$

2° *If k is a positive integer, then $P_{n+k} + (-1)^k P_{n-k} = Q_k P_n$.*
3° $P_{2n} = Q_n P_n$.
4° *If $k, n \in \mathbb{N}$ with k odd, then $Q_n \mid Q_{kn}$.*
5° *If $k, n \in \mathbb{N}$ with k even, then $Q_n P_n \mid P_{kn}$.*
6° *If $d = \gcd(m, n)$ and m/d, n/d are odd, then $\gcd(Q_m, Q_n) = Q_{\gcd(m,n)}$.*

Proof

1° By the identity (2.66) and the relation $Q_k = V_k\left(1+\sqrt{2}, 1-\sqrt{2}\right)$, valid for $k = 0, 1, \ldots$, the desired formula follows.
2° This relation follows from the Binet-type formulae (2.32) and (2.33). Denoting $\alpha = 1+\sqrt{2}$ and $\beta = 1-\sqrt{2}$, we have $\alpha\beta = -1$, hence

$$\begin{aligned} Q_k P_n &= \frac{1}{2\sqrt{2}}\left(\alpha^k + \beta^k\right)\left(\alpha^n - \beta^n\right) = \frac{1}{2\sqrt{2}}\left(\alpha^{n+k} - \beta^{n+k} + \alpha^n\beta^k - \alpha^k\beta^n\right) \\ &= P_{n+k} + \frac{(-1)^k}{2\sqrt{2}}\left(\alpha^{n-k} - \beta^{n-k}\right) = P_{n+k} + (-1)^k P_{n-k}. \end{aligned}$$

3° In the previous relation we take $k = n$.
4° It follows by Lemma 2.1 3° and $Q_s = V_s\left(1+\sqrt{2}, 1-\sqrt{2}\right)$, $s = 0, 1, \ldots$.
5° The proof is similar to Theorem 2.19 5°. Using the notation $k = 2l$, one obtains by the Binet-type formula

$$\begin{aligned} \frac{P_{kn}}{Q_n} &= \frac{1}{2\sqrt{2}}\frac{\alpha^{nk} - \beta^{kn}}{\alpha^n + \beta^n} = \frac{1}{2\sqrt{2}}\frac{(\alpha^{2n})^l - (\beta^{2n})^l}{\alpha^n + \beta^n} \\ &= \frac{1}{2\sqrt{2}}\frac{\alpha^{2n} - \beta^{2n}}{\alpha^n + \beta^n}\left(\alpha^{2n(l-1)} + \alpha^{2n(l-2)}\beta^{2n} + \cdots + \alpha^{2n}\beta^{2n(l-2)} + \beta^{2n(l-1)}\right) \\ &= P_n\left(L_{2n(l-1)} + L_{2n(l-3)} + \cdots\right), \end{aligned}$$

where we used the relation $\alpha\beta = -1$.
6° We apply Theorem 2.17 for $(a, b) = \left(1+\sqrt{2}, 1+\sqrt{2}\right)$ and the conclusion follows since $a + b = 2$ and $ab = -1$ are relatively prime, while $Q_k = V_k\left(1+\sqrt{2}, 1+\sqrt{2}\right)$, for $k = 0, 1, \ldots$.

□

2.2.6 *Zeckendorf's Theorem*

A result concerning the decomposition of integers as a sum of Fibonacci numbers has been established by Zeckendorf.

Theorem 2.22 (Zeckendorf) *Every positive integer can be represented as sum of several distinct Fibonacci numbers.*

Proof We prove by induction that every positive integer k can be written as a sum of several distinct Fibonacci numbers.

The case $k = 1$ is obvious, since we have $1 = F_1$.

Assume now that every positive integer $1, \dots, n-1$ can be written a sum of several distinct Fibonacci numbers, and let $k = n$. Denoting by f_i the greatest Fibonacci number which is less than n, we have

$$n = f_i + (n - f_i).$$

By the induction hypothesis, $n - f_i$ is a positive integer less than n, therefore it can be written as

$$n - f_i = f_{j_1} + f_{j_2} + \cdots + f_{j_t}, \quad j_1 < j_2 < \cdots < j_t.$$

We must have $i > j_t$, as otherwise we would have

$$n = f_i + f_{j_1} + f_{j_2} + \cdots + f_{j_t} > f_i + f_{i-1} = f_{i+1},$$

in contradiction with the choice of i. So, $n = f_{j_1} + f_{j_2} + \cdots + f_{j_t} + f_i$ is a representation of n as a sum of distinct Fibonacci numbers. □

Moreover, this representation is unique. Indeed, assume that the integer n has two Zeckendorf representations, i.e., there are two sets of positive integers T, S containing the indices in the two decompositions as sums of Fibonacci numbers. Denote by $T_1' = T_1 \setminus T_2$ and $T_2' = T_2 \setminus T_1$. Since the sets S' and T' are disjoint, we have

$$\sum_{x \in T_1'} x = \sum_{x \in T_1} x - \sum_{a \in T_1 \cap T_2} a = \sum_{y \in T_1} y - \sum_{b \in T_1 \cap T_2} = \sum_{y \in T_2'} y.$$

If either T_1' or T_2' is empty, then its sum would be 0, as well as the sum of the other set, hence we would have $T_1' = T_2' = \emptyset$, hence $T_1 = T_1 \cap T_2 = T_2$.

If the sets T_1 and T_2 are nonempty, denote by $F_{t_1} = \max T_1$ and $F_{t_2} = \max T_2$. Since the sets are disjoint, one may assume that $F_{t_1} < F_{t_2}$. We can check that

$$\sum_{x \in T_1} x < F_{t_1+1} \leq F_{t_2}.$$

Since the sums over T_1 and T_2 must be equal, this is a contradiction.

One can also define the *Fibonacci multiplication* $\circ : \mathbb{N}^2 \to \mathbb{N}$. Given two numbers with Zeckendorf representations

$$a_1 = \sum_{s=0}^{t_1} F_{j_s} \quad a_2 = \sum_{k=0}^{t_2} F_{i_k},$$

we define

$$a_1 \circ a_2 = \sum_{s=0}^{t_1} \sum_{k=0}^{t_2} F_{j_s + i_k}.$$

Theorem 2.23 $(\mathbb{N}, \circ)$ *forms a commutative semigroup.*

The number of terms involved in the Zeckendorf representation has been established by Deza [60, Theorem 1].

Theorem 2.24

1° *For any positive integer n the number $F_{2n+2} - 1$ is the smallest number which is not representable as a sum of at most n Fibonacci numbers.*
2° *$F_{2n+2} - 1$ may be represented as a sum of $n + 1$ Fibonacci numbers.*
3° *$F_{2n+2} - 2$ is not representable as a sum of fewer than n Fibonacci numbers.*

Proof Indeed, if $n = 1$ the theorem holds true. Assume that the theorem is correct for $n \leq m$. The integers on the segment $[1, F_{2m+2} - 2]$ may be represented from part 1° of the theorem, as a sum of $\leq m$ Fibonacci numbers. Also, number $(F_{2m+2} - 2) + 1 = F_{2m+2} - 1$ may be represented by part 2° of the theorem as a sum of $m + 1$ Fibonacci numbers, while the number $(F_{2m+2} - 2) + 2 = F_{2m+2}$ is a Fibonacci number. The integers in the interval

$$[F_{2m+2} + 1, F_{2m+2} + (F_{2m+1} - 1)], \tag{2.84}$$

are the sums of number F_{2m+2} and of the corresponding numbers on the segment $[1, F_{2m+1} - 1]$, which by part 1° of the theorem are representable as a sum of $\leq m$ Fibonacci numbers (since $F_{2m+1} - 1 \leq F_{2m+2} - 2$). Number $F_{2m+2} + (F_{2m+1} - 1) + 1 = F_{2m+3}$ is a Fibonacci number. We deduce that the numbers of the segment

$$[F_{2m+3} + 1, F_{2m+3} + (F_{2m+2} - 2)],$$

are representable as a sum of $\leq m + 1$ Fibonacci numbers for the same reason as those of segment (2.84). Therefore, the numbers not greater than

$$F_{2m+3} + (F_{2m+2} - 2) = F_{2(m+1)+2} - 2,$$

are representable as sums of $\leq m+1$ Fibonacci numbers. By the induction hypothesis, a correct decomposition of the numbers $F_{2m+2}-2$ and $F_{2m+2}-1$ contains m and $m+1$ terms, respectively. Adding F_{2m+3} on the left-hand side of these two decompositions, we obtain the correct decomposition of the numbers $F_{2m+4}-2$ and $F_{2m+4}-1$. The latter contain $m+1$ and $m+2$ terms, respectively. From this and by Zeckendorf's Theorem, it follows that the numbers $F_{2(m+1)+2}-2$ and $F_{2(m+1)+2}-1$ may be represented as the sums of $m+1$ (but not less) and $m+2$ (but not less) Fibonacci numbers.

Notice that the following identity clearly holds:

$$F_{2n+2}-2=\sum_{i=1}^{2n} F_i=\sum_{i=1}^{n} F_{2i+1}.$$

□

2.3 Homographic Recurrences

In this section we present key definitions and results concerning linear fractional transformations, including some formulae for their composition. These transformations are conformal maps which have many applications in engineering, number theory, geometry, or control theory.

We also present the representation formulae for the general term of homographic recurrent sequences obtained by Andrica and Toader in [22]. We then discuss the composition of sequences of linear fractional transformations given by Jacobsen [87], formulated in terms of tail recurrences.

2.3.1 *Key Definitions*

Definition 2.1 Let a, b, c, d be complex numbers. A map $f:\mathbb{C}\to\mathbb{C}$ is called a **linear fractional transformation** (or **homographic function**) if

$$w=f(z)=\frac{az+b}{cz+d},\quad \Delta=bc-ad\neq 0. \tag{2.85}$$

The transformation can be extended to the entire extended complex plane $\hat{\mathbb{C}}=\mathbb{C}\cup\{\infty\}$, by defining

$$f\left(\frac{-d}{c}\right)=\infty,$$

$$f(\infty)=\frac{a}{c},$$

when the transformation is meromorphic and bijective [87], and analytic everywhere except for the point $z = -d/c$ where it has a simple pole.

The conventions adopted in this case are

$$az + b = b \text{ if } c = 0 \text{ and } z = \infty, \tag{2.86}$$

$$cz + d = d \text{ if } d = 0 \text{ and } z = \infty, \tag{2.87}$$

$$z - w = 0 \text{ if } z = w = \infty. \tag{2.88}$$

Notice that if f is not the identity function, then it has one or two fixed points given by the roots q_1, q_2 of the quadratic equation

$$cq^2 + (d - a)q - b = 0, \tag{2.89}$$

which coincide if and only if f is parabolic (i.e., $D = (d - a)^2 + 4bc = 0$).

The following identities hold:

$$\frac{f(z) - q_1}{f(z) - q_2} = \frac{cq_2 + d}{cq_1 + d} \cdot \frac{z - q_1}{z - q_2}, \quad \text{if } q_1 \neq q_2, \tag{2.90}$$

$$\frac{1}{f(z) - q} = \frac{2c}{a + d} + \frac{1}{z - q}, \quad \text{if } q_1 = q_2 = q. \tag{2.91}$$

2.3.2 *Homographic Recurrent Sequences*

The **matrix of the homographic function** f (2.85) is denoted by

$$A_f = \begin{pmatrix} a & b \\ c & d \end{pmatrix}.$$

Theorem 2.25 *The following properties hold:*

1° *If f and g are homographic functions, then the composition $f \circ g$ is a homographic function (where it is defined) and $A_{f \circ g} = A_f \cdot A_g$.*

2° *If f is a homographic function, then the composite transformation*

$$f^n = \underbrace{f \circ f \circ \cdots \circ f}_{n \text{ times}} \tag{2.92}$$

is a homographic function on its domain and $A_{f^n} = (A_f)^n$, for any integer $n \geq 1$.

Definition 2.2 Let $f : \mathbb{C} \setminus \left\{-\frac{d}{c}\right\} \to \mathbb{C}$ with $f(z) = \dfrac{az+b}{cz+d}$, be a homographic function. We call a **homographic recurrence** the sequence $(z_n)_{n\geq 0}$ given by

$$z_{n+1} = f(z_n) = \frac{az_n + b}{cz_n + d}, \quad n \geq 0, \tag{2.93}$$

where z_0 is a fixed complex number.

Clearly, the sequence $(z_n)_{n\geq 0}$ is well defined if and only if $cz_n + d \neq 0$ for any positive integer n. The set consisting in all $z_0 \in \mathbb{C}$ such that $cz_n + d \neq 0$ for all integers $n \geq 0$ is called the **domain** of the homographic recurrence.

A simple inductive argument shows that for integer $n \geq 0$, one has

$$z_n = f^n(z_0), \tag{2.94}$$

where $f^n = \underbrace{f \circ f \circ \cdots \circ f}_{n\ times}$. Denote

$$(A_f)^n = \begin{pmatrix} a_n & b_n \\ c_n & d_n \end{pmatrix}.$$

From the matrix relation $A_{f^n} = (A_f)^n$ and from $z_n = f^n(z_0)$ we obtain

$$z_n = \frac{a_n z_0 + b_n}{c_n z_0 + d_n}, \quad n \geq 0. \tag{2.95}$$

Remark 2.10 From formula (2.95) it follows that the domain of the homographic recurrence sequence $z_{n+1} = f(z_n)$, $n \geq 0$, is $\mathbb{C} \setminus \left\{-\frac{d_n}{c_n} : n = 1, 2, \ldots\right\}$.

Lemma 2.2 *Consider the linear fractional transformation*

$$f(z) = \frac{az+b}{cz+d}, \quad d \neq 0,$$

having the fixed points q_1, q_2. The following assertions hold true:

1° *If f is nonparabolic, then by the identity* (2.90) *we have*

$$\begin{aligned}\frac{F_n(z) - q_1}{F_n(z) - q_2} &= \frac{f(F_{n-1}(z)) - q_1}{f(F_{n-1}(z)) - q_2} = \frac{cq_2 + d}{cq_1 + d} \cdot \frac{F_{n-1}(z) - q_1}{F_{n-1}(z) - q_2} \\ &= \cdots = \left(\frac{cq_2 + d}{cq_1 + d}\right)^n \frac{z - q_1}{z - q_2}, \quad n = 1, 2, \ldots.\end{aligned} \tag{2.96}$$

2° *If f is parabolic* (2.91), *then by denoting the unique fixed point of f by q, the following relations hold:*

$$\frac{1}{F_n(z)-q}=\frac{1}{f(F_{n-1}(z))-q}=\frac{2c}{a+d}+\frac{1}{F_{n-1}(z)-q} \tag{2.97}$$
$$=\cdots=n\cdot\frac{2c}{a+d}+\frac{1}{z-q},\quad n=1,2,\ldots.$$

This lemma can be used to obtain closed formulae for the general term of the homographic sequence.

We give some examples involving homographic recurrent sequences.

Example 2.20 Let the homographic function $f:\mathbb{C}\setminus\left\{-\frac{3}{2}\right\}\to\mathbb{C}$ be given by

$$f(z)=\frac{4z+1}{2z+3}.$$

Find $f^n(2)$, where n is a positive integer.

Solution The matrix of the homographic function f is

$$A_f=\begin{pmatrix}4&1\\2&3\end{pmatrix},$$

and its characteristic equation is

$$\lambda^2-7\lambda+10=0,$$

with $\lambda_1=2$ and $\lambda_2=5$. From Theorem 2.25 we get

$$(A_f)^n=\lambda_1^nB+\lambda_2^nC=2^nB+5^nC,$$

where

$$B=\frac{1}{\lambda_1-\lambda_2}(A_f-\lambda_2I_2)=-\frac{1}{3}\begin{pmatrix}-1&1\\2&-2\end{pmatrix}$$
$$C=\frac{1}{\lambda_2-\lambda_1}(A_f-\lambda_1I_2)=\frac{1}{3}\begin{pmatrix}2&1\\2&1\end{pmatrix}.$$

Therefore

$$(A_f)^n = \begin{pmatrix} \frac{1}{3}(2^n + 2 \cdot 5^n) & \frac{1}{3}(-2^n + 5^n) \\ \frac{1}{3}(-2^{n+1} + 2 \cdot 5^n) & \frac{1}{3}(2^{n+1} + 5^n) \end{pmatrix},$$

and from formulae (2.94) and (2.95) we get

$$f^n(2) = \frac{\frac{1}{2}(2^n + 2 \cdot 5^n) \cdot 2 + \frac{1}{3}(-2^n + 5^n)}{\frac{1}{3}(-2^{n+1} + 2 \cdot 5^n) \cdot 2 + \frac{1}{3}(2^{n+1} + 5^n)}$$

$$= \frac{2^n + 5^{n+1}}{-2^{n+1} + 5^{n+1}}, \quad n \geq 0.$$

Example 2.21 Let $x_0 > 0$, $m > 0$, and $x_{n+1} = \frac{m}{x_n} - 1$ for all integers $n \geq 0$. If $x_0 \neq m$, prove that the sequence $(x_n)_{n \geq 0}$ is convergent to the negative root of quadratic equation $x^2 + x - m = 0$.

Solution We can write $x_{n+1} = \frac{-x_n + m}{x_n}$, i.e., the sequence $(x_n)_{n \geq 0}$ is defined by a homographic recurrence. Let us consider the matrix

$$A = \begin{pmatrix} -1 & m \\ 1 & 0 \end{pmatrix},$$

with the characteristic equation $\lambda^2 + \lambda - m = 0$ and

$$\lambda_1 = \frac{-1 - \sqrt{1 + 4m}}{2} \text{ and } \lambda_2 = \frac{-1 + \sqrt{1 + 4m}}{2}.$$

By Theorem 2.25 we get

$$A^n = \lambda_1^n B + \lambda_2^n C = \begin{pmatrix} b_{11}\lambda_1^n + c_{11}\lambda_2^n & b_{12}\lambda_1^n + c_{12}\lambda_2^n \\ b_{21}\lambda_1^n + c_{21}\lambda_2^n & b_{22}\lambda_1^n + c_{22}\lambda_2^n \end{pmatrix},$$

and from formula (2.95) it follows that

$$x_n = \frac{(b_{11}x_0 + b_{12})\lambda_1^n + (c_{11}x_0 + c_{12})\lambda_2^n}{(b_{21}x_0 + b_{22})\lambda_1^n + (c_{21}x_0 + c_{22})\lambda_2^n}$$

$$= \frac{(b_{11}x_0 + b_{12}) + (c_{11}x_0 + c_{12})\left(\frac{\lambda_2}{\lambda_1}\right)^n}{(b_{21}x_0 + b_{22}) + (c_{21}x_0 + c_{22})\left(\frac{\lambda_2}{\lambda_1}\right)^n}.$$

Since $\left|\frac{\lambda_2}{\lambda_1}\right| < 1$, we get

$$\lim_{n\to\infty} x_n = \frac{b_{11}x_0 + b_{12}}{b_{21}x_0 + b_{22}},$$

where $b_{11}, b_{12}, b_{21}, b_{22}$ are the entries of matrix B. We have

$$B = \frac{1}{\lambda_1 - \lambda_2}(A - \lambda_2 I_2) = \frac{1}{\lambda_1 - \lambda_2}\begin{pmatrix} -1-\lambda_2 & m \\ 1 & -\lambda_2 \end{pmatrix},$$

hence the desired limit is

$$\frac{(-1-\lambda_2)x_0 + m}{x_0 - \lambda_2} = \frac{-(1+\lambda_2)x_0 + \lambda_2(1+\lambda_2)}{x_0 - \lambda_2} = -(1+\lambda_2) = \lambda_1.$$

2.3.3 Representation Theorems for Homographic Sequences

The general term of (2.93) can be written by roots of the quadratics

$$\begin{aligned} &cq^2 + (d-a)q - b = 0 \\ &p^2 - (a+d)p + ad - bc = 0, \end{aligned} \tag{2.98}$$

with roots q_1, q_2 and p_1, p_2, which share the common discriminant

$$D = (d-a)^2 + 4bc, \tag{2.99}$$

satisfying the identities

$$\begin{aligned} &q_{1,2} = \frac{a - d \pm \sqrt{D}}{2c}; \\ &cq_k + d = p_k, \quad k = 1, 2. \end{aligned} \tag{2.100}$$

Some key results regarding the general term for degenerate/nondegenerate cases, the set of admissible points, as well as the convergence, divergence, and periodicity have been investigated by Andrica and Toader [22].

These results allow one to express the terms of $(z_n)_{n\geq 0}$ as functions of z_0 and q_1, q_2, or p_1, p_2. Notice that similar explicit results can also be derived from the formula (2.96) and (2.97) given by Jacobsen [87].

Theorem 2.26 (nondegenerate case, Theorem 2.1 [22]) *If $D \neq 0$, then the following equivalent formulae hold true:*

$$z_n = \frac{q_1(z_0 - q_2)(cq_1 + d)^n - q_2(z_0 - q_1)(cq_2 + d)^n}{(z_0 - q_2)(cq_1 + d)^n - (z_0 - q_1)(cq_2 + d)^n}; \tag{2.101}$$

$$z_n = \frac{(p_1 - d)(cz_0 - p_2 + d)p_1^n - (p_2 - d)(cz_0 - p_1 + d)p_2^n}{c(cz_0 - p_2 + d)p_1^n - c(cz_0 - p_1 + d)p_2^n}. \tag{2.102}$$

Proof As $q_k = (aq_k + b)/(cq_k + d)$, we have

$$z_{n+1} - q_k = \frac{(z_n - q_k)(ad - bc)}{(cz_n + d)(cq_k + d)}, \quad k = 1, 2,$$

thus

$$\frac{z_{n+1} - q_1}{z_{n+1} - q_2} = \frac{z_n - q_1}{z_n - q_2} \cdot \frac{cq_2 + d}{cq_1 + d}.$$

By induction, we obtain

$$\frac{z_n - q_1}{z_n - q_2} = \frac{z_0 - q_1}{z_0 - q_2}\left(\frac{cq_2 + d}{cq_1 + d}\right)^n,$$

which leads to (2.101). Relation (2.102) can now be derived from (2.101). □

Remark 2.11 From (2.101) and (2.102), z_0 admits the values ($cz_n + d \neq 0$)

$$z_0 \notin \left\{\frac{q_2(cq_1 + d)^n - q_1(cq_2 + d)^n}{(cq_1 + d)^n - (cq_2 + d)^n}, \; n = 0, 1, \ldots\right\},$$

$$z_0 \notin \left\{\frac{(p_2 - d)p_1^n - (p_1 - d)p_2^n}{c(p_1^n - p_2^n)}, \; n = 0, 1, \ldots\right\}.$$

Remark 2.12 The relation (2.102) can also be proved directly, by solving a system of two recurrences. Consider the system

$$\begin{cases} x_{n+1} = ax_n + by_n \\ y_{n+1} = cx_n + dy_n, \end{cases} \tag{2.103}$$

where $z_n = x_n/y_n$ and $x_0 = z_0$, $y_0 = 1$. The system (2.103) is equivalent to

$$\begin{cases} x_n = (y_{n+1} - dy_n)/c \\ y_{n+2} - (a + d)y_{n+1} + (ad - bc)y_n = 0. \end{cases} \tag{2.104}$$

This means that $(y_n)_{n \geq 0}$ satisfies a second-order linear recurrence relation which has (2.98) as characteristic equation. This verifies the initial conditions $y_0 = 1$, $y_1 = cz_0 + d$, hence it can be represented by

$$y_n = \frac{(cz_0 + d - p_2)p_1^n - (cz_0 + d - p_1)p_2^n}{p_1 - p_2}.$$

The result follows from the first relation of (2.104) and $z_n = x_n/y_n$.

Similarly, one can formulate the result for $D = 0$.

Theorem 2.27 (degenerate case, Theorem 2.3 [22]) *If* $D = 0$, *then*

$$z_n = \frac{(a + d)z_0 + n\,[(a - d)z_0 + 2b]}{a + d + n(2cz_0 - a + d)}. \tag{2.105}$$

Proof For $D = 0$, (2.98) has a unique root $p = (a + d)/2$. By (2.104) we have

$$y_n = \frac{a + d + n(2cz_0 + d - a)}{2}\left(\frac{a + d}{2}\right)^{n-1};$$

$$x_n = \frac{2cz_0(a + d) + n(a - d)(2cz_0 - a + d)}{4c}\left(\frac{a + d}{2}\right)^{n-1}.$$

The desired results follows from the relation $z_n = x_n/y_n$. □

Remark 2.13 The admissible set for z_0 from formula (2.105) is in this case

$$z_0 \notin \left\{\frac{n(a - d) - a - d}{2cn},\ n = 1, 2, \ldots\right\}.$$

In this case, the avoidance set can be obtained from the recurrence relation

$$y_{n+1} = \frac{-dy_n + b}{cy_n - a}, \quad y_0 = -\frac{d}{c}.$$

2.3.4 *Convergence and Periodicity*

The representation formulae given above can be used to describe the behavior of the sequence. Based on formulae (2.99) and (2.100) it is convenient to define the following notation:

$$z = \frac{a + d + \sqrt{D}}{a + d - \sqrt{D}} = \frac{p_2}{p_1} = re^{2\pi i\theta}, \quad A = \frac{z_0 - q_2}{z_0 - q_1}.$$

Notice that the initial assumption $ad - bc \neq 0$ implies $a + d - \sqrt{D} \neq 0$.

Theorem 2.28 *If $D = 0$, the sequence $(z_n)_{n \geq 0}$ is convergent to $(a - d)/2c$.*

Proof Indeed, this follows by taking the limit as $n \to \infty$ in Theorem 2.27. □

For $D \neq 0$, formula (2.101) is equivalent to the more compact version

$$z_n = \frac{Aq_1 - q_2 z^n}{A - z^n}. \tag{2.106}$$

Here we reformulate [22, Theorem 3.1], in the single variable z.

Theorem 2.29 *Let $(z_n)_{n \geq 0}$ be a homographic recurrence defined by (2.93), where $D \neq 0$. The following results hold:*

1° *If $z = 1$, then $(z_n)_{n \geq 0}$ converges to $(a - d)/2c = (q_2 + q_1)/2$.*
2° *If $|z| < 1$, then $(z_n)_{n \geq 0}$ converges to q_1.*
3° *If $|z| > 1$, then $(z_n)_{n \geq 0}$ converges to q_2.*
4° *If $|z| = 1$, then the following two distinct cases are possible.*

(a) *For $0 = p/q \in \mathbb{Q}$ irreducible fraction, $(z_n)_{n \geq 0}$ is periodic.*
(b) *For $\theta = p/q \in \mathbb{R} \setminus \mathbb{Q}$, $(z_n)_{n \geq 0}$ is dense within the curve $w(z) = \frac{q_2 z - Aq_1}{z - A}$, defined on the unit circle $S = \{z \in \mathbb{C} : |z| = 1\}$.*

Proof The proofs for 1°, 2°, and 3° follow easily from formula (2.106).

When $|z| = 1$ and θ is rational, it is clear that z^n is a periodic sequence, hence the sequence $(z_n)_{n \geq 0}$ is also periodic.

On the other hand, when θ is irrational, by Kronecker's Lemma, the sequence $(z^n)_{n \geq 0}$ is dense in the unit circle, hence the sequence $(z_n)_{n \geq 0}$ is dense in the image of the function $w : S \to \mathbb{C}$, defined by $w(z) = \frac{q_2 z - Aq_1}{z - A}$. Its image is a circle of radius $|A|$, if $|A| \neq 1$, or a straight line if $|A| = 1$. □

Some examples are shown in Figures 2.1 and 2.2. To express the recurrence relation (2.106) in A and z, one may choose a, b, d, and then compute c from

$$c = \frac{(a + d)^2 \left[(z - 1)(z + 1)^{-1}\right]^2 - (a - d)^2}{4b}. \tag{2.107}$$

In Figure 2.1 are plotted convergent orbits. One may notice that the orbit converges to q_1 for $r < 1$ in Figure 2.1a, and to q_2 for $r > 1$ in Figure 2.1b.

Figure 2.2 shows a periodic orbit with 13 points (a) and a dense orbit (b).

Example 2.22 Let $a \in \mathbb{R}$. Study the convergence of the sequence defined by

$$x_0 = 1, \quad x_{n+1} x_n + a(x_{n+1} - x_n) + 1 = 0, \quad n = 0, 1, \ldots.$$

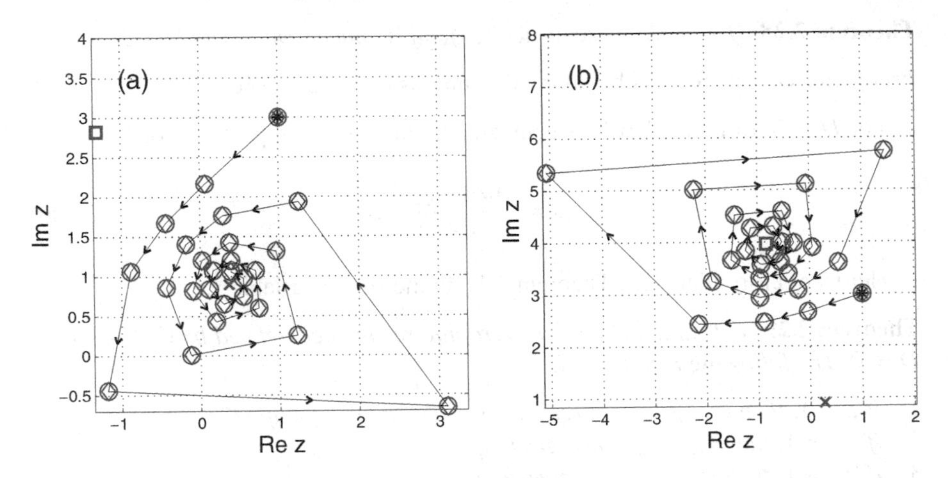

Fig. 2.1 First 25 terms of sequence $(z_n)_{n\geq 0}$ obtained from the relation (2.93) (diamonds) and direct formula (2.106) (circles) for $z_0 = 1 + 3i$ (star) and $z = r^{2\pi i\,x}$, where (**a**) $r = 0.9$, $x = 1/6$; (**b**) $r = 1.1$, $x = 1/6$. For $a = 3 + i$, $d = 1 - 2i$, $b = 2 - 2i$, and c given by (2.107). Also plotted are limits q_1 (cross) and q_2 (square). Arrows show the increase of index

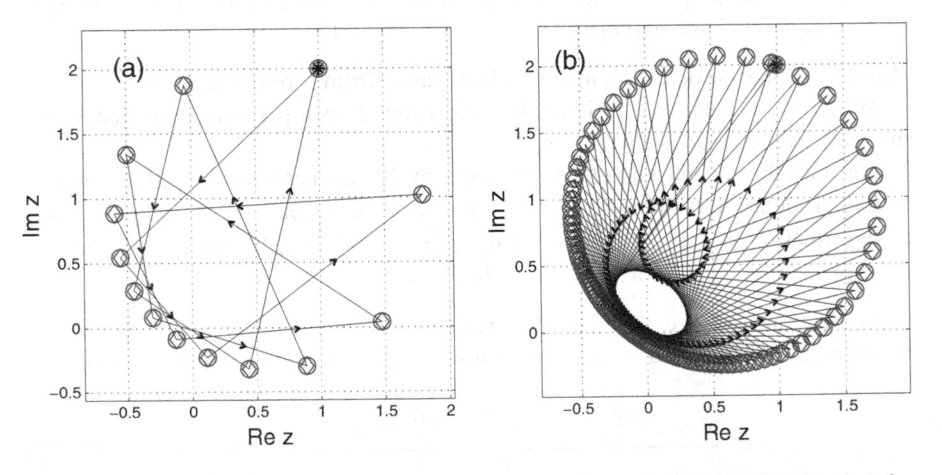

Fig. 2.2 First 100 terms of $(z_n)_{n\geq 0}$ computed by (2.93) (diamonds) and (2.106) (circles) for $z_0 = 1+2i$ (star) and $z = r^{2\pi i\,x}$, where (**a**) $r = 1$, $x = 4/13$; (**b**) $r = 1$, $x = \sqrt{2}/4$. For $a = 3+i$, $d = 1 - 2i$, $b = 2 - 2i$, and c given by (2.107). Arrows indicate the increase of sequence index

Solution We have $x_{n+1} = \frac{ax_n - 1}{x_n + a}$ for $n = 0, 1, \ldots$, hence $x_n = \frac{a_n x_0 + b_n}{c_n x_0 + d_n}$, where

$$\begin{pmatrix} a_n & b_n \\ c_n & d_n \end{pmatrix} = \begin{pmatrix} a & -1 \\ 1 & a \end{pmatrix}^n = \left(\sqrt{1+a^2}\right)^n \begin{pmatrix} \cos nt & -\sin nt \\ \sin nt & \cos nt \end{pmatrix},$$

where $\tan t = \frac{1}{a}$ and $a \neq 0$. If $a = 0$, then the sequence repeats with period 2, i.e., $x_{2n+1} = -1$ and $x_{2n} = 1$, $n = 0, 1, \ldots$. It follows that

$$x_n = \frac{\cos nt - \sin nt}{\cos nt + \sin nt} = \frac{1 - \tan nt}{1 + \tan nt}.$$

If $\frac{t}{\pi} \notin \mathbb{Q}$, then the set $\{\tan nt : n \in \mathbb{N}\}$ is dense in $\mathbb{R}$, and the range of the function $f(x) = \frac{1-x}{1+x}$ is $\mathbb{R} \setminus \{-1\}$, so the sequence $(x_n)_{n\geq 0}$ is dense in $\mathbb{R}$.

The expression $x_{n+1} = \frac{ax_n - 1}{x_n + a}$, is well defined as $x_n \neq -a$ for $n = 0, 1, \ldots$. Otherwise, if $x_n = -a$ for some n, then $-ax_{n+1} + a(x_{n+1} + a) + 1 = 0$, therefore $a^2 + 1 = 0$, which is impossible for $a \in \mathbb{R}$.

2.3.5 *Homographic Recurrences with Variable Coefficients*

In this section we present some results concerning homographic recurrent sequences with variable coefficients, and we investigate some cases when such special recurrent sequences are convergent.

Consider the linear fractional transformations $(f_n)_{n\geq 1}$, defined by

$$f_n(z) = \frac{a_n z + b_n}{c_n z + d_n}, \qquad \Delta_n = a_n d_n - b_n c_n \neq 0, \quad n = 1, 2, \ldots, \tag{2.108}$$

where $(a_n)_{n\geq 0}$, $(b_n)_{n\geq 0}$, $(c_n)_{n\geq 0}$, $(d_n)_{n\geq 0}$ are sequences of complex numbers.

Denoting by $w = f_n(z)$, the inverse transform for f_n is

$$z = f_n^{-1}(w) = -\frac{d_n w - b_n}{c_n w - a_n}.$$

Definition 2.3 Consider the sequence $(f_n)_{n\geq 1}$ of homographic transformations (2.108) We call a **homographic recurrence with variable coefficients** the sequence $(z_n)_{n\geq 0}$ given by

$$z_{n+1} = f_n(z_n) = \frac{a_n z_n + b_n}{c_n z_n + d_n}, \qquad n \geq 0, \quad z_0 \in \mathbb{C}. \tag{2.109}$$

Remark 2.14 Clearly, we have

$$\begin{aligned} z_1 &= f_0(z_0) \\ z_2 &= f_1(z_1) = (f_1 \circ f_0)(z_0) \\ &\vdots \\ z_n &= f_{n-1}(z_{n-1}) = (f_{n-1} \circ f_{n-2})(z_{n-2}) = \cdots = (f_{n-1} \circ \cdots \circ f_0)(z_0). \end{aligned}$$

Remark 2.15 A formula for (2.109) can be found by solving a system of two recurrence relations with variable coefficients. Following the steps outlined in

Remark 2.12, we consider the system

$$\begin{cases} x_{n+1} = a_n x_n + b_n y_n \\ y_{n+1} = c_n x_n + d_n y_n \end{cases}, \tag{2.110}$$

where $z_n = x_n/y_n$ and $x_0 = z_0$, $y_0 = 1$. The system (2.110) is equivalent to

$$\begin{cases} x_n = (y_{n+1} - d_n y_n)/c_n \\ y_{n+2} - (a_n + d_n) y_{n+1} + (a_n d_n - b_n c_n) y_n = 0. \end{cases} \tag{2.111}$$

This means that $(y_n)_{n\geq 0}$ satisfies a second-order linear recurrence relation with variable coefficients, from which one obtains the solution of (2.109).

Jacobsen [87] uses an alternative form of the identities (2.90) and (2.91), to study compositions with variable coefficients (in reverse order) like

$$F_n = f_1 \circ f_2 \circ \cdots \circ f_n, \quad n = 1, 2, \ldots, \tag{2.112}$$

using the concept of tail sequence defined below.

Definition 2.4 (Definition 2.1 [87]) The sequence $(t_n)_{n\geq 0}$, with $t_n \in \hat{\mathbb{C}}$ is called a *tail sequence* for $(f_n)_{n\geq 1}$ if

$$t_{n-1} = f_n(t_n), \quad n = 1, 2, 3, \ldots. \tag{2.113}$$

The following properties can be deduced by simple computations involving the formulae (2.96) and (2.97).

Proposition 2.1 *Let $(s_n)_{n\geq 0}$ and $(t_n)_{n\geq 0}$ be two tail sequences for the sequence $(f_n)_{n\geq 1}$ of linear fractional transformation. The following properties hold:*

1° *If $s_n \neq t_n$ for some $n \geq 0$, then $s_n \neq t_n$ for all $n \geq 0$;*
2° *If $t_{n-1} \neq \infty$ and $t_n \neq \infty$, then for all $z \in \hat{\mathbb{C}}$ we have*

$$\frac{1}{f_n(z) - t_{n-1}} = \frac{c_n(c_n t_n + d_n)}{-\Delta_n} + \frac{(c_n t_n + d_n)^2}{-\Delta_n} \cdot \frac{1}{z - t_n}.$$

3° *If $s_n, s_{n-1}, t_n, t_{n-1} \neq \infty$, and $s_0 \neq t_0$ then for all $z \in \hat{\mathbb{C}}$ we have*

$$\frac{f_n(z) - s_{n-1}}{f_n(z) - t_{n-1}} = \frac{c_n t_n + d_n}{c_n s_n + d_n} \cdot \frac{z - s_n}{z - t_n}.$$

These results can be used to produce a representation of F_n (2.112) and are shown to have applications to continued fractions.

Theorem 2.30 *Let $(s_n)_{n\geq 0}$ and $(t_n)_{n\geq 0}$ be two tail sequences for the sequence $(f_n)_{n\geq 1}$ of linear fractional transformations, and consider the composition F_n, $n = 1, 2, \ldots$ (2.112). For every integer $N \geq 1$, the following statements hold.*

1° *If $t_n \neq 0$ for $n = 0, 1, \ldots, N$, then for all $z \in \hat{\mathbb{C}}$ we have*

$$\frac{1}{F_N(z) - t_0} = \sum_{j=1}^{N} c_j \frac{c_j t_j + d_j}{-\Delta_j} \prod_{k=1}^{j-1} \frac{(c_k t_k + d_k)^2}{-\Delta_k} + \frac{1}{z - t_N} \prod_{k=1}^{N} \frac{(c_k t_k + d_k)^2}{-\Delta_k}.$$

2° *If in addition, $s_n \neq 0$ for $n = 0, 1, \ldots, N$, then for all $z \in \hat{\mathbb{C}}$ we have*

$$\frac{F_N(z) - s_0}{F_N(z) - t_0} = \frac{z - s_N}{z - t_N} \prod_{j=1}^{N} \frac{c_j t_j + d_j}{c_j s_j + d_j}.$$

Proof

1° This follows from Proposition 2.1 2°, applied for $n = 1, \ldots, N$, and the formula (2.113) for the tail recurrence.
2° This follows from Proposition 2.1 3°, applied for $n = 1, \ldots, N$, and the formula (2.113) for the tail recurrence.

□

Proposition 2.2 *Consider the recurrence defined by the formula*

$$x_{n+1} = \frac{a_n x_n}{c_n x_n + a_n}, \qquad n = 0, 1, \ldots,$$

where $(a_n)_{n\geq 0}$, and $(c_n)_{n\geq 0}$ are complex sequences with $a_n c_n \neq 0$ for $n = 0, 1, \ldots$, and x_0 is a complex number. The sequence $(x_n)_{n\geq 0}$ converges if and only if the series $\sum_{n=0}^{\infty} \frac{c_n}{a_n}$ is convergent.

Proof Indeed, one may rewrite the relation as

$$\frac{1}{x_{n+1}} = \frac{c_n}{a_n} + \frac{1}{x_n}, \qquad n = 0, 1, \ldots.$$

If $N \geq 1$ is an integer, by adding the relations for $n = 0, \ldots, N-1$ one obtains

$$\frac{1}{x_N} = \sum_{n=0}^{N-1} \frac{c_n}{a_n} + \frac{1}{x_0},$$

and the conclusion follows by letting N go to infinity. □

Example 2.23 Find the limit of the sequence defined by the recursive formula

$$x_{n+1} = \frac{n^2 x_n}{x_n + n^2}, \qquad n = 1, 2, \ldots,$$

where x_0 is a nonzero real number.

Solution Consider the sequences defined by $a_n = n^2$ and $c_n = 1$, $n = 1, 2, \ldots$, in Proposition 2.2, we deduce that the sequence $(x_n)_{n\geq 0}$ satisfies

$$\frac{1}{x_n} = \sum_{k=1}^{n-1} \frac{1}{k^2} + \frac{1}{x_0},$$

hence

$$\lim_{n\to\infty} x_n = \lim_{n\to\infty} \frac{1}{\sum_{k=1}^{n-1} \frac{1}{k^2} + \frac{1}{x_0}} = \frac{1}{\frac{\pi^2}{6} + \frac{1}{x_0}}.$$

We conclude that the limit of $(x_n)_{n\geq 1}$ is $\frac{6x_0}{6+\pi^2 x_0}$.

Proposition 2.3 *Consider the recursive sequence defined by the formula*

$$x_{n+1} = \frac{x_n + b_{n+1}}{1 - b_{n+1} x_n}, \qquad n = 0, 1, \ldots,$$

where $(b_n)_{n\geq 1}$ is a real sequence and x_0 is a real number. The sequence $(x_n)_{n\geq 0}$ converges if and only if the series $\sum_{n=1}^{\infty} \arctan b_n$ is convergent.

Proof We show by induction that

$$x_n = \tan\left(\arctan x_0 + \sum_{k=1}^{n} \arctan b_k\right). \tag{2.114}$$

When $n = 1$ we get

$$\begin{aligned} x_1 &= \frac{x_0 + b_1}{1 - b_1 x_0} = \frac{\tan(\arctan x_0) + \tan(\arctan b_1)}{1 - \tan(\arctan x_0)\tan(\arctan b_1)} \\ &= \tan\left(\arctan x_0 + \arctan b_1\right), \end{aligned}$$

hence the formula is true. Assuming that (2.114) holds for n, we get

$$x_{n+1} = \frac{x_n + b_{n+1}}{1 - b_{n+1} x_n} = \frac{\tan(\arctan x_0 + \sum_{k=1}^{n} \arctan b_k) + \tan(\arctan b_{n+1})}{1 - \tan(\arctan b_{n+1})\tan(\arctan x_0 + \sum_{k=1}^{n} \arctan b_k)}$$

$$= \tan\left(\arctan x_0 + \sum_{k=1}^{n} \arctan b_k + \arctan b_{n+1}\right),$$

and we are done. □

Example 2.24 Decide if the sequence $(x_n)_{n\geq 0}$ defined by the formula

$$x_{n+1} = \frac{2(n+1)^2 x_n + 1}{2(n+1)^2 - x_n}, \quad n = 0, 1, \dots,$$

where x_0 is a nonzero real number, is convergent.

Solution We have

$$x_{n+1} = \frac{x_n + \frac{1}{2(n+1)^2}}{1 - \frac{1}{2(n+1)^2} x_n}, \quad n = 0, 1, \dots,$$

hence we can consider $b_n = \frac{1}{2n^2}$, in Proposition 2.3. Using the identity

$$\arctan\frac{1}{2k^2} = \arctan\frac{k}{k+1} - \arctan\frac{k-1}{k}, \quad k = 1, 2, \dots,$$

it follows that

$$\sum_{k=1}^{n} \arctan\frac{1}{2k^2} = \arctan\frac{n}{n+1}.$$

From formula (2.114), for each $n = 0, 1, \dots,$ we obtain

$$x_n = \tan\left(\arctan x_0 + \arctan\frac{n}{n+1}\right) = \frac{x_0 + \frac{n}{n+1}}{1 - x_0\frac{n}{n+1}} = \frac{(1+x_0)n + x_0}{(1-x_0)n + 1},$$

therefore $(x_n)_{n\geq 0}$ converges to $\frac{1+x_0}{1-x_0}$ if $x_0 \neq 1$.

If $x_0 = 1$, then it follows that $x_n = 2n + 1$ for $n = 0, 1, \dots$.

Example 2.25 Let $(t_n)_{n\geq 0}$ be a sequence of real numbers such that $t_n \in (0, 1)$ for $n = 0, 1, \dots,$ and there exists $\lim_{n\to\infty} t_n \in (0, 1)$. Prove that the sequences $(x_n)_{n\geq 0}$ and $(y_n)_{n\geq 0}$ defined by the recurrence relations

$$\begin{cases} x_{n+1} = t_n x_n + (1 - t_n) y_n \\ y_{n+1} = (1 - t_n) x_n + t_n y_n, \quad n \geq 0, \end{cases}$$

are convergent and calculate their limits.

Solution Let $U_n = \begin{pmatrix} x_n \\ y_n \end{pmatrix}$ and let $A_n = \begin{pmatrix} t_n & 1-t_n \\ 1-t_n & t_n \end{pmatrix}$. Since $U_{n+1} = A_n U_n$, $n = 0, 1, \ldots$, we have $U_{n+1} = A_n A_{n-1} \cdots A_0 U_0$. We now calculate the matrix product $A_n A_{n-1} \cdots A_0$. The eigenvalues of A_n are $\lambda_1 = 1$ and $\lambda_2 = 2t_n - 1$, and the corresponding eigenvectors are $X_1 = \begin{pmatrix} 1 \\ 1 \end{pmatrix}$ and $X_2 = \begin{pmatrix} -1 \\ 1 \end{pmatrix}$ (which are in fact the same for all $n \geq 0$). Denoting by $P = \begin{pmatrix} 1 & -1 \\ 1 & 1 \end{pmatrix}$, we have

$$A_n = P \begin{pmatrix} 1 & 0 \\ 0 & 2t_n - 1 \end{pmatrix} P^{-1},$$

hence

$$A_n A_{n-1} \cdots A_0 = P \begin{pmatrix} 1 & 0 \\ 0 & s_n \end{pmatrix} P^{-1},$$

where $s_n = \prod_{k=0}^{n}(2t_k - 1)$. If one of the terms of $(t_n)_{n\geq 0}$ is $\frac{1}{2}$ (say $t_{n_0} = \frac{1}{2}$), then $s_n = 0$ for all $n \geq n_0$. If all the terms are different from $\frac{1}{2}$, then we have

$$\lim_{n\to\infty} \frac{s_{n+1}}{s_n} = \lim_{n\to\infty} (2t_{n+1} - 1) \in (-1, 1),$$

hence $\lim_{n\to\infty} s_n = 0$, which implies

$$\lim_{n\to\infty} U_{n+1} = P \begin{pmatrix} 1 & 0 \\ 0 & 0 \end{pmatrix} P^{-1} \begin{pmatrix} x_0 \\ y_0 \end{pmatrix}.$$

We conclude that $\lim_{n\to\infty} x_n = \frac{x_0+y_0}{2} = \lim_{n\to\infty} y_n$.

Example 2.26 Study the convergence of the sequence $(x_n)_{n\geq 0}$ and $(y_n)_{n\geq 0}$ defined by the system of linear recurrences

$$\begin{cases} x_{n+1} = ax_n - by_n \\ y_{n+1} = bx_n + ay_n, \end{cases},$$

where $a, b, x_0, y_0 \in \mathbb{R}$ and $a^2 + b^2 \leq 1$.

Solution Let $U_n = \begin{pmatrix} x_n \\ y_n \end{pmatrix}$ and let $A = \begin{pmatrix} a & -b \\ b & a \end{pmatrix}$. Since $U_{n+1} = AU_n$, for $n \geq 0$, we get $U_n = A^n U_0$. Let $r = \sqrt{a^2 + b^2}$ and let $t \in [0, 2\pi)$ such that $a = r\cos t$ and $b = r\sin t$. It follows that

$$A^n = r^n \begin{pmatrix} \cos nt & -\sin nt \\ \sin nt & \cos nt \end{pmatrix},$$

which implies $x_n = r^n(x_0 \cos nt - y_0 \sin nt)$ and $y_n = r^n(x_0 \sin nt + y_0 \cos nt)$.

- If $r \in [0, 1)$, then $(x_n)_{n\geq 0}$ and $(y_n)_{n\geq 0}$ converge to the common limit zero.
- If $r = 1$ and $\frac{t}{\pi} = \frac{p}{q} \in \mathbb{Q}$ with $(p, q) = 1$, then the sequences $(x_n)_{n\geq 0}$ and $(y_n)_{n\geq 0}$ are periodic, having the same period $2q$.
- If $r = 1$ and $\frac{t}{\pi} \in (\mathbb{R} \setminus \mathbb{Q})$, then the sequences $(x_n)_{n\geq 0}$ and $(y_n)_{n\geq 0}$ are dense within the interval $\left[-\sqrt{x_0^2 + y_0^2}, \sqrt{x_0^2 + y_0^2}\right]$.

Example 2.27 Let $A = \begin{pmatrix} a & b \\ c & d \end{pmatrix} \in \mathcal{M}_2(\mathbb{R})$ be such that $bc \neq 0$ and there exists an integer $n \geq 2$ such that $b_n c_n = 0$, where $A^n = \begin{pmatrix} a_n & b_n \\ c_n & d_n \end{pmatrix}$. Prove that $a_n = d_n$.

Solution By the identity $A A^n = A^n A$ we deduce the relations

$$\begin{aligned} aa_n + bc_n &= a_n a + b_n c \\ ab_n + bd_n &= a_n b + b_n d \\ ca_n + dc_n &= c_n a + d_n c \\ cb_n + dd_n &= c_n b + d_n d, \end{aligned}$$

therefore $bc_n = b_n c$, $(a - d)b_n = (a_n - d_n)b$, $(a - d)c_n = (a_n - d_n)c$, $cb_n = c_n b$. Since $b \neq 0$ and $c \neq 0$, if $b_n = 0$ or $c_n = 0$ we have $a_n = d_n$.

Example 2.28 Let $A = \begin{pmatrix} a & b \\ c & d \end{pmatrix} \in \mathcal{M}_2(\mathbb{Q})$ be such that $bc \neq 0$ and there exists an integer $n \geq 2$ such that $b_n c_n = 0$, where $A^n = \begin{pmatrix} a_n & b_n \\ c_n & d_n \end{pmatrix}$, $n \in \mathbb{N}$.

1° Prove that $a_n = d_n$.
2° Study the convergence of the sequence $(x_n)_{n\geq 0}$ defined by

$$x_{n+1} = \frac{ax_n + b}{cx_n + d}, \quad n \geq 0, \quad x_0 \in \mathbb{R} \setminus \mathbb{Q}.$$

Solution

1° The conclusion follows from the previous example.
2° If $ad = bc$, then $x_0 = x_1 = \cdots$, hence the sequence $(x_n)_{n\geq 0}$ is constant.

Otherwise, we have the relation $x_0 = \frac{-dx_1+b}{-a+cx_1}$. Assuming that x_1 is rational, since $a, b, c, d \in \mathbb{Q}$ we obtain that $x_0 \in \mathbb{Q}$, a contradiction. We deduce that $x_k \in \mathbb{R} \setminus \mathbb{Q}$ for all $k \geq 0$, and $cx_k + d \neq 0$.

Let $f : \mathbb{R} \setminus \mathbb{Q} \to \mathbb{R} \setminus \mathbb{Q}$ be the function defined by $f(x) = \frac{ax+b}{cx+d}$. The recurrence relation $x_{k+1} = f(x_k)$ implies that $x_k = f^k(x_0)$, where $f^k = f \circ \cdots \circ f$. When $k = n$, by part 1° we have $a_n = d_n$, which combined with $b_n c_n = 0$ gives $b_n = c_n = 0$, hence $x_n = \frac{a_n x_0}{a_n} = x_0$. It follows that

$$x_{n+1} = \frac{ax_n + b}{cx_n + d} = \frac{ax_0 + b}{cx_0 + d} = x_1, \quad x_{n+2} = \frac{ax_{n+1} + b}{cx_{n+1} + d} = \frac{ax_1 + b}{cx_1 + d} = x_2,$$

and $x_{n+k} = x_n$, $k \in \mathbb{N}$. This sequence converges if constant, i.e., $x_0 = f(x_0)$. Hence, x_0 satisfies $cx_0^2 + (d-a)x_0 - b = 0$, with the solutions

$$x_0 = \frac{a - d \pm \sqrt{D}}{2c}, \qquad D = (d-a)^2 + 4bc > 0.$$

Thus, the sequence converges for $D > 0$ and $D \neq q^2$, where $q \in \mathbb{Q}$.

Example 2.29 Prove that $(x_n)_{n\geq 0}$ defined below

$$x_{n+1} = 1 + \frac{1}{x_n}, \qquad n = 0, 1, \ldots, \qquad x_0 \in \mathbb{R} \setminus \mathbb{Q}.$$

1° $(x_n)_{n\geq 0}$ is well defined.
2° $(x_n)_{n\geq 0}$ is convergent and its limit does not depend on x_0.

Solution One can show by induction that

$$x_{n+1} = \frac{F_{n+1}x_0 + F_n}{F_n x_0 + F_{n-1}}, \qquad n = 1, 2, \ldots,$$

where $(F_n)_{n\geq 0}$ is the Fibonacci sequence.

1° Since x_0 is irrational, it is clear that $F_n x_0 + F_{n-1} \neq 0$, for all $n = 1, 2, \ldots$.
2° By Theorem 2.14, we deduce that $\lim_{n\to\infty} \frac{F_{n+1}}{F_n} = \varphi$, where $\varphi = \frac{\sqrt{5}+1}{2}$ is the golden ratio, and also that $\lim_{n\to\infty} \frac{F_n}{F_{n+1}} = \frac{1}{\varphi}$. It follows that

$$\lim_{n\to\infty} x_{n+1} = \lim_{n\to\infty} \frac{F_n\left(\frac{F_{n+1}}{F_n}x_0 + 1\right)}{F_n\left(x_0 + \frac{F_{n-1}}{F_n}\right)} = \frac{\varphi x_0 + 1}{x_0 + \frac{1}{\varphi}} = \varphi.$$

Chapter 3
Arithmetic and Trigonometric Properties of Some Classical Recurrent Sequences

The classical second-order recurrent sequences discussed in Chapter 2 have many interesting properties. In Section 3.1 we present arithmetic properties of the Fibonacci and Lucas sequences first found in [23]. Section 3.2 studies properties of Pell and Pell–Lucas sequences. We also discuss about certain notions of Fibonacci, Lucas, Pell, or Pell–Lucas primality. These results have been extended to generalized Lucas and Pell–Lucas sequences in our paper [16]. Section 3.3 presents factorizations of the terms of classical sequences as products of trigonometric expressions. These are derived from the complex factorizations of the general polynomials U_n and V_n (2.60), and some involve the resultant of polynomials.

3.1 Arithmetic Properties of Fibonacci and Lucas Sequences

Here we give some arithmetic properties of Fibonacci and Lucas numbers, focusing on results given in [23].

Theorem 3.1 *Let p be an odd prime, k a positive integer, and r an arbitrary integer. The following relations hold:*

$$2F_{kp+r} \equiv \left(\frac{p}{5}\right) F_k L_r + F_r L_k \pmod{p}, \tag{3.1}$$

and

$$2L_{kp+r} \equiv 5\left(\frac{p}{5}\right) F_k F_r + L_k L_r \pmod{p}, \tag{3.2}$$

where $\left(\frac{p}{5}\right)$ *is the Legendre's symbol.*

D. Andrica, O. Bagdasar, *Recurrent Sequences*, Problem Books in Mathematics,
https://doi.org/10.1007/978-3-030-51502-7_3

Proof We shall prove (3.1) directly from the definition. First, one may write the formula $(1+\sqrt{5})^s = a_s + b_s\sqrt{5}$, where a_s and b_s are positive integers, $s = 0, 1, \ldots$. From Binet's formula, we have

$$F_{kp+r} = \frac{1}{\sqrt{5}}\left[\left(\frac{1+\sqrt{5}}{2}\right)^{kp+r} - \left(\frac{1-\sqrt{5}}{2}\right)^{kp+r}\right]$$

$$= \frac{1}{2^{kp+r}\sqrt{5}}[(a_k + b_k\sqrt{5})^p(a_r + b_r\sqrt{5}) - (a_k - b_k\sqrt{5})^p(a_r - b_r\sqrt{5})]$$

$$= \frac{1}{2^{kp+r}\sqrt{5}}[(a_r + b_r\sqrt{5})\sum_{j=0}^{p}\binom{p}{j}a_k^{p-j}(b_k\sqrt{5})^j$$

$$- (a_r - b_r\sqrt{5})\sum_{j=0}^{p}\binom{p}{j}(-1)^j a_k^{p-j}(b_k\sqrt{5})^j]$$

$$= \frac{1}{2^{kp+r}\sqrt{5}}[a_r\sum_{j=0}^{p}\binom{p}{j}(1-(-1)^j)a_k^{p-j}(b_k\sqrt{5})^j$$

$$+ b_r\sqrt{5}\sum_{j=0}^{p}\binom{p}{j}(1+(-1)^j)a_k^{p-j}(b_k\sqrt{5})^j].$$

Since p divides $\binom{p}{j}$ for $j = 1, 2, \cdots, p-1$, it follows that

$$2^{kp+r-1}F_{kp+r} \equiv \left(a_r b_k^p 5^{\frac{p-1}{2}} + b_r a_k^p\right) \pmod{p}.$$

Using Fermat's Little Theorem and Euler's Criterion, we have further that

$$2^{kp+r-1}F_{kp+r} \equiv \left(\frac{p}{5}\right) a_r b_k + b_r a_k \pmod{p}, \tag{3.3}$$

where by Gauss's Quadratic Reciprocity Law we deduced that $\left(\frac{5}{p}\right) = \left(\frac{p}{5}\right)$.

On the other hand, from the relation $(1+\sqrt{5})^s = a_s + b_s\sqrt{5}$ we get the identity $(1-\sqrt{5})^s = a_s - b_s\sqrt{5}$, hence we have $a_s = 2^{s-1} \cdot L_s$ and $b_s = 2^{s-1} \cdot F_s$ for $s = 0, 1, \ldots$. Substituting this back into (3.3), we obtain

$$2^{kp-k-1}F_{kp+r} \equiv \left(\frac{p}{5}\right) L_r F_k + F_r L_k \pmod{p},$$

and the relation (3.1) follows via Fermat's Little Theorem.

To deduce (3.2), we employ the following well-known formula

$$L_n = F_n + 2F_{n-1},$$

which can be proved either directly from the definition or by noting that the sequences $(L_n)_{n\geq 0}$ and $(F_n + 2F_{n-1})_{n\geq 0}$ satisfy the same initial conditions and the same recursion formula. From this identity we also deduce that

$$L_n + 2L_{n-1} = 5F_n.$$

By (3.1), we have

$$2F_{kp+r} \equiv \left(\frac{p}{5}\right) F_k L_r + F_r L_k \pmod{p},$$

and

$$4F_{kp+r-1} \equiv 2\left(\frac{p}{5}\right) F_k L_{r-1} + 2F_{r-1} L_k \pmod{p}.$$

Adding these two relations yields

$$2F_{kp+r} + 4F_{kp+r-1} \equiv \left(\frac{p}{5}\right) F_k(L_r + L_{r-1}) + L_k(F_r + 2F_{r-1}).$$

Then using the two identities which we mentioned above we deduce that

$$2F_{kp+r} + 4F_{kp+r-1} = L_{kp+r},$$

and

$$\left(\frac{p}{5}\right) F_k(L_r + L_{r-1}) + L_k(F_r + 2F_{r-1}) = 5\left(\frac{p}{5}\right) F_k F_r + L_k L_r,$$

which gives the relation (3.2). □

Some consequences are presented in the examples below.

Example 3.1 Taking $r = 0$ in relation (3.1), we obtain that for any positive integer k one has

$$F_{kp} \equiv \left(\frac{p}{5}\right) F_k \pmod{p}. \tag{3.4}$$

In the special case $k = 1$ we get

$$F_p \equiv \left(\frac{p}{5}\right) \pmod{p}.$$

Taking $k = 1$ and $r = 1$ in relation (3.1) we get

$$2F_{p+1} \equiv \left(\frac{p}{5}\right) + 1 \pmod{p}. \tag{3.5}$$

Taking $k = 1$ and $r = -1$ in relation (3.1) we get

$$2F_{p-1} \equiv -\left(\frac{p}{5}\right) + 1 \pmod{p}. \tag{3.6}$$

If $\left(\frac{p}{5}\right) = -1$, then from (3.5) we have $p \mid F_{p+1}$. In the case $\left(\frac{p}{5}\right) = 1$, from (3.6) one obtains $p \mid F_{p-1}$. We can summarize these in the following known property:

$$p \mid F_{p-\left(\frac{p}{5}\right)}. \tag{3.7}$$

Remark 3.1 A composite integer n is called **Fibonacci pseudoprime** if $n \mid F_{n-\left(\frac{n}{5}\right)}$. Lehmer proved in [112] that there are infinitely many such pseudoprimes. The list of even such pseudoprimes is indexed as A141137 in the OEIS [157]. The list of known odd Fibonacci pseudoprimes is indexed in [157] as to A081264 , and begins with the terms:

$$323, 377, 1891, 3827, 4181, 5777, 6601, 6721, 8149, 10877, 11663, 13201,$$
$$13981, 15251, 17119, 17711, 18407, 19043, 23407, 25877, 27323, 30889,$$
$$34561, 34943, 35207, 39203, 40501, 50183, 51841, 51983, 52701, \ldots.$$

In contrast to (3.7), there is no prime $p < 2.8 \times 10^{16}$ such that $p^2 \mid F_{p-\left(\frac{p}{5}\right)}$. Crandall et al. called in [56] such a prime number p satisfying $p^2 \mid F_{p-\left(\frac{p}{5}\right)}$ a **Wall–Sun–Sun prime**. There is no known example of a Wall–Sun–Sun prime and the congruence $F_{p-\left(\frac{p}{5}\right)} \equiv 0 \pmod{p^2}$, can only be checked through explicit powering computations. Further remarks on this topic can be found in [9] or [73].

Example 3.2 Taking $k = 1$ and $r = 0$ in the relation (3.2), we obtain

$$L_p \equiv 1 \pmod{p}. \tag{3.8}$$

Remark 3.2 A composite number n satisfying $n \mid L_n - 1$ is called a **Bruckman–Lucas pseudoprime**. In 1964, Lehmer [112] proved that the set of Lucas pseudoprimes is infinite. The sequence is indexed in the OEIS [157] as to A005845 , and begins with the terms:

$$705, 2465, 2737, 3745, 4181, 5777, 6721, 10877, 13201, 15251, 24465, 29281,$$
$$34561, 35785, 51841, 54705, 64079, 64681, 67861, 68251, 75077, 80189, 90061,$$
$$96049, 97921, 100065, 100127, 105281, 113573, 118441, 146611, 161027, \ldots.$$

Remark 3.3 A composite number n is called a **Fibonacci–Bruckner–Lucas pseudoprime** if it satisfies simultaneously the properties

$$n \mid F_{n-\left(\frac{p}{5}\right)} \text{ and } n \mid L_n - 1.$$

These numbers produce the sequence [157] as A212424, beginning with

$$4181, 5777, 6721, 10877, 13201, 15251, 34561, 51841, 64079, 64681, \ldots.$$

Bruckman proved in 1994 [42] that there are infinitely numbers n with this property. It was later shown that they correspond to Frobenius pseudoprimes with respect to the Fibonacci polynomial $x^2 - x - 1$ [57, 142].

Example 3.3 From (3.4), it follows that for two positive integers k and s, p divides $F_{kp} - F_{sp}$ if and only if p divides $F_k - F_s$. In particular, since $F_2 = F_1 = 1$, we get

$$p \mid F_{2p} - F_p.$$

Taking $k = 1$ and $r = 1$ in relation (3.2), we get

$$2L_{p+1} \equiv 5\left(\frac{p}{5}\right) + 1 \pmod{p}. \tag{3.9}$$

Taking $k = 1$ and $r = -1$ in relation (3.2) we get

$$2L_{p-1} \equiv 5\left(\frac{p}{5}\right) - 1 \pmod{p}. \tag{3.10}$$

If $\left(\frac{p}{5}\right) = -1$, then from (3.9) we have $p \mid L_{p+1} + 2$. In the case $\left(\frac{p}{5}\right) = 1$, from (3.10) one obtains $p \mid L_{p-1} - 2$.

We can summarize these remarks in the following formula:

$$p \mid L_{p-\left(\frac{p}{5}\right)} - 2\left(\frac{p}{5}\right). \tag{3.11}$$

The relations (3.7) and (3.11) are just the first in a sequences of divisibility relations as we can see from the following result.

Theorem 3.2 *If p is an odd prime and k a positive integer, then we have*

1° $F_{kp-\left(\frac{p}{5}\right)} \equiv F_{k-1} \pmod{p}$.

2° $L_{kp-\left(\frac{p}{5}\right)} \equiv \left(\frac{p}{5}\right) L_{k-1} \pmod{p}$.

Proof

1° Let us consider in (3.1) $r = 1$ and $r = -1$ to deduce the relations $2F_{kp+1} \equiv \left(\frac{p}{5}\right) F_k + L_k \pmod{p}$ and $2F_{kp-1} \equiv -\left(\frac{p}{5}\right) F_k + L_k \pmod{p}$, respectively. These relations can be summarized as

$$2F_{kp-\left(\frac{p}{5}\right)} \equiv L_k - F_k \pmod{p}.$$

The sequences $(L_j - F_j)_{j\geq 0}$ and $(2F_{j-1})_{j\geq 0}$ satisfy the same initial conditions for $j = 0$, $j = 1$ and the same recursive relation, hence $L_j - F_j = 2F_{j-1}$.

2° The argument is similar. Consider $r = 1$ and $r = -1$ in (3.2), to get the relations $2L_{kp+1} \equiv 5\left(\frac{p}{5}\right) F_k + L_k \pmod{p}$ and $2L_{kp-1} \equiv 5\left(\frac{p}{5}\right) F_k - L_k \pmod{p}$, respectively. These relations can be summarized as

$$2L_{kp-\left(\frac{p}{5}\right)} \equiv \left(\frac{p}{5}\right)(5F_k - L_k) \pmod{p}.$$

Now observe that the sequences $(5F_j - L_j)_{j\geq 0}$ and $(2L_{j-1})_{j\geq 0}$ satisfy the same initial conditions for $j = 0$, $j = 1$ and the same recursive relation, therefore $5F_j - L_j = 2L_{j-1}$, and the property is proved.

□

Remark 3.4 The first relation in Theorem 3.2 shows that for every odd prime p, there is an arithmetic progression $a_0, a_1, \ldots$, with ratio p, such that

$$(F_{a_0}, F_{a_1}, F_{a_2}, \ldots) \equiv (F_0, F_1, F_2, \ldots) \pmod{p}.$$

The second relation in the Theorem shows that for every odd prime p, there is an arithmetic progression $a_0, a_1, \ldots$, with ratio p, such that

$$(L_{a_0}, L_{a_1}, L_{a_2}, \ldots) \equiv \left(\frac{p}{5}\right)(L_0, L_1, L_2, \ldots) \pmod{p}.$$

3.2 Arithmetic Properties of Pell and Pell–Lucas Numbers

Here we present some arithmetic properties of Pell and Pell–Lucas numbers.

Theorem 3.3 *Let p be an odd prime, k a positive integer, and r an arbitrary integer. The following relations hold:*

$$2P_{kp+r} \equiv (-1)^{\frac{p^2-1}{8}} P_k Q_r + P_r Q_k \pmod{p} \tag{3.12}$$

$$2Q_{kp+r} \equiv 8(-1)^{\frac{p^2-1}{8}} P_k P_r + Q_k Q_r \pmod{p}. \tag{3.13}$$

Proof We will prove (3.12) using the Binet-type formula for Pell numbers (2.32). Note that one may write $(1+\sqrt{2})^s = u_s + v_s\sqrt{2}$, where u_s and v_2 are positive integers, $s = 0, 1, \ldots$, while $(1-\sqrt{2})^s = u_s - v_s\sqrt{2}$. We have

$$\begin{aligned} P_{kp+r} &= \frac{1}{2\sqrt{2}}\left[(1+\sqrt{2})^{kp+r} - (1-\sqrt{2})^{kp+r}\right] \\ &= \frac{1}{2\sqrt{2}}\left[(u_k + v_k\sqrt{2})^p(u_r + v_r\sqrt{2}) - (u_k - v_k\sqrt{2})^p(u_r - v_r\sqrt{2})\right] \\ &= \frac{1}{2\sqrt{2}}\left[(u_r + v_r\sqrt{2})\sum_{j=0}^{p}\binom{p}{j}u_k^{p-j}\left(v_k\sqrt{2}\right)^j\right. \\ &\quad \left. - (u_r - v_r\sqrt{2})\sum_{j=0}^{p}\binom{p}{j}(-1)^j u_k^{p-j}\left(v_k\sqrt{2}\right)^j\right] \\ &= \frac{1}{2\sqrt{2}}\left[u_r\sum_{j=0}^{p}\left(1-(-1)^j\right)\binom{p}{j}u_k^{p-j}\left(v_k\sqrt{2}\right)^j\right. \\ &\quad \left. + v_r\sqrt{2}\sum_{j=0}^{p}\left(1+(-1)^j\right)\binom{p}{j}u_k^{p-j}\left(v_k\sqrt{2}\right)^j\right]. \end{aligned}$$

Since p divides $\binom{p}{j}$ for $j = 1, \ldots, p-1$, it follows that

$$P_{kp+r} \equiv u_r v_k^p 2^{\frac{p-1}{2}} + u_k^p v_r \pmod{p}.$$

By Fermat's Little Theorem and $2^{\frac{p-1}{2}} \equiv (-1)^{\frac{p^2-1}{8}} \pmod{p}$, we get

$$P_{kp+r} \equiv (-1)^{\frac{p^2-1}{8}} u_r v_k + u_k v_r \pmod{p}. \tag{3.14}$$

On the other hand, from $(1+\sqrt{2})^s = u_s + v_2\sqrt{2}$ and $(1-\sqrt{2})^s = u_s - v_2\sqrt{2}$, we get $2u_s = Q_s$ and $v_s = P_s$, for $s = 0, 1, \ldots$. Substituting into (3.14), we get

$$2P_{kp+r} \equiv (-1)^{\frac{p^2-1}{8}}(2u_r)v_k + (2u_k)v_r \pmod{p},$$

and the relation (3.12) follows.

To prove the relation (3.12), with similar computations we have

$$Q_{kp+r} = u_r\sum_{j=0}^{p}\left(1+(-1)^j\right)\binom{p}{j}u_k^{p-j}\left(v_k\sqrt{2}\right)^j$$

$$+ v_r\sqrt{2}\sum_{j=0}^{p}\left(1-(-1)^j\right)\binom{p}{j}u_k^{p-j}\left(v_k\sqrt{2}\right)^j,$$

hence

$$\begin{aligned} Q_{kp+r} &\equiv 2u_k^p u_r + 4v_k^p v_r 2^{\frac{p-1}{2}} \pmod p \\ &\equiv 2u_k u_r + 4v_k v_r 2^{\frac{p-1}{2}} \pmod p. \end{aligned}$$

Using again the relation $2^{\frac{p-1}{2}} \equiv (-1)^{\frac{p^2-1}{8}} \pmod p$, we obtain

$$2Q_{kp+r} \equiv (2u_k)(2u_r) + (-1)^{\frac{p^2-1}{8}} v_k v_r \pmod p,$$

and the desired relation follows using $2u_s = Q_s$ and $v_s = P_s$, $s = 0, 1, \ldots$. □

Below we present some subsequences of the relations (3.12) and (3.13).

Example 3.4 Taking $r = 0$ in relation (3.12), we obtain that for any positive integer k one has

$$P_{kp} = (-1)^{\frac{p^2-1}{8}} P_k \pmod p. \tag{3.15}$$

The special case $k = 1$ gives

$$P_p = (-1)^{\frac{p^2-1}{8}} \pmod p.$$

Taking $k = 1$ and $r = 1$ in (3.12) we get

$$P_{p+1} = (-1)^{\frac{p^2-1}{8}} + 1 \pmod p. \tag{3.16}$$

Taking $k = 1$ and $r = -1$ in (3.12) we get

$$P_{p-1} = -(-1)^{\frac{p^2-1}{8}} + 1 \pmod p. \tag{3.17}$$

If $p \equiv 5 \pmod 8$, then from (3.16) we have $p \mid P_{p+1}$. In the cases $p \equiv 1, 3, 7 \pmod 8$, from (3.17) one obtains $p | P_{p-1}$. We can summarize these relations in the following property:

$$p \mid P_{p-(-1)^{\frac{p^2-1}{8}}}. \tag{3.18}$$

Remark 3.5 An odd composite number n is called a **Pell pseudoprime** if n divides $P_{n-(-1)^{\frac{n^2-1}{8}}}$. The list of Pell pseudoprimes is indexed as A099011 in the OEIS [157], starting with the terms

$$169, 385, 741, 961, 1121, 2001, 3827, 4879, 5719, 6215, 6265, 6441, 6479, 6601,$$
$$7055, 7801, 8119, 9799, 10945, 11395, 13067, 13079, 13601, 15841, 18241,$$
$$19097, 20833, 20951, 24727, 27839, 27971, 29183, 29953, 31417, 31535, \ldots.$$

Kiss et al. [96] showed that this sequence is infinite.

By analogy with Wall–Sun–Sun primes, we call a prime p **strong Pell prime** if

$$p^2 \mid P_{p-(-1)^{\frac{p^2-1}{8}}}.$$

Finding examples of such primes and algorithms to check the sequence

$$P_{p-(-1)^{\frac{p^2-1}{8}}} \equiv 0 \pmod{p^2},$$

are interesting open problems.

Example 3.5 From relation (3.16) it follows that for two positive integers k and s, p divides $P_{kp} - P_{sp}$ if and only if p divides $P_k - P_s$. Moreover, we have the relation

$$P_{kp} - P_{sp} \equiv (-1)^{\frac{p^2-1}{8}} (P_k - P_s) \pmod{p}.$$

In particular, since $P_2 = 2$ and $P_1 = 1$, we get

$$P_{2p} - P_p \equiv (-1)^{\frac{p^2-1}{8}} \pmod{p}.$$

Example 3.6 Taking $r = 0$ in relation (3.14), we get

$$Q_{kp} \equiv Q_k \pmod{p}.$$

The special case $k = 1$ gives

$$Q_p \equiv 2 \pmod{p}.$$

Taking $k = 1$ and $r = 1$ in relation (3.14) we obtain

$$Q_{p+1} \equiv 4(-1)^{\frac{p^2-1}{8}} + 2 \pmod{p}.$$

Taking $k = 1$ and $r = -1$ in relation (3.14) we obtain

$$Q_{p-1} \equiv 4(-1)^{\frac{p^2-1}{8}} - 2 \pmod{p}.$$

Remark 3.6 We say that an odd composite number n is a **Pell–Lucas pseudoprime** if n divides $Q_n - 2$. The list of Pell–Lucas pseudoprimes starts with

$$169, 385, 961, 1105, 1121, 3827, 4901, 6265, 6441, 6601, 7107, 7801, 8119, 10945,$$
$$11285, 13067, 15841, 18241, 19097, 20833, 24727, 27971, 29953, 31417, 34561,$$
$$35459, 37345, 37505, 38081, 39059, 42127, 45451, 45961, 47321, 49105, \ldots,$$

and was recently indexed in the OEIS as A330276.

Also, it seems that the following property is true, but at this moment we are not aware of the existence of a proof for this result.

Conjecture 3.1 There exist infinitely many Pell–Lucas pseudoprimes.

We now define another pseudoprimality concept for which we formulate a conjecture suggested by numerical experiments.

Remark 3.7 We call an odd composite number n a **Pell–Pell–Lucas pseudoprime** if it satisfies both

$$n \mid P_{n-(-1)^{\frac{n^2-1}{8}}} \text{ and } n \mid Q_n - 2.$$

The list of such pseudoprimes that we know at this moment is

$$169, 385, 961, 1121, 3827, 6265, 6441, 6601, 7801, 8119, 10945, 13067, 15841,$$
$$18241, 19097, 20833, 24727, 27971, 29953, 31417, 34561, 35459, 37345, \ldots.$$

The sequence was recently indexed in the OEIS as A327652.

Conjecture 3.2 There exist infinitely many Pell–Pell–Lucas pseudoprimes.

3.3 Trigonometric Expressions for the Fibonacci, Lucas, Pell, and Pell–Lucas Numbers

The complex factorization in connection with some trigonometric expansions of Fibonacci, Lucas, Pell, and Pell–Lucas numbers have been studied by many authors. We mention the works of Cahill et al. [44], Sury [155, 156], and Wu [167] (see also the monographs of Koshy [101, 102]).

In this section we present unifying methods which allow the derivation of many general trigonometric expansions, including some new formulae.

The polynomial $U_n \in \mathbb{Z}[x, y]$ (2.60) has the complex factorization

$$U_n(x, y) = \prod_{j=1}^{n-1} (x - \omega_j y), \tag{3.19}$$

where ω_j are the nth roots of unity, namely

$$\omega_j = \cos \frac{2j\pi}{n} + i \sin \frac{2j\pi}{n}, \quad j = 1, \ldots, n-1.$$

When n is odd, by grouping the factors in (3.19), we can write

$$U_n(x, y) = \prod_{j=1}^{\frac{n-1}{2}} (x - \omega_j y)(x - \overline{\omega}_j y) = \prod_{j=1}^{\frac{n-1}{2}} \left(x^2 + y^2 - 2xy \cos \frac{2j\pi}{n} \right).$$

Because

$$\begin{aligned} x^2 + y^2 - 2xy \cos \frac{2j\pi}{n} &= (x+y)^2 - 2xy \left(1 + \cos \frac{2j\pi}{n} \right) \\ &= (x+y)^2 - 4xy \cos^2 \frac{j\pi}{n}, \end{aligned}$$

and since $U_n\left(\frac{1+\sqrt{5}}{2}, \frac{1-\sqrt{5}}{2}\right) = F_n$, for $n = 0, 1, \ldots,$ we obtain the following trigonometric expansion of the nth Fibonacci number F_n

$$F_n = \prod_{j=1}^{\frac{n-1}{2}} \left(1 + 4 \cos^2 \frac{j\pi}{n} \right),$$

valid when n is an odd integer.

This formula can be completed for the case when n is even, by the factor corresponding to $j = \frac{n}{2}$, which is 1. Therefore, we have the following result.

Theorem 3.4 *For every positive integer $n \geq 2$, we have*

$$F_n = \prod_{j=1}^{\lfloor \frac{n-1}{2} \rfloor} \left(1 + 4 \cos^2 \frac{j\pi}{n} \right). \tag{3.20}$$

The identity (3.20) was proved in [38] by counting the spanning trees in some classes of graphs. With the same technique, the authors completed the formula by

$$F_n = \prod_{j=1}^{\lfloor \frac{n-1}{2} \rfloor} \left(1 + 4\cos^2 \frac{j\pi}{n}\right) = \prod_{j=1}^{\lfloor \frac{n-1}{2} \rfloor} \left(1 + 4\sin^2 \frac{j\pi}{n}\right).$$

The identity (3.20) is also mentioned in several references, including the paper by Garnier and Ramaré [67].

Using the relation $P_n = U_n \left(1 + \sqrt{2}, 1 - \sqrt{2}\right)$, for $n = 0, 1, \ldots$, we obtain the following trigonometric expansion for P_n, the nth Pell number.

Theorem 3.5 *For every positive integer $n \geq 2$, we have*

$$P_n = 2^{\lfloor \frac{n-1}{2} \rfloor + \lfloor \frac{n}{2} \rfloor} \prod_{j=1}^{\lfloor \frac{n-1}{2} \rfloor} \left(1 + \cos^2 \frac{j\pi}{n}\right). \tag{3.21}$$

Using the relation $2\cos^2 t = 1 + \cos 2t$, formula (3.21) is equivalent to

$$P_n = 2^{\lfloor \frac{n}{2} \rfloor} \prod_{j=1}^{\lfloor \frac{n-1}{2} \rfloor} \left(3 + \cos \frac{2j\pi}{n}\right),$$

which is called the Shapiro's formula (see Koshy [101, pp. 288–290]).

Similarly, the polynomial $V_n \in \mathbb{Z}[x, y]$ defined by (2.60) has the following factorization over complex numbers:

$$V_n(x, y) = \prod_{k=0}^{n-1} (x - \zeta_k y), \tag{3.22}$$

where ζ_k are the nth roots of -1, namely

$$\zeta_k = \cos \frac{(2k+1)\pi}{n} + i \sin \frac{(2k+1)\pi}{n}, \quad k = 0, \ldots, n-1.$$

When n is odd, we can write

$$\begin{aligned} V_n(x, y) &= (x + y) \prod_{k=1}^{\frac{n-1}{2}} (x - \zeta_j y)(x - \overline{\zeta}_j y) \\ &= (x + y) \prod_{k=1}^{\frac{n-1}{2}} \left(x^2 + y^2 - 2xy \cos \frac{(2k+1)\pi}{n}\right). \end{aligned}$$

Using the relation $1 + \cos \frac{(2k+1)\pi}{n} = 2\cos^2 \frac{(2k+1)\pi}{2n}$, one obtains

$$x^2 + y^2 - 2xy\cos\frac{(2k+1)\pi}{n} = (x+y)^2 - 4xy\cos^2\frac{(2k+1)\pi}{2n}.$$

Because $V_n\left(\frac{1+\sqrt{5}}{2}, \frac{1-\sqrt{5}}{2}\right) = L_n$, for $n = 0, 1, \ldots$, we obtain the following trigonometric expansion of the Lucas number L_n

$$L_n = \prod_{k=1}^{\frac{n-1}{2}} \left(1 + 4\cos^2\frac{(2k+1)\pi}{2n}\right), \tag{3.23}$$

valid when n is an odd integer.

When n is even, we have the following factorization for the polynomial V_n

$$V_n = \prod_{k=0}^{\frac{n-2}{2}} \left[(x+y)^2 - 4xy\cos^2\frac{(2k+1)\pi}{2n}\right],$$

and we obtain the formula below for the nth Lucas number L_n.

Theorem 3.6 *For every positive integer $n \geq 2$, the following trigonometric expansion of the Lucas number L_n holds:*

$$L_n = \begin{cases} \prod_{k=1}^{\frac{n-1}{2}} \left(1 + 4\cos^2\frac{(2k+1)\pi}{2n}\right) & \text{if } n \text{ is odd} \\ \prod_{k=0}^{\frac{n-2}{2}} \left(1 + 4\cos^2\frac{(2k+1)\pi}{2n}\right) & \text{if } n \text{ is even.} \end{cases} \tag{3.24}$$

From the relation $V_n\left(1+\sqrt{2}, 1-\sqrt{2}\right) = Q_n$, for $n = 0, 1, \ldots$, we obtain the following trigonometric expansion for Q_n, the nth Lucas–Pell number.

Theorem 3.7 *For every positive integer $n \geq 2$, the following trigonometric expansion of the Lucas number Q_n holds:*

$$Q_n = \begin{cases} 2^n \prod_{k=1}^{\frac{n-1}{2}} \left(1 + \cos^2\frac{(2k+1)\pi}{2n}\right) & \text{if } n \text{ is odd} \\ 2^n \prod_{k=0}^{\frac{n-2}{2}} \left(1 + \cos^2\frac{(2k+1)\pi}{2n}\right) & \text{if } n \text{ is even.} \end{cases} \tag{3.25}$$

3.4 Identities Involving the Resultant of Polynomials

Consider two monic polynomials $f, g \in \mathbb{C}[x]$, given by $f(x) = x^m + a_{m-1}x^{m-1} + \cdots + a_1x + a_0$ and $g(x) = x^n + b_{n-1}x^{n-1} + \cdots + b_1x + b_0$. Recall that the **resultant of the polynomials** f and g is defined as

$$R(f,g)=\prod_{k,j}(\alpha_k-\beta_j),$$

where α_k, $k=1,\ldots,m$, are the roots of f, and β_j, $j=1,\ldots,n$, are the roots of g, not necessarily distinct. A simple but useful observation is that we have

$$R(f,g)=\prod_{k=1}^{m}g(\alpha_k)=(-1)^{mn}\prod_{j=1}^{n}f(\beta_j). \tag{3.26}$$

Many identities can be obtained for different choices of f and g in (3.26).

- In the special case $f(x)=x^m-1$ and g an arbitrary monic polynomial, we denote the mth roots of unity by $\omega_k=\cos\frac{2k\pi}{m}+i\sin\frac{2k\pi}{m}$, $k=0,\ldots,m-1$. The relation (3.26) becomes

$$\prod_{k=0}^{m-1}g(\omega_k)=(-1)^{mn}\prod_{j=1}^{n}(\beta_j^m-1). \tag{3.27}$$

- If $f(x)=x^m+1$ and g is an arbitrary monic polynomial, denote the mth roots of -1 by $\zeta_k=\cos\frac{(2k+1)\pi}{m}+i\sin\frac{(2k+1)\pi}{m}$, $k=0,\ldots,m-1$. By (3.26) we obtain

$$\prod_{k=0}^{m-1}g(\zeta_k)=(-1)^{mn}\prod_{j=1}^{n}(\beta_j^m+1). \tag{3.28}$$

Case 1 The polynomial $g(x)=x^2-x-1$ has the roots $\beta_1=\frac{1+\sqrt{5}}{2}$ and $\beta_2=\frac{1-\sqrt{5}}{2}$. From (3.27) it follows

$$\begin{aligned}\prod_{k=0}^{m-1}\left(\omega_k^2-\omega_k-1\right)&=\left(\beta_1^m-1\right)\left(\beta_2^m-1\right)=(\beta_1\beta_2)^m-\beta_1^m-\beta_2^m+1\\&=(-1)^m-L_m+1,\end{aligned}$$

where L_m is the mth Lucas number. On the other hand, we can write

$$\omega_k^2-\omega_k-1=\omega_k\left(\omega_k-\frac{1}{\omega_k}-1\right)=-\omega_k\left(1-2i\sin\frac{2k\pi}{m}\right),$$

for $k=0,1,\ldots,m-1$. As $\omega_0\omega_1\ldots\omega_{m-1}=(-1)^{m-1}$, we have

$$\prod_{k=0}^{m-1}\left(1-2i\sin\frac{2k\pi}{m}\right)=L_m+(-1)^{m+1}-1.$$

The factor corresponding to $k=0$ is 1, hence considering the cases m odd and m even and grouping the factors in pairs of the form $(z,\overline{z})$, we get

$$\prod_{k=1}^{\lfloor\frac{m-1}{2}\rfloor}\left(1+4\sin^2\frac{2k\pi}{m}\right)=L_m+(-1)^{m+1}-1. \tag{3.29}$$

Now, from (3.28) we have

$$\prod_{k=0}^{m-1}\left(\zeta_k^2-\zeta_k-1\right)=\left(\beta_1^m+1\right)\left(\beta_2^m+1\right)=(\beta_1\beta_2)^m+\beta_1^m+\beta_2^m+1$$
$$=(-1)^m+L_m+1.$$

Also, notice that for $k=0,1,\ldots,m-1$, we have

$$\zeta_k^2-\zeta_k-1=\zeta_k\left(\zeta_k-\frac{1}{\zeta_k}-1\right)=-\zeta_k\left(1-2i\sin\frac{(2k+1)\pi}{m}\right).$$

Therefore, by using $\zeta_0\zeta_1\cdots\zeta_{m-1}=(-1)^m$, one obtains

$$\prod_{k=0}^{m-1}\left(1-2i\sin\frac{(2k+1)\pi}{m}\right)=L_m+(-1)^m+1.$$

If m is odd, then the factor corresponding to $k=\frac{m-1}{2}$ is 1, hence we can group the factors of the above product in pairs of the form $(z,\overline{z})$. Therefore, collecting these terms, we obtain

$$\prod_{k=0}^{\lfloor\frac{m-1}{2}\rfloor}\left(1+4\sin^2\frac{(2k+1)\pi}{m}\right)=L_m+(-1)^m+1. \tag{3.30}$$

Case 2 The polynomial $g(x)=x^2-2x-1$ has the roots $\beta_1=1+\sqrt{2}$ and $\beta_2=1-\sqrt{2}$, hence using the formula (3.27) we get

$$\prod_{k=0}^{m-1}\left(\omega_k^2-2\omega_k-1\right)=\left(\beta_1^m-1\right)\left(\beta_2^m-1\right)=(\beta_1\beta_2)^m-\beta_1^m-\beta_2^m+1$$
$$=(-1)^m-Q_m+1,$$

where Q_m is the mth Pell–Lucas number. Since we have

$$\omega_k^2 - 2\omega_k - 1 = \omega_k \left(\omega_k - \frac{1}{\omega_k} - 2\right) = -2\omega_k \left(1 - i \sin \frac{2k\pi}{m}\right),$$

for $k = 0, 1, \ldots, m-1$ and $\omega_0\omega_1 \ldots \omega_{m-1} = (-1)^{m-1}$, we obtain the identity

$$2^m \prod_{k=0}^{m-1} \left(1 - i \sin \frac{2k\pi}{m}\right) = Q_m + (-1)^{m+1} - 1.$$

Again, the factor corresponding to $k = 0$ is 1, and considering the cases m odd and m even, we obtain the identity

$$\prod_{k=1}^{\lfloor \frac{m-1}{2} \rfloor} \left(1 + \sin^2 \frac{2k\pi}{m}\right) = \frac{1}{2^m} \left(Q_m + (-1)^{m+1} - 1\right). \tag{3.31}$$

Applying formula (3.28), it follows that

$$\begin{aligned}\prod_{k=0}^{m-1} \left(\zeta_k^2 - 2\zeta_k - 1\right) &= \left(\beta_1^m + 1\right)\left(\beta_2^m + 1\right) = (\beta_1\beta_2)^m + \beta_1^m + \beta_2^m + 1 \\ &= (-1)^m + Q_m + 1.\end{aligned}$$

Taking into account that

$$\zeta_k^2 - 2\zeta_k - 1 = \zeta_k \left(\zeta_k - \frac{1}{\zeta_k} - 2\right) = -2\zeta_k \left(1 - i \sin \frac{(2k+1)\pi}{m}\right),$$

for $k = 0, 1, \ldots, m-1$, by using $\zeta_0\zeta_1 \ldots \zeta_{m-1} = (-1)^m$, we obtain the identity

$$2^m \prod_{k=0}^{m-1} \left(1 - i \sin \frac{(2k+1)\pi}{m}\right) = Q_m + (-1)^m + 1.$$

Considering the cases m odd and m even, and grouping the factors in the product, we obtain the identity

$$\prod_{k=0}^{\lfloor \frac{m-1}{2} \rfloor} \left(1 + \sin^2 \frac{(2k+1)\pi}{m}\right) = \frac{1}{2^m} \left(Q_m + (-1)^m + 1\right). \tag{3.32}$$

Remark 3.8 Consider the polynomial $g(x) = x^2 - ax - 1$, where $a \geq 1$ is a positive integer, having the roots

$$\beta_1 = \frac{a+\sqrt{a^2+4}}{2} \qquad \beta_2 = \frac{a-\sqrt{a^2+4}}{2}.$$

With similar computations as before, we can extend the identities (3.29) and (3.31), for $a = 1$ and $a = 2$, respectively. Denoting by

$$A_m = \beta_1^m + \beta_2^m = \left(\frac{a+\sqrt{a^2+4}}{2}\right)^m + \left(\frac{a-\sqrt{a^2+4}}{2}\right)^m,$$

we clearly have

$$\prod_{k=0}^{m-1}(\omega_k^2 - a\omega_k - 1) = (\beta_1\beta_2)^m - \beta_1^m - \beta_2^m + 1 = -A_m + (-1)^m + 1.$$

Notice that

$$\omega_k^2 - a\omega_k - 1 = \omega_k\left(\omega_k - \frac{1}{\omega_k} - a\right) = -\omega_k\left(a - 2i\sin\frac{2k\pi}{m}\right),$$

for $k = 0, 1, \ldots, m-1$. As $\omega_0\omega_1\ldots\omega_{m-1} = (-1)^{m-1}$, we have

$$\prod_{k=0}^{m-1}\left(a - 2i\sin\frac{2k\pi}{m}\right) = A_m + (-1)^{m+1} - 1.$$

The factor corresponding to $k = 0$ is a, hence considering the cases m odd and m even and grouping the factors in pairs of the form $(z, \overline{z})$, we obtain the formula

$$\prod_{k=1}^{\lfloor\frac{m-1}{2}\rfloor}\left(a^2 + 4\sin^2\frac{2k\pi}{m}\right) = A_m + (-1)^{m+1} - 1. \tag{3.33}$$

Furthermore, this is also equivalent to

$$\prod_{k=0}^{m-1}\left(a^2 + 4\sin^2\frac{2k\pi}{m}\right) = \left(A_m + (-1)^{m+1} - 1\right)^2.$$

Similarly, the identities (3.30) and (3.32) can be seen as the particular cases $a = 1$ and $a = 2$, respectively, of the more general formula

$$\prod_{k=0}^{\lfloor \frac{m-1}{2} \rfloor} \left(a^2 + 4\sin^2 \frac{(2k+1)\pi}{m}\right) = A_m + (-1)^m + 1. \tag{3.34}$$

This is also equivalent to

$$\prod_{k=0}^{m-1} \left(a^2 + 4\sin^2 \frac{(2k+1)\pi}{m}\right) = \left(A_m + (-1)^m + 1\right)^2. \tag{3.35}$$

Case 3 The polynomial $g(x) = x^2 - ax + 1$, with $a \geq 2$ integer, has the roots

$$\beta_1 = \frac{a + \sqrt{a^2 - 4}}{2} \qquad \beta_2 = \frac{a - \sqrt{a^2 - 4}}{2}.$$

It is useful to denote

$$C_m = \left(\frac{a + \sqrt{a^2 - 4}}{2}\right)^m + \left(\frac{a - \sqrt{a^2 - 4}}{2}\right)^m.$$

From (3.27) we obtain

$$\prod_{k=0}^{m-1} \left(\omega_k^2 - a\omega_k + 1\right) = \left(\beta_1^m - 1\right)\left(\beta_2^m - 1\right) = (\beta_1\beta_2)^m - \beta_1^m - \beta_2^m + 1$$
$$= -C_m + 2.$$

Notice that we can write

$$\omega_k^2 - a\omega_k + 1 = \omega_k \left(\omega_k + \frac{1}{\omega_k} - a\right) = \omega_k \left(2\cos\frac{2k\pi}{m} - a\right),$$

for $k = 0, 1, \ldots, m-1$. As $\omega_0\omega_1 \ldots \omega_{m-1} = (-1)^{m-1}$, we obtain the identity

$$\prod_{k=0}^{m-1} \left(a - 2\cos\frac{2k\pi}{m}\right) = C_m - 2. \tag{3.36}$$

With similar computations we obtain

$$\prod_{k=0}^{m-1} \left(\zeta_k^2 - a\zeta_k + 1\right) = C_m + 2,$$

and

$$\zeta_k^2 - a\zeta_k + 1 = \zeta_k \left(2\cos\frac{(2k+1)\pi}{m} - a\right), \quad k = 0, 1, \ldots, m-1.$$

Therefore, the following identity is derived:

$$\prod_{k=0}^{m-1} \left(2\cos\frac{(2k+1)\pi}{m} - a\right) = C_m + 2. \tag{3.37}$$

Chapter 4
Generating Functions

Generating functions play an important role in the study of recurrent sequences (see, e.g., [103, 113, 123, 154, 166]). In this chapter we present basic properties, operations, and examples involving ordinary generating functions (Section 4.1), or exponential generating functions (Section 4.2). Then we derive such generating functions for some classical polynomials and integer sequences. In Section 4.3 we give applications of the Cauchy integral formula in the derivation of integral representations for classical number sequences.

4.1 Ordinary Generating Functions

4.1.1 Basic Operations and Examples

The generating function of an infinite sequence

$$a_0,\ a_1,\ \ldots,a_n,\ \ldots,$$

is the infinite series

$$F(z) = a_0 + a_1 z + a_2 z^2 + \cdots + a_n z^n + \cdots. \tag{4.1}$$

Identification Principle Let $F(z) = \sum_{n=0}^{\infty} a_n z^n$ and $G(z) = \sum_{n=0}^{\infty} b_n z^n$ be two generating functions. Then, $F(z) = G(z)$ if and only if $a_n = b_n$, $n \geq 0$.

The following operations hold.

D. Andrica, O. Bagdasar, *Recurrent Sequences*, Problem Books in Mathematics,
https://doi.org/10.1007/978-3-030-51502-7_4

1. Addition

$$\sum_{n=0}^{\infty} a_n z^n + \sum_{n=0}^{\infty} a_n z^n = \sum_{n=0}^{\infty} c_n z^n, \qquad \text{where } c_n = a_n + b_n.$$

2. Multiplication by a constant If $\alpha \in \mathbb{C}$ a scalar, then

$$\alpha \sum_{n=0}^{\infty} a_n z^n = \sum_{n=0}^{\infty} c_n z^n, \qquad \text{where } c_n = \alpha a_n.$$

3. Formal differentiation

$$\frac{\mathrm{d}}{\mathrm{d}z}(F(z)) = \frac{\mathrm{d}}{\mathrm{d}z}\left(\sum_{n=0}^{\infty} a_n z^n\right) = \sum_{n=0}^{\infty} n a_n z^{n-1}.$$

4. Formal integration

$$\int F(z)\mathrm{d}z = \int \sum_{n=0}^{\infty} a_n z^n \mathrm{d}z = \sum_{n=0}^{\infty} \frac{a_n}{n+1} z^{n+1}.$$

5. Multiplication

$$F(z)G(z) = \left(\sum_{n=0}^{\infty} a_n z^n\right)\left(\sum_{n=0}^{\infty} b_n z^n\right) = \sum_{n=0}^{\infty} c_n z^n, \qquad \text{where } c_n = \sum_{k=0}^{n} a_k b_{n-k}.$$

6. Hadamard multiplication

$$F(z) \circ G(z) = \left(\sum_{n=0}^{\infty} a_n z^n\right)\left(\sum_{n=0}^{\infty} b_n z^n\right) = \sum_{n=0}^{\infty} c_n z^n, \qquad \text{where } c_n = a_n b_n.$$

7. Composition If $a_0 = 0$, we have

$$G(F(z)) = \sum_{n=0}^{\infty} b_n [F(z)]^n = \sum_{n=0}^{\infty} b_n \left(\sum_{\substack{j_1+\cdots+j_k=n \\ j_1,\ldots,j_k \geq 1}} a_{j_1} \cdots a_{j_k} \right) z^n,$$

8. Division If $b_0 \neq 0$, then we have

$$\frac{F(z)}{G(z)} = \frac{\sum_{n=0}^{\infty} a_n z^n}{\sum_{n=0}^{\infty} b_n z^n} = \sum_{n=0}^{\infty} c_n z^n,$$

$$c_n = \frac{1}{b_0}\left(a_n - \sum_{k=1}^{n} b_k c_{n-k}\right).$$

9. Inverse The power series $F(z)$ and $G(z)$ are inverse if $F(z)G(z) = 1$, which implies $a_0b_0 = 1$ and

$$b_n = -\frac{1}{a_0}\sum_{k=1}^{n} a_k b_{n-k}, \qquad \text{for } n \geq 1.$$

We now present some illustrative examples of generating functions.

Example 4.1 (finite sequences) For example, a finite sequence

$$a_0, a_1, \ldots, a_n,$$

can be seen as the infinite sequence

$$a_0, a_1, \ldots, a_n, 0, 0, \ldots,$$

whose generating function is the polynomial

$$F(z) = a_0 + a_1 z + a_2 z^2 + \cdots + a_n z^n. \tag{4.2}$$

Example 4.2 (constant sequence) The generating function of the sequence

$$1,\ 1,\ \ldots, 1,\ \ldots,$$

is the function

$$F(z) = 1 + z + z^2 + \cdots + z^n + \cdots = \frac{1}{1-z}, \qquad |z| < 1.$$

Example 4.3 (binomial coefficients) For an integer $n \geq 1$, the generating function for the binomial coefficients

$$\binom{n}{0}, \binom{n}{1}, \binom{n}{2}, \ldots, \binom{n}{n-1}, \binom{n}{n}, 0, \ldots,$$

is the function

$$\sum_{k=0}^{n}\binom{n}{k} z^k = (1+z)^n.$$

Example 4.4 (generalized binomial coefficients) Recall that for a real number α and an integer $n \geq 0$, the generalized binomial coefficient is defined by

$$\binom{\alpha}{n} = \frac{\alpha(\alpha-1)\cdots(\alpha-n+1)}{n!}.$$

The generating function for the generalized binomial coefficients

$$\binom{\alpha}{0}, \binom{\alpha}{1}, \binom{\alpha}{2}, \ldots, \binom{\alpha}{n}, \ldots,$$

is given by

$$\sum_{n=0}^{\infty} \binom{\alpha}{n} z^n = (1+z)^{\alpha}.$$

Example 4.5 Let k be a positive integer and let

$$a_1, a_2, \ldots, a_n, \ldots,$$

be the infinite sequence whose general term a_n is the number of nonnegative integer solutions of the linear diophantine equation

$$x_0 + x_1 + \cdots + x_n = n.$$

Then, the generating function of the sequence $(x_n)_{n\geq 0}$ is

$$\begin{aligned} F(z) &= \sum_{n=0}^{\infty} \left(\sum_{j_1+\cdots+j_k=n} 1 \right) z^n = \sum_{n=0}^{\infty} \sum_{j_1+\cdots+j_k=n} z^{j_1+\cdots+j_k} \\ &= \left(\sum_{j_1=0}^{\infty} z^{j_1} \right) \left(\sum_{j_2=0}^{\infty} z^{j_2} \right) \cdots \left(\sum_{j_k=0}^{\infty} z^{j_k} \right) = \frac{1}{(1-z)^k} \\ &= \sum_{n=0}^{\infty} (-1)^n \binom{-k}{n} z^n = \sum_{n=0}^{\infty} \binom{n+k-1}{n} z^n. \end{aligned}$$

Example 4.6 Let a_n be the number of integer solutions of the equation

$$x_1 + x_2 + x_3 = n,$$

where $0 \leq a_1 \leq 4$, $2 \leq a_2 \leq 3$ and $a_3 \geq 3$. Find the generating function.

Solution The generating function of this sequence is

$$F(z) = \left(1 + z + z^2 + z^3 + z^4\right)\left(z^2 + z^3\right)\left(z^3 + z^4 + \cdots\right)$$
$$= \frac{z^5\left(1 + z + z^2 + z^3 + z^4\right)(1 + z)}{1 - z}.$$

Example 4.7 Find the generating function for the number of n-combinations of red, green, blue, and yellow balls, with the properties: the number of red balls is $0, 1, 2$ or 3, the number of green balls is at least 5, the number of blue balls is odd, while the number of yellow balls is even.

Solution The generating function of this sequence is

$$F(z) = \left(\sum_{k=0}^{3} z^k\right)\left(\sum_{k=5}^{\infty} z^k\right)\left(\sum_{k=0}^{\infty} z^{2k+1}\right)\left(\sum_{k=0}^{\infty} z^{2k}\right)$$
$$= \frac{z^6\left(1 - z^4\right)\left(1 - z^2\right)^2}{(1 - z)^2}.$$

Example 4.8 Let k be a positive integer and let

$$a_0,\ a_1,\ \ldots,\ a_n,\ \ldots,$$

be the infinite sequence whose general term a_n is the number of nonnegative integer solutions of the linear diophantine equation

$$x_1 + x_2 + \cdots + x_k = n.$$

Then, the generating function of the sequence $(x_n)_{n\geq 0}$ is

$$F(z) = \sum_{n=0}^{\infty}\left(\sum_{j_1+\cdots+j_k=n} 1\right) z^n = \sum_{n=0}^{\infty} \sum_{j_1+\cdots+j_k=n} z^{j_1+\cdots+j_k}$$
$$= \left(\sum_{j_1=0}^{\infty} z_{j_1}\right)\left(\sum_{j_2=0}^{\infty} z_{j_2}\right)\cdots\left(\sum_{j_k=0}^{\infty} z_{j_k}\right) = \frac{1}{(1 - z)^k}$$
$$= \sum_{n=0}^{\infty}(-1)^n\binom{-k}{n} z^n = \sum_{n=0}^{\infty}\binom{n + k - 1}{n} z^n.$$

Example 4.9 Let k be a positive integer and let $x_{k,n}$ be the number of integer solutions $(j_1, \ldots, j_k)$ of the equation

$$a_1 + a_2 + \cdots + a_k = n,$$

such that the numbers $j_1, \ldots, j_k$ are odd positive integers.

The generating function of the sequence $(x_{n,k})_{n\geq 0}$ is

$$F(z) = \left(\sum_{j=0}^{\infty} z^{2j+1}\right) \cdots \left(\sum_{j=0}^{\infty} z^{2j+1}\right) = \frac{z^k}{(1-z^2)^k}$$

$$= z^k \sum_{n=0}^{\infty} \binom{n+k-1}{n} z^{2n} = \sum_{n=0}^{\infty} \binom{n+k-1}{n} z^{2n+k}.$$

Example 4.10 Let a_n denote the number of nonnegative integer solutions of the equation

$$3x_1 + 4x_2 + x_3 + 5x_4 = n.$$

The generating function of the sequence $(a_n)_{n\geq 0}$ is

$$F(z) = \left(\sum_{k=0}^{\infty} z^{3k}\right)\left(\sum_{k=0}^{\infty} z^{4k}\right)\left(\sum_{k=0}^{\infty} z^{k}\right)\left(\sum_{k=0}^{\infty} z^{5k}\right)$$

$$= \frac{1}{\left(1-z^3\right)\left(1-z^4\right)(1-z)\left(1-z^5\right)}.$$

More generally, if for an integer $k \geq 1$ and given integers $a_1, \ldots, a_k$, then the number of integer solutions $(x_1, \ldots, x_k)$ of the equation

$$a_1x_1 + a_2x_2 + \cdots + a_kx_k = n,$$

is given by the formula

$$F(z) = \frac{1}{(1-z^{a_1})(1-z^{a_2})\cdots(1-z^{a_k})}.$$

Note that we also have the following useful identities:

$$\frac{1}{(1-z)^n} = \sum_{k=0}^{\infty} \binom{-n}{k} (-z)^k = \sum_{k=0}^{\infty} \binom{n+k-1}{k} z^k, \quad |z| < 1;$$

$$\frac{1}{(1-az)^n} = \sum_{k=0}^{\infty} \binom{-n}{k} (-az)^k = \sum_{k=0}^{\infty} \binom{n+k-1}{k} a^k z^k, \quad |z| < \frac{1}{|a|}.$$

These formulae can be used to find the generating functions associated with numerous sequences.

Example 4.11 Show that the generating function of the sequence

$$0,\ 1,\ 2^2,\ \ldots,\ n^2,\ \ldots,$$

is given by the formula

$$F(z) = \frac{z(1+z)}{(1-z)^3}.$$

Solution Indeed, we have $\frac{1}{1-z} = \sum_{k=0}^{\infty} z^k$, hence

$$\frac{1}{(1-z)^2} = \frac{\mathrm{d}}{\mathrm{d}z}\left(\frac{1}{1-z}\right)\sum_{k=0}^{\infty}\frac{\mathrm{d}}{\mathrm{d}z}\left(z^k\right) = \sum_{k=0}^{\infty} kz^{k-1}.$$

Multiplying both sides by z we obtain $\frac{z}{(1-z)^2} = \sum_{k=0}^{\infty} kz^k$, which by differentiation with respect to z gives

$$\frac{1+z}{(1-z)^3} = \sum_{k-0}^{\infty} k^2 z^{k-1}.$$

The desired result is obtained by multiplication with z.

Similarly one can obtain the generating functions for the sequence

$$0,\ 1,\ 2^k,\ \ldots,\ n^k,\ \ldots,$$

where $k \geq 2$ is a fixed integer.

Example 4.12 (Catalan sequence) Consider the Catalan sequence given by

$$C_{n+1} = \sum_{k=0}^{n} C_k C_{n-k}, \quad C_0 = 1.$$

Using the generating function $F(z) = \sum_{n=0}^{\infty} C_n z^n$, we have

$$\begin{aligned} F(z)F(z) &= \left(\sum_{n=0}^{\infty} C_n z^n\right)\left(\sum_{n=0}^{\infty} C_n z^n\right) = \sum_{n=0}^{\infty}\left(\sum_{k=0}^{\infty} C_k C_{n-k}\right) z^n \\ &= \sum_{n=0}^{\infty} C_{n+1} z^n = \frac{1}{z}\sum_{n=1}^{\infty} C_n z^n = \frac{F(z)}{z} - \frac{1}{z}, \end{aligned}$$

hence the following identity holds:

$$zF(z)^2 - F(z) + 1 = 0.$$

Solving for $F(z)$ we obtain

$$F(z) = \frac{1 \pm \sqrt{1-4z}}{2z}.$$

Since we have

$$\sqrt{1-4z} = 1 + \sum_{n=1}^{\infty} (-1)^n \binom{\frac{1}{2}}{n} 4^n z^n = 1 + \sum_{n=1}^{\infty} a_n z^n,$$

with

$$\begin{aligned} a_n &= (-1)^n \left[\frac{1}{2}\left(\frac{1}{2}-1\right)\cdots\left(\frac{1}{2}-n+1\right)\right] \cdot \frac{2^n}{n!} \cdot 2^n \\ &= (-1)^n \frac{(-1)(-3)\cdots(-2(n-1)+1)}{n!} \cdot 2^n \\ &= -\frac{1\cdot 3\cdot 5\cdots(2(n-1)-1)}{n!} \cdot 2^n = -2\frac{(2(n-1))!}{n!(n-1)!}. \end{aligned}$$

It follows that

$$\sqrt{1-4z} = 1 - 2\sum_{n=0}^{\infty} \frac{(2n)!}{n!(n+1)!} z^{n+1},$$

hence

$$F(z) = \frac{1-\sqrt{1-4z}}{2z} = \sum_{n=0}^{\infty} \frac{(2n)!}{n!(n-1)!} z^n = \sum_{n=0}^{\infty} \frac{1}{n+1}\binom{2n}{n} z^n.$$

The sequence $(C_n)_{n\geq 0}$ given by

$$C_n = \frac{1}{n+1}\binom{2n}{n},$$

is known as the Catalan sequence. One can prove by induction that C_n is the number of ways to add brackets to evaluate the matrix product

$$A_1 A_2 \cdots A_{n+1}, \quad n \geq 0.$$

The number of ways to evaluate the product $A_1 A_2 \cdots A_{n+2}$ is determined by multiplying two matrices at the end, which is given exactly by $n+1$, i.e.,

$$A_1 A_2 \cdots A_{n+2} = (A_1 \cdots A_{k+1})(A_{k+1} \cdots A_{n+1}), \quad 0 \le k \le n.$$

This suggests that the following recurrence relation holds:

$$C_{n+1} = \sum_{k=0}^{n} C_k C_{n-k},$$

which produces Catalan's sequence.

4.1.2 *Generating Functions of Classical Polynomials*

Using the formulae for U_n and V_n (2.60), one obtains

$$\sum_{n=0}^{\infty} U_n(x, y) z^n = \frac{1}{x-y} \sum_{n=0}^{\infty} \left(x^n - y^n\right) z^n = \frac{1}{x-y} \left(\frac{1}{1-xz} - \frac{1}{1-yz} \right)$$

$$= \frac{z}{(1-xz)(1-yz)} = \frac{z}{1-(x+y)z+(xy)z^2}, \tag{4.3}$$

$$\sum_{n=0}^{\infty} V_n(x, y) z^n = \sum_{n=0}^{\infty} \left(x^n + y^n\right) z^n = \frac{1}{1-xz} + \frac{1}{1-yz}$$

$$= \frac{2-(x+y)z}{1-(x+y)z+(xy)z^2}. \tag{4.4}$$

Substituting for the polynomials given in Section 1.2.3, we obtain the generating functions for special classical polynomials.

Fibonacci polynomials Since $f_n(x) = U_n\left(\frac{x+\sqrt{x^2+4}}{2}, \frac{x-\sqrt{x^2+4}}{2}\right)$, we have

$$\sum_{n=0}^{\infty} f_n(x) z^n = \frac{z}{1-xz-z^2}.$$

Lucas polynomials Since $l_n(x) = V_n\left(\frac{x+\sqrt{x^2+4}}{2}, \frac{x-\sqrt{x^2+4}}{2}\right)$, we have

$$\sum_{n=0}^{\infty} l_n(x) z^n = \frac{2-xz}{1-xz-z^2}.$$

Pell polynomials As $p_n(x) = U_n\left(x + \sqrt{x^2+1}, x - \sqrt{x^2+1}\right)$, we have

$$\sum_{n=0}^{\infty} p_n(x)z^n = \frac{z}{1-2xz-z^2}.$$

Pell–Lucas polynomials As $q_n(x) = V_n\left(x + \sqrt{x^2+1}, x - \sqrt{x^2+1}\right)$

$$\sum_{n=0}^{\infty} q_n(x)z^n = \frac{2-2xz}{1-2xz-z^2}.$$

Chebyshev polynomials of the first kind Since we have the relation $T_n(x) = \frac{1}{2}V_n\left(x + \sqrt{x^2-1}, x - \sqrt{x^2-1}\right)$, it follows that

$$\sum_{n=0}^{\infty} T_n(x)z^n = \frac{1-2xz}{1-2xz+z^2}.$$

Chebyshev polynomials of the second kind Since we have the relation $u_n(x) = U_{n+1}\left(x + \sqrt{x^2-1}, x - \sqrt{x^2-1}\right)$, it follows that

$$\sum_{n=0}^{\infty} u_n(x)z^n = \frac{1}{1-2xz+z^2}.$$

Hoggatt–Bicknell–King polynomials of Fibonacci kind Since we have $g_n(x) = U_n\left(\frac{x+\sqrt{x^2-4}}{2}, \frac{x-\sqrt{x^2-4}}{2}\right)$, it follows that

$$\sum_{n=0}^{\infty} g_n(x)z^n = \frac{z}{1-xz+z^2}.$$

Hoggatt–Bicknell–King polynomials of Lucas kind By the relation $h_n(x) = V_n\left(\frac{x+\sqrt{x^2-4}}{2}, \frac{x-\sqrt{x^2-4}}{2}\right)$, one has

$$\sum_{n=0}^{\infty} h_n(x)z^n = \frac{2-xz}{1-xz+z^2}.$$

Jacobsthal polynomials Since $J_n(x) = U_n\left(\frac{1+\sqrt{8x+1}}{2}, \frac{1-\sqrt{8x+1}}{2}\right)$, we have

$$\sum_{n=0}^{\infty} J_n(x)z^n = \frac{z}{1-z-4xz^2}.$$

Morgan–Voyce polynomials As $B_{n-1}(x) = g_n(x+2)$, $g_0(x) = 0$, we have

$$\sum_{n=0}^{\infty} B_n(x)z^n = \sum_{n=0}^{\infty} g_{n+1}(x)z^n = \frac{1}{z}\sum_{n=1}^{\infty} g_n(x+2)z^n = \frac{1}{1-(x+2)z+z^2}.$$

Brahmagupta polynomials For the integer parameter $t > 0$ we have $x_n(x, y) = \frac{1}{2}V_n\left(x + y\sqrt{t}, x - y\sqrt{t}\right)$, $y_n(x, y) = yU_n\left(x + y\sqrt{t}, x - y\sqrt{t}\right)$, so

$$\sum_{n=0}^{\infty} x_n(x, y)z^n = \frac{1-xz}{1-2xz+(x^2-y^2t)z^2}$$

$$\sum_{n=0}^{\infty} y_n(x, y)z^n = \frac{yz}{1-2xz+(x^2-y^2t)z^2}.$$

4.1.3 *Generating Functions of Classical Sequences*

Using $(x, y) = \left(\frac{1+\sqrt{5}}{2}, \frac{1-\sqrt{5}}{2}\right)$ one obtains the generating functions which correspond to Fibonacci and Lucas sequences. In this case, $xy = -1$ and $x + y = 1$, hence by the formulae (4.3) and (4.4) we have

$$\sum_{n=0}^{\infty} F_n z^n = \frac{z}{1-z-z^2}$$

$$\sum_{n=0}^{\infty} L_n z^n = \frac{2-z}{1-z-z^2}.$$

For $(x, y) = \left(1+\sqrt{2}, 1-\sqrt{2}\right)$, the generating functions for Pell and Pell–Lucas sequences are obtained. Indeed, here $xy = -1$ and $x + y = 2$, hence by the formulae (4.3) and (4.4) we have

$$\sum_{n=0}^{\infty} P_n z^n = \frac{z}{1-2z-z^2}$$

$$\sum_{n=0}^{\infty} Q_n z^n = \frac{2-2z}{1-2z-z^2}.$$

4.1.4 The Explicit Formula for the Fibonacci, Lucas, Pell, and Pell–Lucas Polynomials

Using the ordinary generating functions of the Fibonacci, Lucas, Pell, and Pell–Lucas polynomials, we can derive the following algebraic expressions.

Theorem 4.1 *For every positive integer n, we have*

1° $$f_n(x) = \sum_{j=0}^{\lfloor \frac{n-1}{2} \rfloor} \binom{n-j-1}{j} x^{n-2j-1}; \tag{4.5}$$

2° $$l_n(x) = \sum_{j=0}^{\lfloor \frac{n}{2} \rfloor} \frac{n}{n-j} \binom{n-j}{j} x^{n-2j}; \tag{4.6}$$

3° $$p_n(x) = \sum_{j=0}^{\lfloor \frac{n-1}{2} \rfloor} \binom{n-j-1}{j} 2^{n-2j-1} x^{n-2j-1}; \tag{4.7}$$

4° $$q_n(x) = \sum_{j=0}^{\lfloor \frac{n}{2} \rfloor} \frac{n}{n-j} \binom{n-j}{j} 2^{n-2j} x^{n-2j}. \tag{4.8}$$

Proof

1° Using the geometric series we can write

$$\sum_{n=0}^{\infty} f_n(x) z^n = \frac{z}{1-xz-z^2} = \frac{z}{1-(x+z)z} = \sum_{m=0}^{\infty} (x+z)^m z^{m+1}$$
$$= \sum_{n=0}^{\infty} \left(\sum_{j=0}^{\lfloor \frac{n-1}{2} \rfloor} \binom{n-j-1}{j} x^{n-2j-1} \right) z^n,$$

and the formula follows by identification of the corresponding coefficients.

2° Following the same idea as before, we have

$$\sum_{n=0}^{\infty} l_n(x) z^n = \frac{2-xz}{1-xz-z^2} = \frac{2-xz}{1-(x+z)z} = \sum_{m=0}^{\infty} (2-xz)(x+z)^m z^m$$
$$= \sum_{n=0}^{\infty} \left(\sum_{j=0}^{\lfloor \frac{n}{2} \rfloor} \frac{n}{n-j} \binom{n-j}{j} x^{n-2j} \right) z^n,$$

hence the formula (4.6) follows.

3° From the generating function of the Pell numbers, we obtain

$$\sum_{n=0}^{\infty} p_n(x)z^n = \frac{z}{1-2xz-z^2} = \frac{z}{1-(2x+z)z} = \sum_{m=0}^{\infty}(2x+z)^m z^{m+1}$$

$$= \sum_{n=0}^{\infty}\left(\sum_{j=0}^{\lfloor \frac{n-1}{2} \rfloor} \binom{n-j-1}{j} 2^{n-2j-1} x^{n-2j-1}\right) z^n,$$

and the formula follows by identifying the corresponding coefficients.

4° Analogously, for the Pell–Lucas polynomials, we have

$$\sum_{n=0}^{\infty} q_n(x)z^n = \frac{2-2xz}{1-2xz-z^2} = \frac{2-2xz}{1-(2x+z)z} = 2\sum_{m=0}^{\infty}(1-xz)(2x+z)^m z^m$$

$$= \sum_{n=0}^{\infty}\left(\sum_{j=0}^{\lfloor \frac{n}{2} \rfloor} \frac{n}{n-j}\binom{n-j}{j} 2^{n-2j} x^{n-2j}\right) z^n,$$

and the desired formula follows.

□

Because $F_n = f_n(1)$, by (4.5) we obtain the Lucas formula in Theorem 2.15 with a different proof. Similar formulae hold for L_n, P_n, and Q_n.

Corollary 4.1 *The following relations hold:*

$$1^\circ \quad F_n = \sum_{j=0}^{\lfloor \frac{n-1}{2} \rfloor} \binom{n-j-1}{j}; \tag{4.9}$$

$$2^\circ \quad L_n = \sum_{j=0}^{\lfloor \frac{n}{2} \rfloor} \frac{n}{n-j}\binom{n-j}{j}; \tag{4.10}$$

$$3^\circ \quad P_n = \sum_{j=0}^{\lfloor \frac{n-1}{2} \rfloor} \binom{n-j-1}{j} 2^{n-2j-1}; \tag{4.11}$$

$$4^\circ \quad Q_n = \sum_{j=0}^{\lfloor \frac{n}{2} \rfloor} \frac{n}{n-j}\binom{n-j}{j} 2^{n-2j}. \tag{4.12}$$

4.1.5 From Generating Functions to Properties of the Sequence

In many situations, it is possible to obtain the generating function of the sequence $(x_n)_{n\geq 0}$ using only the recurrence relation, and then to derive some properties of the sequence. We illustrate this idea for second-order recurrence relations. Assume that $(x_n)_{n\geq 0}$ is given by $x_{n+2} = ax_{n+1} + bx_n$, $n = 0, 1, \ldots$, where $x_0 = \alpha_0$ and $x_1 = \alpha_1$, while a, b are real (or complex) numbers with $b \neq 0$. We can write

$$\begin{aligned} F(z) &= \sum_{n=0}^{\infty} x_n z^n = x_0 + x_1 z + \sum_{n=2}^{\infty} x_n z^n \\ &= \alpha_0 + \alpha_1 z + \sum_{n=2}^{\infty} (ax_{n-1} + bx_{n-2})\, z^n \\ &= \alpha_0 + \alpha_1 z + az\,(F(z) - \alpha_0) + bz^2 F(z), \end{aligned}$$

and obtain the following relation:

$$F(z) = \frac{\alpha_0 + (\alpha_1 - a\alpha_0)\, z}{1 - az - bz^2}. \tag{4.13}$$

From formula (4.13) we can derive the Binet-type formula for the sequence $(x_n)_{n\geq 0}$. Let t_1, t_2 be the roots of the characteristic equation associated with the sequences, $t^2 - at - b = 0$. Clearly, we have $1 - az - bz^2 = (1 - t_1 z)(1 - t_2 z)$, hence the decomposition of the fraction in (4.13) can be written as

$$\frac{\alpha_0 + (\alpha_1 - a\alpha_0)\, z}{1 - az - bz^2} = \frac{A}{1 - t_1 z} + \frac{B}{1 - t_2 z}, \tag{4.14}$$

where the values of A and B can be determined by identifying the coefficients in the equality $\alpha_0 + (\alpha_1 - a\alpha_0)\, z = A(1 - t_2 z) + B(1 - t_1 z)$. This leads to the system $A + B = \alpha_0$ and $t_2 A + t_1 B = a\alpha_0 - \alpha_1$. If $t_1 \neq t_2$, then we get

$$A = \frac{\alpha_0 t_1 + \alpha_1 - a\alpha_0}{t_1 - t_2}, \qquad B = -\frac{\alpha_0 t_2 + \alpha_1 - a\alpha_0}{t_1 - t_2}.$$

We can derive the explicit formula for x_n by expanding the fractions in (4.14) as geometric series, and the comparing coefficients of z^n on either side. It follows that for $n = 0, 1, \ldots$, we have

$$\begin{aligned} x_n &= At_1^n + Bt_2^n \\ &= \frac{1}{t_1 - t_2}\left[(\alpha_0 t_1 + \alpha_1 - a\alpha_0)\, t_1^n - (\alpha_0 t_2 + \alpha_1 - a\alpha_0)\, t_2^n\right], \quad n = 0, 1, \ldots. \end{aligned} \tag{4.15}$$

If $t_1 = t_2$, then it is possible to derive the formula for x_n directly from (4.15), by taking the limit $t_2 \to t_1$. We obtain

$$\begin{aligned} x_n &= \alpha_0 \lim_{t_2 \to t_1} \frac{t_1^{n+1} - t_2^{n+1}}{t_1 - t_2} + (\alpha_1 - a\alpha_0) \lim_{t_2 \to t_1} \frac{t_1^n - t_2^n}{t_1 - t_2} \\ &= \alpha_0 (n+1) t_1^n + + (\alpha_1 - a\alpha_0)\, n t_1^{n-1} \\ &= [\alpha_0 t_1 + (\alpha_0 t_1 + \alpha_1 - a\alpha_0)\, n]\, t_1^{n-1}, \quad n = 0, 1, \ldots. \end{aligned}$$

Example 4.13 Find the recurrence relation satisfied by the sequence $(F_{rn})_{n \geq 0}$, consisting of those Fibonacci numbers whose index is a multiple of r, where r is a fixed positive integer.

Solution The generating function can be found by Binet's formula

$$\begin{aligned} \sum_{n=0}^{\infty} F_{rn} z^n &= \sum_{n=0}^{\infty} \frac{1}{\sqrt{5}} \left(\alpha^{rn} - \beta^{rn}\right) z^n = \frac{1}{\sqrt{5}} \left(\sum_{n=0}^{\infty} \alpha^{rn} z^n - \sum_{n=0}^{\infty} \beta^{rn} z^n \right) \\ &= \frac{1}{\sqrt{5}} \left(\sum_{n-0}^{\infty} (\alpha^r z)^n - (\beta^r z)^n \right) = \frac{1}{\sqrt{5}} \left(\frac{1}{1 - \alpha^r z} + \frac{1}{1 - \beta^r z} \right), \end{aligned}$$

where $\alpha = \frac{1+\sqrt{5}}{2}$ and $\beta = \frac{1-\sqrt{5}}{2}$. We also obtain the relation

$$\begin{aligned} \sum_{n=0}^{\infty} F_{rn} z^n &= \frac{1}{\sqrt{5}} \left(\frac{1}{1 - \alpha^r z} + \frac{1}{1 - \beta^r z} \right) = \frac{\frac{1}{\sqrt{5}} (\alpha^r - \beta^r) z}{1 - (\alpha^r + \beta^r) z + (\alpha\beta)^r z^2} \\ &= \frac{F_r z}{1 - L_r z + (-1)^r z^2}, \end{aligned}$$

where L_r denotes the rth Lucas number. The denominator leads immediately to the recurrence relation

$$F_{r(n+2)} = L_r F_{r(n+1)} + (-1)^{r+1} F_{r(n-1)}, \quad n = 0, 1, \ldots. \tag{4.16}$$

4.2 Exponential Generating Functions

As seen earlier, the method of ordinary generating functions was very useful for finding sequence terms, especially when these were linked to binomial coefficients. However, in other applications, one might have to consider generating functions with different properties. Such examples are the exponential generating functions.

4.2.1 Basic Operations and Examples

For the sequence $(a_n)_{n\geq 0}$, the exponential generating function is the series

$$E(z) = \sum_{n=0}^{\infty} \frac{a_n}{n!} z^n. \tag{4.17}$$

Identification Principle Let $F(z) = \sum_{n=0}^{\infty} \frac{1}{n!} a_n z^n$ and $G(z) = \sum_{n=0}^{\infty} \frac{1}{n!} b_n z^n$ be two generating functions. Then, $F(z) = G(z)$ if and only if $a_n = b_n$, $n \geq 0$.

The following operations hold.

1. Addition

$$\sum_{n=0}^{\infty} \frac{1}{n!} a_n z^n + \sum_{n=0}^{\infty} \frac{1}{n!} a_n z^n = \sum_{n=0}^{\infty} \frac{1}{n!} c_n z^n, \quad \text{where } c_n = a_n + b_n.$$

2. Multiplication by a constant If $\alpha \in \mathbb{C}$ a scalar, then

$$\alpha \sum_{n=0}^{\infty} \frac{1}{n!} a_n z^n = \sum_{n=0}^{\infty} \frac{1}{n!} c_n z^n, \quad \text{where } c_n = \alpha a_n.$$

3. Formal differentiation

$$\frac{\mathrm{d}}{\mathrm{d}z}(F(z)) = \frac{\mathrm{d}}{\mathrm{d}z}\left(\sum_{n=0}^{\infty} \frac{1}{n!} a_n z^n\right) = \sum_{n=0}^{\infty} \frac{1}{n!} n a_n z^{n-1} = \sum_{n=0}^{\infty} \frac{1}{n!} a_{n+1} z^n.$$

4. Formal integration

$$\int F(z)\mathrm{d}z = \int \sum_{n=0}^{\infty} \frac{1}{n!} a_n z^n \mathrm{d}z = \sum_{n=0}^{\infty} \frac{1}{n!} \frac{a_n}{n+1} z^{n+1} = \sum_{n=1}^{\infty} \frac{1}{n!} a_{n-1} z^n.$$

4. Multiplication

$$F(z)G(z) = \left(\sum_{n=0}^{\infty} \frac{1}{n!} a_n z^n\right)\left(\sum_{n=0}^{\infty} \frac{1}{n!} b_n z^n\right) = \sum_{n=0}^{\infty} \frac{1}{n!} c_n z^n,$$

$$c_n = \sum_{k=0}^{n} \binom{n}{k} a_k b_{n-k}.$$

5. Hadamard multiplication

$$F(z) \circ G(z) = \left(\sum_{n=0}^{\infty} \frac{1}{n!} a_n z^n\right) \left(\sum_{n=0}^{\infty} \frac{1}{n!} b_n z^n\right) = \sum_{n=0}^{\infty} \frac{1}{n!} c_n z^n, \quad c_n = \frac{1}{n!} a_n b_n.$$

6. Composition If $a_0 = 0$, we have

$$G(F(z)) = \sum_{n=0}^{\infty} \frac{1}{n!} b_n [F(z)]^n = \sum_{n=0}^{\infty} \frac{1}{n!} b_n \left(\sum_{\substack{j_1+\cdots+j_k=n \\ j_1,\ldots,j_k \geq 1}} \frac{a_{j_1}}{j_1!} \cdots \frac{a_{j_k}}{j_k!} \right) z^n$$

$$= \sum_{n=0}^{\infty} \frac{1}{n!} b_n \left(\sum_{\substack{j_1+\cdots+j_k=n \\ j_1,\ldots,j_k \geq 1}} \frac{1}{n!} \binom{n}{j_1, \ldots, j_k} a_{j_1} \cdots a_{j_k} \right) z^n.$$

7. Division If $b_0 \neq 0$, then we have

$$\frac{F(z)}{G(z)} = \frac{\sum_{n=0}^{\infty} \frac{1}{n!} a_n z^n}{\sum_{n=0}^{\infty} \frac{1}{n!} b_n z^n} = \sum_{n=0}^{\infty} \frac{1}{n!} c_n z^n, \quad c_n = \frac{1}{b_0} \left(a_n - \sum_{k=1}^{n} \binom{n}{k} b_k c_{n-k} \right).$$

8. Inverse The power series $F(z)$ and $G(z)$ are inverse if $a_0 b_0 = 1$ and

$$b_n = -\frac{1}{a_0} \sum_{k=1}^{n} \binom{n}{k} a_k b_{n-k}, \quad \text{for } n \geq 1.$$

The following exponential generating functions are immediate

- $a_n = 1, n \geq 0$ (constant sequence)

$$E(z) = \sum_{k=0}^{\infty} \frac{z^k}{k!} = e^z.$$

- $a_n = a^n, n \geq 0$ and $a \in \mathbb{C}$ a complex number (geometric sequence)

$$E(z) = \sum_{k=0}^{\infty} \frac{a^k z^k}{k!} = \sum_{k=0}^{\infty} \frac{(az)^k}{k!} = e^{az}.$$

- If $0 \leq k \leq n$ be positive integers and $P(n,k) = \frac{n!}{(n-k)!}$ denotes the number of arrangements of k objects selected out of n, then the exponential generating function of the sequence

$$P(n,0),\ P(n,1),\ P(n,2),\ \ldots,\ P(n,n),\ 0,\ldots,$$

is given by the formula

$$E(z) = \sum_{k=0}^{n} \frac{P(n,k)}{k!} z^k = \sum_{k=0}^{n} \binom{n}{k} z^k = (1+z)^k.$$

Theorem 4.2 (**n-permutations of multisets**) *Let $M = \{n_1\alpha_1, n_2\alpha_2, \ldots, n_k\alpha_k\}$ be a multiset over the set $S = \{\alpha_1, \alpha_2, \ldots, \alpha_k\}$, where n_j is multiplicity of the element α_j, $j = 1, \ldots, k$. Denoting by a_n the number of n-permutations of the multiset M, the exponential generating function of the sequence $(a_n)_{n\geq 0}$ is*

$$E(z) = \left(\sum_{j=0}^{n_1} \frac{z^j}{j!}\right)\left(\sum_{j=0}^{n_2} \frac{z^j}{j!}\right)\cdots\left(\sum_{j=0}^{n_k} \frac{z^j}{j!}\right). \tag{4.18}$$

Proof Notice that for $n \geq n_1 + \cdots + n_k$, we have $a_n = 0$, hence $E(z)$ is a polynomial. Notice also that the right side of (4.18) can be expanded as

$$\sum_{j_1,j_2,\ldots,j_k=0}^{n_1,n_2,\ldots,n_k} \frac{z^{j_1+j_2\cdots+j_k}}{j_1!j_2!\cdots j_k!} = \sum_{n=0}^{n_1+n_2+\cdots+n_k} \frac{z^n}{n!} \sum_{\substack{j_1+\cdots+j_k \\ 0\leq j_1\leq n_1,\ldots,0\leq j_k\leq n_k}} \frac{n!}{j_1!j_2!\cdots j_k!}$$

The number of permutations of M with exactly $j_1\alpha_1's$, $j_2\alpha_2's$, $\ldots$, $j_k\alpha_k's$ such that $j_1 + j_2 + \cdots + j_k = n$ is the multinomial coefficient

$$\binom{n}{j_1, j_2, \ldots, j_k} = \frac{n!}{j_1!j_2!\cdots j_k!}.$$

This shows that a_n is indeed given by

$$a_n = \sum_{\substack{j_1+\cdots+j_k \\ 0\leq j_1\leq n_1,\ldots,0\leq j_k\leq n_k}} \frac{n!}{j_1!j_2!\cdots j_k!}.$$

□

Example 4.14 Determine the number of ways to color the squares of a 1-by-n chessboard using black, white, green, and red, if an even number of squares is colored in black.

Solution Denoting the numbers of colorings by a_n, we have $a_1 = 0$. Each such coloring can be seen as a permutation of three objects b (black), w (white), g (green), and r (red), with repetitions allowed, where b appears an even number of times. This is given by the exponential generating function

$$
\begin{aligned}
E(z) &= \left(\sum_{n=0}^{\infty} \frac{z^{2n}}{(2n)!}\right)\left(\sum_{n=0}^{\infty} \frac{z^n}{n!}\right)^3 \\
&= \frac{e^z + e^{-z}}{2} e^{3z} = \frac{1}{2}\left(e^{4z} + e^{2z}\right) \\
&= \frac{1}{2}\left(\sum_{n=0}^{\infty} \frac{4^n z^n}{n!} + \sum_{n=0}^{\infty} \frac{2^n z^n}{n!}\right) = \frac{1}{2}\sum_{n=0}^{\infty} (4^n + 2^n) \cdot \frac{z^n}{n!}.
\end{aligned}
$$

This shows that $a_n = 2^{n-1}(2^n + 1)$, $n \geq 1$.

Example 4.15 Determine the number a_n of n digit (in base 10) numbers with each digit even, where the digits 0, 2, and 4 occur an even number of times.

Solution Denoting this number by a_n and setting $a_1 = 0$, this is the number of n-permutations of the multiset $M = \{\infty 0, \infty 2, \infty 4, \infty 6, \infty 8\}$ (having infinitely many copies of each element), in which 0, 2, and 4 occur an even number of times. The exponential generating function is

$$
\begin{aligned}
E(z) &= \left(\sum_{n=0}^{\infty} \frac{z^{2n}}{(2n)!}\right)^3 \left(\sum_{n=0}^{\infty} \frac{z^n}{n!}\right)^2 \\
&= \left(\frac{e^z + e^{-z}}{2}\right)^3 e^{2z} = \frac{1}{8}\left(e^{-z} + 3e^z + 3e^{3z} + e^{5z}\right) \\
&= \frac{1}{8}\left(\sum_{n=0}^{\infty} \frac{(-1)^n z^n}{n!} + 3\sum_{n=0}^{\infty} \frac{z^n}{n!} + 3\sum_{n=0}^{\infty} \frac{3^n z^n}{n!} + \sum_{n=0}^{\infty} \frac{5^n z^n}{n!}\right) \\
&= \frac{1}{8}\sum_{n=0}^{\infty} (5^n + 3^n + 1 + (-1)^n) \cdot \frac{z^n}{n!}.
\end{aligned}
$$

This shows that

$$
a_n = \frac{5^n + 3^n + 1 + (-1)^n}{8}, \quad n \geq 0.
$$

Example 4.16 Find the number of ways to color the squares of a 1-by-n board with red, blue, green, and white, where the number of red squares is odd, the number of blue squares is even, and at least one square is white.

Solution The exponential generating function is

$$
\begin{aligned}
E(z) &= \left(\sum_{n=0}^{\infty} \frac{z^{2n+1}}{(2n+1)!}\right)\left(\sum_{n=0}^{\infty} \frac{z^{2n}}{(2n)!}\right)\left(\sum_{n=0}^{\infty} \frac{z^n}{n!}\right)\left(\sum_{n=1}^{\infty} \frac{z^n}{n!}\right) \\
&= \left(\frac{e^z - e^{-z}}{2}\right)\left(\frac{e^z + e^{-z}}{2}\right) e^z(e^z - 1) \\
&= \frac{1}{4}\left(e^{4z} - e^{3z} - 1 + e^{-z}\right) \\
&= -\frac{1}{4} + \frac{1}{4}\sum_{n=0}^{\infty} (4^n - 3^n + (-1)^n) \cdot \frac{z^n}{n!}.
\end{aligned}
$$

This shows that

$$
a_n = \frac{4^n - 3^n + (-1)^n}{4}, \qquad n \geq 1,
$$

and $a_0 = 0$.

4.2.2 Generating Functions for Polynomials U_n and V_n

Using the formulae for the polynomials U_n and V_n (2.60), we have

$$
\begin{aligned}
\sum_{n=0}^{\infty} \frac{1}{n!} U_n(x, y) z^n &= \frac{1}{x - y} \sum_{n=0}^{\infty} \frac{1}{n!} \left(x^n - y^n\right) z^n \\
&= \frac{1}{x - y} \left[\sum_{n=0}^{\infty} \frac{1}{n!} (xz)^n - \sum_{n=0}^{\infty} \frac{1}{n!} (yz)^n\right] \\
&= \frac{e^{xz} - e^{yz}}{x - y}.
\end{aligned}
\tag{4.19}
$$

and

$$
\sum_{n=0}^{\infty} \frac{1}{n!} V_n(x, y) z^n = \sum_{n=0}^{\infty} \frac{1}{n!} \left(x^n + y^n\right) z^n = e^{xz} + e^{yz}. \tag{4.20}
$$

4.2.3 *Generating Functions of Classical Polynomials*

Here we use the hyperbolic functions $\sinh u = \frac{e^u - e^{-u}}{2}$ and $\cosh u = \frac{e^u + e^{-u}}{2}$.

Fibonacci polynomials Since $f_n(x) = U_n\left(\frac{x+\sqrt{x^2+4}}{2}, \frac{x-\sqrt{x^2+4}}{2}\right)$, we have

$$\sum_{n=0}^{\infty} \frac{1}{n!} f_n(x) z^n = \frac{1}{\sqrt{x^2+4}} \left(e^{\frac{x+\sqrt{x^2+4}}{2} \cdot z} - e^{\frac{x-\sqrt{x^2+4}}{2} \cdot z} \right)$$

$$= \frac{2e^{\frac{xz}{2}}}{\sqrt{x^2+4}} \frac{e^{\frac{1}{2}z\sqrt{x^2+4}} - e^{-\frac{1}{2}z\sqrt{x^2+4}}}{2} = \frac{2e^{\frac{xz}{2}}}{\sqrt{x^2+4}} \sinh \frac{z\sqrt{x^2+4}}{2}.$$

Lucas polynomials Since $l_n(x) = V_n\left(\frac{x+\sqrt{x^2+4}}{2}, \frac{x-\sqrt{x^2+4}}{2}\right)$, we have

$$\sum_{n=0}^{\infty} \frac{1}{n!} l_n(x) z^n = e^{\frac{x+\sqrt{x^2+4}}{2} \cdot z} + e^{\frac{x-\sqrt{x^2+4}}{2} \cdot z}$$

$$= 2e^{\frac{xz}{2}} \frac{e^{\frac{1}{2}z\sqrt{x^2+4}} + e^{-\frac{1}{2}z\sqrt{x^2+4}}}{2} = 2e^{\frac{xz}{2}} \cosh \frac{z\sqrt{x^2+4}}{2}.$$

Pell polynomials As $p_n(x) = U_n\left(x + \sqrt{x^2+1}, x - \sqrt{x^2+1}\right)$, we have

$$\sum_{n=0}^{\infty} \frac{1}{n!} p_n(x) z^n = \frac{2e^{xz}}{\sqrt{x^2+1}} \sinh z\sqrt{x^2+1}.$$

Pell–Lucas polynomials As $q_n(x) = V_n\left(x + \sqrt{x^2+1}, x - \sqrt{x^2+1}\right)$

$$\sum_{n=0}^{\infty} \frac{1}{n!} q_n(x) z^n = 2e^{xz} \cosh z\sqrt{x^2+1}.$$

Chebyshev polynomials of the first kind Since we have the relation $T_n(x) = \frac{1}{2} V_n\left(x + \sqrt{x^2-1}, x - \sqrt{x^2-1}\right)$, it follows that

$$\sum_{n=0}^{\infty} \frac{1}{n!} T_n(x) z^n = e^{xz} \cosh z\sqrt{x^2-1}.$$

Chebyshev polynomials of the second kind We have the relation $u_n(x) = U_{n+1}\left(x + \sqrt{x^2-1}, x - \sqrt{x^2-1}\right)$, while by (4.19) we deduce that

$$\sum_{n=0}^{\infty} \frac{1}{n!} U_{n+1}(x, y) z^n = \frac{x e^{xz} - y e^{yz}}{x - y}.$$

It follows that

$$\begin{aligned}\sum_{n=0}^{\infty} \frac{1}{n!} u_n(x) z^n &= \frac{\left(x + \sqrt{x^2-1}\right) e^{\left(x+\sqrt{x^2-1}\right)z} - \left(x - \sqrt{x^2-1}\right) e^{\left(x-\sqrt{x^2-1}\right)z}}{2\sqrt{x^2-1}} \\ &= \frac{e^{xz}}{\sqrt{x^2-1}} \left(x \sinh z\sqrt{x^2-1} + \sqrt{x^2-1} \cosh z\sqrt{x^2-1}\right).\end{aligned}$$

Hoggatt–Bicknell–King polynomials of Fibonacci kind Since we have $g_n(x) = U_n\left(\frac{x+\sqrt{x^2-4}}{2}, \frac{x-\sqrt{x^2-4}}{2}\right)$, it follows that

$$\sum_{n=0}^{\infty} \frac{1}{n!} g_n(x) z^n = \frac{2e^{\frac{xz}{2}}}{\sqrt{x^2-4}} \sinh \frac{z\sqrt{x^2-4}}{2}.$$

Hoggatt–Bicknell–King polynomials of Lucas kind By the relation $h_n(x) = V_n\left(\frac{x+\sqrt{x^2-4}}{2}, \frac{x-\sqrt{x^2-4}}{2}\right)$, we have

$$\sum_{n=0}^{\infty} \frac{1}{n!} h_n(x) z^n = 2e^{\frac{xz}{2}} \cosh \frac{z\sqrt{x^2-4}}{2}.$$

Jacobsthal polynomials Since $J_n(x) = U_n\left(\frac{1+\sqrt{8x+1}}{2}, \frac{1-\sqrt{8x+1}}{2}\right)$, we have

$$\sum_{n=0}^{\infty} \frac{1}{n!} J_n(x) z^n = \frac{2e^{\frac{z}{2}}}{\sqrt{8x+1}} \sinh \frac{z\sqrt{8x+1}}{2}.$$

Morgan–Voyce polynomials Since $B_{n-1}(x) = g_n(x+2)$ and $g_0(x) = 0$, by the same argument as for Chebysev polynomials of the second kind, it follows that

$$\sum_{n=0}^{\infty} \frac{1}{n!} B_n(x) z^n = \frac{\frac{x+\sqrt{x^2-4}}{2} e^{\frac{x+\sqrt{x^2-4}}{2} z} - \frac{x-\sqrt{x^2-4}}{2} e^{\frac{x-\sqrt{x^2-4}}{2} z}}{\sqrt{x^2-4}}$$

$$= \frac{e^{\frac{xz}{2}}}{\sqrt{x^2-4}} \left(x \sinh \frac{z\sqrt{x^2-4}}{2} + \sqrt{x^2-4} \cosh \frac{z\sqrt{x^2-4}}{2} \right).$$

Brahmagupta polynomials For the integer parameter $t > 0$ we have $x_n(x, y) = \frac{1}{2} V_n \left(x + y\sqrt{t}, x - y\sqrt{t}\right)$, $y_n(x, y) = yU_n \left(x + y\sqrt{t}, x - y\sqrt{t}\right)$, so

$$\sum_{n-0}^{\infty} \frac{1}{n!} x_n(x, y) z^n = e^{xz} \cosh yz\sqrt{t}.$$

$$\sum_{n=0}^{\infty} \frac{1}{n!} y_n(x, y) z^n = \frac{e^{xz}}{\sqrt{t}} \sinh yz\sqrt{t}.$$

4.2.4 *Generating Functions of Classical Sequences*

Substituting $x - 1$ in the formulae for the exponential generating functions for Fibonacci, Lucas, Pell, and Lucas–Pell polynomials, we obtain

$$\sum_{n=0}^{\infty} \frac{1}{n!} F_n z^n = \frac{2e^{\frac{z}{2}}}{\sqrt{5}} \sinh \frac{z\sqrt{5}}{2};$$

$$\sum_{n=0}^{\infty} \frac{1}{n!} L_n z^n = 2e^{\frac{z}{2}} \cosh \frac{z\sqrt{5}}{2};$$

$$\sum_{n=0}^{\infty} \frac{1}{n!} P_n z^n = \frac{2e^{z}}{\sqrt{2}} \sinh z\sqrt{2};$$

$$\sum_{n=0}^{\infty} \frac{1}{n!} Q_n z^n = 2e^{z} \cosh z\sqrt{2}.$$

4.3 The Cauchy Integral Formula

The Cauchy integral formula is one of the fundamental results in complex analysis, having a long history and applications in numerous fields of mathematics, including complex analysis, combinatorics, discrete mathematics, or number theory.

Here we will use a version of Cauchy's formula derived in [12], to derive integral representations for the terms of some classical integer sequences. Recently, this approach was used to compute exact integral formulae for the coefficients of cyclotomic polynomials [11], Gaussian and multinomial polynomials [13], polygonal polynomials [14], and other general classes of polynomials [15].

4.3.1 A Useful Version of the Cauchy Integral Formula

Recall that a function $h : \Omega \to \mathbb{C}$ is said to be *meromorphic* at a point $z_0 \in \Omega$, if h can be written as a quotient of two analytic functions f, g in a neighborhood $\mathcal{U} \subset \Omega$ of z_0

$$\forall z \in \mathcal{U} \setminus \{z_0\} \; : \; h(z) = \frac{f(z)}{g(z)}.$$

In this case we also have the expansion

$$h(z) = \sum_{n \geq -m} c_n (z - z_0)^n, \tag{4.21}$$

for all $z \neq z_0$ in a disk centered at z_0. If m is the largest number for which $c_{-m} \neq 0$ in (4.21), then z_0 is called a *pole* of order m. The coefficient c_{-1} of $(z - z_0)^{-1}$ in (4.21) is denoted by $\mathrm{Res}(h, z_0)$, and called the *residue* of h at the point z_0.

Cauchy's Residue Theorem relates global properties of a meromorphic function and its integral along closed curves, to local characteristics, i.e., the residues at poles.

Theorem 4.3 (Residue Theorem) *Let $h : \Omega \to \mathbb{C}$ be a meromorphic function in a domain Ω, and λ be a simple loop in Ω along which the function is analytic. Then*

$$\frac{1}{2\pi i} \int_\lambda h(z)\, \mathrm{d}z = \sum_s Res(h(z), z = s),$$

where the sum is over all poles s of h enclosed by λ.

Through this formula one can obtain a unitary formula for the coefficients of an analytic function [70].

Theorem 4.4 (Cauchy integral formula) *Let $f(z) = \sum_{n \geq 0} c_n z^n$ be an analytic function in a disk centered at 0, and let Γ be a curve in the interior of this disk, which winds around the origin exactly once in positive orientation, that is the winding number with respect to 0 is equal to 1. Then we have*

$$c_n = \frac{1}{2\pi i} \int_\Gamma \frac{f(z)}{z^{n+1}}\, \mathrm{d}z, \quad n = 0, 1, \ldots. \tag{4.22}$$

Letting Γ be the circle of radius $R > 0$ centered at 0 and $z = R(\cos t + i \sin t)$, $t \in [0, 2\pi]$, we have $\mathrm{d}z = R(-\sin t + i \cos t)\mathrm{d}t = iR(\cos t + i \sin t)\mathrm{d}t$. Then, formula (4.22) can be written as

$$c_n = \frac{1}{2\pi i}\int_0^{2\pi} \frac{iR(\cos t + i \sin t) f(R(\cos t + i \sin t))}{R^{n+1}(\cos t + i \sin t)^{n+1}} dt$$
$$= \frac{1}{2\pi R^n}\int_0^{2\pi} \frac{f(R(\cos t + i \sin t))}{\cos nt + i \sin nt} dt,$$

therefore

$$c_n = \frac{1}{2\pi R^n}\int_0^{2\pi} (\cos nt - i \sin nt) f(R(\cos t + i \sin t))\, dt, \quad n = 0, 1, \ldots. \tag{4.23}$$

If the coefficients c_n are real numbers, then from (4.23) we obtain

$$c_n = \frac{1}{2\pi R^n}\int_0^{2\pi} \mathrm{Re}\,[(\cos nt - i \sin nt) f(R(\cos t + i \sin t))]\, dt, \quad n = 0, 1, \ldots. \tag{4.24}$$

In addition, in this case the following formula can be deduced:

$$\int_0^{2\pi} \mathrm{Im}\,[(\cos nt - i \sin nt) f(R(\cos t + i \sin t))]\, dt = 0, \quad n = 0, 1, \ldots.$$

Remark 4.1 When f is a polynomial, we can give a direct proof to formula (4.23). Indeed, assuming that $f(z) = \sum_{k=0}^{m} c_k z^k$, we obtain

$$z^{-n} f(z) = c_n + \sum_{k=0, k\neq n}^{m} c_k z^{k-n}.$$

Let $z = R(\cos t + i \sin t)$, $t \in [0, 2\pi]$, and consider the integral over the interval $[0, 2\pi]$. Clearly, the integral of z^{k-n} vanishes for $k \neq n$, hence

$$\int_0^{2\pi} R(\cos t + i \sin t)^{-n} f(R(\cos t + i \sin t))\, dt = 2\pi c_n, \quad n = 0, 1, \ldots, \tag{4.25}$$

therefore formula (4.23) follows. Notice that the above argument can also be applied for Laurent polynomials.

Example 4.17 Let with $0 < R < 1$ and consider the geometric series $\sum_{n\geq 0} z^n$. Clearly, we have $f(z) = \frac{1}{1-z}$ and

$$f(R(\cos t + i \sin t)) = \frac{1}{1 - R(\cos t + i \sin t)} = \frac{1 - R(\cos t - i \sin t)}{1 + R^2 - 2R\cos t}.$$

By formula (4.23) it follows that for every $n \geq 0$, we have

$$1 = c_n = \frac{1}{2\pi R^n} \int_0^{2\pi} \frac{(\cos nt - i \sin nt)\,[1 - R(\cos t - i \sin t)]}{1 + R^2 - 2R\cos t}\, dt$$
$$= \frac{1}{2\pi R^n} \int_0^{2\pi} \frac{\cos nt - R\cos(n+1)t - i\,[\sin nt - R\sin(n+1)t]}{1 + R^2 - 2R\cos t}\, dt.$$

Because the coefficients c_n are real numbers, from (4.24) we obtain

$$1 = \frac{1}{2\pi R^n} \int_0^{2\pi} \frac{\cos nt - R\cos(n+1)t}{1 + R^2 - 2R\cos t}\, dt,$$

and

$$0 = \frac{1}{2\pi R^n} \int_0^{2\pi} \frac{\sin nt - R\sin(n+1)t}{1 + R^2 - 2R\cos t}\, dt.$$

4.3.2 The Integral Representation of Classical Sequences

We begin with the integral representation of $U_n(x, y)$, where x, y are nonzero real numbers with $x \neq y$. Assume that $R < \min\left\{\frac{1}{|x|}, \frac{1}{|y|}\right\}$. The ordinary generating function of $U_n(x, y)$ is given in formula (4.3) as

$$\sum_{n=0}^{\infty} U_n(x, y) z^n = \frac{1}{x - y} \sum_{n=0}^{\infty} \left(\frac{1}{1 - xz} - \frac{1}{1 - yz}\right).$$

The power series in the left-hand side is convergent for $|z| < R$, since $|xz| < 1$ and $|yz| < 1$. Applying formula (4.23) we obtain

$$U_n(x, y) = \frac{1}{2\pi(x - y)R^n} \int_0^{2\pi} \left[\frac{\cos nt - i\sin nt}{1 - xR(\cos t + i\sin t)} - \frac{\cos nt - i\sin nt}{1 - yR(\cos t + i\sin t)}\right] dt.$$

In order to calculate the first expression in the integral, we have

$$\frac{\cos nt - i\sin nt}{1 - xR(\cos t + i\sin t)} = \frac{(\cos nt - i\sin nt)\,[1 - xR(\cos t - i\sin t)]}{x^2R^2 + 1 - 2xR\cos t}$$
$$= \frac{\cos nt - i\sin nt - xR\,[\cos(n+1)t - i\sin(n+1)t]}{x^2R^2 + 1 - 2xR\cos t},$$

and similarly

$$\frac{\cos nt - i\sin nt}{1 - yR(\cos t + i\sin t)} = \frac{(\cos nt - i\sin nt)\,[1 - yR(\cos t - i\sin t)]}{y^2R^2 + 1 - 2xR\cos t}$$
$$= \frac{\cos nt - i\sin nt - yR\,[\cos(n+1)t - i\sin(n+1)t]}{y^2R^2 + 1 - 2yR\cos t}.$$

Because $U_n(x, y)$ is a real number, it follows that

$$U_n(x, y) = \frac{1}{2\pi(x - y)R^n}\int_0^{2\pi}\left[\frac{\cos nt - xR\cos(n+1)t}{x^2R^2 + 1 - 2xR\cos t} - \frac{\cos nt - yR\cos(n+1)t}{y^2R^2 + 1 - 2yR\cos t}\right]\mathrm{d}t.$$

After simple computations we obtain the integral formula

$$\begin{aligned} U_n(x, y) &= \frac{1}{2\pi R^{n-1}}\int_0^{2\pi}\frac{xyR^2\cos(n+1)t}{a + b\cos t + c\cos^2 t}\,\mathrm{d}t \\ &- \frac{1}{2\pi R^{n-1}}\int_0^{2\pi}\frac{(x+y)R\cos nt}{a + b\cos t + c\cos^2 t}\,\mathrm{d}t \\ &+ \frac{1}{2\pi R^{n-1}}\int_0^{2\pi}\frac{\cos(n-1)t}{a + b\cos t + c\cos^2 t}\,\mathrm{d}t, \end{aligned} \tag{4.26}$$

where

$$\begin{aligned} a &= a(x, y, R) = (xy)^2R^4 + \left(x^2 + y^2\right)R^2 + 1, \\ b &= b(x, y, R) = -2(x + y)(xyR^2 + 1)R, \\ c &= c(x, y, R) = 4xyR^2. \end{aligned} \tag{4.27}$$

Similarly, the ordinary generating function of $V_n(x, y)$ is given by

$$\sum_{n=0}^{\infty} V_n(x, y)z^n = \frac{1}{1 - xz} + \frac{1}{1 - yz}.$$

Applying formula (4.23) it follows that

$$V_n(x, y) = \frac{1}{2\pi R^n}\int_0^{2\pi}\left[\frac{\cos nt - xR\cos(n+1)t}{x^2R^2 + 1 - 2xR\cos t} + \frac{\cos nt - yR\cos(n+1)t}{y^2R^2 + 1 - 2yR\cos t}\right]\mathrm{d}t,$$

hence

$$V_n(x, y) = \frac{1}{2\pi R^n}\int_0^{2\pi}\frac{2xyR^2\cos(n+2)t}{a + b\cos t + c\cos^2 t}\,\mathrm{d}t$$

$$- \frac{1}{2\pi R^n} \int_0^{2\pi} \frac{(x+y)\left(xyR^2+2\right) R\cos(n+1)t}{a+b\cos t+c\cos^2 t} \mathrm{d}t$$
$$+ \frac{1}{2\pi R^n} \int_0^{2\pi} \frac{\left[(x+y)^2 R^2+2\right]\cos nt}{a+b\cos t+c\cos^2 t} \mathrm{d}t$$
$$- \frac{1}{2\pi R^n} \int_0^{2\pi} \frac{(x+y)R\cos(n-1)t}{a+b\cos t+c\cos^2 t} \mathrm{d}t, \tag{4.28}$$

where a, b, c are defined by (4.27).

Remark 4.2 Considering the integral

$$I_k = I_k(x, y, R) = \int_0^{2\pi} \frac{\cos kt}{a+b\cos t+c\cos^2 t} \mathrm{d}t,$$

formula (4.26) shows that $U_n(x, y)$ is a linear combination of the integrals I_{n+1}, I_n and I_{n-1}, i.e., we have

$$U_n(x, y) = \frac{xy}{2\pi R^{n-3}} I_{n+1} - \frac{x+y}{2\pi R^{n-2}} I_n + \frac{1}{2\pi R^{n-1}} I_{n-1}.$$

Similarly, from (4.28), we obtain $V_n(x, y)$ as a linear combination of the terms I_{n+2}, I_{n+1}, I_n and I_{n-1}, given explicitly by

$$V_n(x, y) = \frac{xy}{\pi R^{n-2}} I_{n+2} - \frac{(x+y)\left(xyR^2+2\right)}{2\pi R^{n-1}} I_{n+1} + \frac{(x+y)^2 R^2+2}{2\pi R^n} I_n - \frac{x+y}{2\pi R^{n-1}} I_{n-1}.$$

Integral formula for Fibonacci numbers Because $F_n = U_n\left(\frac{1+\sqrt{5}}{2}, \frac{1-\sqrt{5}}{2}\right)$, for Fibonacci numbers we have $x+y=1$ and $xy=-1$, therefore

$$F_n = \frac{1}{2\pi R^{n-1}} \int_0^{2\pi} \frac{-R^2\cos(n+1)t - R\cos nt + \cos(n-1)t}{R^4+3R^2+1+2(R^3-R)\cos t - 4R^2\cos^2 t} \mathrm{d}t, \tag{4.29}$$

for every positive real number $R < \min\left\{\frac{2}{1+\sqrt{5}}, \frac{2}{\sqrt{5}-1}\right\} = \frac{2}{1+\sqrt{5}} = \frac{\sqrt{5}-1}{2}$.

Using the notation

$$I_k = I_k\left(\frac{1+\sqrt{5}}{2}, \frac{1-\sqrt{5}}{2}, R\right)$$
$$= \int_0^{2\pi} \frac{\cos kt}{R^4+3R^2+1+2(R^3-R)\cos t - 4R^2\cos^2 t} \mathrm{d}t,$$

this can be further written as

$$F_n = -\frac{1}{2\pi R^{n-3}} I_{n+1} - \frac{1}{2\pi R^{n-2}} I_n + \frac{1}{2\pi R^{n-1}} I_{n-1}.$$

Integral formula for Lucas numbers Because $L_n = V_n\left(\frac{1+\sqrt{5}}{2}, \frac{1-\sqrt{5}}{2}\right)$, and $x^2 + y^2 = 3$, we obtain the following integral formula for Lucas numbers

$$\begin{aligned} L_n = &- \frac{1}{\pi R^{n-2}} \int_0^{2\pi} \frac{\cos(n+2)t}{R^4 + 3R^2 + 1 + 2(R^3 - R)\cos t - 4R^2\cos^2 t}\, dt \\ &+ \frac{R^2 - 2}{2\pi R^{n-1}} \int_0^{2\pi} \frac{\cos(n+1)t}{R^4 + 3R^2 + 1 + 2(R^3 - R)\cos t - 4R^2\cos^2 t}\, dt \\ &+ \frac{R^2 + 2}{2\pi R^{n}} \int_0^{2\pi} \frac{\cos nt}{R^4 + 3R^2 + 1 + 2(R^3 - R)\cos t - 4R^2\cos^2 t}\, dt \\ &- \frac{1}{2\pi R^{n-1}} \int_0^{2\pi} \frac{\cos(n-1)t}{R^4 + 3R^2 + 1 + 2(R^3 - R)\cos t - 4R^2\cos^2 t}\, dt, \end{aligned} \tag{4.30}$$

for every positive real number $R < \min\left\{\frac{2}{1+\sqrt{5}}, \frac{2}{\sqrt{5}-1}\right\} = \frac{2}{1+\sqrt{5}} = \frac{\sqrt{5}-1}{2}$.

With the I_k's used for Fibonacci numbers, for Lucas numbers we have

$$L_n = -\frac{1}{\pi R^{n-2}} I_{n+2} + \frac{R^2 - 2}{2\pi R^{n-1}} I_{n+1} + \frac{R^2 + 2}{2\pi R^{n}} I_n - \frac{1}{2\pi R^{n-1}} I_{n-1}.$$

Integral formula for Pell numbers Because $P_n = U_n\left(1 + \sqrt{2}, 1 - \sqrt{2}\right)$, for Pell numbers we have $x + y = 2$ and $xy = -1$, therefore

$$P_n = \frac{1}{2\pi R^{n-1}} \int_0^{2\pi} \frac{-R^2\cos(n+1)t - 2R\cos nt + \cos(n-1)t}{R^4 + 6R^2 + 1 + 4(R^3 - R)\cos t - 4R^2\cos^2 t}\, dt, \tag{4.31}$$

for every positive real number $R < \min\left\{\frac{1}{1+\sqrt{2}}, \frac{1}{\sqrt{2}-1}\right\} = \sqrt{2} - 1$.

Using the notation

$$\begin{aligned} I_k &= I_k\left(1 + \sqrt{2}, 1 - \sqrt{2}, R\right) \\ &= \int_0^{2\pi} \frac{\cos kt}{R^4 + 6R^2 + 1 + 4(R^3 - R)\cos t - 4R^2\cos^2 t}\, dt, \end{aligned}$$

we have

$$P_n = -\frac{1}{2\pi R^{n-3}} I_{n+1} - \frac{1}{\pi R^{n-2}} I_n + \frac{1}{2\pi R^{n-1}} I_{n-1}.$$

Integral formula for Pell–Lucas numbers Since $Q_n = V_n\left(1+\sqrt{2}, 1-\sqrt{2}\right)$ and $x^2 + y^2 = 6$, for Pell–Lucas numbers we obtain

$$\begin{aligned} Q_n = &- \frac{1}{\pi R^{n-2}} \int_0^{2\pi} \frac{\cos(n+2)t}{R^4 + 6R^2 + 1 + 4(R^3 - R)\cos t - 4R^2\cos^2 t}\,dt \\ &+ \frac{R^2 - 2}{\pi R^{n-1}} \int_0^{2\pi} \frac{\cos(n+1)t}{R^4 + 6R^2 + 1 + 4(R^3 - R)\cos t - 4R^2\cos^2 t}\,dt \\ &+ \frac{2R^2 + 1}{\pi R^n} \int_0^{2\pi} \frac{\cos nt}{R^4 + 6R^2 + 1 + 4(R^3 - R)\cos t - 4R^2\cos^2 t}\,dt \\ &- \frac{1}{\pi R^{n-1}} \int_0^{2\pi} \frac{\cos(n-1)t}{R^4 + 6R^2 + 1 + 4(R^3 - R)\cos t - 4R^2\cos^2 t}\,dt, \end{aligned} \tag{4.32}$$

for every positive real number $R < \min\left\{\frac{1}{1+\sqrt{2}}, \frac{1}{\sqrt{2}-1}\right\} = \sqrt{2} - 1$.

With the I_k's used for Pell numbers, for Pell–Lucas numbers we have

$$Q_n = -\frac{1}{\pi R^{n-2}} I_{n+2} + \frac{R^2 - 2}{\pi R^{n-1}} I_{n+1} + \frac{2R^2 + 1}{\pi R^n} I_n - \frac{1}{\pi R^{n-1}} I_{n-1}.$$

Chapter 5
More on Second-Order Linear Recurrent Sequences

The sequence $(x_n)_{n\geq 0} = (x_n(a, b; p, q))_{n\geq 0}$ defined by

$$x_{n+2} = px_{n+1} + qx_n, \quad x_0 = a, x_1 = b, \tag{5.1}$$

with a, b, p, and q arbitrary complex numbers is called a Horadam sequence. For simplicity, we shall denote $(x_n(a, b; p, q))_{n\geq 0}$ by $(x_n)_{n\geq 0}$ hereafter. The recurrent sequence (5.1) was named to honor A.F. Horadam, who initiated the study of this general recursion in its general form since the 1960s in [78–82]. Many classical sequences are particular instances: when $(a, b) = (0, 1)$, for $(p, q) = (1, 1)$ one gets the Fibonacci sequence, while for $(p, q) = (1, -1)$ the Lucas sequence, respectively.

In Section 5.1 we discuss the formula of the general term of the Horadam sequence, which is used to characterize periodic orbits in Section 5.2. The periodic orbits are enumerated in Section 5.3, and their geometric structure is discussed in Section 5.4. Non-periodic Horadam patterns are detailed in Sections 5.5 and 5.6, with applications discussed in Section 5.7. Certain examples of nonhomogeneous orbits are presented in Section 5.8.

5.1 Preliminary results

Historical details about this sequence are given in the survey paper of Larcombe et al. [107], or [126]. Many properties and identities are shown in [2, 47, 93, 94, 104–106, 108, 109, 111, 148, 150, 159, 171–173, 176, 177]. One can find links to other related sequences in [75, 77, 140], results related to generating functions in [95, 113, 154, 170], and many generalizations and extensions in [83, 149, 169].

D. Andrica, O. Bagdasar, *Recurrent Sequences*, Problem Books in Mathematics,
https://doi.org/10.1007/978-3-030-51502-7_5

The characteristic equation associated with the recurrence (5.1) is

$$P(x) = x^2 - px - q = 0, \tag{5.2}$$

whose roots, termed *generators*, are denoted by z_1 and z_2. By Vieta's relations for the polynomial P we have the relations

$$-p = z_1 + z_2, \quad q = z_1 z_2, \tag{5.3}$$

showing that the recurrence (5.1) may be defined either through the coefficients p, q, or equivalently, by the generators z_1, z_2.

5.1.1 General Sequence Term

We start with formulae for the general term x_n of the sequence (5.1), when the characteristic polynomial (5.2) has distinct, or equal roots. These formulae are equivalent to the Binet-type formulae discussed in Chapter 2.

5.1.1.1 Nondegenerate Case: Distinct Roots ($z_1 \neq z_2$)

The general term of the sequence $(x_n)_{n\geq 0}$, when (5.2) has the distinct roots z_1 and z_2 is given by (see, e.g., [1, Chapter 7], [62, Chapter 1], or [89])

$$x_n = Az_1^n + Bz_2^n, \tag{5.4}$$

where the constants A and B obtained from the initial conditions are

$$A = \frac{az_2 - b}{z_2 - z_1}, \quad B = \frac{b - az_1}{z_2 - z_1}. \tag{5.5}$$

The general Horadam term can be written explicitly as

$$x_n = \frac{1}{z_2 - z_1}\left[(az_2 - b)z_1^n + (b - az_1)z_2^n\right]. \tag{5.6}$$

5.1.1.2 Degenerate Case: Equal Roots ($z_1 = z_2$)

When $z_1 = z_2 = z$, the general term of the Horadam sequence is given by

$$x_n = \left[A + Bn\right]z^n = \left[a + \left(\frac{b}{z} - a\right)n\right]z^n. \tag{5.7}$$

5.1.2 Ratios of Horadam Sequences

Here we explore ratios of consecutive terms of Horadam sequences, first considered by [138]. We extend these with analysis of non-convergent cases, as seen in [34]. The results extend naturally to higher order recurrences.

Extending the concept of Golden ratio for a general Horadam sequence $(x_n)_{n\geq 0}$, we define the following sequence of ratios of consecutive terms

$$q_n = \frac{x_{n+1}}{x_n}, \quad n \in \mathbb{N}, \tag{5.8}$$

well defined if $x_n \neq 0$ for $n \in \mathbb{N}$. We have two main cases to analyze.

Case 1. Nondegenerate case ($z_1 \neq z_2$) In this case we have

$$q_n = \frac{x_{n+1}}{x_n} = \frac{Az_1^{n+1} + Bz_2^{n+1}}{Az_1^n + Bz_2^n}. \tag{5.9}$$

The condition $x_n \neq 0$ gives $(az_2 - b)z_1^n \neq (az_1 - b)z_2^n$. The limit satisfies

1° If $|z_1| < |z_2|$, then $\lim_{n\to\infty} q_n = z_2$;
2° If $|z_1| > |z_2|$, then $\lim_{n\to\infty} q_n = z_1$;
3° If $|z_1| = |z_2|$, then $\lim_{n\to\infty}$ is indeterminate.

While 1° and 2° have been investigated in the literature (see, e.g., [138]), item 3° presents special interest. The orbit produced can be periodic if z_1 and z_2 are roots of unity (Figure 5.1a), or dense in a circle (Figure 5.1b).

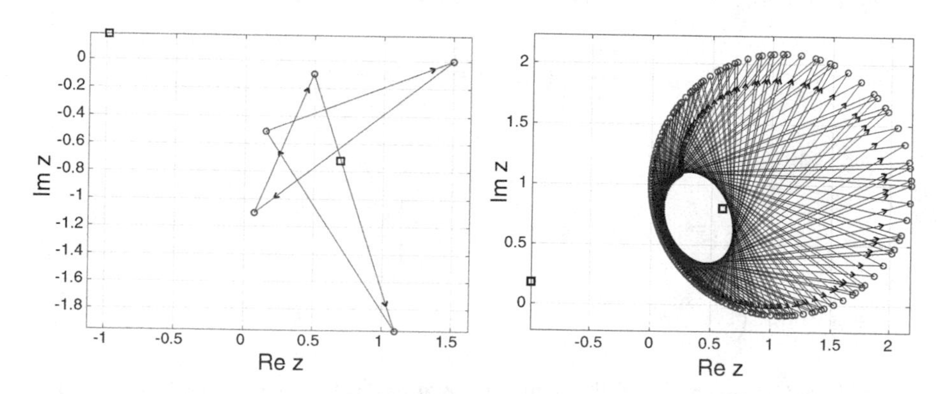

Fig. 5.1 Terms $q_0, \ldots, q_N$ (circles) obtained from (5.5), for $a = 2 + \frac{2}{3}i$, $b = 3 + i$ (stars) and (**a**) $z_1 = e^{2\pi i \frac{\sqrt{2}}{3}}$, $z_2 = e^{2\pi i \frac{\sqrt{2}}{3} + \frac{2}{5}}$, $N = 200$; (**b**) $z_1 = e^{2\pi i \frac{\sqrt{2}}{3}}$, $z_2 = e^{2\pi i \frac{\sqrt{5}}{15}}$, $N = 2000$. Arrows indicate the direction of the orbit

Indeed, if $|z_1| = |z_2|$, denoting by $z = z_2/z_1 = e^{i\theta}$, one can write

$$q_n = \frac{x_{n+1}}{x_n} = z_1 \frac{A + Be^{i(n+1)x}}{A + Be^{inx}}. \tag{5.10}$$

If $\theta \in \mathbb{Q}$, then the sequence $(q_n)_{n\geq 0}$ is periodic, while for $\theta \in \mathbb{R} \setminus \mathbb{Q}$ the orbit of $(q_n)_{n\geq 0}$ is dense in a circle. Note that z_1 and z_2 must not be roots of unity.

Case 2. Degenerate case ($z_1 = z_2 = z$) The condition $x_n \neq 0$ reduces here to $C \neq -nD$ for all $n \in \mathbb{N}$, and the limit is

$$\lim_{n\to\infty} q_n = \lim_{n\to\infty} \frac{x_{n+1}}{x_n} = \lim_{n\to\infty} \frac{(C + D(n+1))z^{n+1}}{(C + Dn)z^n} = z. \tag{5.11}$$

5.1.3 Particular Horadam Orbits

Here we explore some particular orbits produced by distinct generators [32].

5.1.3.1 Orbits Produced by Conjugate Generators (if $a, b \in \mathbb{R}$)

Theorem 5.1 *Let $(x_n)_{n\geq 0}$ be a Horadam sequence of general term (5.6). If $z_1 = \overline{z}_2$, then the sequence $(x_n)_{n\geq 0}$ is a subset of the real line whenever $a, b \in \mathbb{R}$. In this case we obtain the classical Horadam sequence for real numbers.*

Proof Denoting $z = z_2 = \overline{z}_1$, the general sequence term is

$$x_n = \frac{1}{z - \overline{z}} \left[(az - b)\overline{z}^n + (b - a\overline{z})z^n \right],$$

which is equivalent to

$$\begin{aligned} x_n &= \frac{1}{z - \overline{z}} \left[az\overline{z} \left(\overline{z}^{n-1} - z^{n-1} \right) + b(z^n - \overline{z}^n) \right] \\ &= \left[-az\overline{z} \left(\overline{z}^{n-2} + z\overline{z}^{n-3} + \cdots + \overline{z}z^{n-3} + z^{n-2} \right) + b(z^{n-1} + \cdots + \overline{z}^{n-1}) \right]. \end{aligned}$$

As a, b and $z\overline{z}$ are real and the sums are symmetric in z and $\overline{z}$ this indicates that $x_n = \overline{x_n}$, therefore x_n is real. This ends the proof. □

5.1.3.2 Concentric Orbits Produced by Opposite Entries

Theorem 5.2 *Let k be an even natural number, z_1, z_2 opposite primitive kth roots satisfying $z_2 = -z_1$ and a, b arbitrary complex numbers. The orbit of the sequence $(x_n)_{n\geq 0}$ defined in (5.1) is formed from two concentric regular k/2-gons, whose nodes represent a bipartite graph.*

Proof Considering $z = z_2 = -z_1$, from (5.4) we obtain

$$x_n = \frac{1}{2z}\left[(az - b)\,(-z)^n + (b + az)\,z^n \right],$$

therefore the even and odd terms of the sequence are given by

$$x_{2n} = az^{2n}, \quad x_{2n+1} = bz^{2n}, \quad n \in \mathbb{N}.$$

Hence, the even and odd terms of the sequence represent the vertices of the polygons generated by z^2 (the root $z = e^{2\pi i p/k}$ generates a $k/\gcd(2p, k)$-gon). These are on the circles of radii $|a|$ and $|b|$, respectively, as shown in Figure 5.2. □

5.1.3.3 Conjugate Orbits Produced by Conjugate Parameters

Theorem 5.3 *Let $(x_n)_{n\geq 0}$ be the sequence defined in (5.4) for generators $z_1 \neq z_2$ and initial conditions a and b. The sequence $(x_n^*)_{n\geq 0}$ generated by the conjugate*

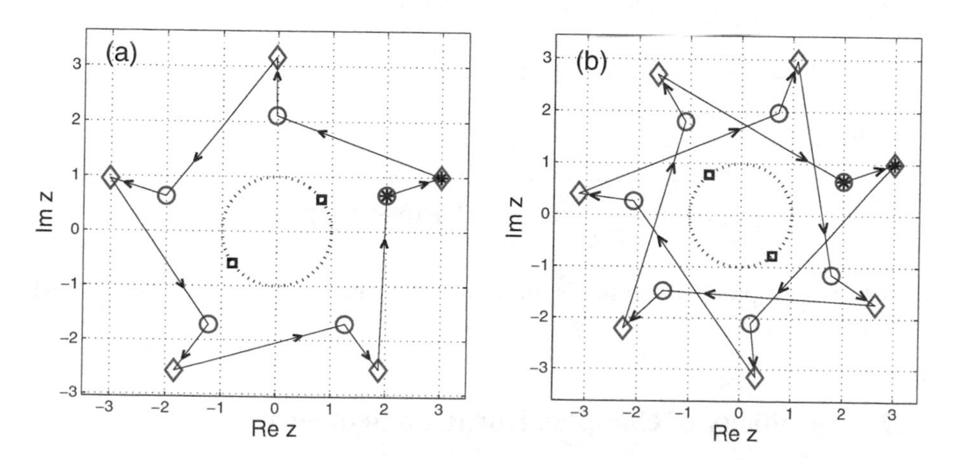

Fig. 5.2 First $N = 100$ orbit terms of sequence $(x_n)_{n\geq 0}$ obtained from (5.4), computed for pairs of opposite roots **(a)** $k = 10$, $z_1 = e^{2\pi i \frac{1}{10}}$, $z_1 = e^{2\pi i \frac{6}{10}}$; **(b)** $k = 14$, $z_1 = e^{2\pi i \frac{5}{14}}$, $z_1 = e^{2\pi i \frac{12}{14}}$. Arrows indicate the direction of the orbit from one term to the next. The dotted line is the unit circle. The initial conditions $x_0 = a = 2 + 2/3i$ and $x_1 = b = 3 + i$ shown as stars

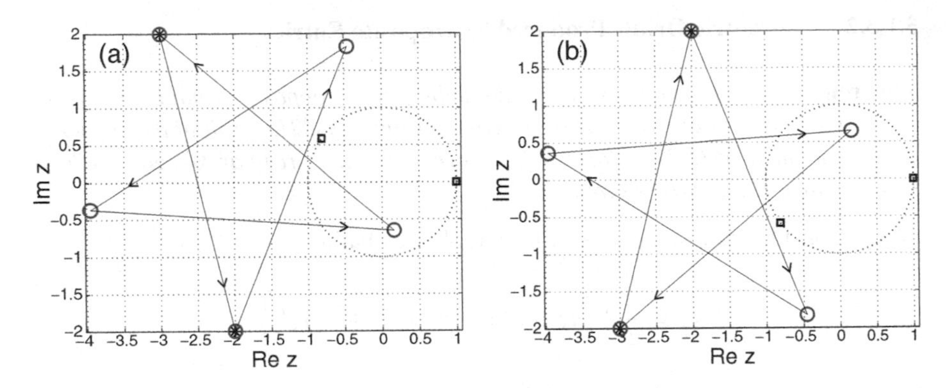

Fig. 5.3 Orbit of sequences (**a**) $(x_n)_{n\geq 0}$ defined by (5.4) and (**b**) $(X_n)_{n\geq 0}$ defined by (5.13), computed for $z_1 = e^{2\pi i \frac{2}{5}}$, $z_2 = 1$ and initial conditions $a = -3 + 2i$, $b = -2 - 2i$. Arrows indicate orbit's direction, while the dotted line is the unit circle

generators $\overline{z_1}, \overline{z_2}$ *and initial conditions* $\overline{a}, \overline{b}$ *satisfies*

$$X_n = \overline{x}_n, \quad n \in \mathbb{N}. \tag{5.12}$$

Proof From (5.4) for $(x_n)_{n\geq 0}$ we obtain

$$X_n = \frac{1}{\overline{z_2} - \overline{z_1}} \left[(\overline{a}\,\overline{z_2} - \overline{b})\overline{z_1}^n + (\overline{b} - \overline{a}\,\overline{z_1})\overline{z_2}^n \right]. \tag{5.13}$$

Using basic properties of the complex conjugate (see [6, 168])

$$\overline{uv} = \overline{u}\,\overline{v}; \quad \overline{u+v} = \overline{u} + \overline{v}; \quad \overline{uv^{-1}} = \overline{u}\,\overline{v}^{-1}\,(u \in \mathbb{C}, v \in \mathbb{C}^*),$$

one obtains

$$X_n = \overline{\frac{1}{z_2 - z_1}\left[(az_2 - b)z_1^n + (b - az_1)z_2^n \right]} = \overline{x_n},$$

which ends the proof. The result is illustrated in Figure 5.3. □

5.2 Periodicity of Complex Horadam Sequences

In an early article [79], Horadam remarked two periodic instances of the sequence $(x_n(a, b; \pm 1, 1))_{n\geq 0}$. Here we present sufficient conditions for the periodicity of the sequence $(x_n)_{n\geq 0}$ are established when the roots z_1, z_2 of (5.2) are distinct or identical. The results were published in [27].

5.2.1 Geometric Progressions of Complex Argument

The behavior of the sequence $(z^n)_{n\geq 0}$ for an arbitrary value of $z \in \mathbb{C}$ is essential for the study of the periodicity of Horadam sequences. This is related to Weyl's criterion (see, e.g., [165], [65, Chapter 2], or [72]). We state it for convenience. Full details of the proof are given in [27]. Denote the floor and fractional part of x by $\lfloor x \rfloor = \max\{m \in \mathbb{Z} : m \leq x\}$ and $\{x\} = x - \lfloor x \rfloor$, respectively.

Lemma 5.1 *The set $M = \{\{n\theta\} : dn \in \mathbb{N}\}$ is dense in $[0, 1]$ for every $\theta \in \mathbb{R}\backslash\mathbb{Q}$.*

The behavior of sequence $(z^n)_{n\geq 0}$ for arbitrary $z \in \mathbb{C}$ is described below.

Lemma 5.2 *Let $z = re^{2\pi i\theta} \in \mathbb{C}$ be a complex number $(r > 0)$. Then $(z^n)_{n\geq 0}$ is*

(i) a regular k-gon if $r = 1$, and $\theta = j/k \in \mathbb{Q}$ with $gcd(j, k) = 1$;
(ii) a dense subset of the unit circle for $r = 1$ and $\theta \in \mathbb{R} \setminus \mathbb{Q}$;
(iii) an inward spiral for $r < 1$;
(iv) an outward spiral for $r > 1$.

Proof

(i) For $r = 1$ the terms of the sequence $(z^n)_{n\geq 0}$ are located on the unit disk. When $\theta = j/k \in \mathbb{Q}$ is an irreducible fraction, z is a primitive kth root of unity. As $z^k = 1$, the sequence $(z^n)_{n\geq 0} = \{1, z, \ldots, z^{k-1}, \ldots\}$ is periodic and describes a closed finite orbit.
[The result is illustrated in Figure 5.4 for (a) $k = 5$ and (b) $k = 8$.]

(ii) As the argument of z^n is $2\pi nx$, the principal arguments of the terms in the sequence $(z^n)_{n\geq 0}$ form the set $\{2\pi\{nx\}\}$ which, from Lemma 5.1, is dense in the interval $[0, 2\pi]$.
Thus, the orbit of $(z^n)_{n\geq 0}$ is a dense subset of the unit circle.

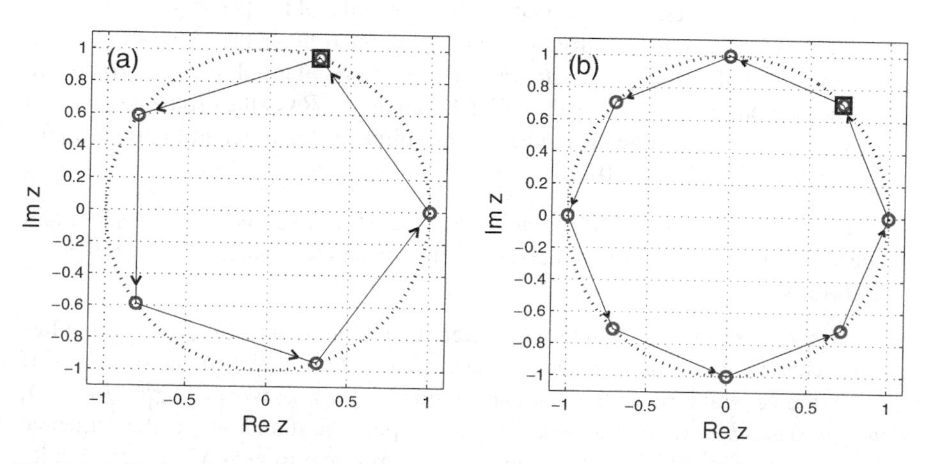

Fig. 5.4 Orbit of $(z^n)_{n\geq 0}$ obtained for $r = 1$ and (**a**) $x = 1/5$; (**b**) $x = 1/8$. Arrows indicate orbit's direction, dotted line the unit circle, while $z = r\exp(2\pi ix)$ is shown as a square

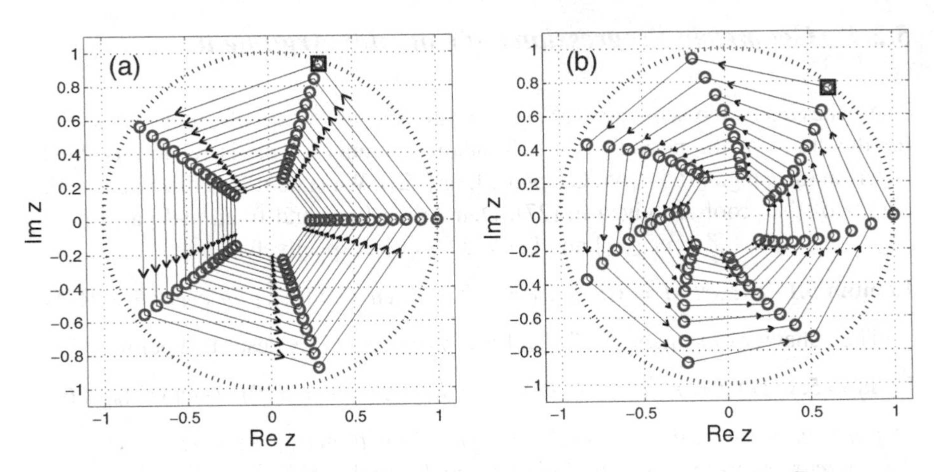

Fig. 5.5 First 71 terms of $(z^n)_{n\geq 0}$ obtained for $r = 0.98$ and **(a)** $x = 1/5$; **(b)** $x = \sqrt{2}/10$. Orbit's direction shown by arrows, unit circle by a dotted line, $z = r\exp(2\pi i x)$ by a square

(iii) For $r < 1$ we have $\lim_{n\to\infty} |z^n| = \lim_{n\to\infty} |z|^n = \lim_{n\to\infty} r^n = 0$, therefore the sequence $(z^n)_{n\geq 0}$ converges to the origin.
[The terms of $(z^n)_{n\geq 0}$ are all real for $\{x\} \in \{0, 1/2\}$. When $\theta = j/k \in \mathbb{Q}$ is an irreducible fraction ($k \geq 3$) the orbit of $(z^n)_{n\geq 0}$ forms an inward spiral whose points are aligned with the vertices of a regular polygon, as shown in Figure 5.5a for $\theta = 1/5$ and $r = 0.98$. When $\theta \in \mathbb{R}\backslash\mathbb{Q}$ the orbit also converges to the origin, this time the resulting points form a spiral, as in Figure 5.5b for $x = \sqrt{2}/10$ and $r = 0.98$.]

(iv) For $r > 1$ we have $\lim_{n\to\infty} |z^n| = \lim_{n\to\infty} |z|^n = \lim_{n\to\infty} r^n = \infty$, hence the sequence of absolute values $(|z^n|)_{n\geq 0}$ diverges to infinity.
[Notice that the terms of sequence $(z^n)_{n\geq 0}$ are all real for $\{x\} \in \{0, 1/2\}$. When fraction $\theta = j/k \in \mathbb{Q}$ is irreducible ($k \geq 3$), the orbit of the sequence $(z^n)_{n\geq 0}$ forms a set of rays aligned with the vertices of a regular polygon, as depicted in Figure 5.6a for $\theta = 1/10$ and $r = 1.01$. When $\theta \in \mathbb{R}\backslash\mathbb{Q}$ the orbit also diverges to infinity but this time the points form a spiral, as illustrated in Figure 5.6b for $\theta = \sqrt{2}/10$ and $r = 1.01$.] □

If one of the roots of the characteristic polynomial (5.2) is zero, the Horadam sequence is in fact a first-order recurrence relation, and the analysis of periodicity is straightforward.

Remark 5.1 For roots z_1, z_2 of the characteristic polynomial (5.2), $z_1 z_2 = 0$ implies $q = 0$, when the Horadam sequence reduces to a first-order recurrence sequence. If $z_2 = 0$, the general term of the Horadam sequence $(x_n)_{n\geq 0}$ is $x_n = bz_1^{n-1}$ ($n \geq 1$). For $z_1 \neq 0$ and $b \neq 0$, the sequence $(x_n)_{n\geq 0}$ is periodic if and only if the sequence $(z_1^n)_{n\geq 0}$ is periodic, which from Lemma 5.2 is equivalent to z_1 is a kth root of unity

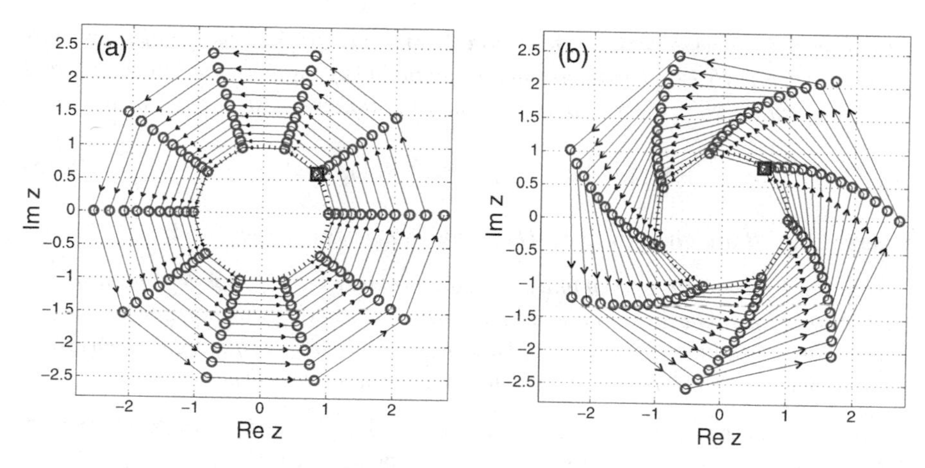

Fig. 5.6 First 101 terms of $(z^n)_{n\geq 0}$ obtained for $r = 1.01$ and (**a**) $x = 1/10$; (**b**) $x = \sqrt{2}/10$. Orbit's direction shown by arrows, unit circle by a dotted line, $z = r\exp(2\pi i x)$ by a square

for some $k \subset \mathbb{N}$. When $b = 0$ or z_1 also vanish, the recurrence relation is satisfied by $x_n = 0$ $(n \geq 2)$, with the only nonzero terms being potentially x_0, x_1.

5.2.2 Nondegenerate Case

The conditions for periodicity of the recurrence sequence (5.1) are examined when the roots z_1, z_2 of the characteristic polynomial (5.2) are distinct.

Theorem 5.4 (sufficient condition for periodicity) *Let $z_1 \neq z_2$ be distinct kth roots of unity $(k \geq 2)$, and let the polynomial $P(x)$ be*

$$P(x) = (x - z_1)(x - z_2), \quad x \in \mathbb{C}. \tag{5.14}$$

The recurrence sequence $(x_n)_{n\geq 0}$ generated by the characteristic polynomial (5.14), and the arbitrary initial values $x_0 = a$, $x_1 = b$, is periodic.

Proof As $z_1^k = z_2^k = 1$, the sequence $(x_n)_{n\geq 0}$ is periodic. The period is a divisor of k, generally $\text{lcm}(\text{ord}(z_1), \text{ord}(z_2))$ (where $\text{ord}(z)$ is the order of z).

From (5.4), the following degenerate (i.e., not both generators contribute to x_n) periodic cases are possible. When $B = 0$ one obtains $b = az_1$, therefore $x_n = az_1^n$ $(n \geq 0)$, while $A = 0$ gives $x_n = az_2^n$ $(n \geq 0)$. When $A = B = 0$ we have $b = az_1 = az_2$, therefore $a = b = 0$ and $x_n = 0$ $(n \geq 0)$. □

By Lemma 5.2, periodic sequences can only occur when z_1 and z_2 are roots of unity, as otherwise the orbits of $(z_1^n)_{n\geq 0}$ and $(z_2^n)_{n\geq 0}$ have infinitely many distinct terms. A necessary condition is presented next.

Theorem 5.5 (necessary periodicity condition) *Let $z_1 \neq z_2$ be the distinct roots of the quadratic (5.14). The recurrence sequence $(x_n)_{n\geq 0}$ generated by z_1, z_2, and the initial values $x_0 = a$, $x_1 = b$, is periodic only if there exists $k \in \mathbb{N}$ such that*

$$A(z_1^k - 1)z_1 = 0, \quad B(z_2^k - 1)z_2 = 0, \tag{5.15}$$

where A and B are given by (5.5). The following cases emerge.

(i) z_1 and z_2 are kth roots of unity (for some integer $k \geq 2$) (nondegenerate);
(ii) z_1 or z_2 is a kth root of unity and the other is zero (regular polygon);
(iii) z_1 or z_2 is a kth root of unity and $b = az_1$ or $b = az_2$, resp. (regular polygon);
(iv) z_1 and z_2 are arbitrary, and $a = b = 0$ (degenerate orbit).

Proof Let us assume that the sequence is periodic, and let $k \in \mathbb{N}$ be the period. Under this assumption the periodicity can be expressed trivially as

$$x_n = x_{n+k}, \quad \forall n \in \mathbb{N}. \tag{5.16}$$

As $z_1 \neq z_2$ relations (5.4) and (5.16) give

$$x_n = Az_1^n + Bz_2^n = Az_1^{n+k} + Bz_2^{n+k} = x_{n+k}, \quad \forall n \in \mathbb{N},$$

which can further be written as

$$A(z_1^k - 1)z_1^n + B(z_2^k - 1)z_2^n = 0, \quad \forall n \in \mathbb{N}. \tag{5.17}$$

The case when $0 = z_1 z_2$ (discussed in Remark 5.1) implies (5.15).

Assuming that $z_1, z_2 \neq 0$ and that there exist nonzero numbers α and β such that, $\forall n \in \mathbb{N}$, $\alpha z_1^n + \beta z_2^n = 0$, one can write (5.17) for $n = 1, 2$ to obtain

$$-\frac{\alpha}{\beta} = \frac{z_2^2}{z_1^2} = \frac{z_2}{z_1},$$

equivalent to $z_1 = z_2$, a contradiction. The periodicity of $(x_n)_{n\geq 0}$ requires

$$A(z_1^k - 1) = 0, \quad B(z_2^k - 1) = 0,$$

which implies (5.15). This ends the proof.

The only nondegenerate solution of (5.17) (i.e., both generators have a nonzero contribution) requires $z_1^k = z_2^k = 1$, which is seen in Case (i). If either A or B is zero (but not both), (5.4) shows that the sequence is determined by only one of the generators which leads to Cases (ii) and (iii). Finally, if both A and B are zero we obtain a degenerate orbit (Case (iv)). □

Theorems 5.4 and 5.5 highlight the importance of both the generators and initial values to the periodicity of the orbits of Horadam sequences; we see this again in the degenerate roots instance.

5.2.3 Degenerate Case

Here the conditions for periodicity are examined in the case when the characteristic solutions z_1, z_2 are equal.

Theorem 5.6 (sufficient condition for periodicity) *Let $k \geq 2$ be an integer, z a kth root of unity, and let the polynomial $P(x)$ be defined by*

$$P(x) = (x - z)^2, \quad x \in \mathbb{C}. \tag{5.18}$$

The sequence $(x_n)_{n\geq 0}$ having the characteristic polynomial (5.18), and arbitrary initial values $x_0 = a$, $x_1 = b$, is periodic when $b = az$, and divergent otherwise.

Proof The closed form (5.7) shows that $(x_n)_{n\geq 0}$ diverges (and is clearly not periodic) whenever $B \neq 0$. For $B = 0$ one obtains $b = az$, and in turn a periodic orbit with general term $x_n = az^n$ $(n \geq 0)$, while $A = 0$ gives $x_n = bnz^{n-1}$ $(n \geq 1)$ for which the sequence $(x_n)_{n\geq 0}$ represents a divergent spiral. When $A = B = 0$ we have $a = 0 = b$ and the trivial sequence $x_n = 0$ $(n \geq 0)$. □

Proposition 5.1 *When generated by a repeated kth root of unity, the terms of the divergent subsequence $(x_{Nk+j})_{n\geq 0}$ are collinear for each $j \in \{0, \ldots, k-1\}$.*

Proof For a fixed $j \in \{0, \ldots, k-1\}$, the general term of $(x_{Nk+j})_{n\geq 0}$ (5.7) is

$$x_{Nk+j} = [A + B(Nk + j)]z^{Nk+j} = [A + B(Nk + j)]z^j, \quad \forall N \in \mathbb{N},$$

therefore

$$x_{Nk+j} - x_j = NkBz^j,$$

whose argument is independent of N, hence the terms of the subsequence $(x_{Nk+j})_{n\geq 0}$ are collinear for every $j \in \{0, \ldots, k-1\}$, as shown in Figure 5.7, where sequence terms are calculated from (5.7). □

For each $j \in \{0, \ldots, k-1\}$, there are $k/\gcd(j, k)$ rays of aligned sequence terms, as seen in Figure 5.7 for (a) $k = 6$, $j = 1$ and (b) $k = 6$, $j = 2$.

Theorem 5.7 (necessary condition for periodicity) *The recurrence sequence $(x_n)_{n\geq 0}$ generated by the characteristic polynomial (5.18), and arbitrary initial values $x_0 = a$, $x_1 = b$, is periodic only if one of the following conditions is true*

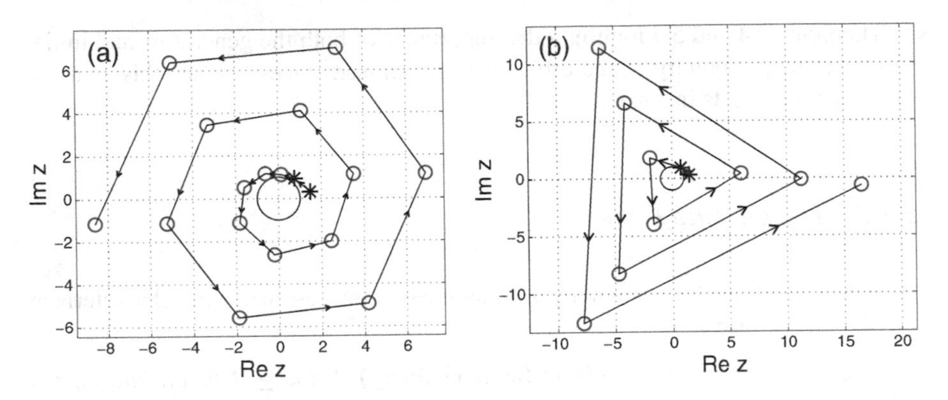

Fig. 5.7 First N terms of the sequence $(x_n)_{n\geq 0}$ for the initial values $a = 1.5\exp(2\pi i/30)$, $b = 1.2\exp(2\pi i/7)$ (shown as stars) for (**a**) $N = 18$, $z = e^{2\pi i \frac{1}{6}}$; (**b**) $N = 11$, $z = e^{2\pi i \frac{2}{6}}$. Arrows indicate the direction of the sequence trajectory

$$\begin{aligned} &(1)\ z = 0, \\ &(2)\ z^k - 1 = 0, \quad B = 0, \\ &(3)\ z^k - 1 \neq 0, \quad A = B = 0. \end{aligned} \tag{5.19}$$

The following orbit types emerge.

(i) $z = 0$ *(degenerate orbit);*
(ii) z *is a kth root of unity (for some integer* $k \geq 2$*) and* $b = az$ *(regular polygon);*
(iii) z *is arbitrary and* $a = b = 0$ *(degenerate orbit).*

Proof For $z_1 = z_2 = z$, the general term formula (5.7) and periodicity property given by (5.16), together give the condition

$$x_n = Az^n + Bnz^n = Az^{n+k} + B(n+k)z^{n+k} = x_{n+k}, \quad \forall n \in \mathbb{N}.$$

This can be written as

$$\left[A(z^k - 1) + Bkz^k + Bn(z^k - 1)\right] z^n = 0, \quad \forall n \in \mathbb{N}.$$

The case $z = 0$ discussed in Remark 5.1 implies (5.19). When $z \neq 0$ one has

$$A(z^k - 1) + Bkz^k + Bn(z^k - 1) = 0, \quad \forall n \in \mathbb{N}. \tag{5.20}$$

As this holds of all n, the linear polynomial in n is null, whence

$$A(z^k - 1) + Bkz^k = 0,$$
$$B(z^k - 1) = 0.$$

It follows readily that if $z^k - 1 = 0$ we have $B = 0$, while $z^k - 1 \neq 0$ implies $A = B = 0$. The only nondegenerate case occurs, therefore, when $b = az$, with z a kth root of unity ($k \geq 2$). This ends the proof. □

5.3 The Geometry of Periodic Horadam Orbits

As discussed in [27], nontrivial periodic orbits of $(x_n)_{n\geq 0}$ are obtained when the two roots z_1 and z_2 of the quadratic (5.2) are distinct roots of unity, i.e., $z_1 = e^{2\pi i p_1/k_1}$ and $z_2 = e^{2\pi i p_2/k_2}$, where p_1, p_2, k_1, k_2 are positive integers. A classification of orbits based on the properties of the numbers p_1, p_2, k_1, k_2 was given by Bagdasar et al. in [33].

We also discuss the geometric patterns that can be recovered from the (finite) *orbit* of the periodic sequence $(x_n)_{n\geq 0}$ defined as

$$W = \{z \in \mathbb{C} : \exists n \in \mathbb{N} \text{ such that } x_n = z\}.$$

To avoid trivial cases, we consider distinct and nonzero generators z_1, z_2.

The classification of periodic Horadam orbits is based on the orders of z_1 and z_2. Taking z_1 and z_2 as primitive roots (i.e., $\gcd(p_1, k_1) = \gcd(p_2, k_2) = 1$), while the period of the sequence $(x_n)_{n\geq 0}$ can be shown to be given by

$$k_1 k_2/\gcd(k_1, k_2). \tag{5.21}$$

This orbit may exhibit a rich variety of patterns, as illustrated below.

First, for $k_2 = 1$ the orbit is the regular star polygon $\{k_1/p_1\}$, while for $k_2 = 2$ the orbit is a bipartite graph. In general, for integers k_1, k_2 satisfying $d = \gcd(k_1, k_2)$, one obtains a multipartite graph whose vertices can be divided into either k_2 regular k_1/d-gons, or into k_1 regular k_2/d-gons.

In the end of this section we show that periodic Horadam orbits are located within an annulus, whose inner and outer boundaries we derive.

The close relationship with the Fibonacci numbers/patterns suggests that Horadam sequences can be used to generate structures with optimal properties, by appropriate choices of the initial conditions a, b and generators z_1, z_2. A first application was the design of a pseudo-random number generator with geometric structure produced by aperiodic Horadam sequences, proposed in [26].

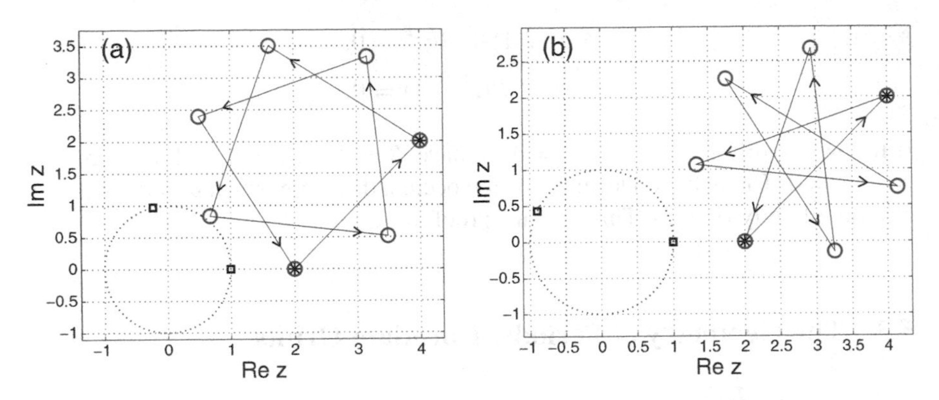

Fig. 5.8 The first $N = 100$ orbit terms of sequence $(x_n)_{n\geq 0}$ obtained from (5.22), for $z_2 = 1$ and (**a**) $k = 7$, $z_1 = e^{2\pi i \frac{2}{7}}$; (**b**) $k = 7$, $z_1 = e^{2\pi i \frac{3}{7}}$; when $a = 2$ and $b = 4 + 2i$. Arrows indicate the orbit's direction visiting x_0, x_1 (star), $x_2, \ldots, x_N$ (circle). The dotted line represents the unit circle

5.3.1 Regular Star Polygons

Theorem 5.8 *If $z_1 = e^{2\pi i p/k}$ is a primitive kth root ($k \geq 2$) and $z_2 = 1$, then the orbit of $(x_n)_{n\geq 0}$ is the regular star polygon $\{k/p\}$, as shown in Figure 5.8.*

Proof In this case, the general formula (5.4) gives

$$x_n = Az_1^n + B, \tag{5.22}$$

where A and B are constant expressions of z_1. The sequence $(x_n)_{n\geq 0}$ can be obtained from $(z_1^n)_{n\geq 0}$ by rotating with $\arg(A)$, scaling with $|A|$ and translating with B, therefore the shape of the orbit is a regular k-gon, similar to the orbit of $(z_1^n)_{n\geq 0}$ ([27, Lemma 2.2]).

There is a close relation between these orbits and regular star polygons. As from x_n to x_{n+1} there is a jump of p adjacent vertices in the k-gon, the directed orbit of $(x_n)_{n\geq 0}$ for $\gcd(p, k) = 1$ generates the regular star polygon $\{k/p\}$. In Figure 5.8 are depicted the star polygons (a) $\{7/2\}$ and (b) $\{7/3\}$. □

5.3.2 Bipartite Graphs

Here we show bipartite periodic orbits obtained for $z_1 = e^{2\pi i p/k}$ and $z_2 = -1$.

Theorem 5.9 **(k odd)** *Let $k \geq 2$ be an odd number, z_1 a primitive kth root and $z_2 = -1$. The orbit of sequence $(x_n)_{n\geq 0}$ is a 2k-gon, whose nodes can be divided into two regular k-gons representing a bipartite graph, as illustrated in Figure 5.9a.*

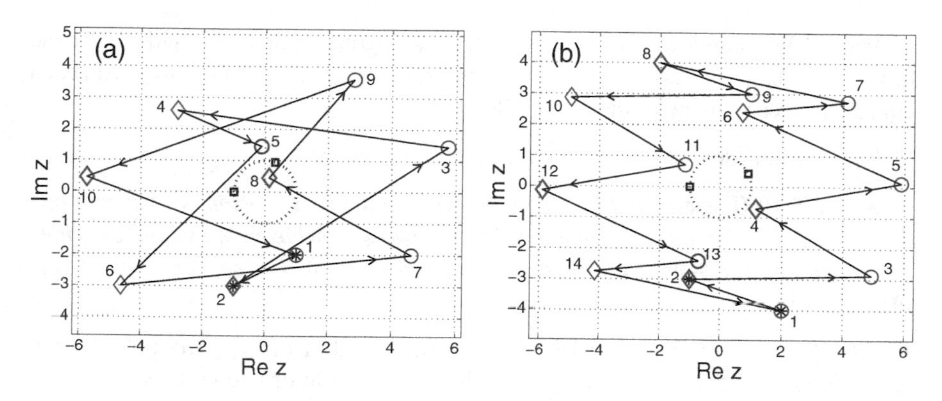

Fig. 5.9 First $N = 100$ orbit terms of $(x_n)_{n\geq 0}$ obtained from (5.23), for $z_2 = -1$ and **(a)** $k = 5$ (odd), $z_1 = e^{2\pi i \frac{1}{5}}$ and $a = 2 - 4i$, $b = -1 - 3i$; **(b)** $k = 14$ (even), $z_1 = e^{2\pi i \frac{1}{14}}$ and $a = 1 - 2i$, $b = -1 - 3i$. Arrows indicate the direction of the orbit visiting x_0, x_1 (star), $x_2, \ldots, x_N$. The sets W_0 and W_1 defined in (5.24) and (5.26) are represented by circles and diamonds respectively. The dotted line represents the unit circle

Proof In this case, the general formula (5.4) gives

$$x_n = Az_1^n + (-1)^n B, \tag{5.23}$$

where A, B are constants. As z_1 is a primitive kth root and $\gcd(k, 2) = 1$, the period of sequence $(x_n)_{n\geq 0}$ computed by (5.21) is $2k$ and the orbit is

$$W = \{x_0, x_1, \ldots, x_{2k-1}\}.$$

The set W can be partitioned into the disjoint ordered sets x_0 and x_1

$$\begin{aligned} W_0 &= \{A + B, Az_1^2 + B, \ldots, Az_1^{k-1} + B, Az_1^{k+1} + B, \ldots, Az_1^{2k-2} + B\}, \\ W_1 &= \{Az_1 - B, Az_1^3 - B, \ldots, Az_1^{k-2} - B, Az_1^k - B, \ldots, Az_1^{2k-1} - B\}, \end{aligned} \tag{5.24}$$

containing the terms of even and odd index, respectively. As $z_1^k = 1$, we have

$$\begin{aligned} W_0 &= \{A + B, Az_1^2 + B, \ldots, Az_1^{k-1} + B, Az_1^1 + B, \ldots, Az_1^{k-2} + B\}, \\ W_1 &= \{Az_1 - B, Az_1^3 - B, \ldots, Az_1^{k-2} - B, Az_1^0 - B, \ldots, Az_1^{k-1} - B\}. \end{aligned} \tag{5.25}$$

Finally, the sets W_0 and W_1 contain the points

$$\{Az_1^j + B, j = 0, \ldots, k-1\}, \quad \{Az_1^j - B, j = 0, \ldots, k-1\},$$

so W_0 and W_1 represent vertices of two regular k-gons, visited alternatively by the terms of $(x_n)_{n\geq 0}$. Moreover, by (5.25), x_n jumps over $2p$ adjacent vertices with each consecutive visit in any of the k-gons W_0 or W_1. □

By a similar idea, one can prove the result for k even.

Theorem 5.10 (k even) *Let $k \geq 2$ be an even number, z_1 a primitive kth root and $z_2 = -1$. The orbit of the sequence $(x_n)_{n\geq 0}$ is a k-gon, whose nodes can be divided into two regular $k/2$-gons representing a bipartite graph. The property is shown in Figure 5.9b.*

Proof When k is even, z_2 is a kth root and the period of $(x_n)_{n\geq 0}$ from (5.21) is k. As in Theorem 5.5, W can be partitioned into the disjoint ordered sets W_0 and W_1

$$W_0 = \{A + B, Az_1^2 + B, \ldots, Az_1^{k-2} + B\},$$
$$W_1 = \{Az_1 - B, Az_1^3 - B, \ldots, Az_1^{k-1} - B\}, \tag{5.26}$$

containing the terms of even and odd index, respectively. The number $Z = z_1^2$ is a primitive $k/2$th root, therefore the set $\{1, Z, \ldots, Z^{k/2-1}\}$ represents the vertices of a $k/2$-gon. The sets W_0 and W_1 can be written using Z as

$$W_0 = \{AZ^j + B, j = 0, \ldots, k/2 - 1\},$$
$$W_1 = \{Az_1Z^j - B, j = 0, \ldots, k/2 - 1\},$$

so W_0 and W_1 are the vertices of two regular $k/2$-gons visited alternatively by the terms of $(x_n)_{n\geq 0}$. Also, (5.26) shows that x_n jumps over p adjacent vertices each time it visits W_0 or W_1 (as $z_1^{n+2} = z_1^n e^{2\pi i \frac{2p}{k}} = z_1^n e^{2\pi i \frac{p}{k/2}}$). □

5.3.3 *Multipartite Graphs*

Here we present some multipartite periodic orbits of $(x_n)_{n\geq 0}$ obtained for generic distinct roots of unity $z_1 = e^{2\pi i p_1/k_1}$ and $z_2 = e^{2\pi i p_2/k_2}$.

Theorem 5.11 *Let $k_1, k_2, d \geq 2$ be natural numbers such that $\gcd(k_1, k_2) = d$ and z_1, z_2 be k_1th and k_2th primitive roots, respectively. The orbit of the sequence $(x_n)_{n\geq 0}$ is then a k_1k_2/d-gon, whose nodes can be divided into k_1 regular k_2/d-gons representing a multipartite graph. By duality, the nodes of the orbit can also be divided into k_2 regular k_1/d-gons. The property is shown in Figures 5.10 and 5.11.*

Proof As $\gcd(k_1, k_2) = d$ the period of $(x_n)_{n\geq 0}$ from (5.21) is k_1k_2/d, and the orbit consists of the first k_1k_2/d terms

$$W = \{x_0, x_1, \ldots, x_{k_1k_2/d-1}\}. \tag{5.27}$$

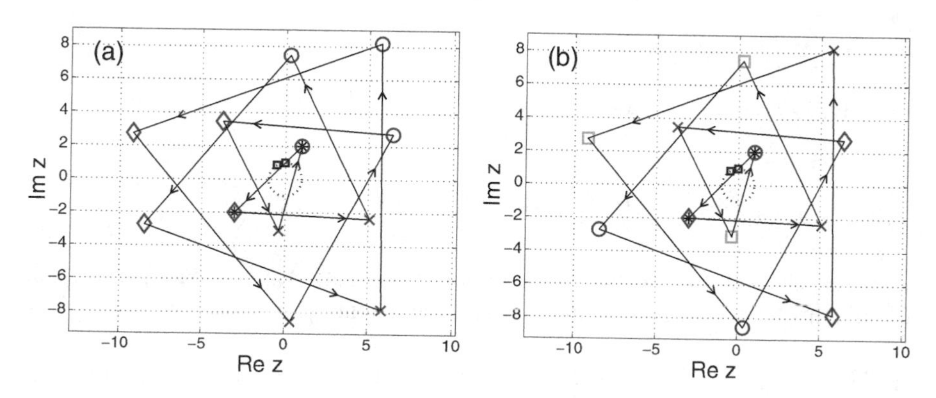

Fig. 5.10 First $N = 100$ terms of $(x_n)_{n\geq 0}$ given by (5.4), for $k_1 = 3$, $k_2 = 4$. We compute $x_0, \ldots, x_N$ for $z_1 = e^{2\pi i \frac{1}{4}}$, $z_2 = e^{2\pi i \frac{1}{3}}$ and initial conditions $a = 1+2i$, $b = -3-2i$. The orbits are partitioned into (**a**) three squares; (**b**) four equilateral triangles; arrows indicate the direction of the orbit from one term to the next. The dotted line represents the unit circle

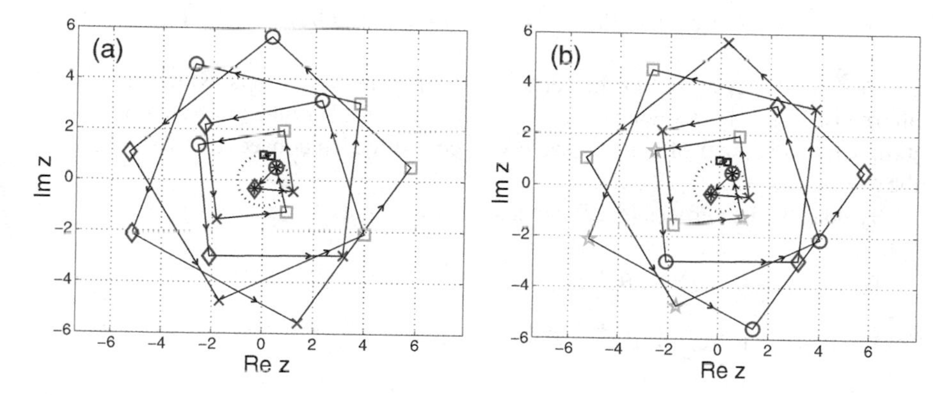

Fig. 5.11 First $N = 100$ terms of $(x_n)_{n\geq 0}$ given by (5.4), for $k_1 = 4$, $k_2 = 5$. We compute $x_0, \ldots, x_N$ for $z_1 = e^{2\pi i \frac{1}{5}}$, $z_2 = e^{2\pi i \frac{1}{4}}$ and initial conditions $a = (1+i)/2$, $b = -(1+i)/3$. Orbits are partitioned into (**a**) four regular pentagons; (**b**) five squares; arrows indicate orbit's direction from one term to the next. The dotted line is the unit circle

Clearly, W can be partitioned into k_2 disjoint sets $W_0, \ldots, W_{k_2-1}$ defined by

$$W_0 = \{A + B, Az_1^{k_2} + Bz_2^{k_2}, \ldots, Az_1^{(k_1/d-1)k_2} + Bz_2^{(k_1/d-1)k_2}\},$$

$$W_1 = \{Az_1 + Bz_2, Az_1^{k_2+1} + Bz_2^{k_2+1}, \ldots, Az_1^{(k_1/d-1)k_2+1} + Bz_2^{(k_1/d-1)k_2+1}\},$$

$$\ldots$$

$$W_{k_2-1} = \{Az_1^{k_2-1} + Bz_2^{k_2-1}, Az_1^{2k_2-1} + Bz_2^{2k_2-1}, \ldots, Az_1^{(k_1/d)k_2-1} + Bz_2^{(k_1/d)k_2-1}\}.$$

As $z_2^{k_2} = 1$ the above sets can be simplified to

$$W_0 = \{A + B, Az_1^{k_2} + B, \ldots, Az_1^{(k_1/d-1)k_2} + B\}, \tag{5.28}$$
$$W_1 = \{Az_1 + Bz_2, Az_1^{k_2+1} + Bz_2, \ldots, Az_1^{(k_1/d-1)k_2+1} + Bz_2\},$$
$$\ldots$$
$$W_{k_2-1} = \{Az_1^{k_2-1} + Bz_2^{k_2-1}, Az_1^{2k_2-1} + Bz_2^{k_2-1}, \ldots, Az_1^{(k_1/d)k_2-1} + Bz_2^{k_2-1}\}.$$

It can be checked that no two terms in any of the sets $W_0, \ldots, W_{k_2-1}$ are equal. Should this be the case, we would have the coefficients j, r and q such that

$$Az_1^{rk_2+j} + Bz_2^j = Az_1^{qk_2+j} + Bz_2^j, \quad j \in \{0, \ldots, k_2-1\}, \quad r, q \in \{0, \ldots, (k_1/d)-1\}.$$

For $A \neq 0$ this is only possible when $z_1^{(r-q)k_2} = 1$, which only happens for $r = q$, as $\gcd(k_2, k_1) = d$ and z_1 is a primitive k_1th root. This shows that

$$W_j = \{Az_1^r + Bz_2^j, r = 0, \ldots, (k_1/d) - 1\}, \quad j = 0, \ldots, k_2 - 1, \tag{5.29}$$

so $W_0, \ldots, W_{k_2-1}$ represent the vertices of k_2 regular k_1/d-gons, which are visited alternatively by the sequence $(x_n)_{n\geq 0}$. Moreover, (5.28) and (5.29) show that x_n jumps over a number of $k_2 p_1$ adjacent vertices with each consecutive visit in any of the k_1/d-gons $W_0, \ldots, W_{k_2-1}$.

By duality, the vertices (5.27) can also be divided as

$$W_1' = \{A + B, Az_1^{k_1} + Bz_2^{k_1}, \ldots, Az_1^{(k_2/d-1)k_1} + Bz_2^{(k_2/d-1)k_1}\}, \tag{5.30}$$
$$W_2' = \{Az_1 + Bz_2, Az_1^{k_1+1} + Bz_2^{k_1+1}, \ldots, Az_1^{(k_2/d-1)k_1+1} + Bz_2^{(k_2/d-1)k_1+1}\},$$
$$\ldots$$
$$W_{k_1-1}' = \{Az_1^{k_1-1} + Bz_2^{k_1-1}, Az_1^{2k_1-1} + Bz_2^{2k_1-1}, \ldots, Az_1^{(k_2/d)k_1-1} + Bz_2^{(k_2/d)k_1-1}\},$$

and using the same argument as above $W_0', \ldots, W_{k_1-1}'$ represent the vertices of k_1 regular k_2/d-gons, which are visited alternatively by the sequence $(x_n)_{n\geq 0}$. Moreover, (5.30) shows that x_n jumps over a number of $k_1 p_2$ adjacent vertices with each consecutive visit in any of the k_2/d-gons $W_0', \ldots, W_{k_1-1}'$. □

Some direct consequences of Theorem 5.11 present particular interest.

Corollary 5.1 *Let $k_1, k_2 \geq 2$ be such that $\gcd(k_1, k_2) = 1$ and z_1, z_2 be k_1th and k_2th primitive roots, respectively. The orbit of the sequence $(x_n)_{n\geq 0}$ is then a k_1k_2-gon, whose nodes can be divided into k_1 regular k_2-gons representing a multipartite graph. By duality, the orbit can also be divided into k_2 regular k_1-gons. The property is illustrated in Figure 5.11, where the twenty points of the orbit can be decomposed in either four regular pentagons, or five squares.*

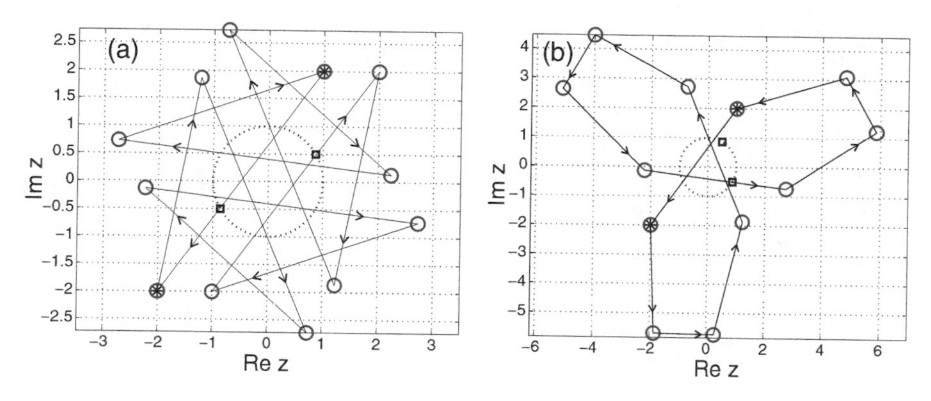

Fig. 5.12 First $N = 100$ orbit terms of sequence $(x_n)_{n\geq 0}$ obtained from (5.4), for initial conditions are $a = 1 + 2i$ and $b = -2 - 2i$. We compute $x_0, \ldots, x_N$ for **(a)** $k_1 = k_2 = 12$, $z_1 = e^{2\pi i \frac{1}{12}}$, $z_2 = e^{2\pi i \frac{7}{12}}$; **(b)** $k_1 = 2k_2 = 12$, $z_1 = e^{2\pi i \frac{1}{6}}$, $z_2 = e^{2\pi i \frac{11}{12}}$. Arrows indicate the direction of the orbit from one term to the next. The dotted line is the unit circle

Corollary 5.2 *If $k_2|k_1$ the orbit is a k_1-gon whose nodes can be divided into k_2 regular k_1/k_2-gons. The asymmetry in Figure 5.12 illustrates this, as the only regular polygons in the periodic orbit of $(x_n)_{n\geq 0}$ are 1- and 2-gons.*

5.3.4 Geometric Bounds of Periodic Orbits

Periodic Horadam orbit are confined to an annulus, whose boundaries are easily obtained. This feature is displayed in the orbit diagrams thereafter.

Theorem 5.12 *When the Horadam sequence $(x_n)_{n\geq 0}$ is periodic, the orbit is subject to the following geometric boundaries*

(i) For $z_1 \neq z_2$ the orbit is located inside the annulus

$$\{z \in \mathbb{C} : \big||A| - |B|\big| \leq |x_n| \leq |A| + |B|\}, \quad \forall n \in \mathbb{N},$$

where the constants A and B are given by (5.5);

(ii) For $z_1 = z_2 = z$ the orbit is either a subset (regular k-gon) of the circle

$$S(0, |a|) = \{z \in \mathbb{C} : |z| = |a|\},$$

for $a \neq 0$, or else the zero set $\{0\}$ for $a = 0$.

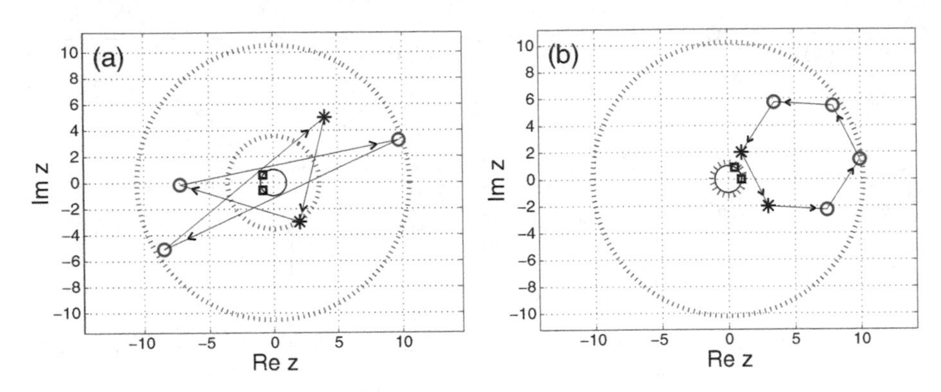

Fig. 5.13 Orbit of a periodic Horadam sequence $(x_n)_{n\geq 0}$ computed for (**a**) $z_1 = e^{2\pi i \frac{2}{5}}$, $z_2 = e^{2\pi i \frac{3}{5}}$, $a = 4 + 5i$, $b = 2 - 3i$; (**b**) $z_1 = e^{2\pi i \frac{1}{6}}$, $z_2 = e^{2\pi i \frac{5}{6}}$, $a = 1 + 2i$, $b = 3 - 2i$. Also plotted are the initial values a, b (stars), the generators z_1, z_2 (squares), the unit circle $S(0, 1)$ (solid line) and boundaries of the annulus $U(0, ||A| - |B||, |A| + |B|)$ (dashed lines)

Proof The complex numbers u and v satisfy the triangle inequalities

$$||u| - |v|| \leq |u + v| \leq |u| + |v|. \tag{5.31}$$

(i) For $z_1 \neq z_2$ the general term of (5.4), combined with (5.31), gives

$$||Az_1^n| - |Bz_2^n|| \leq |x_n| = |Az_1^n + Bz_2^n| \leq |Az_1^n| + |Bz_2^n|,$$

which as $|z_1| = |z_2| = 1$ is equivalent to

$$||A| - |B|| \leq |x_n| = |Az_1^n + Bz_2^n| \leq |A| + |B|.$$

[We illustrate this result in Figure 5.13 for sequences obtained when z_1 and z_2 are (a) 5th roots and (b) 6th roots of unity. One should note that in the proof we have only used the fact that z_1 and z_2 lie on the unit circle.]

(ii) When $z_1 = z_2 = z$ then, from Theorem 5.7, the sequence can be periodic only when $B = 0$ and z is a kth primitive root, when $x_n = az^n$ $(n \geq 0)$ and the orbit is a regular k-gon with $|x_n| = |a|$; when a is also zero we have $x_n = 0$.

□

5.3.5 "Masked" Periodicity

As shown in [27, Theorem 3.1], a sufficient condition for periodicity is that generators are distinct roots of unity. Here we discuss the possible periods yielded by self-repeating Horadam sequences obtained from a fixed generator pair, and for different initial conditions a, b, based on [29].

We denote by $\text{lcm}(k_1, k_2)$ the least common multiple of k_1 and k_2. In general, two primitive roots of orders k_1, k_2 are expected to produce an orbit of length $\text{lcm}(k_1, k_2)$. However, this is not always true, as the following theorem shows.

Theorem 5.13 *Consider the distinct primitive roots of unity* $z_1 = e^{2\pi i p_1/k_1}$ *and* $z_2 = e^{2\pi i p_2/k_2}$, *where* p_1, p_2, k_1, k_2 *are positive integers and let the polynomial* $P(x)$ *be defined by*

$$P(x) = (x - z_1)(x - z_2), \quad x \in \mathbb{C}. \tag{5.32}$$

The recurrence sequence $(x_n)_{n\geq 0}$ *generated by the characteristic polynomial (5.32) and the arbitrary initial values* $x_0 = a$, $x_1 = b$ *is periodic. The following periods are possible*

(a) $AB \neq 0$*: the period is* $\text{lcm}(k_1, k_2)$*;*
(b) $A = 0$, $B \neq 0$*: the period is* k_2*;*
(c) $A \neq 0$, $B = 0$*: the period is* k_1*;*
(d) $A = B = 0$*: the period is* 1 *(constant sequence).*

Proof

(a) For $AB \neq 0$, the general term of the sequence given by formula (5.1) is a nondegenerated linear combination of z_1^n and z_2^n, hence the period is $\text{lcm}(k_1, k_2)$.
(b) Here Az_1^n does not feature in (5.1), hence the orbit is the regular star polygon $\{k_2/p_2\}$.
(c) Similar to point (b), here the orbit is the regular star polygon $\{k_1/p_1\}$.
(d) When $A = B = 0$, one has $b = az_1 = az_2$, hence $a(z_2 - z_1) = 0$. Since z_1 and z_2 are distinct, this implies $a = b = 0$ and the sequence is constant.

□

The only possible masked periods for $k_1 = k_2 = k$ have lengths k and 1.

The following example illustrates the result of Theorem 5.13.

Example 5.1 Consider the primitive roots of unity $z_1 = e^{2\pi i 2/3}$, $z_2 = e^{2\pi i 1/4}$. By Theorem 5.13, the periodic Horadam orbit produced by formula (5.1) may have the lengths 1, 3, 4, or 12. The distinct periodic orbits produced in this example are sketched in Figure 5.14a–d.

Theorem 5.13 also has the following consequence.

Proposition 5.2 *Let* $k, k_1, k_2 \geq 2$ *be natural numbers such that* $\text{lcm}(k_1, k_2) = k$. *There is a periodic Horadam recurrent sequence* $(x_n)_{n\geq 0}$ *of length* k, *which for different initial conditions* $x_0 = a$, $x_1 = b$ *masks orbits of period* k_1 *and* k_2, *respectively.*

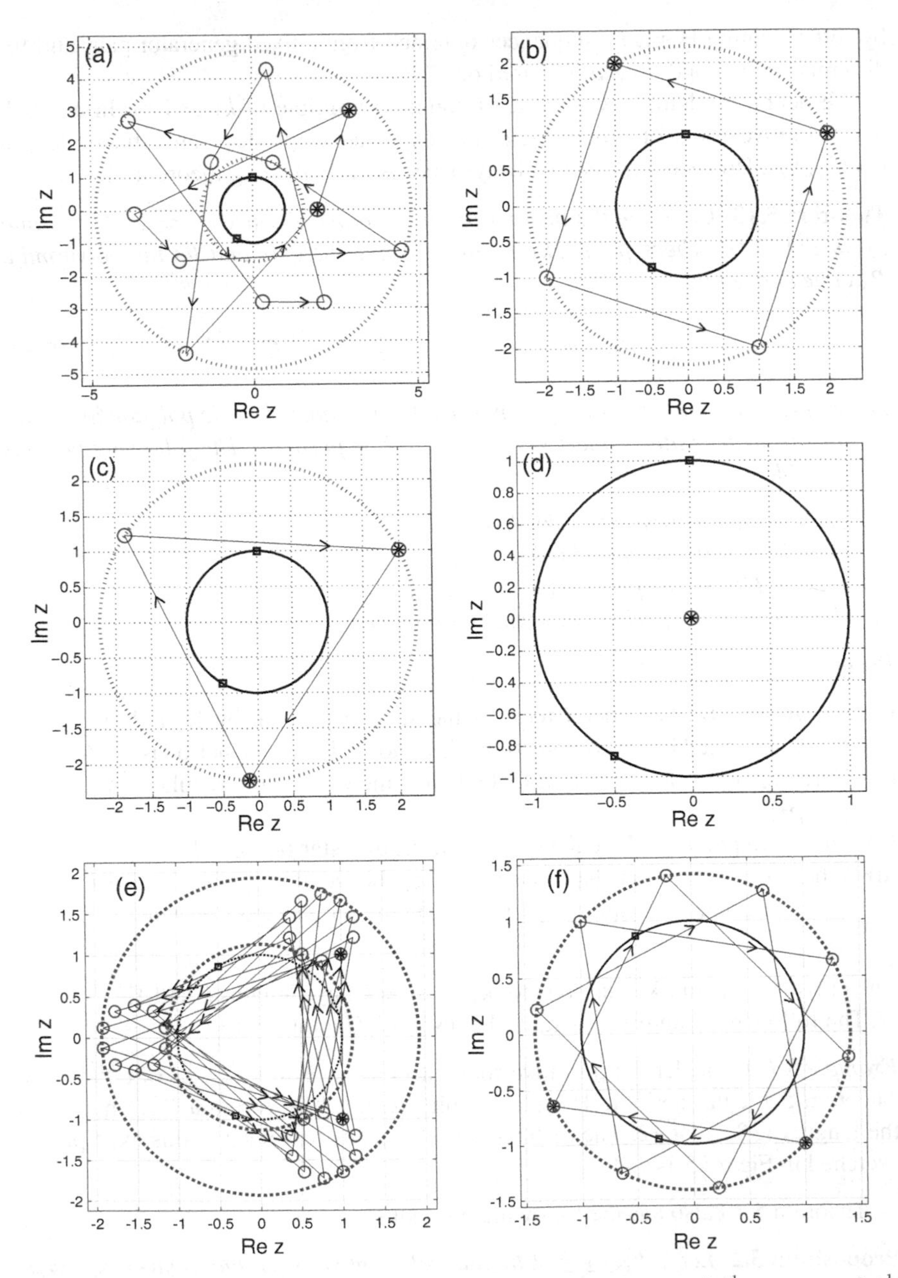

Fig. 5.14 Terms $x_0, \dots, x_N$ (circles) from (5.1) for the generators $z_1 = e^{2\pi i \frac{1}{3}}$ and $z_2 = e^{2\pi i \frac{1}{4}}$ (squares) and seeds (stars) **(a)** $a = 2$, $b = 3 + 3i$ $(AB \neq 0)$; **(b)** $a = 2$, $b = 2e^{2\pi i \frac{1}{4}}$ $(A = 0)$; **(c)** $a = 2$, $b = 2e^{2\pi i \frac{2}{3}}$ $(B = 0)$; **(d)** $a = b = 0$ $(A = B = 0)$; then, $z_1 = e^{2\pi i \frac{1}{3}}$, $z_2 = e^{2\pi i \frac{7}{10}}$ and **(e)** $a = 1 - i$, $b = 1 + i$ $(AB \neq 0)$; **(f)** $a = 1 - i$, $b = az_2$ $(A = 0)$; finally, $z_1 = e^{2\pi i \frac{1}{2}}$, $z_2 = e^{2\pi i \frac{11}{15}}$ and **(g)** $a = 1 - i$, $b = 1 + i$ $(AB \neq 0)$; **(h)** $a = 1 - i$, $b = az_2$ $(A = 0)$. Arrows indicate the direction of the orbit. Boundaries of annulus $U(0, ||A| - |B||, |A| + |B|)$ (solid line) with A, B from ((5.5)) and the unit circle (dotted line) are also plotted

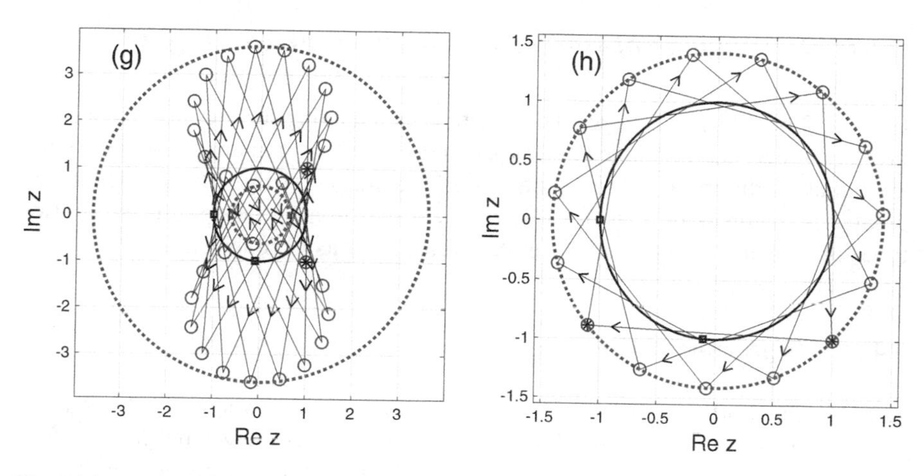

Fig. 5.14 (continued)

Proof One may select the distinct roots $z_1 = e^{2\pi i p_1/k_1}$ and $z_2 = e^{2\pi i p_2/k_2}$, where p_1, p_2, k_1, k_2 are integers. Following Theorem 5.13, the sequence generated by (5.1) will have length k for $AB \neq 0$, with A, B given by (5.5). □

The property in Proposition 5.2 is illustrated by the following example.

Example 5.2 For $k = 30$, the primitive generator pair $z_1 = e^{2\pi i \frac{1}{3}}$ and $z_2 = e^{2\pi i \frac{7}{10}}$ can mask periods of length 3 and 10, while the pair $z_1 = e^{2\pi i \frac{1}{2}}$ and $z_2 = e^{2\pi i \frac{11}{15}}$ can mask periods of length 2 or 15, respectively. These are sketched in Figure 5.14e–h.

5.4 The Enumeration of Periodic Horadam Patterns

Here we present the number of distinct Horadam sequences which (for arbitrary initial conditions) have a fixed period, giving enumeration formulae in both degenerate and nondegenerate cases. These were given in [28] and involve Euler's totient function φ and the number of divisors function ω.

Recall that for equal roots $z_1 = z_2 = z$ of (5.2), the general term of the Horadam sequence $(x_n)_{n \geq 0}$ is given by $x_n = (A + Bz)z^n$ (5.7). This can only be periodic when $b = az$ and z is a root of unity.

For distinct roots $z_1 \neq z_2$ of (5.2), the general term of Horadam's sequence $(x_n)_{n \geq 0}$ is $x_n = Az_1^n + Bz_2^n$ (5.4), where A and B are given by (5.5). When $AB = 0$, at least one of the generators z_1 and z_2 does not appear explicitly in x_n, and the sequence orbit degenerates to either a regular polygon centered in 0, or to a point. For $AB \neq 0$, the sequence is periodic when the distinct generators z_1 and z_2 are roots of unity.

5.4.1 The Number of Horadam Patterns of Fixed Length

Let $k \geq 2$ be a positive integer. The function enumerating the Horadam sequences $(x_n)_{n\geq 0}$ of period k is denoted by $H_P(k)$, and depends on the generators z_1, z_2 and the initial conditions a, b. For k_1 and k_2 positive integers, let $\gcd(k_1, k_2)$ denote the greatest common divisor, and $\operatorname{lcm}(k_1, k_2)$ the least common multiple. There are two types of (degenerate and nondegenerate) periodic orbits to consider.

5.4.1.1 Degenerate Orbits

A degenerated orbit is a regular polygon centered in 0 or a singleton. As detailed in [27], this happens when the Horadam sequences $(x_n)_{n\geq 0}$ given by (5.7) or (5.4) depend on only one of the generators and this is a root of unity, say $z_1 = e^{2\pi i p_1/k_1}$. The number of distinct sequences having period k is given by

$$H_P(k) = |\{(p_1, k_1) : \gcd(p_1, k_1) = 1, \ k_1 = k\}| = \varphi(k),$$

where φ is Euler's well-known totient function [125].

If no generator appears explicitly in (5.7) or (5.4) (i.e., $z_1 \neq z_2$, $A = 0$, $B = 0$ or $z_1 = z_2 = z$, $a = 0$, $b = 0$), then the periodic sequence is constant and there are no generator configurations leading to periodicity $k \geq 2$.

5.4.1.2 Nondegenerate Orbits

Nondegenerated periodic orbits are obtained when the generators are distinct roots of unity $z_1 = e^{2\pi i p_1/k_1}$ and $z_2 = e^{2\pi i p_2/k_2}$, and the arbitrary initial conditions a, b are such that $AB \neq 0$ for A, B defined in (5.5).

As established in [27], the period of the Horadam sequence delivered by a generator pair z_1, z_2 is $\operatorname{lcm}(\operatorname{ord}(z_1), \operatorname{ord}(z_2))$, where $\operatorname{ord}(z)$ is the order of z. Representing the pair (z_1, z_2) by the quadruple (p_1, k_1, p_2, k_2), we want to select those producing a sequence having period k. To ensure that the enumeration formula generates all the distinct periodic sequences, we shall assume without loss of generality that z_1, z_2 are primitive roots of unity and $k_1 \leq k_2$.

The distinct sequences of period k are enumerated from the quadruples

$$H_P(k) = |\{(p_1, k_1, p_2, k_2) : \gcd(p_1, k_1) = \gcd(p_2, k_2) = 1, \operatorname{lcm}(k_1, k_2) = k, \ k_1 \leq k_2\}|. \tag{5.33}$$

Some formulae for this expression are identified, based on the properties of pairs (k_1, k_2) satisfying $\operatorname{lcm}(k_1, k_2)$, and corresponding generators $z_1 = e^{2\pi i p_1/k_1}$ and $z_2 = e^{2\pi i p_2/k_2}$.

5.4.2 A First Formula

To derive this formula we first generate the pairs (k_1, k_2) satisfying $\mathrm{lcm}(k_1, k_2) = k$ and then count the pairs (p_1, p_2) such that (p_1, k_1, p_2, k_2) satisfies (5.33).

We first count the quadruples (p_1, k_1, p_2, k_2) in (5.33) for which $k_1 = k_2$.

Lemma 5.3 *If* $k_1 = k_2$ *and* $lcm(k_1, k_2) = k$, *then* $k_1 = k_2 = k$.

The result is not difficult to prove and shows that the only pair (k_1, k_2) such that $k_1 = k_2$ is (k, k). The number of quadruples produced in this case is

$$H'_P(k) = |\{(p_1, p_2) : \gcd(p_1, k) = \gcd(p_2, k) = 1, p_1 < p_2\}| = \frac{1}{2}\varphi(k)(\varphi(k) - 1),$$

as the number of choices for each of p_1 and p_2 is $\varphi(k)$ and $p_1 < p_2$. We then count the quadruples (p_1, k_1, p_2, k_2) when $k_1 \neq k_2$ and $\mathrm{lcm}(k_1, k_2) = k$.

Lemma 5.4 *If* $lcm(k_1, k_2) = k$ *and* $k_1 \neq k_2$, *then we have*

$$H''_P(k) = |\{(p_1, k_1, p_2, k_2) : \gcd(p_1, k_1) = \gcd(p_2, k_2) = 1, \mathrm{lcm}(k_1, k_2) = k\}| = \varphi(k_1)\varphi(k_2),$$

quadruples (p_1, k_1, p_2, k_2) *satisfying* (5.33).

Proof As $k_1 \neq k_2$ the primitive roots z_1 and z_2 are distinct for all combinations p_1 and p_2. This means that any combination pairs satisfying the relation $\gcd(p_1, k_1) = \gcd(p_2, k_2) = 1$ may be considered. There are $\varphi(k_1)$ pairs (p_1, k_1) and $\varphi(k_2)$ pairs (p_2, k_2), which confirms the result. □

Theorem 5.14 *The number of distinct Horadam sequences of period* $k \geq 2$ *is*

$$H_P(k) = \sum_{[k_1,k_2]=k,\ k_1<k_2} \varphi(k_1)\varphi(k_2) + \frac{1}{2}\varphi(k)(\varphi(k) - 1). \tag{5.34}$$

To evaluate this formula one needs to generate all ordered pairs (k_1, k_2), whose lcm is k. Some special formulae are obtained for periods with particular prime decompositions.

The first few terms of the number sequence $H_P(k)$

$$1, 1, 3, 5, 10, 11, 21, 22, 33, 34, 55, 46, 78, 69, 92, 92, 136, 105, \ldots,$$

recover the OEIS sequence A102309, providing its first enumerative context.

The values of the function for prime numbers, powers of primes, or products of two prime numbers are given below.

Example 5.3 (prime numbers) When k is a prime, we have $\varphi(k) = k-1$. In this case we have two divisor pairs $(k_1, k_2) \in \{(1, k), (k, k)\}$, with multiplicities $\varphi(1)\varphi(k) = k - 1$ and $\varphi(k)(\varphi(k) - 1)/2 = (k - 1)(k - 2)/2$, giving

$$H_P(k) = k(k-1)/2. \tag{5.35}$$

When $k = 23$ there is a total of $23 \cdot 22/2 = 253$ distinct solutions, while for $k = 11$ there is a total of $11 \cdot 10/2 = 55$ distinct solutions. Explicitly, for $k = 5$ there are 10 solutions given by the fraction pairs

$$\left(\frac{p_1}{k_1}, \frac{p_2}{k_2}\right) \in \left\{\left(\frac{1}{1}, \frac{1}{5}\right), \left(\frac{1}{1}, \frac{2}{5}\right), \left(\frac{1}{1}, \frac{3}{5}\right), \left(\frac{1}{1}, \frac{4}{5}\right), \left(\frac{1}{5}, \frac{2}{5}\right),\right.$$
$$\left.\left(\frac{1}{5}, \frac{3}{5}\right), \left(\frac{1}{5}, \frac{4}{5}\right), \left(\frac{2}{5}, \frac{3}{5}\right), \left(\frac{2}{5}, \frac{4}{5}\right), \left(\frac{3}{5}, \frac{4}{5}\right)\right\}.$$

Example 5.4 (powers of a prime) If p is prime and $k = p^m$ with $m \geq 2$, then $\varphi(k) = p^m(1 - 1/p) = p^m - p^{m-1}$. The pairs of divisors (k_1, k_2) in the set $\{(1, k), (p, k), \ldots, (p^{m-1}, k), (k, k)\}$, have multiplicities $\varphi(p^j)\varphi(k)$, $j = 0, \ldots, m-1$, and $\varphi(k)(\varphi(k) - 1)/2 = (k - k/p)(k - k/p - 1)/2$. We obtain a telescopic sum in which the consecutive terms (up to the last two) cancel out

$$H_P(k) = \left(1 + (p-1) + \cdots + (p^{m-1} - p^{m-2}) + (p^m - p^{m-1} - 1)/2\right)\varphi(k)$$
$$= \frac{k^2 - k^2/p^2 - k + k/p}{2} = \frac{\varphi(k)[2k - \varphi(k) - 1]}{2}. \tag{5.36}$$

For example, when $k = 9 = 3^2$ one obtains $H_P(k) = \frac{6[18-6-1]}{2} = 33$ while for $k = 4$ one obtains $H_P(k) = \frac{2[8-2-1]}{2} = 5$ and the distinct solutions are

$$\left(\frac{p_1}{k_1}, \frac{p_2}{k_2}\right) \in \left\{\left(\frac{1}{1}, \frac{1}{4}\right), \left(\frac{1}{1}, \frac{3}{4}\right), \left(\frac{1}{2}, \frac{1}{4}\right), \left(\frac{1}{2}, \frac{3}{4}\right), \left(\frac{1}{4}, \frac{3}{4}\right)\right\}.$$

Example 5.5 (product of two primes) When $k = pq$ $(p < q)$ is the product of two prime numbers, $\varphi(k) = \varphi(p)\varphi(q)$, we have five divisor pairs

$$(k_1, k_2) \in \{(1, k), (p, q), (p, k), (q, k), (k, k)\},$$

with multiplicities

$$\varphi(1)\varphi(k),\ \varphi(p)\varphi(q),\ \varphi(p)\varphi(k),\ \varphi(q)\varphi(k),\ \varphi(k)(\varphi(k) - 1)/2.$$

This gives the following formula

$$H_P(k) = (p-1)(q-1)(pq + p + q)/2.$$

For example, when $k = 6 = 2 \cdot 3$ the solutions are

$$\left(\frac{p_1}{k_1}, \frac{p_2}{k_2}\right) \in \left\{\left(\frac{1}{1}, \frac{1}{6}\right), \left(\frac{1}{1}, \frac{5}{6}\right), \left(\frac{1}{2}, \frac{1}{3}\right), \left(\frac{1}{2}, \frac{2}{3}\right), \left(\frac{1}{2}, \frac{1}{6}\right), \left(\frac{1}{2}, \frac{5}{6}\right), \left(\frac{1}{3}, \frac{1}{6}\right), \left(\frac{1}{3}, \frac{5}{6}\right), \left(\frac{2}{3}, \frac{1}{6}\right), \left(\frac{2}{3}, \frac{5}{6}\right), \left(\frac{1}{6}, \frac{5}{6}\right)\right\},$$

for a total of 11 distinct solutions. Some of the orbits realized for $k = 6$ are plotted in Figure 5.15. One can notice the geometric variety of shapes produced even for small values of k, which range from regular polygons in Figure 5.15a to more complex orbits in Figure 5.15b, c, or d.

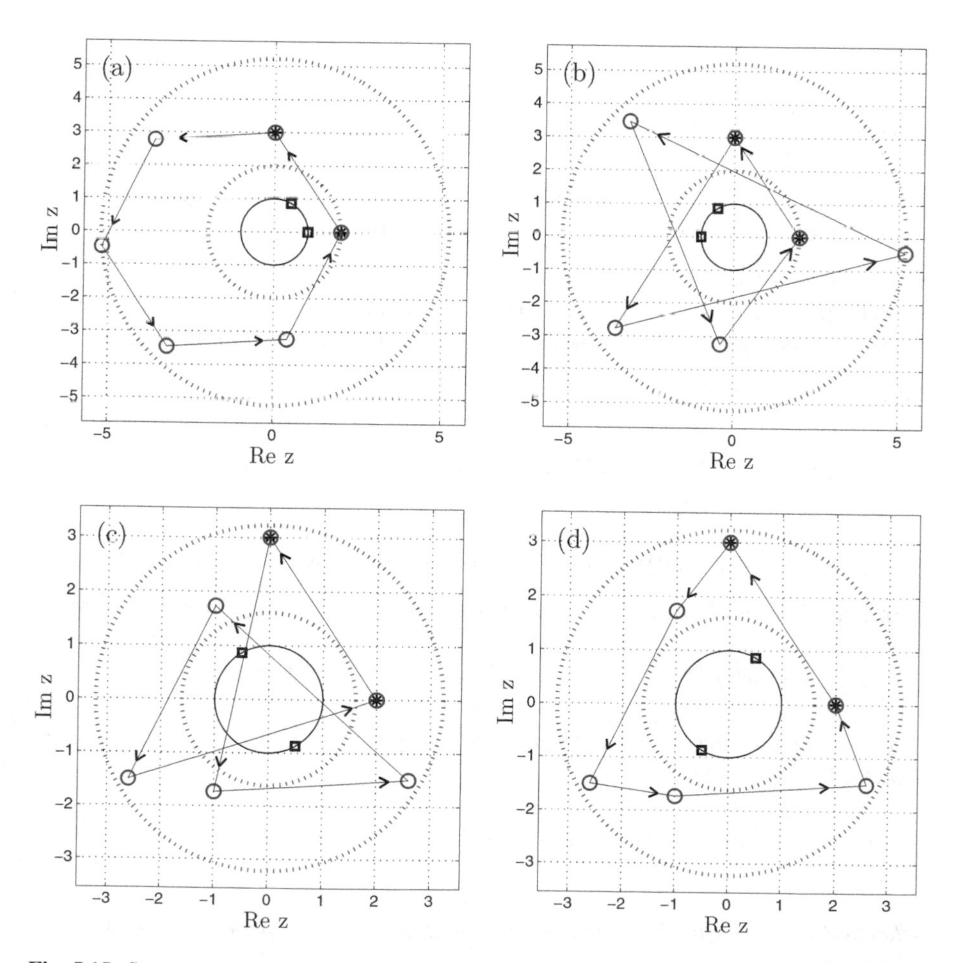

Fig. 5.15 Sequence terms $(x_n)_{n\geq 0}$ obtained from (5.4) for the pairs $(\frac{p_1}{k_1}, \frac{p_2}{k_2})$ **(a)** $(\frac{1}{1}, \frac{1}{6})$; **(b)** $(\frac{1}{2}, \frac{1}{3})$; **(c)** $(\frac{1}{3}, \frac{5}{6})$; **(d)** $(\frac{2}{3}, \frac{1}{6})$ when $a = 2$ and $b = 3i$ (stars). Arrows indicate the orbit's direction $x_0, x_1, \ldots, x_6 = x_0$ (circles). We also plot generators z_1, z_2 (squares), unit circle (solid line), and boundaries of $U(0, ||A| - |B||, |A| + |B|)$ (dotted line) with A, B from (5.5)

Example 5.6 (composite integers) For $k = 12$, we have the divisor pairs

$$(k_1, k_2) \in \{(1, 12), (2, 12), (3, 4), (3, 12), (4, 6), (4, 12), (6, 12), (12, 12)\},$$

with multiplicities $\varphi(p)\varphi(q)$ for each pair (p, q) in the list satisfying $p < q$, and finally, $\varphi(12)(\varphi(12) - 1)/2$ for the pair $(12, 12)$. This gives the formula

$$H_P(12) = 4 + 4 + 4 + 8 + 4 + 8 + 8 + 4 \cdot 3/2 = 46.$$

An equivalent (but more direct) formula for $H_P(k)$, which does not require the generation of all quadruples (p_1, k_2, p_2, k_2), is described below.

5.4.3 A Second Formula

By the formula $\varphi(\gcd(k_1, k_2)) \cdot \varphi(\text{lcm}(k_1, k_2)) = \varphi(k_1) \cdot \varphi(k_2)$, we can produce an algorithmic version of (5.34), consisting of the following steps:

- Choose a divisor d of k, such that $1 \leq d < k$;
- Estimate how many pairs k_1, k_2 satisfy $d = \gcd(k_1, k_2)$ and $k = \text{lcm}(k_1, k_2)$;
- Sum all the terms $\varphi(d)\varphi(k)$ over d, with the corresponding multiplicity.

Formula (5.34) becomes

$$H_P(k) = \left[\sum_{d|k,\, d<k} \varphi(d)\text{GL}(d, k) \right] \varphi(k) + \frac{1}{2} \varphi(k) \left(\varphi(k) - 1 \right), \tag{5.37}$$

where the arithmetic function $\text{GL}(d, k)$ is computed in the following lemma.

Lemma 5.5 *If d, k are natural numbers such that $d|k$, whose factorizations are*

$$d = p_1^{d_1} p_2^{d_2} \cdots p_n^{d_n}, \quad k = p_1^{m_1} p_2^{m_2} \cdots p_n^{m_n}, \quad (1 \leq d_i \leq m_i).$$

The number of pairs of natural numbers k_1, k_2 with $d = \gcd(k_1, k_2)$ and $k = \text{lcm}(k_1, k_2)$ is

$$GL(d, k) = |\{(k_1, k_2) : d = \gcd(k_1, k_2) \text{ and } k = \text{lcm}(k_1, k_2)\}| = 2^{\omega(k/d)-1}, \tag{5.38}$$

where $\omega(x)$ represents the number of distinct prime divisors for the integer x.

Proof Let the numbers k_1 and k_2 be written as

$$k_1 = p_1^{\alpha_1} p_2^{\alpha_2} \cdots p_n^{\alpha_n}, \quad k_2 = p_1^{\beta_1} p_2^{\beta_2} \cdots p_n^{\beta_n}.$$

When $d = \gcd(k_1, k_2)$ and $k = \operatorname{lcm}(k_1, k_2)$, for each index $i \in \{1, \dots, n\}$, we have

$$\min\{\alpha_i, \beta_i\} = d_i, \quad \max\{\alpha_i, \beta_i\} = m_i.$$

Number k_1 and k_2 are either distinct, or $d = k$. There are two possibilities.

If $d_i = m_i$, one has $\alpha_i = \beta_i = d_i = m_i$. Each $i \in I = \{i \in \{1, \dots, n\} : d_i < m_i\}$ generates two possible pairs $(\alpha_i, \beta_i) \in \{(d_i, m_i), (m_i, d_i)\}$, giving a total of $2^{|I|}$ distinct pairs of powers, hence we have $2^{|I|-1}$ pairs with $k_1 < k_2$. As the prime decomposition of k/d is

$$k/d = p_1^{m_1-d_1} p_2^{m_2-d_2} \cdots p_n^{m_n-d_n} = \prod_{i \in I} p_i^{m_i-d_i},$$

one obtains that $|I| = \omega(k/d)$. This ends the proof. □

Theorem 5.15 *Using formula (5.37), $H_P(k)$ can be written more compactly as*

$$H_P(k) = \left[\sum_{d|k,\, d<k} \varphi(d) 2^{\omega(k/d)} + \varphi(k) - 1 \right] \frac{\varphi(k)}{2}. \tag{5.39}$$

Example 5.7 (Example 5.6 using (5.39)) The divisors of 12 smaller than 12 are

$$1,\ 2 = 2^1,\ 3 = 3^1,\ 4 = 2^2,\ 6 = 2 \cdot 3.$$

Writing the terms in formula (5.39) explicitly one obtains

$$\left[\varphi(1)2^1 + \varphi(2)2^1 + \varphi(3)2^0 + \varphi(4)2^0 + \varphi(6)2^0 \right] \varphi(12) + \frac{\varphi(12)\left(\varphi(12) - 1\right)}{2} = 46.$$

Example 5.8 (square-free integers) When $k = p_1 p_2 \dots p_m$, $m \geq 2$, and $p_1, \cdots, p_m$ prime numbers, a compact formula for $H_P(k)$ can be given. Each divisor d of k can be written as $p_{i_1} p_{i_2} \cdots p_{i_j}$, where $1 \leq i_1 \leq \cdots \leq i_j \leq m$ for $j = 0, \dots, m$. The corresponding term in (5.39) can be written as

$$\varphi(d) 2^{\omega(k/d)} = \varphi(p_{i_1}) \varphi(p_{i_2}) \cdots \varphi(p_{i_j}) 2^{m-j}.$$

Summing over all possible divisors d of k one obtains the formula

$$H_P(k) = \left[\sum_{j=0}^{m-1} \left(\sum_{1 \leq i_1 \leq \cdots \leq i_j \leq m} \varphi(p_{i_1}) \cdots \varphi(p_{i_j}) \right) 2^{m-j} + \varphi(p_1) \cdots \varphi(p_m) - 1 \right] \frac{\varphi(k)}{2}$$

$$= \left[(\varphi(p_1) + 2) \cdots (\varphi(p_m) + 2) - 1 \right] \frac{\varphi(k)}{2}$$

$$= \left[(p_1 + 1) \cdots (p_m + 1) - 1 \right] \frac{(p_1 - 1) \cdots (p_m - 1)}{2}, \tag{5.40}$$

by the identities $\varphi(k) = \varphi(p_1) \cdots \varphi(p_m)$ and $\varphi(p) = p - 1$ for any prime p.

For example, when $k = 30 = 2 \cdot 3 \cdot 5$ the number of periodic orbits is

$$H_P(k) = \left[3 \cdot 4 \cdot 6 - 1 \right] \frac{1 \cdot 2 \cdot 4}{2} = 284.$$

Remark 5.2 An alternative formula for $H_P(k)$ is obtained using the generator pairs $z_1 = e^{2\pi i p_1/k}$ and $z_2 = e^{2\pi i p_2/k}$ with $1 \leq p_1 < p_2 \leq k$, when these are not primitive roots of unity. Clearly, $\text{ord}(z_1) = k/\gcd(p_1, k)$ and $\text{ord}(z_2) = k/\gcd(p_2, k)$. The sequence generated by z_1 and z_2 has period k if $\text{lcm}(\text{ord}(z_1), \text{ord}(z_2)) = k$. Using the well-known property $\text{lcm}(x, y)\gcd(x, y) = xy$ (for $x, y \in \mathbb{N}$) for the positive integers $\text{ord}(z_1)$ and $\text{ord}(z_2)$, one obtains the condition

$$k \gcd\left(\frac{k}{(p_1, k)}, \frac{k}{(p_2, k)} \right) = \frac{k}{\gcd(p_1, k)} \frac{k}{\gcd(p_2, k)}$$

$$\iff \gcd(p_1, k)\gcd(p_2, k)\gcd\left(\frac{k}{\gcd(p_1, k)}, \frac{k}{\gcd(p_2, k)} \right) = k.$$

Using $x\gcd(y, z) = \gcd(xy, xz)$ (for $x, y, z \in \mathbb{N}$), the above relations give

$$\gcd\Big(\gcd(p_2, k)k, \gcd(p_1, k)k \Big) = k \iff \gcd((p_1, k), \gcd(p_2, k)) = 1.$$

All periodic orbits can be generated from the pairs (p_1, p_2) satisfying

$$H_P(k) = \left| \{(p_1, p_2) : \gcd(\gcd(p_1, k), \gcd(p_2, k)) = 1, \ 1 \leq p_1 < p_2 \leq k\} \right|. \tag{5.41}$$

When written explicitly, this formula yields a result similar to (5.34).

5.4.4 Computational Complexity

To evaluate $H_P(k)$ by (5.34), we enumerate the ordered pairs of positive integers (k_1, k_2) such that $\text{lcm}(k_1, k_2) = k$, that is $\max\{\alpha_i, \beta_i\} = m_i$ for all $i \in \{1, \ldots, n\}$. As $0 \leq \alpha_i, \beta_i \leq m_i$, there are $(m_i + 1)^2$ pairs (α_i, β_i), of which m_i^2 satisfy $0 \leq \alpha_i, \beta_i \leq m_i - 1$. The number of pairs (α_i, β_i) satisfying the identity $\max\{\alpha_i, \beta_i\} =$

m_i is $(m_i + 1)^2 - m_i^2 = 2m_i + 1$. For $i \in \{1, \ldots, n\}$, the number of divisor pairs (k_1, k_2) is $(2m_1 + 1)(2m_2 + 1) \cdots (2m_n + 1)$. Apart from (k, k) each pair appeared twice, so the number of ordered pairs in (5.34) is

$$[(2m_1 + 1)(2m_2 + 1) \cdots (2m_n + 1) + 1]/2.$$

In formula (5.37) one must find all distinct divisors d of k, whose number is

$$(m_1 + 1)(m_2 + 1) \cdots (m_n + 1),$$

and multiply them by the appropriate weights $\mathrm{GL}(d, k)$ defined in (5.38).

This suggests that for numbers with many different prime divisors the second formula provides the value $H_P(k)$ in fewer steps.

5.4.5 Asymptotic Bounds

The first few terms of the sequence $H_P(k)$ are plotted in Figure 5.16a, along with some lower and upper boundaries given by the expressions

$$\frac{\varphi(k)k}{2} \leq H_P(k) \leq \frac{(k-1)k}{2},$$

which can be derived from formulae (5.39) and (5.41) as detailed below.

Formula $k(k-1)/2$ is the number of pairs (p_1, p_2) satisfying $1 \leq p_1 < p_2 \leq k$ in (5.41), so this is an upper bound for $H_P(k)$, which only holds when k is prime (5.35).

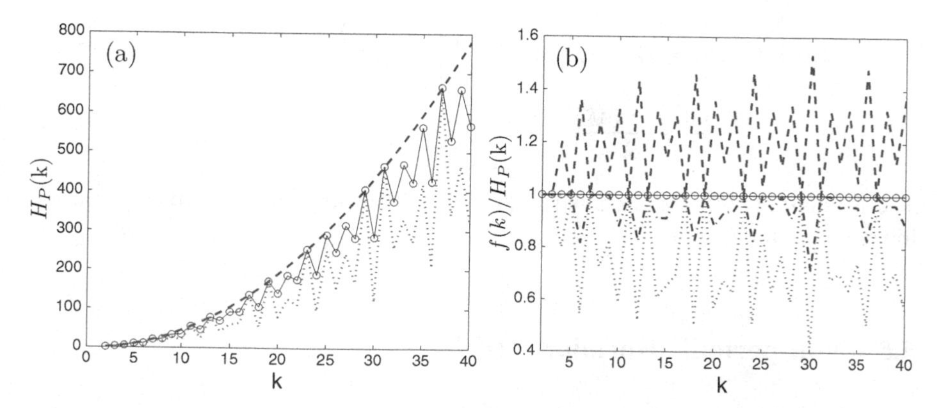

Fig. 5.16 First 40 terms of the sequences **(a)** $H_P(k)$ (circles), $(k-1)k/2$ (dashed) and $\frac{\varphi(k)k}{2}$ (dotted); **(b)** $f(k)/H_P(k)$, where $f(k)$ is $H_P(k)$ (circles), $(k-1)k/2$ (dashed), $\frac{\varphi(k)k}{2}$ (dotted) and $\frac{\varphi(k)[2k-\varphi(k)-1]}{2}$ (dash-dotted)

Whenever $H_P(k) = k(k-1)/2$, all the pairs (p_1, p_2) such that $1 \le p_1 < p_2 \le k$ have to satisfy the relation $\gcd(\gcd(p_1, k), \gcd(p_2, k)) = 1$ (as shown in (5.41)). If k has a proper divisor $1 < d < k$, then the pair $(p_1, p_2) = (d, k)$ has the property $\gcd(\gcd(d, k), \gcd(k, k)) = d$, which is a contradiction.

The lower bound can be obtained from (5.39) by writing

$$H_P(k) = \left[\sum_{d|k,\, 1 \le d < k} \varphi(d) \left(2^{\omega(k/d)} - 1 \right) + \sum_{d|k,\, d<k} \varphi(d) + \varphi(k) - 1 \right] \frac{\varphi(k)}{2}$$
$$\ge \left[1 + \sum_{d|k,\, 1<d<k} \varphi(d) \left(2^{\omega(k/d)} - 1 \right) \right] \frac{\varphi(k)}{2} + (k-1)\frac{\varphi(k)}{2} \ge \frac{\varphi(k)}{2} k.$$

We have used the relation $\sum_{d|k,\, 1<d<k} \varphi(d) = k$ and that $2^{\omega(k)} \ge 2$. From (5.35), the lower bound is attained when k is a prime number.

As illustrated in Figure 5.16b, a better lower bound for $H_P(k)$ seems to be given by formula $\frac{\varphi(k)[2k-\varphi(k)-1]}{2}$ (5.36), which attains equality when k is a prime power. Here we prove that this is a lower bound for $H_P(k)$ whenever $k = p_1 p_2 \cdots p_m$ is square-free, by using the inequality

$$(p_1+1)(p_2+1)\cdots(p_m+1) + (p_1-1)(p_2-1)\cdots(p_m-1) \ge 2p_1 p_2 \cdots p_m, \tag{5.42}$$

which holds as the terms with minus sign cancel out. One can write

$$(p_1+1)(p_2+1)\cdots(p_m+1) \ge 2k - \varphi(k), \tag{5.43}$$

which by (5.40) gives the following inequality, valid when is k square-free

$$H_P(k) = \left[(p_1+1)\cdots(p_m+1) - 1 \right] \frac{(p_1-1)\cdots(p_m-1)}{2} \tag{5.44}$$
$$\ge \frac{\varphi(k)[2k - \varphi(k) - 1]}{2}. \tag{5.45}$$

The inequalities for $\varphi(k)$ detailed in [125] could perhaps be used to improve the lower bound for $H_P(k)$.

5.5 Non-periodic Horadam Orbits

In this section we investigate orbits produced by non-periodic Horadam sequences. We first study the degenerate case (equal generators), where the orbits are either trivial or simple spirals. Then, we present an atlas of nondegenerate Horadam patterns, including convergent, divergent, quasi-convergent, discrete, or dense orbits.

Some patterns inspired the design a Horadam-based random-number generator are also presented [25].

5.5.1 *Degenerate Orbits*

We first examine the geometric patterns produced by some degenerate orbits. Denote by $S = S(0; 1)$, $U = U(0; 1)$, $S(z_0, r)$, and $U(0; r_1, r_2)$ the unit circle, unit disc, circle of center z_0 and radius r, and annulus of radii r_1 and r_2 centered in the origin.

Let $(x_n)_{n\geq 0}$ be a Horadam sequence satisfying the recurrence relation

$$x_{n+2} = px_{n+1} + qx_n, \quad x_0 = a, x_1 = b,$$

with a, b, p, q are complex numbers.

For equal roots $z_1 = z_2$ of (5.2), the general term is

$$x_n = \left[a + \left(\frac{b}{z} - a\right)n\right]z^n = \left[az + (b - az)n\right]z^{n-1}. \tag{5.46}$$

For distinct roots $z_1 \neq z_2$ of (5.2), the general term of $(x_n)_{n\geq 0}$ is

$$x_n = Az_1^n + Bz_2^n, \tag{5.47}$$

where the constants A and B are obtained from initial conditions $x_0 = a$ and $x_1 = b$. To avoid trivial orbits, it will be assumed from now on that $AB \neq 0$.

5.5.1.1 Behavior of Sequence $(z^n)_{n\geq 0}$

As linear combinations of $(z_1^n)_{n\geq 0}$ and $(z_2^n)_{n\geq 0}$, Horadam patterns depend on the behavior of $(z^n)_{n\geq 0}$, where $z \in \mathbb{C}$ [27, Lemma 2.1].

Lemma 5.6 *Let $z = re^{2\pi i\theta} \in C$ ($r \geq 0, \theta \in \mathbb{R}$). The orbit of $(z^n)_{n\geq 0}$ is*

(*a*) *a regular k-gon if z is a primitive kth root of unity;*
(*b*) *a dense subset of the unit circle if $r = 1$ and $\theta \in \mathbb{R} \setminus \mathbb{Q}$;*
(*c*) *an inward spiral for $r < 1$;*
(*d*) *an outward spiral for $r > 1$.*

When $\theta = j/k \in \mathbb{Q}$ is irreducible, the spirals in (c) and (d) align along k rays.

5.5.1.2 Patterns Produced by Identical Generators

We first examine the orbits produced by repeated roots of the quadratic (5.2).

Theorem 5.16 *Let $(x_n)_{n\geq 0}$ be the sequence defined by (5.47) for $x_0 = a$, $x_1 = b$ and assume that the polynomial (5.2) has a repeated root $z = z_1 = z_2 = re^{2\pi i\theta}$. The orbit of $(x_n)_{n\geq 0}$ has a single point if $|a|+|b| = 0$. Otherwise, this represents*

(a) the vertices of a k-gon when $b = az$ and z is a primitive kth root of unity;
(b) a dense subset of S when $b = az$ and $|z| = 1$ and $\theta \in \mathbb{R} \setminus \mathbb{Q}$;
(c) a convergent spiral collapsing onto the origin for $|z| < 1$;
(d) a divergent spiral for $|z| > 1$ and $|a| + |b| > 0$, or for $|z| = 1$ and $b \neq az$.

Proof From (5.46), the general sequence term is $x_n = a(1-n)z^n + bz^{n-1}$, so for $a = b = 0$ one obtains $x_n = 0$ for $n \in \mathbb{N}$. When $b = az$ one has $x_n = az^n$, therefore the orbit of $(x_n)_{n\geq 0}$ represents a scaled (with $|a|$) and rotated (with angle $\arg(a)$) version of the orbit of sequence $(z^n)_{n\geq 0}$ described in Lemma 5.6, which proves (a) and (b).

When $b \neq az$, the general term is the product of z^n and the linear polynomial $a+(b/z-a)n$. For $|z| < 1$, the power z^n dominates the polynomial and the sequence converges to 0, which proves (c). Because of a, b and n, the convergence may be non-monotonic in absolute value. Finally, for $|z| > 1$ both the power and polynomial diverge, while for $|z| = 1$ and $b \neq az$ one has $|x_n| = |az + (b - az)n|$, which diverges. This ends the proof of (d). □

5.6 An Atlas of Horadam Patterns

In this section we characterize the orbits of Horadam sequences obtained for arbitrary generators and initial conditions. The distinct generators are denoted by

$$z_1 = r_1 e^{2\pi i\theta_1}, \quad z_2 = r_2 e^{2\pi i\theta_2}, \tag{5.48}$$

where $r_1, r_2, \theta_1, \theta_2$ are real numbers. We may assume that $0 \leq r_1 \leq r_2$.

Horadam patterns produced by formula (5.47) can be summarized below:

1. Stable for $r_1 = r_2 = 1$;
2. Quasi-convergent for $0 \leq r_1 < r_2 = 1$;
3. Convergent for $0 \leq r_1 \leq r_2 < 1$;
4. Divergent for $r_2 \geq 1$.

The geometric patterns obtained in each case are presented below.

5.6.1 *Stable Orbits: $r_1 = r_2 = 1$*

We first examine orbits produced by generators on the unit circle. These are either finite sets (periodic), or sets dense within certain 1D curves, or 2D annuli. Moreover, as shown before, all sequence terms $x_n = Az_1^n + Bz_2^n$ of stable orbits are located inside the annulus

$$\{z \in \mathbb{C} : ||A| - |B|| \leq |z| \leq |A| + |B|\}. \tag{5.49}$$

The following theorem characterizes the stable orbits.

Theorem 5.17 *Let $(x_n)_{n\geq 0}$ be the sequence defined by (5.47). If the distinct roots (5.48) satisfy $r_1 = r_2 = 1$, the following orbit patterns emerge.*

(a) If $\theta_1, \theta_2 \in \mathbb{Q}$, then the orbit is periodic (finite). (see Figures 5.11, 5.12 or 5.15);
(b) If $\theta_1 = p/k \in \mathbb{Q}$ (irreducible), and $\theta_2 \in \mathbb{R} \setminus \mathbb{Q}$, then the orbit is a dense subset of the union of k distinct circles (Figure 5.17);
(c) If $\theta_1, \theta_2 \in \mathbb{R} \setminus \mathbb{Q}$, then we have three distinct types of behavior

(c1) If $\theta_2 - \theta_1 = q \in \mathbb{Q}$, then the orbit's closure is a finite set of concentric circles (Figure 5.18);
(c2) If $\theta_2 = \theta_1 q$ and $q \in \mathbb{Q}$, then the orbit's closure is a stable closed curve (flower) (Figure 5.19);
(c3) If $1, \theta_1, \theta_2$ are linearly independent over $\mathbb{Q}$, then the orbit is dense within the annulus $U(0, ||A| - |B||, |A| + |B|)$ (Figure 5.20).

Proof The proof uses the dimension of the orbits' closure. This is zero for finite orbits, one for orbits dense in closed curves, and two for orbits dense in an annulus. □

5.6.1.1 (a) Stable Periodic (Finite) Orbits

When $\theta_1, \theta_2 \in \mathbb{Q}$ the orbit is finite. Indeed, when $\theta_1 = p_1/k_1$ and $\theta_2 = p_2/k_2$ are irreducible, one has $z_1^{k_1} = z_2^{k_2} = 1$. From (5.47) and since $AB \neq 0$, the sequence terms $(x_n)_{n\geq 0}$ repeat with period $\mathrm{lcm}(k_1, k_2)$.

The properties of periodic orbits have been shown previously, including necessary and sufficient conditions for the periodicity of complex Horadam [27], enumeration formulae for periodic sequences of fixed length [28], or the geometric structure of periodic Horadam orbits [25].

5.6.1.2 (b) Stable Orbits Dense Within 1D Curves

Some Horadam orbits are dense in 1D curves. When $\theta_1 = p/k$ is irreducible, z_1 is a kth primitive root of unity, therefore $(z_1^n)_{n\geq 0}$ only takes the distinct values $1, z_1, \ldots, z_1^{k-1}$ representing the vertices of a regular k-gon.

The subsequences $(x_{nk+j})_{n\geq 0}$, $j = 0, \ldots, k-1$, have the general terms

$$x_{nk+j} = Az_1^{nk+j} + Bz_2^{nk+j} = Az_1^j + Bz_2^{nk+j}. \tag{5.50}$$

The points Az_1^j are situated at distance $|A|$ from the origin and are vertices of a regular k-gon. For each value of $j = 0, \ldots, k-1$, one can write

$$Bz_2^{nk+j} = Bz_2^j(z_2^k)^n = Be^{2\pi(j\theta_2)i}e^{[2\pi(k\theta_2)i]n}, \quad n \geq 0.$$

As θ_2 is irrational, $k\theta_2$ is also irrational. From Lemma 5.6(b), the orbit of $(Bz_2^{nk+j})_{n\geq 0}$ is a dense subset of a circle of radius $|B|$, hence the subsequence $(x_{nk+j})_{n\geq 0}$ given in (5.50) is a dense subset of the circle $S(Az_1^j, |B|)$.

This property is illustrated for $\theta_1 = 1/3$ in Figure 5.17, where $k = 3$. The k circles are disjoint whenever $|A| > |B|$ as seen in Figure 5.17a, and intersect for $|A| < |B|$ as illustrated in Figure 5.17b, respectively. As expected, the sequence terms are located within the annulus $U(0, ||A| - |B||, |A| + |B|)$.

5.6.1.3 (c1) Orbits Dense Within Concentric Circles

Assume that $\theta_1, \theta_2 \in \mathbb{R} \setminus \mathbb{Q}$ and $\theta_2 - \theta_1 = q \in \mathbb{Q}$ (irreducible). By (5.47), x_n is given by

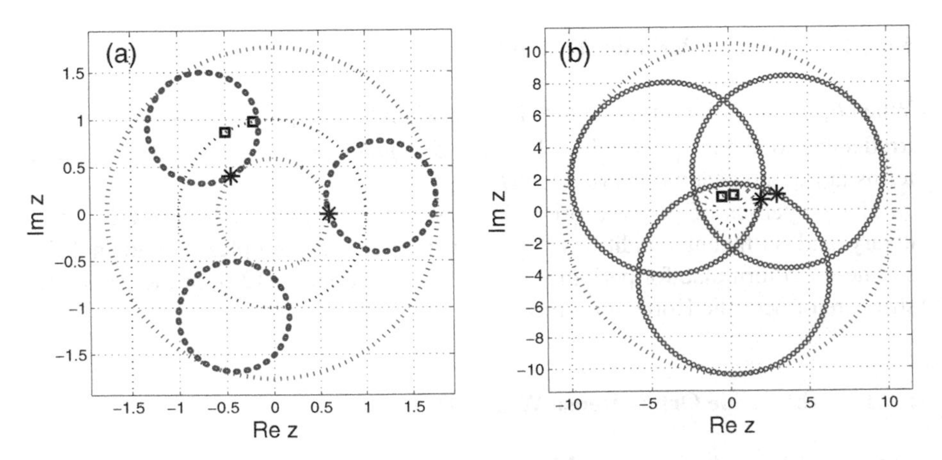

Fig. 5.17 First 500 terms of the sequence $(x_n)_{n\geq 0}$ obtained from (5.47) for $r_1 = r_2 = 1$ and **(a)** $\theta_1 = 1/3$, $\theta_2 = \sqrt{2}/5$, when $a = 0.6$ and $b = 0.6\exp(2\pi i(\sqrt{2}/5 + 0.1))$; **(b)** $\theta_1 = \pi/15$, $\theta_2 = 1/3$, when $a = 2 + 2/3i$ and $b = 3 + i$. Stars represent the initial terms a, b, squares the generators z_1, z_2, and the dotted line the unit circle. Inner and outer circles represent the boundaries of the annulus $U(0, ||A| - |B||, |A| + |B|)$ defined in (5.49)

$$x_n = A z_1^n \left[1 + A^{-1} B \left(z_2 z_1^{-1} \right)^n \right].$$

As $e^{2\pi i(\theta_2-\theta_1)}$ is a kth primitive root of unity, the subsequences $(x_{nk+j})_{n\geq 0}$ defined for $j = 0, \ldots, k-1$ have the general terms

$$x_{nk+j} = A z_1^{nk+j} \left[1 + A^{-1} B e^{2\pi i(\theta_2-\theta_1)(nk+j)} \right] = A z_1^{nk+j} \left[1 + A^{-1} B e^{2\pi i(\theta_2-\theta_1)j} \right]. \tag{5.51}$$

Denoting $x_j = 1 + A^{-1} B e^{2\pi i(\theta_2-\theta_1)j}$ for $j = 0, \ldots, k-1$, one may write

$$x_{nk+j} = A z_1^{nk+j} x_j = \left(A z_1^j x_j \right) \left(z_1^k \right)^n, \quad n \geq 0.$$

Since θ_1 is irrational, $k\theta_1$ irrational, hence by Lemma, the orbit of $((z_1^k)^n)_{n\geq 0}$ is dense within the unit circle. For each value $j = 0, 1, \ldots, k-1$, the orbit of the subsequence $(x_{nk+j})_{n\geq 0}$ defined by (5.51) represents a dense subset of the circle $S(0, |Ax_j|)$. This property is illustrated in Figure 5.18 for the case when θ_1 and θ_2 are irrational and $\theta_2 - \theta_1 = 1/2$ in (a) and $\theta_2 - \theta_1 = 1/3$ in (b), respectively.

In particular, when $\theta_2 - \theta_1$ is an integer, one has

$$x_n = A z_1^n \left[1 + A^{-1} B e^{2\pi i(\theta_2-\theta_1)n} \right] = z_1^n [A + B],$$

therefore the orbit is a dense subset of the circle $S(0; |A + B|)$.

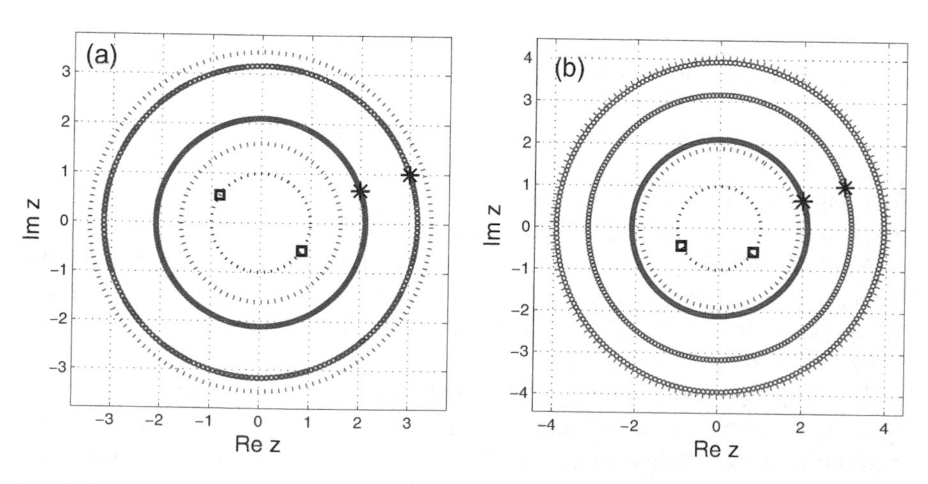

Fig. 5.18 First 1000 terms of the sequence $(x_n)_{n\geq 0}$ obtained from (5.47) for $r_1 = r_2 = 1$ and **(a)** $\theta_1 = e/3$, $\theta_2 = e/3 + 1/2$; **(b)** $\theta_1 = \pi/2$, $\theta_2 = \pi/2 + 1/3$, when $a = 2 + 2/3i$ and $b = 3 + i$. Stars represent the initial terms a, b, squares the generators z_1, z_2, and the dotted line the unit circle. Inner and outer circles represent the orbit boundaries of the annulus $U(0, ||A| - |B||, |A| + |B|)$

5.6.1.4 (c2) Orbits Dense Within Closed 1D Curves

Assume that $\theta_1, \theta_2 \in \mathbb{R} \setminus \mathbb{Q}$ and $\theta_2 = \theta_1 q$, $q \in \mathbb{Q}$. From (5.47), we can write x_n as

$$x_n = Az_1^n + Bz_2^n = Az_1^n + B(z_1^q)^n.$$

From Lemma 5.6, the orbit of $(z_1^n)_{n\geq 0}$ is dense in the unit circle, so $(x_n)_{n\geq 0}$ is dense in the graph of the complex function $f : S \to \mathbb{C}$ defined as

$$f(z) = Az + Bz^q.$$

This property is illustrated in Figure 5.19. As shown in (a), $q = 6$ gives 5 self-intersections and winding number one, while for the case $q = 3/2$ illustrated in (d), winding number is three with two self-intersections. For an irreducible fraction $q = m/k$ one can show that the curve $g(x) = Ae^{2\pi xi} + Bz^{2\pi qxi}$, has m self-intersections and winding number k.

5.6.1.5 (c3) Stable Orbits Dense Within 2D Annuli

Whenever $1, \theta_1, \theta_2$ are linearly independent over $\mathbb{Q}$ (or $\mathbb{Z}$), the resulting Horadam orbit is dense within an annulus.

Let us denote $A = R_1 e^{i\phi_1}$ and $B = R_2 e^{i\phi_2}$. For any point $x^* = re^{i\theta} \in D$, one can find complex arguments φ_1, φ_2 satisfying the identity

$$x^* = re^{i\theta} = R_1 e^{i(\phi_1 + 2\pi\varphi_1)} + R_2 e^{i(\phi_2 + 2\pi\varphi_2)}.$$

The nth term x_n of the Horadam sequence in polar form can be written as

$$x_n = Az_1^n + Bz_2^n = R_1 e^{i(\phi_1 + 2\pi n\theta_1)} + R_2 e^{i(\phi_2 + 2\pi n\theta_2)}.$$

By continuity, for any $\varepsilon > 0$ one has $|x - x^*| < \varepsilon$ provided that $|n\theta_1 - \varphi_1| < \delta$ and $|n\theta_2 - \varphi_2| < \delta$, for sufficiently small $\delta > 0$.

As $(1, \theta_1, \theta_2)$ are linearly independent over $\mathbb{Q}$, Proposition A.3 ensures that the sequence $(\{n\theta_1\}, \{n\theta_2\})_{n\geq 0}$ is dense within $[0, 1] \times [0, 1]$. Hence we can find terms x_n arbitrarily close to x^*. This indicates that $(x_n)_{n\geq 0}$ is a dense subset in the annulus $U(0, ||A| - |B||, |A| + |B|)$.

The property is illustrated in Figure 5.20a for $|A| \neq |B|$, obtained for $r_1 = r_2 = 1, \theta_1 = \frac{\sqrt{2}}{3}, \theta_2 = \frac{\sqrt{5}}{15}$ and $a = 2 + \frac{2}{3}i$, $b = 3 + i$. When $|A| = |B|$, the orbit is dense within the circle $U(0, 2|A|)$, as depicted in Figure 5.20b for the parameter values $r_1 = r_2 = 1, \theta_1 = \exp(1)/2, \theta_2 = \exp(2)/4$ and $a = 1 + 1/3i$, $b = 1.5a\exp(\pi(\theta_1 + \theta_2))$. These types of dense orbits are used to design a random number generator in Section 3.3.

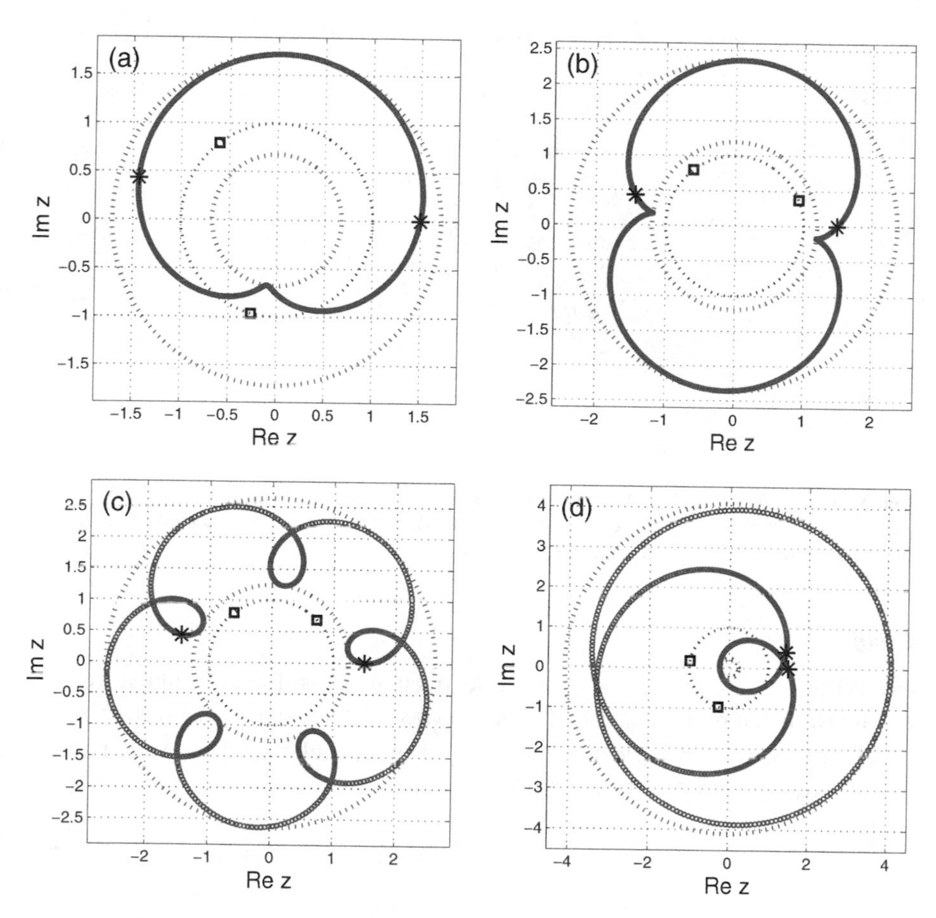

Fig. 5.19 First 1000 terms of $(x_n)_{n\geq 0}$ obtained from (5.47) for $r_1 = r_2 = 1$, $\theta_2 = q\theta_1$ with **(a)** $q = 2$, $\theta_2 = \sqrt{2}/4$; **(b)** $q = 3$, $\theta_2 = \sqrt{2}/4$; **(c)** $q = 6$, $\theta_2 = \sqrt{2}/4$; and $a = 1.5$, $b = 1.5\exp(2\pi i(\sqrt{2}/4 + 1/10))$ and for **(d)** $q = 2/3$, $\theta_2 = \sqrt{2}/3$, and $a = 1.5$, $b = 1.5\exp(2\pi i(2\sqrt{2}/3+1/10))$. Starting values a, b shown by stars, generators z_1, z_2 by squares, unit circle by dotted line. Inner and outer circles show $U(0, ||A| - |B||, |A| + |B|)$

5.6.2 *Quasi-Convergent Orbits for* $0 \leq r_1 < r_2 = 1$

Here orbits collapse to the vertices of a polygon, or onto circles.

Theorem 5.18 *Let* $(x_n)_{n\geq 0}$ *be the sequence defined by (5.47) for initial conditions* $x_0 = a$, $x_1 = b$ *and let us assume that the polynomial (5.2) has distinct roots* $z_1 \neq z_2$ *such that* $r_1 < r_2 = 1$. *The orbit of* $(x_n)_{n\geq 0}$ *has the following patterns*

(a) For $\theta_2 \in \mathbb{Q}$ *the orbits are attracted onto discrete finite sets (vertices of regular polygons) along rays* ($\theta_1 \in \mathbb{Q}$) *or spirals* ($\theta_1 \in \mathbb{R} \setminus \mathbb{Q}$) *(Figure 5.21).*
(b) For $\theta_2 \in \mathbb{R} \setminus \mathbb{Q}$ *the orbits converge to a circle of radius* $|B|$ *(Figure 5.22).*

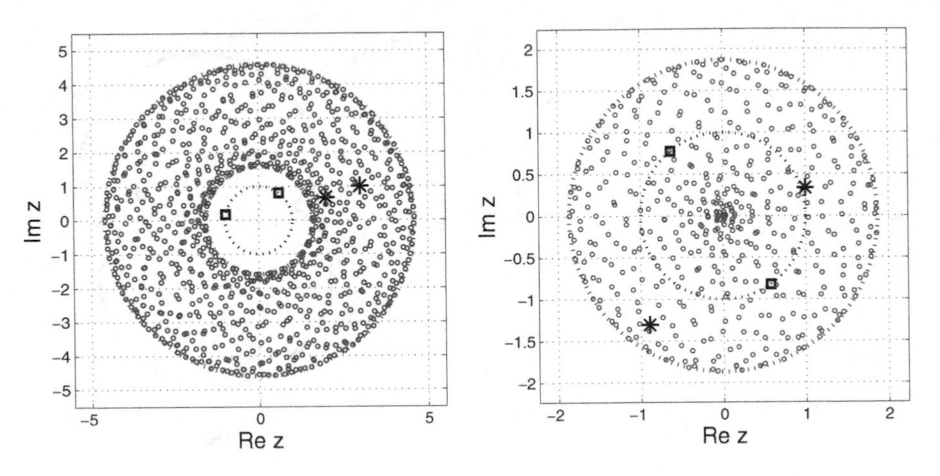

Fig. 5.20 First N terms of $(x_n)_{n\geq 0}$ obtained from (5.4). (**a**) $N = 1000$, $|A| \neq |B|$; (**b**) $N = 500$, $|A| = |B|$. Also plotted, initial conditions a, b (stars), generators z_1, z_2 (squares), unit circle S (solid line), and circle $U(0, |A| + |B|)$ (dotted line)

Proof We have

(a) When $\theta_2 = p_2/k_2$ is an irreducible fraction, z_2 is a k_2th primitive root of unity, therefore the sequence $({z_2}^n)_{n\geq 0}$ is periodic and has just k_2 distinct terms $1, z_2, \ldots, z_2^{k_2-1}$ representing the vertices of a regular k_2-gon. We define the subsequences $(x_{nk_2+j})_{n\geq 0}$ for $j = 0, \ldots, k_2 - 1$, by

$$x_{nk_2+j} = Az_1^{nk_2+j} + Bz_2^{nk_2+j} = Az_1^{nk_2+j} + Bz_2^j. \tag{5.52}$$

The points Bz_2^j are situated at distance $|B|$ from the origin and represent the vertices of the regular k_2-gon. As the sequence $(A{z_1}^n)_{n\geq 0}$ converges to zero, $(x_{nk+j})_{n\geq 0}$ converges to Bz_2^j. This property is illustrated for $\theta_1 = 1/7$ in Figure 5.21.

When $\theta_1 = p_1/k_1 \in \mathbb{Q}$ is irreducible, the subsequences $(x_{nk_2+j})_{n\geq 0}$ approach their limit along rays, as depicted in Figure 5.21a. The number of rays is the same for each attractor. For a fixed value of j one can write (5.52) as

$$x_{nk_2+j} = Az_1^j z_1^{nk_2} + Bz_2^j,$$

which indicates that the number of rays coming out of the jth vertex is equal to the number of distinct principal arguments of $(z_1^{nk_2})_{n\geq 0}$. As the arguments are np_1k_2/k_1, the number of rays is $k_1/\gcd(k_1, k_2)$. When θ_1 is irrational the spirals collapse on the vertices of the k_2-gon, as illustrated in Figure 5.21b.

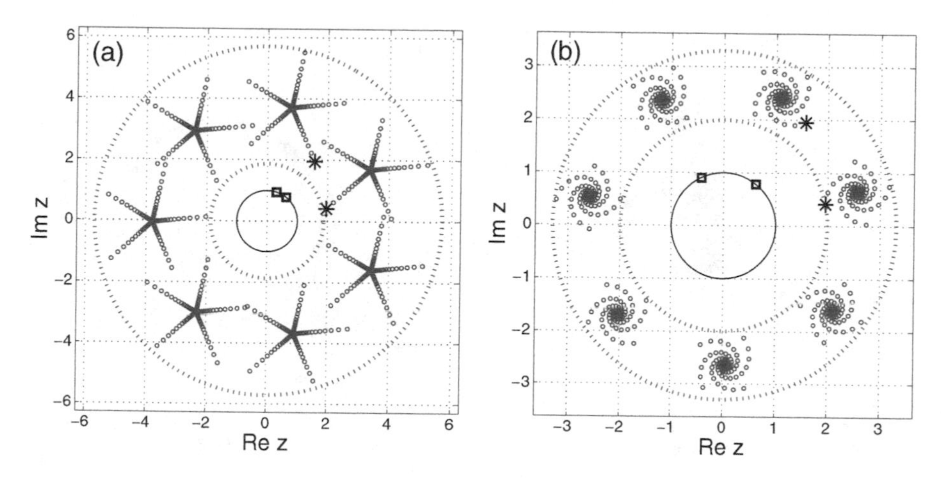

Fig. 5.21 First 1000 terms of sequence $(x_n)_{n\geq 0}$ obtained from (5.47) for $r_1 = 0.995$, $r_2 = 1$ and (**a**) $\theta_1 = 1/5$, $\theta_2 = 1/7$; (**b**) $\theta_1 = \sqrt{5}/7$, $\theta_2 = 1/7$, when $a = 2\exp(2\pi i/30)$ and $2.5\exp(2\pi i/7)$. Stars depict initial conditions a, b, squares the generators z_1, z_2, and the solid line the unit circle. Inner and outer circles represent the boundaries of the annulus $U(0, ||A| - |B||, |A| + |B|)$

(b) When θ_2 is irrational, Lemma 5.6 indicates that the sequence $(Bz_2^n)_{n\geq 0}$ is a dense subset of the circle of radius $|B|$ centered in the origin. As the sequence $(Az_1{}^n)_{n\geq 0}$ converges to zero, the orbit of sequence $(x_n)_{n\geq 0}$ collapses onto the circle of radius $|B|$ centered in the origin, as illustrated in Figure 5.22. There is a distinction between the regular five petal pattern obtained when $\theta_1 = 1/5 \in \mathbb{Q}$ (Figure 5.22a) and the erratic orbit obtained for irrational $\theta_1 = \sqrt{2}/2$ (Figure 5.22b).

□

Note that sequence terms are inside the circle $U(0, |A| + |B|)$, but not inside annulus $U(0, ||A| - |B||, |A| + |B|)$, as seen in Figure 5.22a. With A and B solved from the initial conditions and generators, we can tell when sequence terms might be located outside the annulus. If $|A| > |B|$ we have

$$\lim_{n\to\infty} |x_n| = |B| < |A| - |B| = ||A| - |B||,$$

therefore such a condition is $|A| > 2|B|$, which requires $|az_2 - b| > 2|b - az_1|$. In the example shown in Figure 5.22a we have $|A| = 3.4669$ and $|B| = 1.5254$.

5.6.3 *Convergent Orbits for* $0 \leq r_1 \leq r_2 < 1$

When both generators are located inside the unit circle, the sequence $(x_n)_{n\geq 0}$ defined by (5.47) converges to the origin. However, a surprising variety of patterns

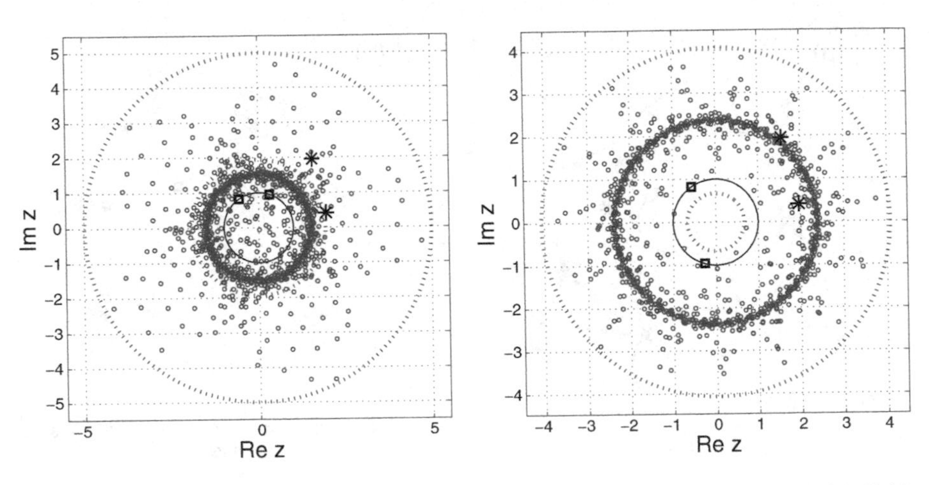

Fig. 5.22 First 1000 sequence terms $(x_n)_{n\geq 0}$ given by (5.47) for $r_1 = 0.995$, $r_2 = 1$ and (**a**) $\theta_1 = 1/5$, $\theta_2 = \sqrt{3}/5$; (**b**) $\theta_1 = \sqrt{2}/2$, $\theta_2 = \sqrt{3}/5$, when $a = 2\exp(2\pi i/30)$ and $2.5\exp(2\pi i/7)$. Stars represent initial conditions a, b, squares generators z_1, z_2, and the solid line the unit circle. Inner and outer circles represent the orbit boundaries $U(0, ||A| - |B||, |A| + |B|)$

is still generated, as detailed below. The cases when r_1 and r_2 are equal, or distinct have to be analyzed separately.

Theorem 5.19 ($r_1 < r_2$) *Let $(x_n)_{n\geq 0}$ be the sequence defined by (5.47) for initial conditions $x_0 = a$, $x_1 = b$ and assume that the polynomial (5.2) has distinct roots $z_1 \neq z_2$ such that $0 \leq r_1 < r_2 < 1$. The following orbit patterns emerge.*

(a) For $\theta_1 = p_1/k_1, \theta_2 = p_2/k_2 \in \mathbb{Q}$ the orbit has $\mathrm{lcm}(k_2, k_1)$ branches merging onto k_2 rays, which converge to the origin (Figure 5.23).
(b) For $\theta_1 \in \mathbb{R} \setminus \mathbb{Q}$, $\theta_2 = p/k \in \mathbb{Q}$ the orbit has k spirally perturbed arms. (Figure 5.24).
(c) For $\theta_1 = p/k \in \mathbb{Q}$, $\theta_2 \in \mathbb{R}\setminus\mathbb{Q}$ the orbit converges concentrically to the origin, as a set of k petals or branches (Figure 5.25).
(d) For $\theta_1, \theta_2 \in \mathbb{R} \setminus \mathbb{Q}$ the orbit is an erratic or ordered spiral (Figure 5.26).

Proof From Lemma 5.6, $(Az_1^n)_{n\geq 0}$ and $(Bz_2^n)_{n\geq 0}$ converge to the origin, therefore $(x_n)_{n\geq 0}$ converges to the origin as well. As $r_1 < r_2$, $(x_n)_{n\geq 0}$ behaves as $(Bz_2^n)_{n\geq 0}$ for larger values of n. This observation allows us to treat the orbits of $(x_n)_{n\geq 0}$ as perturbations of $(Bz_2^n)_{n\geq 0}$ by elements of sequence $(Az_1^n)_{n\geq 0}$. The nature and intensity of the perturbation are determined by the relations between θ_1, θ_2 and r_1 and r_2, respectively.

(a) For $\theta_1 = p_1/k_1$ and $\theta_2 = p_2/k_2$ the general term of sequence $(x_n)_{n\geq 0}$ given in (5.47) can be written as

$$x_n = Az_1^n + Bz_2^n = Ar_1^n \exp\left(2\pi i \frac{np_1}{k_1}\right) + Br_2^n \exp\left(2\pi i \frac{np_2}{k_2}\right). \tag{5.53}$$

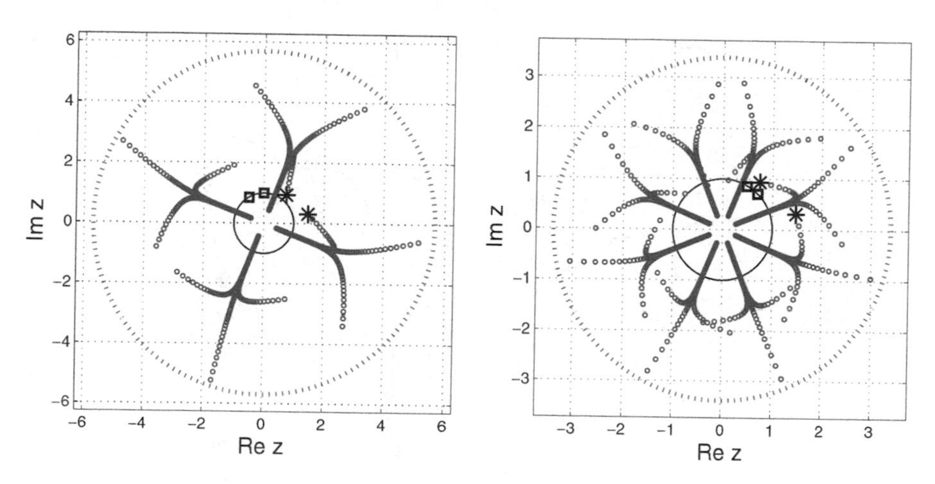

Fig. 5.23 First 2000 terms of sequence $(x_n)_{n\geq 0}$ obtained from (5.47) for $r_1 = 0.99$, $r_2 = 0.999$, and **(a)** $\theta_1 = 1/3$, $\theta_2 = 1/4$; **(b)** $\theta_1 = 1/6$, $\theta_2 = 1/8$; when $a = 1.5\exp(2\pi i/30)$ and $b = 1.2\exp(2\pi i/7)$. Also plotted, initial conditions x_0, x_1 (stars), generators z_1, z_2 (squares), unit circle S (solid line), and circle $U(0, |A| + |B|)$ (dotted line)

Denoting the least common multiple of numbers k_1 and k_2 by $K = \text{lcm}(k_1, k_2)$, one can define the subsequences $(x_{nK+j})_{n\geq 0}$ of (5.53) for $j = 0, \dots, K-1$ as

$$x_{nK+j} = Ar_1^{nK+j}\exp\left(2\pi i\frac{(nK+j)p_1}{k_1}\right) + Br_2^{nK+j}\exp\left(2\pi i\frac{(nK+j)p_2}{k_2}\right)$$
$$= Ar_1^{nK+j}\exp\left(2\pi i\frac{jp_1}{k_1}\right) + Br_2^{nK+j}\exp\left(2\pi i\frac{jp_2}{k_2}\right).$$

For a fixed value of j we denote by $l = j \mod k_2$. The K/k_2 sequences $(x_{nK+mk_2+l})_{n\geq 0}$ defined for $m = 0, \dots, K/k_2 - 1$ are all converging asymptotically to the ray $(x_{nK+l})_{n\geq 0}$. The property is illustrated in Figure 5.23.

(b) For $\theta_1 \in \mathbb{R} \setminus \mathbb{Q}$ and $\theta_2 = p_2/k_2$ one can use (5.47) to define the subsequences $(x_{nk_2+j})_{n\geq 0}$, for $j = 0, \dots, k_2 - 1$ as

$$x_{nk_2+j} = Ar_1^{nk_2+j}\exp\left(2\pi i(nk_2+j)\theta_1\right) + Br_2^{nk_2+j}\exp\left(2\pi i\frac{(nk_2+j)p_2}{k_2}\right)$$
$$= Ar_1^{nk_2+j}\exp\left(2\pi i(nk_2+j)\theta_1\right) + Br_2^{nk_2+j}\exp\left(2\pi i\frac{jp_2}{k_2}\right).$$

As $r_1 < r_2$, for each value of $j = 0, \dots, k_2$ the orbit of $(x_{nk_2+j})_{n\geq 0}$ is a perturbation of the convergent ray $Bz_2^{nk_2+j}$ by the convergent spiral $Az_1^{nk_2+j}$. The amplitude of the perturbation for $r_2 = 0.999$ is depicted for $r_1 = 0.99$ and $r_1 = 0.997$ in Figure 5.24.

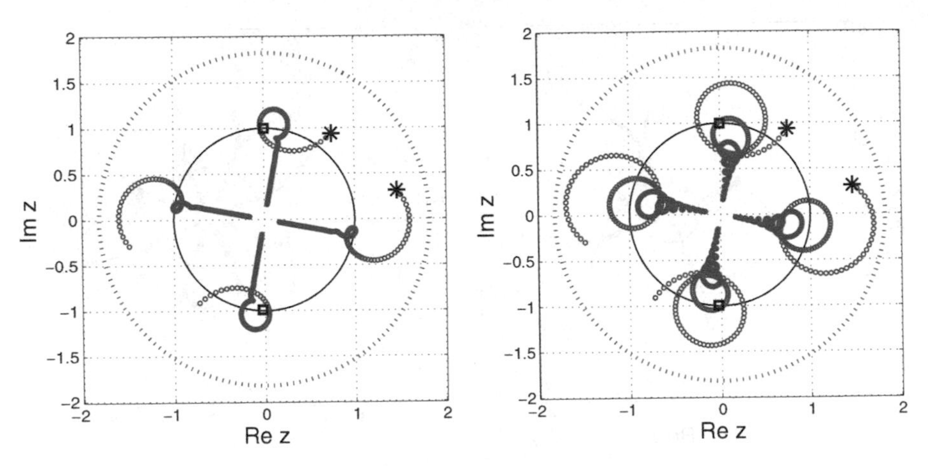

Fig. 5.24 First 2000 terms of sequence $(x_n)_{n\geq 0}$ obtained from (5.47) for $\theta_1 = \sqrt{5}/3$, $\theta_2 = 1/4$ and $r_2 = 0.999$ for (**a**) $r_1 = 0.99$, (**b**) $r_1 = 0.997$, when $a = 1.5\exp(2\pi i/30)$ and $b = 1.2\exp(2\pi i/7)$. Also plotted, initial conditions x_0, x_1 (stars), generators z_1, z_2 (squares), unit circle S (solid line), and circle $U(0, |A|+|B|)$ (dotted line)

(c) For $\theta_1 = p_1/k_1$ and $\theta_2 \in \mathbb{R}\setminus\mathbb{Q}$, $(x_{nk_1+j})_{n\geq 0}$, $j = 0,\ldots,k_1-1$ satisfy

$$x_{nk_1+j} = Ar_1^{nk_1+j}\exp\left(2\pi i\frac{jp_1}{k_1}\right) + Br_2^{nk_1+j}\exp\left(2\pi i(nk_1+j)\theta_2\right).$$

As $r_1 < r_2$, for each value of $j = 0,\ldots,k_1$ the orbit of $(x_{nk_1+j})_{n\geq 0}$ can be viewed as a perturbation of the convergent spiral $Bz_2^{nk_1+j}$ by the convergent ray $Az_1^{nk_1+j}$. For large values of n we have $r_1^n < r_2^n$, therefore the orbit of $(x_n)_{n\geq 0}$ asymptotically converges towards the spiral Bz_2^n, as in Figure 5.25.

(d) For irrational θ_1 and θ_2 the initial part of the orbit is erratic, and then converges asymptotically towards the spiral Bz_2^n, as in Figure 5.26. □

Theorem 5.20 ($r_1 = r_2 = r$) *Let $(x_n)_{n\geq 0}$ be the sequence defined by (5.47) for initial conditions $x_0 = a$, $\theta_1 = b$ and assume that the polynomial (5.2) has distinct roots $z_1 \neq z_2$ such that $0 \leq r_1 = r_2 = r < 1$. The following orbit patterns emerge.*

(a) When $\theta_1 = p_1/k_1, \theta_2 = p_2/k_2 \in \mathbb{Q}$ the orbit consists of $\mathrm{lcm}(k_1,k_2)$ *rays converging to origin (Figure 5.27a).*

(b) When $\theta_1 \in \mathbb{Q}$ and $\theta_2 \in \mathbb{R}\setminus\mathbb{Q}$ the orbit consists of k_2 perturbed spirals (Figure 5.27b–d).

(c) When $\theta_1, \theta_2 \in \mathbb{R}\setminus\mathbb{Q}$, the following patterns are possible

(c1) If $\theta_2 - \theta_1 = q \in \mathbb{Q}$, then the orbit is a multiple spiral (Figure 5.28a).

(c2) If $\theta_1 = \theta_2 q$, $q \in \mathbb{Q}$, then the orbit consists of multi-chamber contours (Figure 5.28b and c).

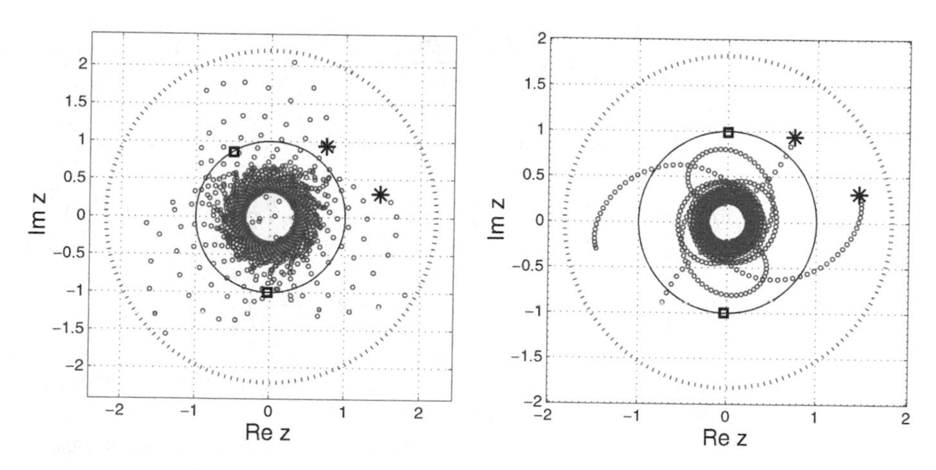

Fig. 5.25 First 2000 terms of $(x_n)_{n\geq 0}$ obtained from (5.47) for $r_1 = 0.99$, $r_2 = 0.999$ and **(a)** $\theta_1 = 1/3$, $\theta_2 = \sqrt{5}/3$; **(b)** $\theta_1 = 1/4$, $\theta_2 = \sqrt{5}/3$; when $a = 1.5\exp(2\pi i/30)$ and $b = 1.2\exp(2\pi i/7)$. Also plotted, initial conditions x_0, x_1 (stars), generators z_1, z_2 (squares), unit circle S (solid line), and circle $U(0, |A| + |B|)$ (dotted line)

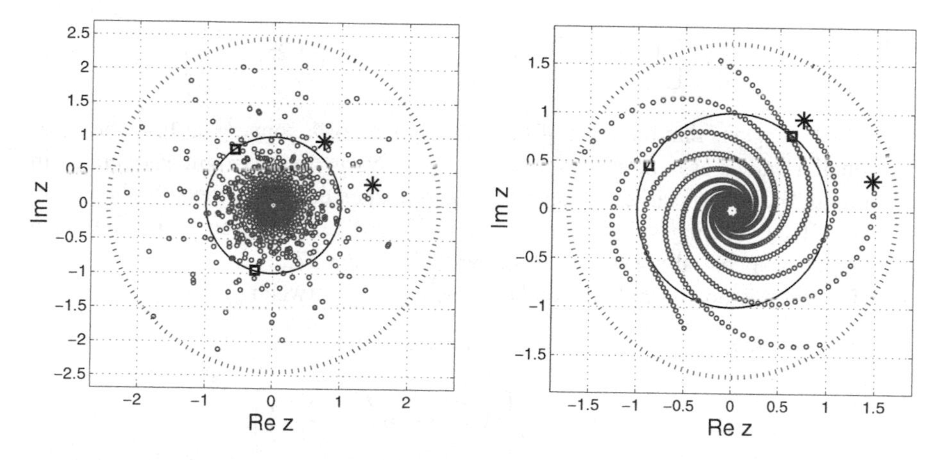

Fig. 5.26 First 2000 terms of $(x_n)_{n\geq 0}$ from (5.47) for $r_1 = 0.99$, $r_2 = 0.997$ and **(a)** $\theta_1 = \sqrt{2}/2$, $\theta_2 = \sqrt{5}/3$; **(b)** $\theta_1 = 3\sqrt{2}/10$, $\theta_2 = \sqrt{2}/10$, when $a = 1.5\exp(2\pi i/30)$ and $b = 1.2\exp(2\pi i/7)$. Also plotted, initial conditions x_0, x_1 (stars), generators z_1, z_2 (squares), unit circle S (solid line), and circle $U(0, |A| + |B|)$ (dotted line)

(c3) *If* $1, \theta_1, \theta_2$ *are linearly independent over* $\mathbb{Q}$, *then the orbit is a convergent spiral (Figure 5.28d).*

Proof From Lemma 5.6, $(Az_1^n)_{n\geq 0}$ and $(Bz_2^n)_{n\geq 0}$ converge to the origin, therefore $(x_n)_{n\geq 0}$ converges to the origin as well.

(a) For $\theta_1 = p_1/k_1$ and $\theta_2 = p_2/k_2$ the general term of sequence $(x_n)_{n\geq 0}$ is

$$x_n = Az_1^n + Bz_2^n = r^n\left[A\exp\left(2\pi i\frac{np_1}{k_1}\right) + B\exp\left(2\pi i\frac{np_2}{k_2}\right)\right]. \quad (5.54)$$

For $K = \mathrm{lcm}(k_1, k_2)$ define the subsequences $(x_{nK+j})_{n\geq 0}$ of (5.54) as

$$\begin{aligned} x_{nK+j} &= r^{nK+j}\left[A\exp\left(2\pi i\frac{(nK+j)p_1}{k_1}\right) + B\exp\left(2\pi i\frac{(nK+j)p_2}{k_2}\right)\right] \\ &= r^{nK+j}\left[A\exp\left(2\pi i\frac{jp_1}{k_1}\right) + B\exp\left(2\pi i\frac{jp_2}{k_2}\right)\right]. \end{aligned}$$

For each $j = 0, \ldots, K-1$ the orbit of $(x_{nK+j})_{n\geq 0}$ represents a ray converging to the origin. The property is illustrated in Figure 5.27a.

(b) When θ_1 is irrational and $\theta_2 = p/k$ is an irreducible fraction (the other case is symmetric for $r_1 = r_2 = r$) one can define the subsequences $(x_{nk+j})_{n\geq 0}$ as

$$\begin{aligned} x_{nk+j} &= r^n\left[A\exp\left(2\pi i(nk+j)\theta_1\right) + B\exp\left(2\pi i(nk+j)\frac{p}{k}\right)\right] \\ &= r^n\left[A\exp\left(2\pi i(nk+j)\theta_1\right) + B\exp\left(2\pi i\frac{jp}{k}\right)\right], \end{aligned}$$

which suggests that the orbit of $(x_n)_{n\geq 0}$ is made of k convergent branches. For smaller values of r these branches rapidly converge to the origin as depicted in Figure 5.27b. For larger values of r the branches converge towards the origin as self-intersecting spirals, as plotted in Figure 5.27c. When θ_1 is small, the branches appear almost solid, as shown in Figure 5.27d.

(c) Here orbit geometry only depends on the relation between θ_1 and θ_2.

(c1) When $\theta_1 - \theta_2 = q = p/k \in \mathbb{Q}$ we can write

$$x_n = z_1^n\left[A + B\exp\left(2\pi i\frac{np}{k}\right)\right],$$

hence $(x_n)_{n\geq 0}$ is composed of k spirals, similar in appearance to z_1^n, The orbit is a double spiral as in Figure 5.28a, obtained for $\theta_2 = 5\sqrt{2}/2$ and $q = 1/2$.

(c2) When $\theta_1 = \theta_2 q$ for $q = p/k \in \mathbb{Q}$, the general term of $(x_n)_{n\geq 0}$ is

$$x_n = Az_1^n + Bz_2^n = A(z_2^n)^q + Bz_2^n.$$

In this case, one can identify a number of interesting orbit patterns. In Figure 5.28b one can see a collapsible contour with four petals obtained

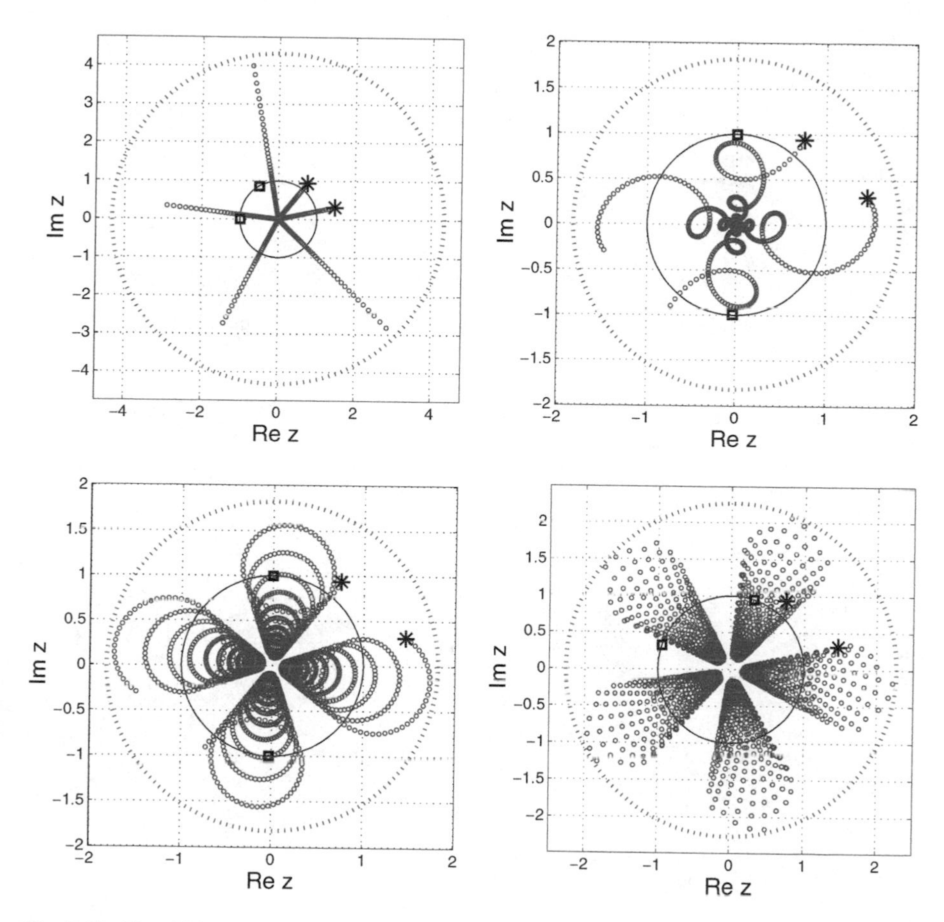

Fig. 5.27 First 2000 terms of $(x_n)_{n\geq 0}$ given by (5.47) for (**a**) $\theta_1 = 1/3$, $\theta_2 = 1/2$, $r = 0.99$; (**b**) $\theta_1 = \sqrt{5}/3$, $\theta_2 = 1/4$, $r = 0.995$; (**c**) $\theta_1 = \sqrt{5}/3$, $\theta_2 = 1/4$, $r = 0.999$; (**d**) $\theta_1 = \sqrt{5}/5$, $\theta_2 = 1/5$, $r = 0.999$ when $a = 1.5\exp(2\pi i/30)$ and $b = 1.2\exp(2\pi i/7)$ (stars). Also plotted, generators z_1, z_2 (squares), unit circle S (solid line), and circle $U(0, |A| + |B|)$ (dotted line)

for $q = 5$ and $\theta_2 = \sqrt{2}/10$, while in Figure 5.28c the orbit is a double-hole contour containing the origin, obtained for $q = 2$ and $\theta_2 = \sqrt{3}/4$.

(c3) In this case sequences $(n\theta_1)_{n\geq 0}$ and $(n\theta_2)_{n\geq 0}$ are linearly independent and dense subsets of the unit interval. Using Lemma 5.1, the arguments of sequence $(x_n)_{n\geq 0}$ are uniformly distributed, and the orbit is an erratic spiral converging to the origin. This situation is illustrated in Figure 5.28d.

□

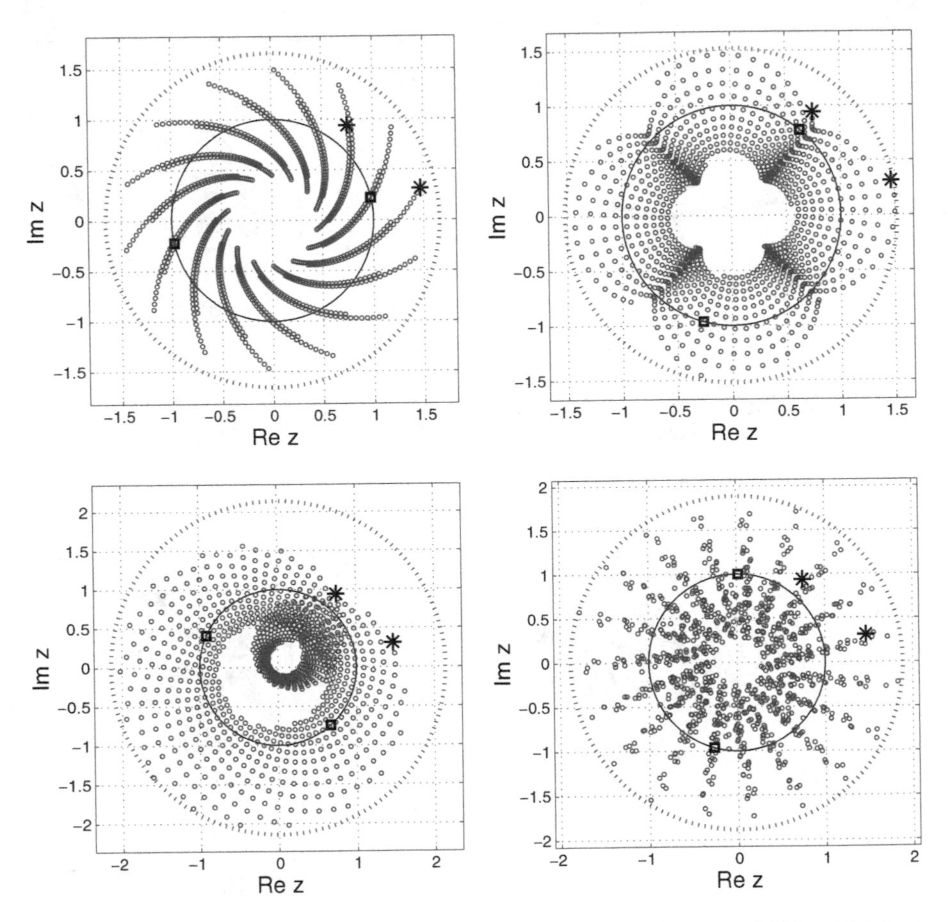

Fig. 5.28 First 1000 terms of sequence $(x_n)_{n\geq 0}$ obtained from (5.47) for $r = 0.999$ and (**a**) $\theta_1 = 5\sqrt{2}/2 + 1/2$, $\theta_2 = 5\sqrt{2}/2$, (**b**) $\theta_1 = 5\sqrt{2}/10$, $\theta_2 = \sqrt{2}/10$; (**c**) $\theta_1 = \sqrt{3}/2$, $\theta_2 = \sqrt{3}/4$; (**d**) $\theta_1 = \sqrt{2}/2$, $\theta_2 = \sqrt{3}/7$; when $a = 1.5\exp(2\pi i/30)$ and $1.2\exp(2\pi i/7)$. Also plotted, initial conditions x_0, x_1 (stars), generators z_1, z_2 (squares), unit circle S (solid line), and circle $U(0, |A| + |B|)$ (dotted line)

5.6.4 *Divergent Orbits for* $1 < r_2$

When one or both generators are located outside the unit circle, the sequence $(x_n)_{n\geq 0}$ defined by (5.47) diverges to infinity. However, a surprising variety of patterns emerges, including different types of spirals, patterns converging to a finite number of rays coming from the origin, or orbits resembling water ripples. The cases when r_1 and r_2 are equal or distinct have to be analyzed separately. The problem is dual to the convergent case presented in the previous section.

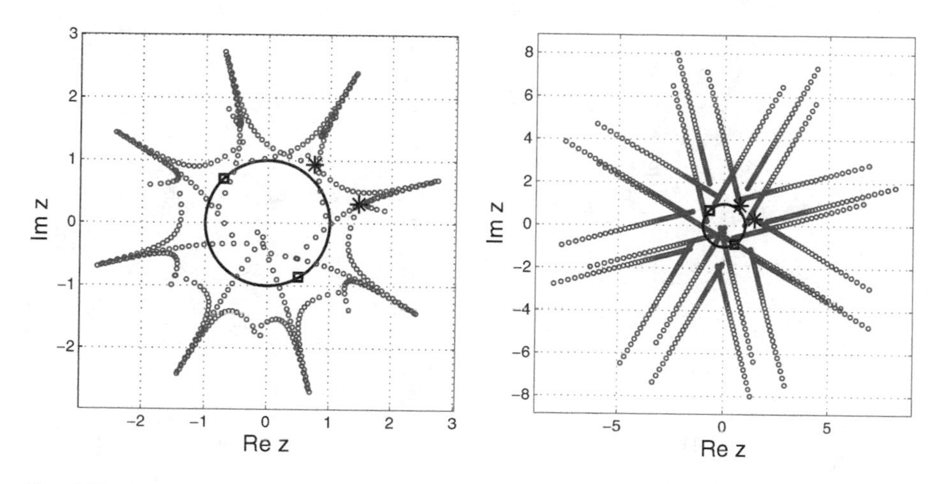

Fig. 5.29 First N terms of sequence $(x_n)_{n\geq 0}$ obtained from (5.47) for $r_2 = 1.002$, $\theta_1 = 5/6$, $\theta_2 = 3/8$ and (**a**) $N = 500$ and $r_2 = 0.999$; (**b**) $N = 1000$ and $r_2 = 1$, when $a = 1.5\exp(2\pi i/30)$ and $b = 1.2\exp(2\pi i/7)$. Also plotted, initial conditions x_0, x_1 (stars), generators z_1, z_2 (squares), and unit circle S (solid line)

Theorem 5.21 $(1 < r_2, r_1 < r_2)$ *Let $(x_n)_{n\geq 0}$ be the sequence defined by (5.47) for initial conditions $x_0 = a$, $\theta_1 = b$ and assume that the polynomial (5.2) has distinct roots $z_1 \neq z_2$ such that $0 < r_1 < r_2$ and $1 < r_2$. The following patterns emerge.*

(a) For $\theta_1 = p_1/k_1, \theta_2 = p_2/k_2 \in \mathbb{Q}$, the orbit has $lcm(k_2, k_1)$ branches merging onto k_2 rays which diverge away from the origin to infinity (Figure 5.29).
(b) For $\theta_1 \in \mathbb{R} \setminus \mathbb{Q}$, $\theta_2 = p/k \in \mathbb{Q}$, the orbit has k divergent spiral arms (Figure 5.30).
(c) For $\theta_1 = p/k \in \mathbb{Q}$ (irreducible) and $\theta_2 \in \mathbb{R} \setminus \mathbb{Q}$, then the diverges concentrically to infinity, as a set of k petals or branches (Figure 5.31).
(d) When $\theta_1, \theta_2 \in \mathbb{R} \setminus \mathbb{Q}$, the orbit is an erratic or ordered spiral (Figure 5.32).

Remark 5.3 The perturbations are stable for $r_1 = 1$, and increase for $r_1 > 1$.

Proof From Lemma 5.6, $(Bz_2^n)_{n\geq 0}$ diverges to infinity, therefore $(x_n)_{n\geq 0}$ diverges as well. As $r_1 < r_2$, $(x_n)_{n\geq 0}$ behaves as $(Bz_2^n)_{n\geq 0}$ for larger n. This observation allows us to treat the orbits of $(x_n)_{n\geq 0}$ as perturbations of $(Bz_2^n)_{n\geq 0}$ by elements of sequence $(Az_1^n)_{n\geq 0}$. The nature and intensity of the perturbation are determined by the relations between θ_1, θ_2 and r_1 and r_2, respectively.

(a) As seen previously, the orbits resemble those obtained for $r_1 < r_2 < 1$, with the notable difference that here the branches diverge to infinity along rays. As one can see in Figure 5.29a, the convergence is evident after a relatively small number of terms for $r_1 < 1$, as in this case the contribution due to r_1^n vanishes. For $r_1 \geq 1$ it takes a larger amount of terms for the branches to approach the rays, as shown in Figure 5.29b.

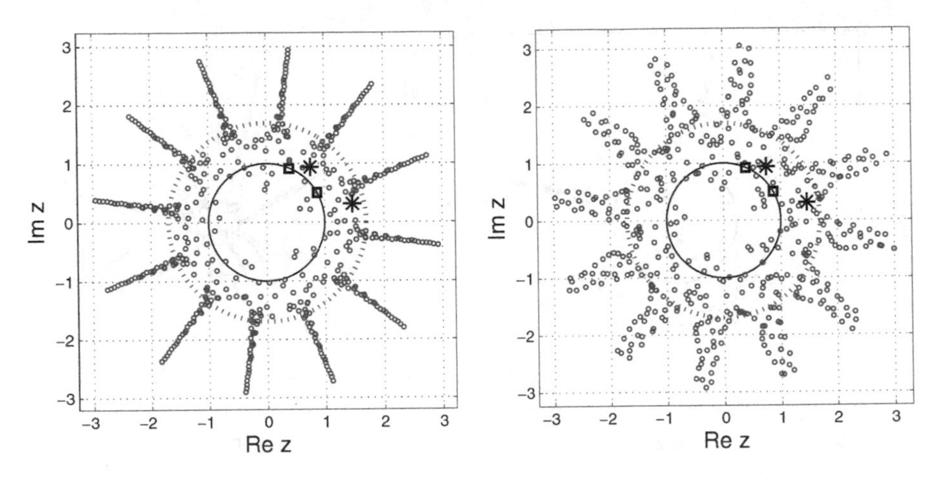

Fig. 5.30 First 500 terms of sequence $(x_n)_{n\geq 0}$ obtained from (5.47) for $r_2 = 1.002$, $\theta_1 = \sqrt{5}/12$, $\theta_2 = 1/12$ and **(a)** $r_1 = 0.99$; **(b)** $r_1 = 0.997$, when $a = 1.5\exp(2\pi i/30)$ and $b = 1.2\exp(2\pi i/7)$. Also plotted, initial conditions x_0, x_1 (stars), generators z_1, z_2 (squares) and unit circle S (solid line), and circle $U(0, |A| + |B|)$ (dotted line)

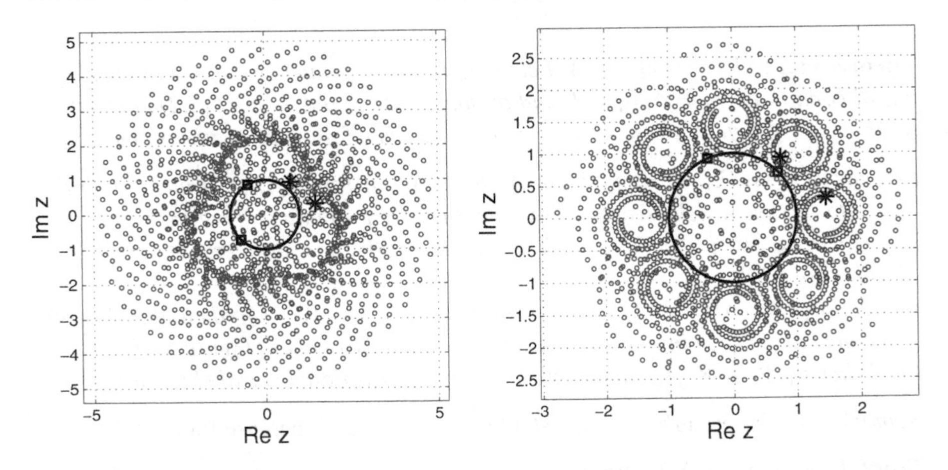

Fig. 5.31 First 1500 terms of sequence $(x_n)_{n\geq 0}$ obtained from (5.47) for $r_2 = 1.001$, and **(a)** $\theta_1 = 1/3$, $\theta_2 = \pi/5$, $r_1 = 0.997$; **(b)** $\theta_1 = 1/8$, $\theta_2 = \pi/10$, $r_1 = 1$, when $a = 1.5\exp(2\pi i/30)$ and $b = 1.2\exp(2\pi i/7)$. Also plotted, initial conditions x_0, x_1 (stars), generators z_1, z_2 (squares), and unit circle S (solid line)

(b) The number of terms stays the same while we modify r_1. The convergence along rays is clearly faster for smaller values of r_1, as in Figure 5.30.

(c) The orbit converges to a spiral in the long term. The contribution of r_1^n fades with the increase of n for $r_1 < 1$, as shown in Figure 5.31a. On the other hand, this contribution stays the same or increases for $r_1 \geq 1$ as illustrated in Figure 5.31b.

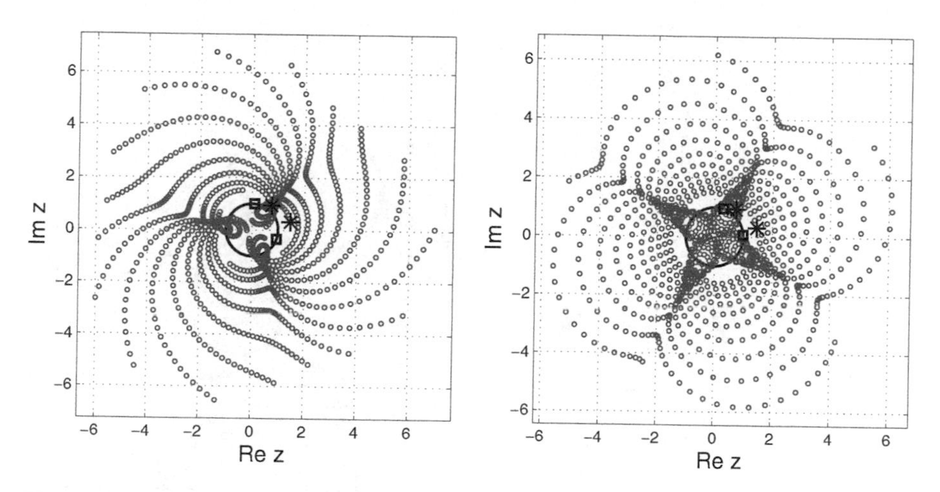

Fig. 5.32 First 1000 terms of sequence $(x_n)_{n\geq 0}$ obtained from (5.47) for $r_2 = 1$, $r_2 = 1.002$, and (**a**) $\theta_1 = 4\sqrt{2}/6$, $\theta_2 = \sqrt{2}/6$; (**b**) $\theta_1 = 5\sqrt{2}/7$, $\theta_2 = \sqrt{2}/7$ when $a = 1.5\exp(2\pi i/30)$ and $b = 1.2\exp(2\pi i/7)$. Also plotted, initial conditions x_0, x_1 (stars), generators z_1, z_2 (squares), and unit circle S (solid line)

(d) In the case depicted in Figure 5.32, all arguments are irrational. However, one may note that $\theta_1 = \theta_2 q$ for $q = p/k \in \mathbb{Q}$, so the result is dual to that in Theorem 5.21(c2). As $r_1 = 1$, the perturbation does not vanish, therefore the spirals may initially exhibit patterns related to the ratio θ_1/θ_2. □

Theorem 5.22 ($r_1 = r_2 = r$) *Let $(x_n)_{n\geq 0}$ be the sequence defined by (5.47) for initial conditions $x_0 = a$, $x_1 = b$ and assume that the polynomial (5.2) has distinct roots $z_1 \neq z_2$ such that $1 < r_1 = r_2 = r$. The following patterns emerge.*

(a) If $\theta_1 = p_1/k_1, \theta_2 = p_2/k_2 \in \mathbb{Q}$, then the orbit consists of $\text{lcm}(k_1, k_2)$ *divergent rays (Figure 5.33a).*
(b) If $\theta_1 \in \mathbb{Q} = p_1/k_1$ and $\theta_2 \in \mathbb{R} \setminus \mathbb{Q}$, then the orbit has k_1 divergent spirals (Figure 5.33b).
(c) If $\theta_1, \theta_2 \in \mathbb{R} \setminus \mathbb{Q}$, then we identify the following distinct cases.

(c1) If $\theta_2 - \theta_1 = q \in \mathbb{Q}$, then the orbit consists of multiple divergent spirals (Figure 5.33c).
(c2) If $\theta_1 = \theta_2 q$, $q \in \mathbb{Q}$, then the orbit has multi-chamber divergent contours (Figure 5.33d).
(c3) If $1, \theta_1, \theta_2$ are linearly independent over $\mathbb{Q}$, then the orbit consists of erratic divergent spirals (Figure 5.34).

Proof As both $(Az_1^n)_{n\geq 0}$ and $(Bz_2^n)_{n\geq 0}$ diverge, $(x_n)_{n\geq 0}$ diverges as well.

(a) The graph is made of $\text{lcm}(k_1, k_2)$ rays, which diverge to infinity.
(b) The situation is dual to that for convergent spirals (Figure 5.33b).

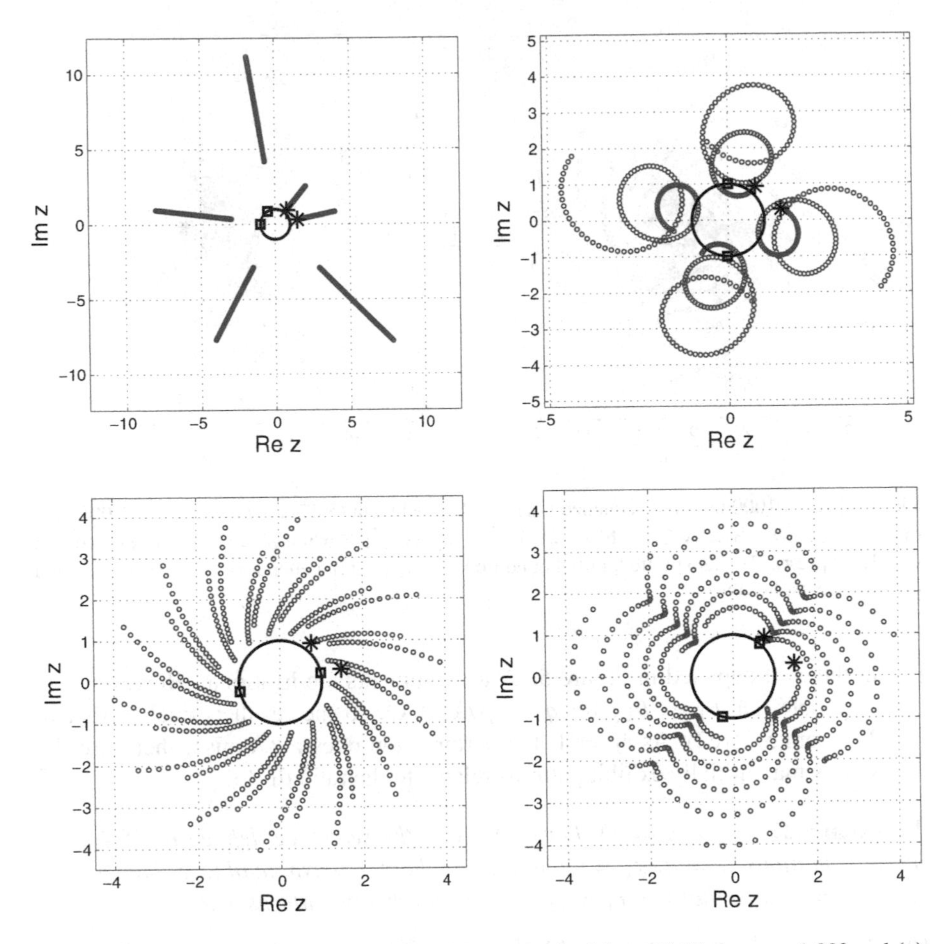

Fig. 5.33 First 500 terms of the sequence $(x_n)_{n \geq 0}$ obtained from (5.47) for $r = 1.002$ and (**a**) $\theta_1 = 1/3$, $\theta_2 = 1/2$; (**b**) $\theta_1 = \sqrt{5}/3$, $\theta_2 = 1/4$; (**c**) $\theta_1 = 5\sqrt{2}/2 + 1/2$, $\theta_2 = 5\sqrt{2}/2$; (**d**) $\theta_1 = \sqrt{2}/2$, $\theta_2 = \sqrt{2}/10$, when $a = 1.5\exp(2\pi i/30)$ and $1.2\exp(2\pi i/7)$. Also plotted, initial conditions x_0, x_1 (stars), generators z_1, z_2 (squares), and unit circle S (solid line)

(c1) As in Figure 5.33c, the orbits have two distinct spirals for $\theta_1 - \theta_2 = 1/2$.
(c2) The graph may have multiple chamber domains when $\theta_1 = q\theta_2$. In Figure 5.33d we have $q = 5$ and the domain made of four petals.
(c3) When θ_1 and θ_2 are not linearly independent over $\mathbb{Q}$ we have Figure 5.34, for $\theta_1 = \exp(1)/1000$, $\theta_2 = \pi/1000$, $r = 1.001$. The orbit looks like a regular spiral up to $N = 1500$ terms Figure 5.34a. However, the pattern becomes irregular if $N = 2400$ terms are plotted, as shown in Figure 5.34b.

□

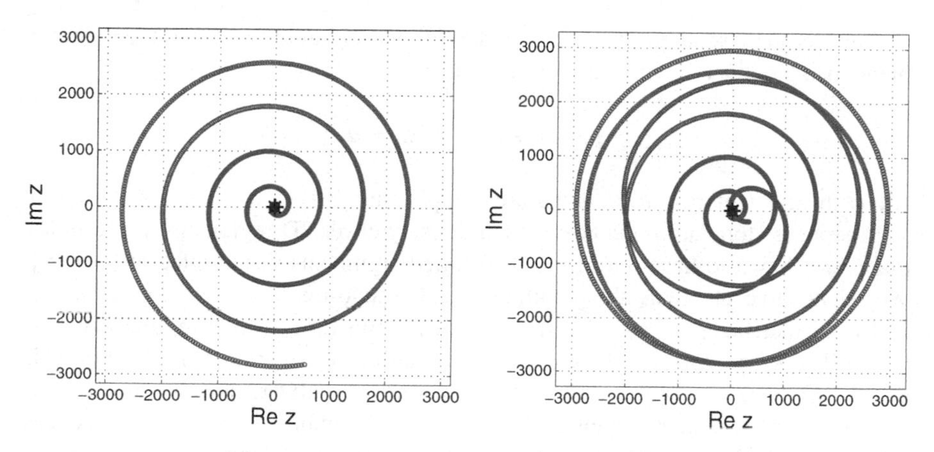

Fig. 5.34 First N terms of the sequence $(x_n)_{n\geq 0}$ obtained from (5.47) for (**a**) $\theta_1 = \exp(1)/1000$, $\theta_2 = \pi/1000$, $r = 1.001$, $N = 1500$; (**b**) $\theta_1 = \exp(1)/1000$, $\theta_2 = \pi/1000$, $r = 1.001$, $N = 2400$; when $a = 1.5\exp(2\pi i/30)$ and $1.2\exp(2\pi i/7)$. Also plotted, initial conditions x_0, x_1 (stars), generators z_1, z_2 (squares), and unit circle S (solid line)

5.7 A Horadam-Based Pseudo-Random Number Generator

Uniformly distributed pseudo-random number generators are a common feature in many numerical simulations. We here present some a random number generation algorithm based on the geometric properties of complex Horadam sequences. For certain parameters, the sequence exhibits uniformity in the distribution of arguments. This feature is exploited to design a pseudo-random number generator which is evaluated using Monte Carlo π estimations, and is found to perform comparatively with commonly used generators like Multiplicative Lagged Fibonacci and the "twister" Mersenne.

5.7.1 *Pseudo-Random Generators and Horadam Sequences*

A random number generator is a core component of statistical sampling algorithms involving and simulations. For example, Monte Carlo methods are widely used for numerically solving integrals, when it is desirable to have a random sequence that closely resembles the underlying differential equations. Current implementations of pseudo-random number generator are based on classical methods including *Linear Congruences* and *Lagged Fibonacci Sequences* [132]. Linear recurrent sequences are commonly used in the generation of pseudo random numbers [136]

We will discuss and evaluate the properties of a random generator based on Horadam sequences. Our examination focused on the following requirements [76]: period, uniformity, and correlation.

The Horadam sequence $(x_n)_{n\geq 0}$ extends Fibonacci numbers [79] to the complex plane, being defined by the recurrence relation

$$x_{n+2} = px_{n+1} + qx_n, \quad x_0 = a, x_1 = b,$$

where the parameters a, b, p, and q are complex numbers.

Aperiodic Horadam sequences which densely cover 2D regions in the complex plane are shown to have uniformly distributed arguments in the interval $[-\pi, \pi]$. We use this to evaluate π, along with classical pseudo-random number generators.

Recall that for $z = re^{2\pi i\theta}$ and $r = 1$, the orbit $(z^n)_{n\geq 0}$ is a regular k-gon if $\theta = j/k \in \mathbb{Q}$, $\gcd(j, k) = 1$, or a dense (and uniformly distributed) subset of S, if $\theta \in \mathbb{R}\setminus\mathbb{Q}$. For $r < 1$ or $r > 1$, one obtains inward, or outward spirals, respectively. The orbits of Horadam sequences (5.4) are linear combinations of the sequences $(z_1^n)_{n\geq 0}$ and $(z_2^n)_{n\geq 0}$ for certain generators z_1, z_2 and coefficients A and B [27].

It is known that for an irrational number θ, the sequence $(n\theta)_{n\geq 0}$ is equidistributed mod 1 (Weyl's criterion). This property represents the basis for a novel random number generator, which is used to evaluate the value of π.

5.7.2 Complex Arguments of 2D Dense Horadam Orbits

Some orbits are dense within a circle or an annulus centered in the origin. Specifically, if $r_1 = r_2 = 1$ with the generators $z_1 = e^{2\pi i\theta_1} \neq z_2 = e^{2\pi i\theta_2}$ such that $1, \theta_1, \theta_2$ are linearly independent over $\mathbb{Q}$, then the orbit of the sequence $(x_n)_{n\geq 0}$ is dense in $U(0, ||A| - |B||, |A| + |B|)$, as in Figure 5.35a, for the parameters $r_1 = r_2 = 1, \theta_1 = \frac{\sqrt{2}}{3}, \theta_2 = \frac{\sqrt{5}}{15}$ and $a = 2 + \frac{2}{3}i$, $b = 3 + i$. These dense orbits are used to design a random number generator.

In the case $|A| = |B|$ shown in Figure 5.35b, for the parameters $r_1 = r_2 = 1$, $\theta_1 = \exp(1)/2$, $\theta_2 = \exp(2)/4$, and $a = 1 + 1/3i$, $b = 1.5a\exp(\pi(\theta_1 + \theta_2))$, we examine the periodicity, uniformity, and autocorrelation for the arguments of terms of sequence $(x_n)_{n\geq 0}$ defined by (5.4).

5.7.2.1 Argument of Horadam Sequence Terms

If $A = R_1e^{i\phi_1}$, $B = R_2e^{i\phi_2}$, $z_1 = e^{2\pi i\theta_1}$, $z_2 = e^{2\pi i\theta_2}$, the term x_n in polar form is

$$re^{i\theta} = Az_1^n + Bz_2^n = R_1e^{i(\phi_1+2\pi n\theta_1)} + R_2e^{i(\phi_2+2\pi n\theta_2)}.$$

Denoting $\varphi_1 = \phi_1 + 2\pi n\theta_1$, $\varphi_2 = \phi_2 + 2\pi n\theta_2$, one can write

$$re^{i\theta} = R_1e^{i\varphi_1} + R_2e^{i\varphi_2}, \quad \varphi_1, \varphi_2 \in \mathbb{R}, \quad R_1, R_2 > 0. \tag{5.55}$$

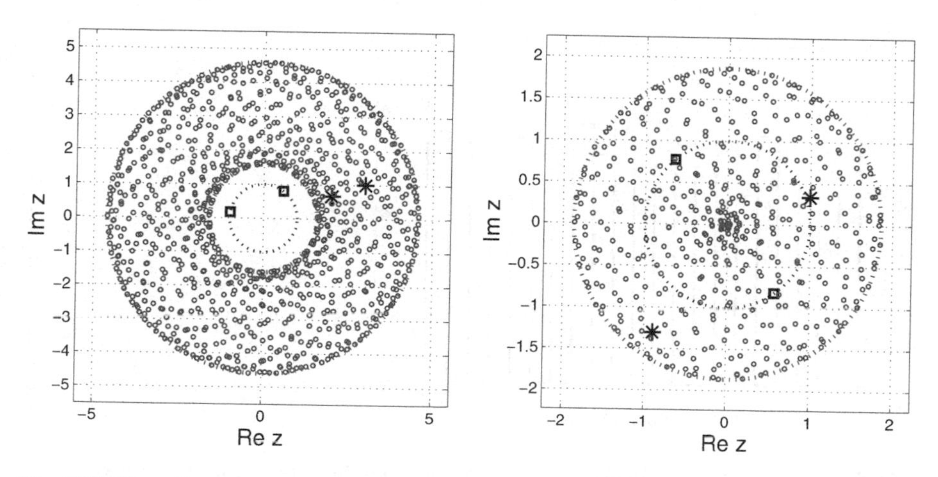

Fig. 5.35 First N sequence terms of $(x_n)_{n\geq 0}$ given by (5.4). (**a**) $N = 1000$, $|A| \neq |B|$; (**b**) $N = 500$, $|A| = |B|$. Also plotted, initial conditions a, b (stars), generators z_1, z_2 (squares), unit circle S (solid line), and circle $U(0, |A| + |B|)$ (dotted line)

For $R = R_1 = R_2$, we obtain $\theta = \frac{1}{2}(\varphi_1 + \varphi_2)$.

5.7.2.2 Periodicity of Arguments

Using the formula for θ, one obtains $\arg(x_n) = \frac{\phi_1+\phi_2}{2} + 2\pi n(\theta_1+\theta_2)$. When $1, \theta_1, \theta_2$ are linearly independent over $\mathbb{Q}$, the sequence $\arg(x_n)$ is aperiodic. This property is also valid for the sequence of normalized arguments $(\arg(x_n) + \pi)/(2\pi)$.

5.7.2.3 Uniform Distribution of Arguments

The sequence of arguments produced by certain Horadam sequences is uniformly distributed in the interval $[-\pi, \pi]$.

When both θ_1 and θ_2 are irrational, the arguments $\varphi_1 = \phi_1 + 2\pi n\theta_2$, and $\varphi_2 = \phi_2 + 2\pi n\theta_2 \in [-\pi, \pi]$ are uniformly distributed modulo 2π, hence θ is uniformly distributed on $[-\pi, \pi]$. The normalized argument $(\theta + \pi)/(2\pi)$ is then uniformly distributed in $[0, 1]$, as seen in Figure 5.36a.

5.7.2.4 Autocorrelation of Arguments

A test for the quality of pseudo-random number generators is the autocorrelation test [76]. For a good quality generator, the 2D diagrams of normalized arguments $((\arg(x_n) + \pi)/(2\pi), (\arg(x_{n+1}) + \pi)/(2\pi))$ should uniformly cover the unit square. The plot depicted in Figure 5.36b suggests that consecutive arguments are

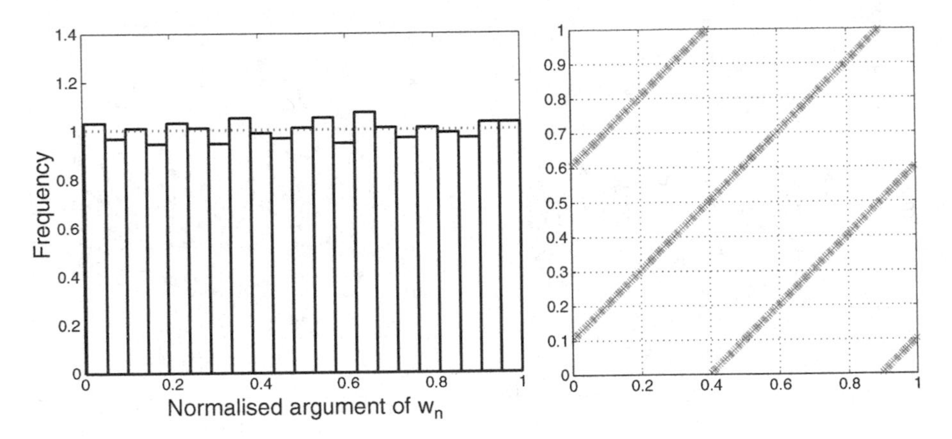

Fig. 5.36 (**a**) Histogram of $\frac{\arg(x_n)+\pi}{2\pi}$ vs uniform density on $[0, 1]$. (**b**) Normalized angle correlations: $\left[\frac{\arg(x_n)+\pi}{2\pi}, \frac{\arg(x_{n+1})+\pi}{2\pi}\right]$

very correlated. This issue can be addressed in several ways. One is to combine arguments produced by two distinct Horadam sequences, while the other approach is to use generalized Horadam orbits.

5.7.3 *Monte Carlo Simulations for Mixed Arguments*

Here we test the performance of the uniform distribution of normalized Horadam angles as a pseudo-random number generator, evaluating the results against those produced by Multiplicative Lagged Fibonacci and Mersenne Twister (the latter being the default random number generator implemented in Matlab).

A Monte Carlo simulation approximating the value of π could involve randomly selecting points $(X_n, Y_n)_{n=1}^N$ in the unit square and determining the ratio $\rho = m/N$, where m is number of points that satisfy $X_n^2 + Y_n^2 \leq 1$. Here we use two sequences (w_n^1) and (w_n^2) computed from formula (5.4).

The parameters are $\theta_1 = \frac{e}{2}, \theta_2 = \frac{e^2}{4}$ for (w_n^1), and $\theta_1 = \frac{e}{10}, \theta_2 = \frac{\pi}{10}$ for (w_n^2), with initial conditions $a = 1 + \frac{1}{3}i$, $b = 1.5a \exp(\pi(\theta_1 + \theta_2))$. The 2D coordinates plotted in Figure 5.37 represent normalized arguments of Horadam sequence terms, given by the formula

$$(X_n, Y_n) = \left(\frac{\arg(w_n^1) + \pi}{2\pi}, \frac{\arg(w_n^2) + \pi}{2\pi}\right). \tag{5.56}$$

In the simulation shown in Figure 5.37a, the sample size is $N = 1000$ and there are 792 points satisfying $X_n^2 + Y_n^2 \leq 1$. Using this data, one obtains

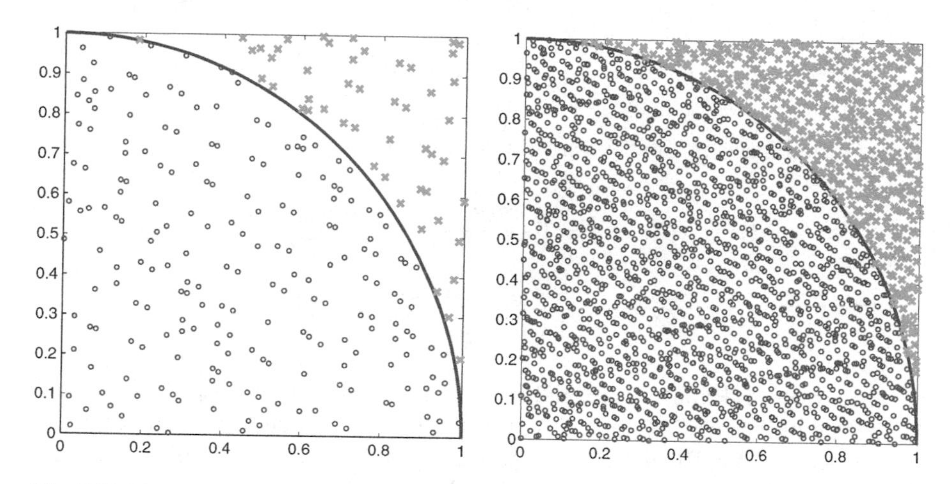

Fig. 5.37 First (**a**) 1000; (**b**) 10000 points having coordinates (X_n, Y_n) given by (5.56). Also represented, points inside (circles) and outside (crosses) the unit circle (solid line)

Table 5.1 10^N is the sample size used in each simulation

10^N	H1	H2	F1	F2	MT1	MT2
1	0.0584	0.0584	−0.3297	−0.3297	−0.7258	0.8584
2	0.2584	−0.0216	0.0985	−0.0615	−0.0215	0.0985
3	0.0784	−0.0016	−0.0136	0.0304	−0.0456	0.0264
4	0.0104	0.0004	0.0092	0.0200	0.0036	0.0096
5	0.0012	−0.0006	−0.0016	−0.0018	0.0004	−0.0034
6	0.0003	0.0000	−0.0001	−0.0010	−0.0026	−0.0015
7	0.0000	0.0000	0.0003	−0.0006	−0.0002	0.0004

$$\rho = \frac{792}{1000} = 0.792 \text{ and } \pi \sim 4\rho = 3.1680.$$

The value improves with the increase in the number of sequence terms, to 3.1420 for $N = 10^4$ (depicted in Figure 5.37b) and to 3.141888 for $N = 10^6$.

A more detailed illustration of this convergence is shown in Table 5.1. Sequences H1 and H2 represent simulations for π obtained from pairs of Horadam sequences. In particular, H1 was obtained from sequences x_n^1 and x_n^2, while H2 from sequences x_n^1 and x_n^3 given below by $(\theta_1, \theta_2, a, b)$

$$x_n^1 : \left(\frac{e}{2}, \frac{e^2}{4}, 1 + \frac{1}{3}i, 1.5ae^{\pi(\theta_1+\theta_2)}\right)$$

$$x_n^2 : \left(\frac{e}{10}, \frac{\sqrt{5}}{15}, 1 + \frac{2}{3}i, 1.5ae^{\pi(\theta_1+\theta_2)}\right)$$

$$x_n^3 : \left(\frac{\sqrt{2}}{3}, \frac{e}{15}, 1 + \frac{2}{3}i, 1.5ae^{\pi(\theta_1+\theta_2)} \right).$$

The 2D coordinates in Table 5.1 are obtained for the sequences

$$\text{H1}: \quad (X_n, Y_n) = \left(\frac{\arg(w_n^1) + \pi}{2\pi}, \frac{\arg(w_n^2) + \pi}{2\pi} \right),$$

$$\text{H2}: \quad (X_n, Y_n) = \left(\frac{\arg(w_n^1) + \pi}{2\pi}, \frac{\arg(w_n^3) + \pi}{2\pi} \right).$$

The sequences F1 and F2 are produced using two coordinates (X_n, Y_n) simulated by the "multFibonacci" Multiplicative Lagged Fibonacci pseudo-random generator. The generators for F1 and F2 had periodicities 2^{31} and 2^{16}, respectively. The difference in the convergence rate is noticeable.

The sequences MT1, MT2 are produced using two coordinates (X_n, Y_n) simulated by the "twister" Mersenne Twister pseudo-random generator. Both MT1 and MT2 used the default seed value as implemented in Matlab.

The convergence is non-monotonic for all methods, although it appears more rapid for our Horadam based generator. This is dependent on the choice of initial parameters. Further examination is needed to fully describe the relationship between seeds and the convergence rate.

5.8 Nonhomogeneous Horadam Sequences

Here we study perturbed versions of the Horadam sequence (5.1), given by

$$x_{n+2} = px_{n+1} + qx_n + u_n, \quad x_0 = a, x_1 = b, \tag{5.57}$$

where a, b, p, and q are complex numbers, while the sequence $(u_n)_{n\geq 0}$ will represent a constant or periodic perturbation, based on [31]. When $(u_n)_{n\geq 0}$ is self-repeating, we show that $(x_n)_{n\geq 0}$ is periodic, and discuss the value of the period. For simplicity, we assume that generators z_1, z_2 are distinct.

5.8.1 Constant Perturbation

Here the perturbation sequence $(u_n)_{n\geq 0}$ is constant.

Theorem 5.23 *Let a, b, c, p, and q be complex numbers, and consider the recurrence relation*

$$x_{n+2} = px_{n+1} + qx_n + c, \quad x_0 = a, x_1 = b. \tag{5.58}$$

If 1, z_1, *and* z_2 *are distinct, the general term of the recurrence sequence is*

$$x_n = Az_1^n + Bz_2^n + C, \tag{5.59}$$

where the constants A, B, C *can be computed explicitly from*

$$A = \frac{\tilde{a}z_2 - \tilde{b}}{z_2 - z_1}, \quad B = \frac{\tilde{b} - \tilde{a}z_1}{z_2 - z_1}, \quad C = \frac{c}{(z_1 - 1)(z_2 - 1)}, \tag{5.60}$$

while $\tilde{a} = a - C$ *and* $\tilde{b} = b - C$.

Theorem 5.24 *When* $1, z_1, z_2$ *are distinct, the sequence given by* (5.58) *is periodic if and only if* z_1 *and* z_2 *are distinct roots of unity. The periodic orbit is contained within the annulus* $U(C, ||A| - |B||, |A| + |B|)$ *with* A, B, C *given by (5.60). This property is illustrated in Figure 5.38.*

5.8.2 Periodic Perturbations

When the sequence $(u_n)_{n\geq 0}$ has period k, there is a finite set

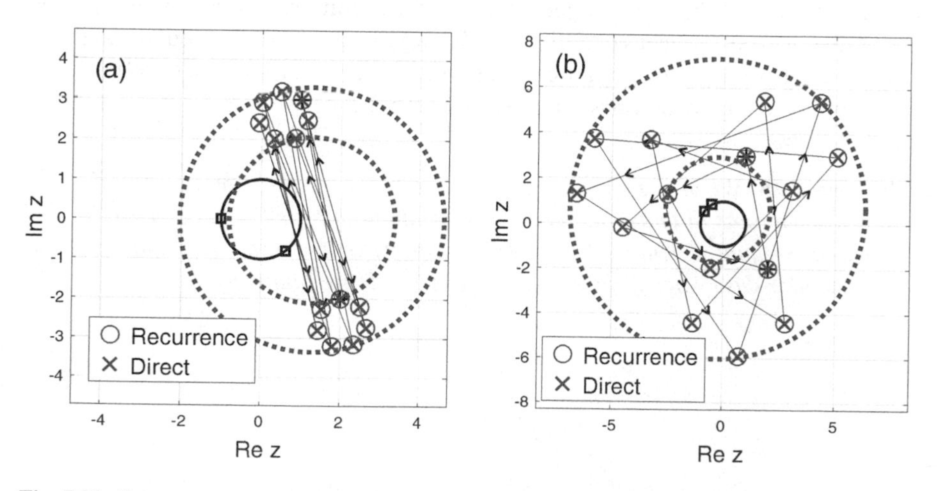

Fig. 5.38 Terms $x_0, x_1, \ldots, x_N$ obtained from recurrence equation (5.58) (circles) and direct formula (5.59) (crosses) for initial conditions $a = 2 - 2i$, $b = 1 + 3i$, (stars), when $c = 1 + 2i$ and generator pairs (squares) (**a**) $z_1 = e^{2\pi i \frac{1}{2}}$ and $z_2 = e^{2\pi i \frac{6}{7}}$; (**b**) $z_1 = e^{2\pi i \frac{1}{3}}$ and $z_2 = e^{2\pi i \frac{2}{5}}$. Arrows indicate the direction of the orbit. Boundaries of $U(C, ||A| - |B||, |A| + |B|)$ (dotted line) with A, B, C from (5.60) and unit circle (solid line) are also shown

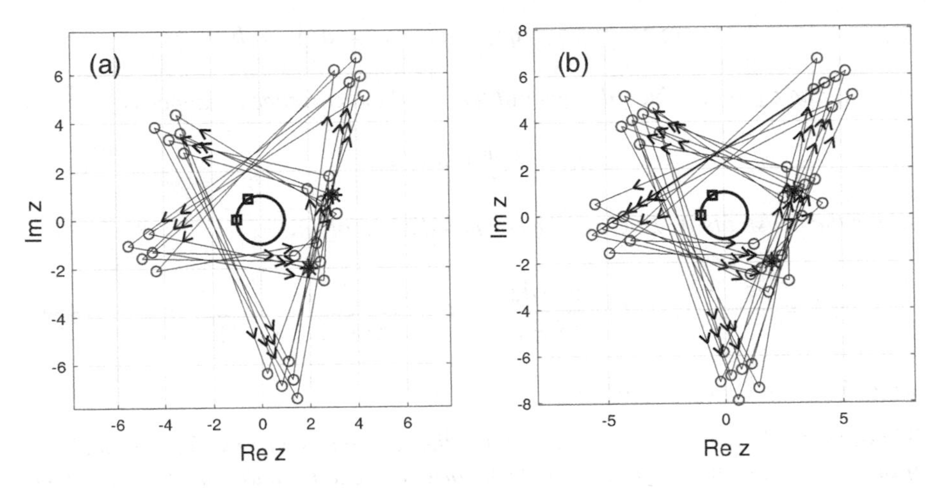

Fig. 5.39 First N terms of sequence $(x_n)_{n\geq 0}$ (circles) from (5.57) for the initial conditions $a = 2-2i$, $b = 3+i$ (stars) and the generator pair (squares) $z_1 = e^{2\pi i \frac{1}{2}}$ and $z_2 = e^{2\pi i \frac{1}{3}}$, while the sequence U given by (5.61) is (**a**) $U = \{1, 2, 3, 4, 5\}$ and (**b**) $U = \{1, 2, 3, 4, 5, 6, 7\}$. Arrows indicate the direction of the orbit

$$U = \{u_1, \ldots, u_{k-1}\}, \tag{5.61}$$

such that $u_{n+k} = u_n$ for all $n \geq 0$.

A closed formula for the general sequence term can be obtained, based on the argument used for the constant perturbations. The orbit $(x_n)_{n\geq 0}$ can be partitioned into k distinct subsequences $(x_{j+nk})_{n\geq 0}$, whose general term can be found explicitly. The techniques are similar to those used in the derivation of formula (5.59).

The periodicity conditions in this case can be derived from those for the homogeneous problem.

Theorem 5.25 *When* $1, z_1, z_2$ *are distinct and* $z_1 = e^{2\pi i p_1/k_1}$ *and* $z_2 = e^{2\pi i p_2/k_2}$ *are primitive roots of unity, the recurrence relation given by* (5.57) *is periodic and its period is equal to* $\mathrm{lcm}(k_1, k_2, k)$. *This property is illustrated in Figure 5.39, in the particular case when* $k_1 = 2$, $k_2 = 3$, *and* $k = 5$ *are relatively prime.*

Chapter 6
Higher Order Linear Recurrent Sequences

This chapter presents key properties of the complex linear recurrent sequences of arbitrary order, including general term formulae, periodicity, or the asymptotic behavior in some particular cases. We also give results for systems of linear recurrent sequences. Following the work of Horadam who investigated second-order general recurrent sequences of complex numbers in general [79], the sequence arising from (6.1) has also been called a generalized Horadam sequence. Many well-known sequences are obtained for particular choices of the initial conditions and recurrence relation coefficients. A history of Horadam sequences is given by Larcombe et al. [107].

In Section 6.1 we provide some general context and exact formulae for the general term of a linear recurrent sequence, and we show that the order of a linear recurrent relation can be reduced at the cost of linearity, following Andrica and Buzeţeanu in [17, 18].

In Section 6.2 we present results concerning systems of linear recurrent sequences, based on the works of Andrica and Toader [21], or Cobzaş [53, Chapter 6, Theorem 2.8]. We first deduce the general solution, then we examine applications involving systems of two and three linear recurrent sequences. Some of these are related to Diophantine Equations [7].

In Section 6.3 we discuss necessary and sufficient periodicity conditions for generalized complex Horadam sequences. The results use formulae for the general term of linear recurrent sequences with arbitrary coefficients, involving the initial conditions $\alpha_1, \ldots, \alpha_m$ and the generators $z_1, \ldots, z_m$, representing the nonzero roots (distinct or equal) of the characteristic equation (6.1).

In Section 6.4 we present the geometric structure of periodic orbits, and enumerate the self-repeating orbits of complex linear recurrent sequences of fixed length. Finally, in Section 6.5 we discuss some asymptotic properties of orbits whose generators are roots of unity.

Numerical implementations of the general term of complex linear recurrent sequences are detailed in Appendix A.3.

D. Andrica, O. Bagdasar, *Recurrent Sequences*, Problem Books in Mathematics,
https://doi.org/10.1007/978-3-030-51502-7_6

6.1 Linear Recurrent Sequences (LRS)

Here we present basic results in the theory of linear recurrent sequences. We first discuss the general term for recurrent sequences of higher order defined for arbitrary initial values and coefficients, based on [21] and the monograph of Everest et al. [62]. Some recent progress concerning key problems are presented based on [133], then we show how to reduce the order of a linear recurrence relation as in [17, 18].

6.1.1 Definition and General Term

Let $m \geq 2$ be a natural number. A *linear recurrence sequence* (LRS) is an infinite sequence $(x_n)_{n\geq 0}$ satisfying the recurrence relation

$$x_n = a_1 x_{n-1} + a_2 x_{n-2} + \cdots + a_m x_{n-m}, \quad m \leq n \in \mathbb{N}. \tag{6.1}$$

If a_i, $i = 1, \ldots, m$ (recurrence coefficients), and α_i, $i = 1, \ldots, m$ (initial conditions) are fixed complex numbers satisfying $a_m \neq 0$ and $x_{i-1} = \alpha_i$, $i = 1, \ldots, m$, then the recurrence relation has order m and is uniquely defined.

The polynomial defined by

$$f(x) = x^m - a_1 x^{m-1} - \cdots - a_{m-1} x - a_m, \tag{6.2}$$

is called the *characteristic polynomial* of the LRS (6.1). The general term of the recursion is given by

$$x_n = P_1(n) z_1^n + \cdots + P_m(n) z_m^n,$$

where $z_1, \ldots, z_m$ are roots of (6.2), and P_i, $i = 1, \ldots, m$ are polynomials in n. If $z_1, \ldots, z_m$ are distinct, then the LRS is *simple* and $P_1, \ldots, P_m$ are constant.

The theory of LRS is a vast subject with extensive applications in mathematics and other sciences. Many properties and results are presented in the monograph of Everest et al. [62]. Decision problems involving LRS with rational terms are discussed by Ouaknine in [133] and other of his papers.

Problem 1 (Skolem) Does $x_n = 0$ for some n?

Problem 2 Is $x_n = 0$ for infinitely many n?

Problem 3 (Positivity) Does $x_n \geq 0$ for all n?

Problem 4 (Ultimate Positivity) Does $x_n \geq 0$ for all but finitely many n?

Berstel and Mignotte showed that Problem 2 is decidable [37]. Ouaknine made progress on Problem 3 for simple LRS of order $m \le 9$ [134], or arbitrary LRS of order $m \le 5$ [135], respectively. Problem 4 holds for simple LRS [133].

6.1.2 The Solution of a Linear Recurrence Equation

Let the sequence $(x_n)_{n\ge 0}$ be defined by the linear recurrence relation (6.1), where $a_1, \dots, a_m \in \mathbb{C}$, $a_m \ne 0$, and $x_0, \dots, x_{m-1} \in \mathbb{C}$ are given. We want to find the general term x_n for $n \ge m$. For this we consider the generating function of the sequence $(x_n)_{n\ge 0}$

$$F(z) = x_0 + x_1 z + \cdots + x_n z^n + \cdots . \tag{6.3}$$

The following result presents the relationship between homogeneous linear recurrent sequences and their generating function (see, e.g., [114]).

Theorem 6.1 *Assume that the sequence $(x_n)_{n\ge 0}$ satisfies the homogeneous linear recurrence relation (6.1). The associated generating function F can be written as a rational fraction $F(z) = \frac{P(z)}{Q(z)}$, where Q is a polynomial of degree m with nonzero constant term and P is a polynomial of degree strictly less than m.*

Conversely, for any such polynomials P and Q, there exists a unique sequence $(x_n)_{n\ge 0}$ satisfying the linear homogeneous recurrence relation (6.1), having the generating function given by the rational function P/Q.

Proof From (6.1) and (6.3), the generating function F for the sequence $(x_n)_{n\ge 0}$ can be written as

$$\begin{aligned}
F(z) &= \sum_{j=0}^{m-1} x_j z^j + \sum_{n=m}^{\infty} x_n z^n = \sum_{j=0}^{m-1} x_j z^j + \sum_{n=m}^{\infty} \left(\sum_{j=1}^{m} a_j x_{n-j} \right) z^n \\
&= \sum_{j=0}^{m-1} x_j z^j + \sum_{j=1}^{m} a_j \sum_{n=k}^{\infty} x_{n-j} z^n = \sum_{j=0}^{m-1} x_j z^j + \sum_{j=1}^{m} a_j \sum_{n=m-j}^{\infty} x_n z^{n+j} \\
&= \sum_{j=0}^{m-1} x_j z^j + a_m z^m \sum_{n=0}^{\infty} x_n z^n + \sum_{j=1}^{m-1} a_j z^j \left(\sum_{n=0}^{\infty} x_n z^n - \sum_{i=0}^{m-j-1} x_i z^i \right) \\
&= \sum_{j=0}^{m-1} x_j z^j + F(z) \sum_{j=1}^{m} a_j z^j - \sum_{j=1}^{m-1} a_j z^j \sum_{i=0}^{m-j-1} x_i z^i \\
&= F(z) \sum_{j=1}^{m} a_j z^j + \sum_{j=0}^{m-1} x_j z^j - \sum_{s=1}^{m-1} z^s \sum_{j=1}^{s} a_j x_{s-j} .
\end{aligned}$$

It follows that

$$F(z)\left(1-\sum_{j=1}^{m}a_jz^j\right)=\sum_{j=0}^{m-1}x_jz^j-\sum_{s=1}^{m-1}z^s\sum_{j=1}^{s}a_jx_{s-j}$$
$$=x_0+\sum_{s=1}^{m-1}\left(x_s-\sum_{j=1}^{s}a_jx_{s-j}\right)z^s.$$

We obtain the polynomials $P\in C_{m-1}[z]$ and $Q\in C_m[z]$ defined by

$$P(z)=x_0+\sum_{s=1}^{m-1}\left(x_s-\sum_{j=1}^{s}a_jx_{s-j}\right)z^s$$
$$Q(z)=1-\sum_{j=1}^{m}a_jz^j.$$

Conversely, if $(x_n)_{n\geq 0}$ is the sequence having the generating function $F(z)=P(z)/Q(z)$, we can write

$$F(z)=\sum_{n=0}^{\infty}x_nz^n,\quad P(z)=\sum_{j=0}^{k}b_jz^j,\quad Q(z)=1-\sum_{j=1}^{m}a_jz^j.$$

Clearly, from $F(z)=\frac{P(z)}{Q(z)}$ we obtain the formula

$$\left(1-\sum_{j=1}^{m}a_jz^j\right)\left(\sum_{n=0}^{\infty}x_nz^n\right)=\sum_{j=0}^{m}b_jz^j.$$

Writing polynomial Q as an infinite series with $a_j=0$ for $j>m$, we have

$$\sum_{n=0}^{\infty}x_nz^n-\sum_{n=0}^{\infty}\left(\sum_{j=1}^{n}a_jx_{n-j}\right)=\sum_{j=0}^{m}b_jz^j.$$

Identifying the coefficients of z^n, one obtains the recurrence relation

$$x_n=\sum_{j=1}^{m}a_jx_{n-j},\quad n\geq m.$$

□

Remark 6.1 We can write F as a sum of simple fractions. If the polynomial

$$f^*(z) = 1 - a_1 z - \cdots - a_k z^k,$$

has the distinct roots $z_1^*, \ldots, z_m^*$, with multiplicities $q_1, \ldots, q_m$, then

$$F(z) = \frac{P(z)}{f^*(z)} = \sum_{i=1}^{m} \sum_{j=1}^{q_i} \frac{f_{ij}}{\left(z - z_i^*\right)^j}. \tag{6.4}$$

Lemma 6.1 *The coefficients f_{ij} in formula* (6.4) *is*

$$f_{ij} = \frac{1}{(q_i - j)!} \left[(z - z_i^*)^{q_i} F(z)\right]_{z=z_i^*}^{(q_i - j)}, \quad j = 1, \ldots, q_i, \quad i = 1, \ldots, m. \tag{6.5}$$

Proof By (6.4) we have for $i = 1, \ldots, m$

$$F(z) = \frac{f_{i1}}{z - z_i^*} + \cdots \frac{f_{iq_i}}{\left(z - z_i^*\right)^{q_i}} + h_i(z),$$

or

$$\left(z - z_i^*\right)^{q_i} F(z) = f_{i1} \left(z - z_i^*\right)^{q_i - 1} + \cdots + f_{iq_i} + \left(z - z_i^*\right)^{q_i} h_i(t),$$

which gives (6.5) by successive differentiation. □

Lemma 6.2 *If $u \neq 0$ and $\left|\frac{z}{u}\right| < 1$, then for all integers $j \geq 1$ we have*

$$\frac{1}{(z - u)^j} = (-1)^j u^{-j} \sum_{n \geq 0} \binom{n + j - 1}{j - 1} \left(\frac{z}{u}\right)^n. \tag{6.6}$$

Proof The result follows from operations with formal series given in [48], by differentiating the identity below $j - 1$ times with respect to z

$$\frac{1}{z - u} = -\frac{1}{u} \sum_{n \geq 0} \left(\frac{z}{u}\right)^n.$$

□

Theorem 6.2 *The sequence $(x_n)_{n \geq 0}$ defined by* (6.1) *is given by*

$$x_n = P_1(n) z_1^n + \cdots + P_m(n) z_m^n, \quad n \geq k, \tag{6.7}$$

where the $z_1, \ldots, z_m$ are the distinct roots of the polynomial

$$f(z) = z^k - a_1 z^{k-1} - \cdots - a_k,$$

with the orders of multiplicity $q_1, \ldots, q_m$, *and*

$$P_i(n) = \sum_{j=1}^{q_i} f_{ij} z_i^j \binom{n+j-1}{j-1}, \quad i = 1, \ldots, m. \tag{6.8}$$

Proof As $z_i = \frac{1}{z_i^*}$, from (6.4) and (6.6) we get

$$F(z) = \sum_{i=1}^{m} \sum_{j=1}^{q_i} \frac{f_{ij}}{\left(z - z_i^*\right)^j} = \sum_{i=1}^{m} \sum_{j=1}^{q_i} \sum_{n \geq 0} (-1)^j f_{ij} z_i^j \binom{n+j-1}{j-1} \left(\frac{z}{z_i^*}\right)^n$$

$$= \sum_{n \geq 0} \sum_{i=1}^{m} \sum_{j=1}^{q_i} (-1)^j z_i^{n+j} f_{ij} \binom{n+j-1}{j-1} z^n = \sum_{n \geq 0} \left(\sum_{i=1}^{m} P_i(n) z_i^n \right) z^n,$$

which is exactly (6.7). □

Example 6.1 The sequence $(x_n)_{n \geq 1}$ is defined as $x_1 = 20$, $x_2 = 12$, and

$$x_{n+2} = x_n + x_{n+1} + 2\sqrt{x_n x_{n+1} + 121}, \quad n \geq 1.$$

1° Compute x_{10};
2° Determine with justification if every term in the sequence is an integer.

Solution It is clear that $x_3 = 20 + 12 + 2 \cdot 19 = 70$. Notice that

$$\begin{aligned}
x_{n+3} &= x_{n+1} + x_{n+2} + 2\sqrt{x_{n+1}x_{n+2} + 121} \\
&= x_{n+1} + x_{n+2} + 2\sqrt{x_{n+1}\left(x_n + x_{n+1} + 2\sqrt{x_n x_{n+1} + 121}\right) + 121} \\
&= x_{n+1} + x_{n+2} + 2\sqrt{x_{n+1}^2 + 2\sqrt{x_n x_{n+1} + 121} + x_n x_{n+1} + 121} \\
&= x_{n+1} + x_{n+2} + 2\left(x_{n+1} + \sqrt{x_n x_{n+1} + 121}\right) \\
&= 3x_{n+1} + x_{n+2} + x_{n+2} - x_n - x_{n+1} \\
&= 2x_{n+2} + 2x_{n+1} - x_n.
\end{aligned}$$

Therefore, it follows that $(x_n)_{n \geq 1}$ is an integer sequence and we can compute $x_4 = 144$, $x_5 = 416$, $x_6 = 1010$, $x_7 = 2788$, $x_8 = 7260$, $x_9 = 19046$, $x_{10} = 49824$.

The characteristic equation of $(x_n)_{n \geq 1}$ is $t^3 - 2t^2 - 2t + 1 = 0$, with the roots $t_1 = -1$, $t_2 = \frac{3+\sqrt{5}}{2}$, $t_3 = \frac{3-\sqrt{5}}{2}$. It follows that

$$x_n = c_1(-1)^n + c_2\left(\frac{3+\sqrt{5}}{2}\right)^n + c_3\left(\frac{3-\sqrt{5}}{2}\right)^n, \quad n = 1, 2, \ldots.$$

Solving the system $x_1 = 20$, $x_2 = 12$, $x_3 = 70$, we obtain the coefficients

$$c_1 = -\frac{54}{5}, \quad c_2 = \frac{12}{5} + \frac{2\sqrt{5}}{5}, \quad c_3 = \frac{12}{5} - \frac{2\sqrt{5}}{5}.$$

Furthermore, one can write

$$x_n = \frac{54}{5}(-1)^{n+1} + \frac{1}{5}\left(6+6+2\sqrt{5}\right)\left(\frac{6+2\sqrt{5}}{4}\right)^n$$

$$+\frac{1}{5}\left(6+6-2\sqrt{5}\right)\left(\frac{6-2\sqrt{5}}{4}\right)^n = \frac{54}{5}(-1)^{n+1} + \frac{6}{5}L_{2n} + \frac{4}{5}L_{2n+2},$$

where L_m is the mth Lucas number.

6.1.3 The Space of Solutions for Linear Recurrence Equations

A solution of the recurrence equation (6.1) is any function $x : \mathbb{N} \to \mathbb{C}$ satisfying

$$x(n) = a_1x(n-1) + a_2x(n-2) + \cdots + a_mx(n-m), \quad n \geq m. \tag{6.9}$$

(The notation $x(n)$ is only used in this section, while for the rest of the paper the more compact subscript notation x_n is preferred.) As each solution is completely determined by m initial conditions $x(0), x(1), \ldots, x(m-1)$, the set containing the solutions of (6.9) forms a vector space V of dimension m over $\mathbb{C}$. The general term of the sequence satisfying the recurrence relation (6.9) is then a linear combination of m functions which form a basis for V.

For $\lambda \neq 0$, the characteristic equation (6.10) is equivalent to

$$\lambda^m = a_1\lambda^{m-1} + a_2\lambda^{m-2} + \cdots + a_{m-1}\lambda + a_m.$$

It may be assumed without loss of generality that the order of the recurrence relation cannot be reduced, therefore $c_m \neq 0$. For finding a base of the vector space V, one may first check that the functions $w(n) = \lambda^n$ ($\lambda \neq 0$) are a solution of (6.9), whenever λ is a zero of the *characteristic polynomial*

$$f(x) = x^m - a_1x^{m-1} - a_2x^{m-2} - \cdots - a_{m-1}x - a_m. \tag{6.10}$$

As a complex polynomial, $f(x)$ has exactly m roots. Examples of bases for V for the cases when the roots of (6.10) are all distinct, all equal, or distinct with arbitrary multiplicities are presented below.

Proposition 6.1 *If the polynomial f (6.10) has m distinct roots $z_1, \ldots, z_m$, then*

$$f_1(n) = z_1^n, \; f_2(n) = z_2^n, \; \ldots, \; f_m(n) = z_m^n,$$

form a basis of the vector space V of solutions for the recurrence relation (6.9).

The proof is based on two facts. First, each function $f_i(n)$ is a solution of (6.9). Second, the Vandermonde determinant involving the first m values $1, z_i, \ldots, z_i^{m-1}$ of each function $f_i(n)$ is nonzero for distinct values $z_1, \ldots, z_m$. A detailed proof is presented in [86, Theorem 1].

Proposition 6.2 *If the polynomial f (6.10) has m roots equal to z, then the m sequences*

$$f_1(n) = z^n, \; f_2(n) = nz^n, \; \ldots, \; f_m(n) = n^{m-1}z^n,$$

form a basis of the vector space V of solutions for the recurrence relation (6.9).

The idea of the argument is that a multiple root of polynomial (6.10) is also a root of its derivative. A detailed proof is given in [86, Corollary 1].

Proposition 6.3 *If a characteristic polynomial of a linear recurrence relation of order d has m distinct roots $z_1, \ldots, z_m$ having multiplicities $d_1, \ldots, d_m$ such that $d_1 + \cdots + d_m = d$, then the d sequences*

$$f_{ij}(n) = n^{j-1} z_i^n, \quad 1 \le i \le m, \quad 1 \le j \le d_i, \tag{6.11}$$

form a basis of the vector space V of solutions for the recurrence relation (6.9).

A proof of this results is given in [86, Theorem 2].

6.1.4 Reduction of Order for LRS

It was shown in Chapter 2 that as suggested by Andrica and Buzeţeanu in 1982 [17], a second-order linear recurrence equation can be reduced to a nonlinear recurrence equation of the first order (see Theorem 2.4). In particular, such first-order formulae obtained for Fibonacci, Lucas, Pell, and Pell–Lucas numbers were presented in Theorem 2.9.

Here we present a more general result, concerning the reduction of order for higher order linear recurrence relation, based on results proved by Andrica and Buzeţeanu in 1985 [18]. Specifically, we show that if a sequence $(x_n)_{n\ge 0}$ satisfies a linear recurrence of order $m \ge 2$, then there exists a polynomial relation between any m consecutive terms. This shows that the linear recurrence relation of order m is in fact reduced to a nonlinear recurrence relation of order $m-1$.

Let $m \geq 2$ be an integer and let the sequence $(x_n)_{n\geq 1}$ satisfying

$$x_n = \sum_{r=1}^{m} a_r x_{n-r}, \quad n > m, \tag{6.12}$$

where the starting values $x_i = \alpha_i$, and recurrence coefficients a_i, $i = 1, \ldots, m$, are given real (or complex) numbers such that $a_m \neq 0$.

For $n \geq m$, let us consider the determinant

$$D_n = \begin{vmatrix} x_{n+m-1} & x_{n+m-2} & \cdots & x_{n+1} & x_n \\ x_{n+m-2} & x_{n+m-3} & \cdots & x_n & x_{n-1} \\ & & \cdots\cdots\cdots\cdots & & \\ x_{n+1} & x_n & \cdots & x_{n-m+3} & x_{n-m+2} \\ x_n & x_{n-1} & \cdots & x_{n-m+2} & x_{n-m+1} \end{vmatrix}, \tag{6.13}$$

for which we can obtain a recursive formula.

Theorem 6.3 *Let $(x_n)_{n\geq 1}$ be a sequence given by* (6.12) *and let D_n be defined by formula* (6.13). *For any $n \geq m$, the following relation holds:*

$$D_n = (-1)^{(m-1)(n-m)} a_m^{n-m} D_m. \tag{6.14}$$

Proof Following the steps outlined in [98, 117, 152] (for $m = 2$), we introduce the matrix

$$A_n = \begin{pmatrix} x_{n+m-1} & x_{n+m-2} & \cdots & x_{n+1} & x_n \\ x_{n+m-2} & x_{n+m-3} & \cdots & x_n & x_{n-1} \\ & & \cdots\cdots\cdots\cdots & & \\ x_{n+1} & x_n & \cdots & x_{n-m+3} & x_{n-m+2} \\ x_n & x_{n-1} & \cdots & x_{n-m+2} & x_{n-m+1} \end{pmatrix}. \tag{6.15}$$

It is easy to see that

$$A_{n+1} = \begin{pmatrix} 0 & 1 & 0 & 0 & \cdots & 0 & 0 & 0 \\ 0 & 0 & 1 & 0 & \cdots & 0 & 0 & 0 \\ & & & & \cdots\cdots\cdots & & & \\ 0 & 0 & 0 & 0 & \cdots & 0 & 1 & 0 \\ 0 & 0 & 0 & 0 & \cdots & 0 & 0 & 1 \\ a_1 & a_2 & a_3 & a_4 & \cdots & a_{m-2} & a_{m-1} & a_m \end{pmatrix} A_n, \tag{6.16}$$

hence

$$A_n = \begin{pmatrix} 0 & 1 & 0 & 0 & \dots & 0 & 0 & 0 \\ 0 & 0 & 1 & 0 & \dots & 0 & 0 & 0 \\ & & & \dots & \dots & & & \\ 0 & 0 & 0 & 0 & \dots & 0 & 1 & 0 \\ 0 & 0 & 0 & 0 & \dots & 0 & 0 & 1 \\ a_1 & a_2 & a_3 & a_4 & \dots & a_{m-2} & a_{m-1} & a_m \end{pmatrix}^{n-m} A_m. \tag{6.17}$$

By taking determinants in (6.17), we obtain

$$\left((-1)^{m-1}a_m\right)^{n-m} D_m = D_n,$$

for $n \geq m$, which ends the proof. □

Theorem 6.4 *Let $(x_n)_{n\geq 1}$ be a sequence given by* (6.12). *There exists a polynomial function of degree m defined by $F_m : \mathbb{C}^m \to \mathbb{C}$, so that the following relation holds:*

$$F_m(x_n, x_{n-1}, \dots, x_{n-m+1}) = (-1)^{(m-1)(n-m)} a_m^{n-m} F_m(\alpha_m, \alpha_{m-1}, \dots, \alpha_1). \tag{6.18}$$

Notice that from the recurrence relation (6.12) one can compute D_m as a function of the known values $\alpha_1, \alpha_2, \dots, \alpha_m$. For the same reason, one can also express D_n as a function of the terms $x_n, x_{n-1}, \dots, x_{n-m+1}$. Thus, there exists a polynomial function of degree m, such that the relation (6.18) is true. If we suppose that the equation (6.18) can be solved with respect to x_n, this results in an expression involving only the terms $x_{n-1}, x_{n-2}, \dots, x_{n-m+1}$. However, the resulting expression is in general very complicated, as shown below.

Example 6.2 For $m = 2$ we obtain

$$F_2(x, y) = x^2 - a_1xy - a_2y^2, \tag{6.19}$$

and the sequence $(x_n)_{n\geq 1}$ is given by

$$x_n = a_1x_{n-1} + a_2x_{n-2}, \quad n \geq 3, \ x_1 = \alpha_1, \ x_2 = \alpha_2, \tag{6.20}$$

where the relation $F_2(x_n, x_{n-1}) = (-1)^n a_2^{n-2} F_2(\alpha_2, \alpha_1)$ holds. This relation was proved by induction in [35]. Writing the relation explicitly we have

$$(2x_n - a_1x_{n-1})^2 = (a_1^2 + 4a_2)x_{n-1}^2 + 4(-1)^{n-1}a_2^{n-2}\left(a_2\alpha_1^2 + a_1\alpha_1\alpha_2 - \alpha_2^2\right). \tag{6.21}$$

Under some special conditions on $(x_n)_{n\geq 1}$, we can express x_n as an explicit formula of x_{n-1}. Moreover, if the sequence satisfies the second-order recurrence equation

(6.20) with a_1, a_2 and α_1, α_2 integers, by (6.21) then

$$(a_1^2 + 4a_2)x_{n-1}^2 + 4(-1)^{n-1}a_2^{n-2}\left(a_2\alpha_1^2 + a_1\alpha_1\alpha_2 - \alpha_2^2\right),$$

is a perfect square. This result extends [90].

Example 6.3 Simple computations show that for $m = 3$ we have

$$\begin{aligned} F_3(x, y, z) = &- x^3 - (a_3 + a_1a_2)y^3 - a_3^2z^3 + 2a_1x^2y + a_2x^2z - (a_2^2 + a_1a_3)y^2z \\ &- (a_1^2 - a_2)xy^2 - a_1a_3xz^2 - 2a_2a_3yz^2 + (3a_3 - a_1a_2)xyz. \end{aligned}$$

By (6.18) we obtain that for the linear recurrence relation

$$x_n = a_1x_{n-1} + a_2x_{n-2} + a_3x_{n-3}, \quad n \geq 4,\ x_1 = \alpha_1, x_2 = \alpha_2, x_3 = \alpha_3, \tag{6.22}$$

one has the relation $F_3(x_n, x_{n-1}, x_{n-2}) = a_3^{n-3}F_3(\alpha_3, \alpha_2, \alpha_1)$.

Example 6.4 The Tribonacci numbers are defined by the recurrence relation

$$T_n = T_{n-1} + T_{n-2} + T_{n-3}, \quad n \geq 4, \tag{6.23}$$

where $T_1 = 1$, $T_2 = 1$ and $T_3 = 2$. In our notation, we have $\alpha_1 = 1$, $\alpha_2 = 1$, $\alpha_3 = 2$, and $a_1 = a_2 = a_3 = 1$. Substituting in Example 6.3, we obtain

$$F_3(x, y, z) = -x^3 - 2y^3 - z^3 + 2x^2y + x^2z - 2y^2z - xz^2 - 2yz^2 + 2xyz.$$

The nonlinear relation satisfied by Tribonacci numbers is

$$F_3(T_n, T_{n-1}, T_{n-2}) = F_3(\alpha_3, \alpha_2, \alpha_1),$$

where $F_3(\alpha_3, \alpha_2, \alpha_1) = -8 - 2 - 1 + 8 + 4 - 2 - 2 - 2 + 4 = -1$.

6.2 Systems of Recurrent Sequences

Here we present some results by Andrica and Toader [21], which explore the convergence of solutions of a system of linear recurrent equations. The paper starts with the case of weighted means, and then deals with a more general case, giving solutions of linear recurrence relations and for the powers of matrices. Some applications concerning systems or three recurrence relations, obtained by Andrica and Marinescu in [20] are also given, where the authors use the dynamics of an iterative process to obtain refinements of geometric inequalities.

6.2.1 *Weighed Arithmetic Mean*

The p-mean is a continuous function $M : \mathbb{R}_+^p \to \mathbb{R}_+$ with the properties

$$\min\{x_1, \dots, x_p\} \le M\left(x_1, \dots, x_p\right) \le \max\{x_1, \dots, x_p\},$$

and

$$M\left(x_1, \dots, x_p\right) = \min\{x_1, \dots, x_p\} \implies x_1 = \dots = x_p.$$

Starting with the p-means $M_1, \dots, M_p$ and the initial values $x_0^1, \dots, x_0^p$, we can construct p sequences $(x_n^i)_{n\ge 0}$, for $i = 1, \dots, p$, by

$$x_{n+1}^i = M_i\left(x_n^1, \dots, x_n^p\right), \quad i = 1, \dots, p.$$

As seen in [158], the sequences $(x_n^i)_{n\ge 0}$, $i = 1, \dots, p$ converge to a common limit $l = l(M_1, \dots, M_p; x_0^1, \dots, x_0^p)$, calculated for two weighted arithmetic means. Here we find the common limit of p arithmetic means, defined by

$$A_i\left(x_1, \dots, x_p\right) = a_{i1}x_1 + \dots + a_{ip}x_p,$$

where $a_{ij} > 0$ for $i, j = 1, \dots, p$ and $a_{i1} + \dots + a_{ip} = 1$ for $i = 1, \dots, p$. Given the initial values $x_0^1, \dots, x_0^p$ we define the sequences $(x_n^i)_{n\ge 0}$ for $i = 1, \dots, p$, as

$$x_{n+1}^i = a_{i1}x_n^1 + \dots + a_{ip}x_n^p, \quad i = 1, \dots, p. \tag{6.24}$$

By the result mentioned above these sequences are all convergent to a common limit $l = l(A_1, \dots, A_p; x_0^1, \dots, x_0^p)$, which we want to determine. For this, we look for the numbers $\lambda_1, \dots, \lambda_p$ with the property

$$\lambda_1 x_n^1 + \dots + \lambda_p x_n^p = \lambda_1 x_{n+1}^1 + \dots + \lambda_p x_{n+1}^p, \quad n \ge 0, \tag{6.25}$$

that is

$$\sum_{i=1}^p \lambda_i x_n^i = \sum_{i=1}^p \lambda_i \sum_{j=1}^p a_{ij} x_n^j = \sum_{j=1}^p \left(\sum_{i=1}^p a_{ij}\lambda_i\right) x_n^j.$$

This happens for every n, if and only if

$$\lambda_i = a_{1i}\lambda_1 + \dots + a_{pi}\lambda_p, \quad i = 1, \dots, p, \tag{6.26}$$

i.e., this homogeneous system of linear equations has nontrivial solutions, because, considering the matrix $A = (a_{ji})_{1\le j,i\le p}$, we have $\det(A - I_p) = 0$. By a theorem of Perron [36], the matrix A has 1 as a simple eigenvalue (which is also greatest in module), and to it corresponds a positive eigenvector. Hence, every solution of (6.26) has the form $\lambda_i = c\lambda_i^0$, $i = 1, \ldots, p$, where

$$\lambda_i^0 \geq 0, \quad i = 1, \ldots, p, \quad \lambda_1^0 + \cdots + \lambda_p^0 = 1. \tag{6.27}$$

By (6.25) we have, step by step

$$\lambda_1^0 x_{n+1}^1 + \cdots + \lambda_p^0 x_{n+1}^p = \lambda_1^0 x_n^1 + \cdots + \lambda_p^0 x_n^p = \cdots = \lambda_1^0 x_0^1 + \cdots + \lambda_p^0 x_0^p,$$

and passing to the limit, we get

$$\lambda_1^0 l + \cdots + \lambda_p^0 l = \lambda_1^0 x_0^1 + \cdots + \lambda_p^0 x_0^p,$$

thus

$$l(A_1, \ldots, A_p; x_0^1, \ldots, x_0^p) = \lambda_1^0 l + \cdots + \lambda_p^0 l = \lambda_1^0 x_0^1 + \cdots + \lambda_p^0 x_0^p, \tag{6.28}$$

where $\lambda_1^0, \ldots, \lambda_p^0$ is the unique solution of (6.26) which satisfies (6.27).

Remark 6.2 It is obvious that a necessary condition for the convergence of the sequences given by (6.24) to a common limit is

$$a_{i1} + \cdots + a_{ip} = 1, \quad i = 1, \ldots, p.$$

6.2.2 The Solution of a System of Linear Recurrence Relations

In what follows we shall consider the system of linear recurrence relations

$$x_{n+1}^i = a_{i1} x_n^1 + \cdots + a_{ip} x_n^p, \quad n \geq p, \quad i = 1, \ldots, p, \tag{6.29}$$

where $a_{ij} \in \mathbb{C}$, for $i, j = 1, \ldots, p$ and $x_0^1, \ldots, x_0^p$ are given. Introducing the matrices $A = (a_{ij})_{1 \leq i, j \leq p}$ and $X_n = (x_n^1, \ldots, x_n^p)$, then (6.29) can be written as

$$X_{n+1} = A \cdot X_n, \quad n \geq 0, \tag{6.30}$$

and so we have

$$X_n = A^n \cdot X_0, \quad n \geq 0, \tag{6.31}$$

This relation shows that the solution of the system (6.29) is equivalent to determining the powers A^n of the matrix A. We shall denote $A^n = (a_{ij}^n)_{1 \leq i, j \leq p}$, for $n \geq 1$. Considering the characteristic polynomial of the matrix A

$$f(t) = \det(t \cdot I_p - A) = t^p - a_1 t^{p-1} - \cdots - a_p,$$

with distinct roots $t_1, \ldots, t_m$ and multiplicities $q_1, \ldots, q_m$, by the theorem of Cayley–Hamilton–Frobenius (see [36]), we have

$$A^p = a_1 A^{p-1} + \cdots + a_p I_p,$$

that is

$$A^n = a_1 A^{n-1} + \cdots + a_p A^{n-p}.$$

It follows that the elements a_{hl}^n verify the linear recurrence relations

$$a_{hl}^n = a_1 \cdot a_{hl}^{n-1} + \cdots + a_p \cdot a_{hl}^{n-p}, \quad h, l = 1, \ldots, p, \quad n \geq p.$$

By (6.7) we get

$$a_{hl}^n = P_{hl}^1(n)\,(t_1)^n + \cdots + P_{hl}^m(n)\,(t_m)^n\,, \quad n \geq p,$$

with

$$P_{hl}^i(n) = \sum_{j=1}^{q_i} (-1)^j f_{hl}^{ij}\,(t_i)^j \binom{n+j-1}{j-1}, \quad i = 1, \ldots, m,$$

where

$$f_{hl}^{ij} = \frac{1}{(q_i - j)!} \left[(t - t_i^*)^{q_i} F_{hl}(t)\right]_{t=t_i^*}^{(q_i - j)}, \quad j = 1, \ldots, q_i, \quad i = 1, \ldots, m,$$

F_{hl} being the generating function of the sequence $(a_{hl}^n)_{n \geq 0}$.

Theorem 6.5 *The solution of the system* (6.29) *is given by*

$$x_n^k = \sum_{i=1}^{m} \left(\sum_{j=1}^{p} P_{kj}^i(n) x_0^j \right) (t_i)^n\,, \quad k = 1, \ldots, p. \tag{6.32}$$

Proof Using the relation (6.31) we get

$$x_n^k = \sum_{j=1}^{p} a_{kj}^n x_0^j = \sum_{j=1}^{p} \left(\sum_{i=1}^{m} P_{kj}^i(n)\,(t_i)^n \right) x_0^j = \sum_{i=1}^{m} \left(\sum_{j=1}^{p} P_{kj}^i(n) x_0^j \right) (t_i)^n\,.$$

□

From the relation (6.32) we deduce the following result.

Theorem 6.6 *If*

$$\sum_{j=1}^{p} P_{kj}^{h}(n)\, x_0^j = \begin{cases} c_k, \text{ if } & t_h = 1 \\ 0, \text{ if } & |t_h| \geq 1, t_h \neq 1, \end{cases}$$

then

$$\lim_{n\to\infty} x_n^k = \begin{cases} c_k, \text{ if } t_h = 1 \text{ for some } h \\ 0, \text{ if } t_h \neq 1 \text{ for every } h. \end{cases}$$

Corollary 6.1 *If A is a Markov matrix [36], i.e., $a_{ij} \geq 0$ for $i, j = 1, \ldots, p$ and $a_{i1} + \cdots + a_{ip} = 1$ for $i = 1, \ldots, p$, then for every $k = 1, \ldots, p$, the sequence $(x_n^k)_{n\geq 0}$ is convergent to the limit $c_k = \sum_{j=1}^{p} f_{kj} x_0^j$, where $f_{kj} = \left((1-t)F_{kj}(t)\right)_{t=1}$, and F_{kj} is the generating function of the sequence $(a_{kj}^n)_{n\geq 0}$.*

Proof From the theorem of Perron cited above (see [36]), it results that A has $t_h = 1$ as a simple eigenvalue, which is the greatest in module. So

$$c_k = \sum_{j=1}^{p} P_{kj}^{h}(n) \cdot x_0^j,$$

with

$$P_{kj}^{h}(n) = -f_{kj}^{h} = -\left((1-t)F_{kj}(t)\right)_{t=1}.$$

We remark that the result is another form of (6.28). □

Remark 6.3 In some cases, we can also resolve the nonhomogeneous system

$$x_{n+1}^i = a_{i1}x_n^1 + \cdots + a_{ip}x_n^p + b^i, \quad n \geq p, \quad i = 1, \ldots, p,$$

or, denoting $B = (b^1, \ldots, b^p)'$, we can write

$$X_{n+1} = A \cdot X_n + B, \quad n \geq 0. \tag{6.33}$$

For example, it has a constant solution T, that is $T = AT + B$, if A does not have 1 as an eigenvalue and in this case $T = (I_p - A)^{-1}B$. The general solution of (6.33) is $X_n = T + Y_n$, where Y_n is the general solution of (6.30).

Remark 6.4 For the determination of A^n, a formula was given by Perron

$$A^n = \sum_{i=1}^{p} \frac{1}{(q_i - 1)!} \left[\frac{(t - t_i)^{q_i}}{f(t)} \cdot \left(tI_p - A\right)^* t^n \right]_{t=t_i}^{(q_i-1)},$$

where X^* denotes the adjoint of matrix X. For some proofs, one can consult [58, 59, 71, 145]. Another formula is given in [49] (which also contains an extensive bibliography). Here the ideas from [50] have been exploited.

Example 6.5 Let p be a prime. At any vertex of a regular polygon with p sides it is written an integer. For any vertex of the polygon we compute the difference between the sum of the integers written at his neighbors and his number. After that we delete all the initial integers and replace them by the newly obtained integers. Prove that the integers obtained after p such steps are the same modulo p with the initial integers.

Solution The transformation is described by the following matrix relation:

$$\begin{pmatrix} a_0^{(1)} \\ \vdots \\ a_{p-1}^{(1)} \end{pmatrix} = \begin{pmatrix} -1 & 1 & 0 & \cdots & 0 & 1 \\ 1 & -1 & 1 & \cdots & 0 & 0 \\ 0 & 1 & -1 & \cdots & 0 & 0 \\ \cdots & \cdots & \cdots & \cdots & \cdots & \cdots \\ 1 & 0 & 0 & \cdots & 1 & -1 \end{pmatrix} \begin{pmatrix} a_0 \\ \vdots \\ a_{p-1} \end{pmatrix} = A \begin{pmatrix} a_0 \\ \vdots \\ a_{p-1} \end{pmatrix},$$

where A is a $p \times p$ matrix. After p steps we obtain

$$\begin{pmatrix} a_0^{(p)} \\ \vdots \\ a_{p-1}^{(p)} \end{pmatrix} = A^p \begin{pmatrix} a_0 \\ \vdots \\ a_{p-1} \end{pmatrix}. \tag{6.34}$$

We can write $A = -I_p + X + X^{p-1}$, where X is the permutation matrix

$$\begin{pmatrix} 0 & 0 & 0 & \cdots & 0 & 1 \\ 1 & 0 & 0 & \cdots & 0 & 0 \\ 0 & 1 & 0 & \cdots & 0 & 0 \\ \cdots & \cdots & \cdots & \cdots & \cdots & \cdots \\ 0 & 0 & 0 & \cdots & 1 & 0 \end{pmatrix}.$$

In $M_p(\mathbb{Z}_p)$ we have

$$\begin{aligned} A^p &= (-I_p + X + X^{p-1})^p = (-I_p)^p + X^p + X^{p(p-1)} \\ &= -I_p + I_p + I_p = I_p, \end{aligned}$$

since $X^p = I_p$. From (6.34) it follows that in $M_{p,1}(\mathbb{Z}_p)$ we have

$$\begin{pmatrix} a_0^{(p)} \\ \vdots \\ a_{p-1}^{(p)} \end{pmatrix} = \begin{pmatrix} a_0 \\ \vdots \\ a_{p-1} \end{pmatrix}, \tag{6.35}$$

which ends the proof.

6.2.3 Systems of Two Linear Recurrent Sequences

Let $A \in M_2(\mathbb{C})$, $A = \begin{pmatrix} a & b \\ c & d \end{pmatrix}$. We denote by $Tr(A) = a + d$ the trace of A, and by $\det A = ad - bc$ the determinant of A.

The quadratic equation $\det(A - \lambda I_2) = 0$ called the *characteristic equation* of matrix A can be written in the explicit form

$$\lambda^2 - Tr(A)\lambda + \det A = 0, \tag{6.36}$$

whose roots are called the *eigenvalues* of matrix A.

The following result is fundamental for the algebra of square matrices. It holds for matrices of order two, but it can also be generalized.

Theorem 6.7 (Hamilton–Cayley for 2×2 matrices) *Any square matrix of order two satisfies its characteristic equation, i.e., the following relation holds:*

$$A^2 - Tr(A) \cdot A + (\det A) \cdot I_2 = O_2.$$

Denoting the eigenvalues of matrix A by λ_1, λ_2, one can obtain exact formulae for the powers of the matrix.

Theorem 6.8 *The following relations hold for all nonnegative integers:*

$$A^n = \begin{cases} \lambda_1^n B + \lambda_2^n C & \text{if } \lambda_1 \neq \lambda_2 \\ \lambda_1^n B + n\lambda_1^{n-1} C & \text{if } \lambda_1 = \lambda_2, \end{cases}$$

for some matrices $B, C \in M_2(\mathbb{C})$.

6.2.3.1 Linear Systems of Two Sequences

Let $(x_n)_{n\geq 0}$ and $(y_n)_{n\geq 0}$ be sequences of complex numbers defined by

$$\begin{cases} x_{n+1} = ax_n + by_n \\ y_{n+1} = cx_n + dy_n, \quad n \geq 0, \end{cases} \tag{6.37}$$

where a, b, c, d, c_0, y_0 are fixed complex numbers. In matrix form we have

$$\begin{pmatrix} x_{n+1} \\ y_{n+1} \end{pmatrix} = \begin{pmatrix} a & b \\ c & d \end{pmatrix} \begin{pmatrix} x_n \\ y_n \end{pmatrix} \text{ i.e., } \begin{pmatrix} x_{n+1} \\ y_{n+1} \end{pmatrix} = A \begin{pmatrix} x_n \\ y_n \end{pmatrix}, \quad n \geq 0,$$

where $A = \begin{pmatrix} a & b \\ c & d \end{pmatrix}$ is the matrix of coefficients.

It follows that

$$\begin{pmatrix} x_n \\ y_n \end{pmatrix} = A^n \begin{pmatrix} x_0 \\ y_0 \end{pmatrix}, \quad n \geq 0, \tag{6.38}$$

i.e., the computation of the general terms x_n and y_n is reduces to the calculation of the powers of matrix A (see Theorem 6.8).

Example 6.6 Find the limits of sequences $(x_n)_{n\geq 0}$ and $(y_n)_{n\geq 0}$, where

$$\begin{cases} x_{n+1} = (1-\alpha)x_n + \alpha y_n \\ y_{n+1} = \beta x_n + (1-\beta)y_n, \end{cases}$$

and α and β are complex numbers with $|1-\alpha-\beta| < 1$.

Solution The matrix of coefficients is

$$A = \begin{pmatrix} 1-\alpha & \alpha \\ \beta & 1-\beta \end{pmatrix},$$

and its characteristic equation is $\lambda^2 - (2-\alpha-\beta)\lambda + 1 - \alpha - \beta = 0$, with $\lambda_1 = 1$ and $\lambda_2 = 1-\alpha-\beta$. we have $\lambda_1 \neq \lambda_2$ and $|\lambda_2| < 1$. By Theorem 6.8, it follows that

$$A^n = \lambda_1^n B + \lambda_2^n C = B + (1-\alpha-\beta)^n C,$$

hence

$$\lim_{n\to\infty} A^n = B + \lim_{n\to\infty} (1-\alpha-\beta)^n C = B,$$

since

$$\lim_{n\to\infty} (1-\alpha-\beta)^n = 0.$$

From (6.38) it follows that sequences $(x_n)_{n\geq 0}$ and $(y_n)_{n\geq 0}$ are convergent. Let $x = \lim_{n\to\infty} x_n$ and $y = \lim_{n\to\infty} y_n$. Also, from (2) we get

$$\begin{pmatrix} x \\ y \end{pmatrix} = B \begin{pmatrix} x_0 \\ y_0 \end{pmatrix},$$

where

$$B = \frac{1}{\lambda_1 - \lambda_2}(A - \lambda_2 I_2) = \frac{1}{\alpha + \beta}\begin{pmatrix} \beta & \alpha \\ \beta & \alpha \end{pmatrix}.$$

We conclude that

$$\lim_{n\to\infty} x_n = \lim_{n\to\infty} y_n = \frac{\beta x_0 + \alpha y_0}{\alpha + \beta}.$$

Example 6.7 Define the sequences $(x_n)_{n\geq 0}$ and $(y_n)_{n\geq 0}$ by

$$\begin{cases} 2x_{n+1} = \sqrt{3}x_n + y_n \\ 2y_{n+1} = -x_n + \sqrt{3}y_n, \quad n \geq 0, \end{cases}$$

where x_0 and y_0 are fixed real numbers. Show that $(x_n)_{n\geq 0}$ and $(y_n)_{n\geq 0}$ are periodic of the same period.

Solution Clearly, the matrix of coefficients is

$$A = \begin{pmatrix} \frac{\sqrt{3}}{2} & \frac{1}{2} \\ -\frac{1}{2} & \frac{\sqrt{3}}{2} \end{pmatrix} = \begin{pmatrix} \cos\frac{\pi}{6} & \sin\frac{\pi}{6} \\ -\sin\frac{\pi}{6} & \cos\frac{\pi}{6} \end{pmatrix}.$$

A simple inductive argument shows that for all positive integers n

$$A^n = \begin{pmatrix} \cos\frac{n\pi}{6} & \sin\frac{n\pi}{6} \\ -\sin\frac{n\pi}{6} & \cos\frac{n\pi}{6} \end{pmatrix}.$$

It is clear that $A^{12} = I_2$ and 12 is the smallest nonnegative integer with this property. Then $A^n = A^n I_2 = A^n A^{12} = A^{n+12}$ and from (6.38) we obtain $x_{n+12} = x_n$ and $y_{n+12} = y_n$ and the desired conclusion follows.

Remark 6.5 If the sequences $(x_n)_{n\geq 0}$ and $(y_n)_{n\geq 0}$ are defined by

$$\begin{cases} x_{n+1} = (\cos t)x_n + (\sin t)y_n \\ y_{n+1} = -(\sin t)x_n + (\cos t)y_n, \; n \geq 0, \end{cases}$$

where t is a fixed real number, then the matrix of coefficients satisfies for all positive integers n

$$A^n = \begin{pmatrix} \cos nt & \sin nt \\ -\sin nt & \cos nt \end{pmatrix},$$

hence from (6.38) we get

$$x_n = (\cos nt)x_0 + (\sin nt)y_0 \quad \text{and} \quad y_n = -(\sin nt)x_0 + (\cos nt)y_0.$$

The sequences $(x_n)_{n\geq 0}$ and $(y_n)_{n\geq 0}$ are periodic if and only if t is rational. If $t = \dfrac{p}{q} \in Q$ is an irreducible fraction, then the common period is $2p$.

Example 6.8 Let k be a positive integer. Prove that there exist integers x, y, neither of which is divisible by 7, such that $x^2 + 6y^2 = 7^k$.

Solution Take $x_1 = y_1 = -1$, and consider the following relations

$$x_{k+1} = x_k - 6y_k, \quad y_{k+1} = x_k + y_k, \quad k = 1, 2, 3, \ldots.$$

One can check that

$$x_{k+1}^2 + y_{k+1}^2 = 7(x_k^2 + 6y_k^2),$$

and

$$x_k \equiv y_k \equiv (-1)^k \pmod 7.$$

Hence, the integers $x = x_k$, $y = y_k$ satisfy the given condition.

Example 6.9 Prove that

(a) For each positive integer n there is a unique positive integer a_n such that

$$(1+\sqrt{5})^n = \sqrt{a_n} + \sqrt{a_n + 4^n}.$$

(b) For n even, a_n is divisible by $5 \cdot 4^{n-1}$ and find the quotient.

Solution Let $(1+\sqrt{5})^n = x_n + y_n\sqrt{5}$, where x_n and y_n, $1, 2, \ldots$, are positive integers, and $x_1 = 1$, $y_1 = 1$. We clearly have

$$\begin{aligned} x_{n+1} + y_{n+1}\sqrt{5} &= (1+\sqrt{5})^{n+1} = (1+\sqrt{5})(1+\sqrt{5})^n \\ &= (1+\sqrt{5})(x_n + y_n\sqrt{5}) = x_n + 5y_n + (x_n + y_n)\sqrt{5}, \end{aligned}$$

hence $(x_n)_{n\geq 1}$ and $(y_n)_{n\geq 1}$ satisfy the system

$$\begin{cases} x_{n+1} = x_n + 5y_n \\ y_{n+1} = x_n + y_n. \end{cases}$$

(a) Clearly, $(1-\sqrt{5})^n = x_n - y_n\sqrt{5}$, $n = 1, 2, \ldots$, hence

$$x_n^2 - 5y_n = (-4)^n, \quad n = 1, 2, \ldots.$$

If n is even, consider $a_n = x_n^2 - 4^n$ and we have

$$\sqrt{a_n} + \sqrt{a_n + 4^n} = \sqrt{x_n^2 - 4^n} + \sqrt{x_n^2} = \sqrt{5y_n^2} + \sqrt{x_n^2}$$
$$= y_n\sqrt{5} + x_n = (1+\sqrt{5})^n.$$

If n is odd, consider $a_n = 5y_n^2 - 4^n$ and we have

$$\sqrt{a_n} + \sqrt{a_n + 4^n} = \sqrt{5y_n^2 - 4^n} + \sqrt{5y_n^2} = \sqrt{x_n^2} + \sqrt{5y_n^2}$$
$$= x_n + y_n\sqrt{5} = (1+\sqrt{5})^n.$$

(b) If n is even, then we have $a_n = x_n^2 - 4^n = 5y_n^2$, where

$$y_n = \frac{1}{2\sqrt{5}}\left[(1+\sqrt{5})^n - (1-\sqrt{5})^n\right]$$
$$= \frac{2^n}{2\sqrt{5}}\left[\left(\frac{1+\sqrt{5}}{2}\right)^n - \left(\frac{1-\sqrt{5}}{2}\right)^n\right] = 2^{n-1}F_n,$$

where F_n is the nth Fibonacci number. In this case we get $a_n = 5 \cdot 4^{n-1}F_n^2$, hence $5 \cdot 4^{n-1} \mid a_n$ and the quotient is F_n^2.

6.2.3.2 Second-Order Linear Recurrent Sequences

Second-order linear recurrent sequences can be written conveniently as systems. Let $(x_n)_{n\geq 0}$ be a sequence of complex numbers defined by

$$x_{n+1} = ax_n + bx_{n-1}, \quad n \geq 1, \tag{6.39}$$

where a, b, x_0, x_1 are fixed complex numbers. Relation (6.39) is equivalent to

$$\begin{cases} x_{n+1} = ax_n + by_n \\ y_{n+1} = x_n \end{cases}, \quad n \geq 0,$$

where $y_0 = \frac{1}{b}(x_1 - ax_0)$ if $b \neq 0$. The matrix of coefficients is

$$A = \begin{pmatrix} a & b \\ 1 & 0 \end{pmatrix},$$

with the characteristic equation $\lambda^2 - a\lambda - b = 0$ having the roots λ_1 and λ_2.

If $\lambda_1 \neq \lambda_2$, from Theorem 6.8.1) we get

$$A^n = \lambda_1^n B + \lambda_2^n C,$$

where

$$B = \frac{1}{\lambda_1 - \lambda_2}(A - \lambda_2 I_2) = \frac{1}{\lambda_1 - \lambda_2}\begin{pmatrix} a - \lambda_2 & b \\ 1 & -\lambda_2 \end{pmatrix},$$

and

$$C = \frac{1}{\lambda_2 - \lambda_1}(A - \lambda_1 I_2) = \frac{1}{\lambda_2 - \lambda_1}\begin{pmatrix} a - \lambda_1 & b \\ 1 & -\lambda_1 \end{pmatrix}.$$

In this case after an elementary computation we obtain

$$x_n = \frac{x_1 - x_0\lambda_2}{\lambda_1 - \lambda_2}\lambda_1^n - \frac{x_1 - x_0\lambda_1}{\lambda_1 - \lambda_2}\lambda_2^n, \quad n \geq 0,$$

i.e., x_n is a linear combination of λ_1^n and λ_2^n. It follows that

$$x_n = \alpha_1\lambda_1^n + \alpha_2\lambda_2^n, \quad n \geq 0, \tag{6.40}$$

where coefficients α_1, α_2 are obtained from the system

$$\alpha_1 + \alpha_2 = x_0, \quad \lambda_1\alpha_1 + \lambda_2\alpha_2 = x_1.$$

If $\lambda_1 = \lambda_2$, in similar way from Theorem 6.8.2 we get

$$A^n = \lambda_1^n B + n\lambda_1^{n-1} C.$$

In this case it follows

$$x_n = (\alpha n + \beta)\lambda_1^n, \quad n \geq 0,$$

where $\beta = x_0$ and $(\alpha + \beta)\lambda_1 = x_1$.

6.2.3.3 Matrices and Pell's Type Equations

Let $D \geq 2$ be a positive integer, which is not a perfect square. *Pell's equation* is defined by the following Diophantine equation [7]

$$x^2 - Dy^2 = 1, \tag{6.41}$$

It is well known that the equation (6.41) has nontrivial solutions in positive integers, i.e., solutions $(x, y) \neq (1, 0)$. For the pair (x, y) of positive integers we consider the matrix

$$A_{(x,y)} = \begin{pmatrix} x & Dy \\ y & x \end{pmatrix}.$$

Denote by S_D the set of all positive integer solutions of (6.41). Then $(x, y) \in S_D$ if and only if $\det A_{(x,y)} = 1$. Also, $(x, y) \neq (1, 0)$ if and only if $A_{(x,y)} \neq I_2$. In what follows we want to describe the set S_D of all solutions to (6.41) by using matrices. The following relation holds:

$$A_{(x,y)} \cdot A_{(u,v)} = A_{(xu+Dyv,\ xv+yu)}. \tag{6.42}$$

Taking determinants in (6.42) we get the following *multiplication principle*: If $(x, y), (u, v) \in S_D$, then $(xu + Dyv,\ xv + yu) \in S_D$, i.e., if $(x, y), (u, v) \in S_D$, the product $A_{(x,y)} \cdot A_{(u,v)}$ generates the solution $(xu + Dyv,\ xv + yu)$.

Let $(x_1, y_1) \in S_D$ be the *fundamental solution* of (6.41), i.e., $(x_1, y_1) \neq (1, 0)$ and x_1 is minimal. For any positive integer n, we have

$$\det A^n_{(x_1,y_1)} = (\det A_{(x_1,y_1)})^n = 1.$$

By the multiplication principle, $A^n_{(x_1,y_1)}$ generates a sequence (x_n, y_n) of solutions, where

$$A^n_{(x_1,y_1)} = \begin{pmatrix} x_n & Dy_n \\ y_n & x_n \end{pmatrix}, \quad n \geq 1.$$

From relation $A^{n+1}_{(x_1,y_1)} = A_{(x_1,y_1)} A^n_{(x_1,y_1)}$, it follows the recursive system

$$\begin{cases} x_{n+1} = x_1 x_n + D y_1 y_n \\ y_{n+1} = y_1 x_n + x_1 y_n, \quad n \geq 1. \end{cases} \tag{6.43}$$

The system (6.43) can be written in the following equivalent matrix form

$$\begin{pmatrix} x_{n+1} \\ y_{n+1} \end{pmatrix} = \begin{pmatrix} x_1 & Dy_1 \\ y_1 & x_1 \end{pmatrix} \begin{pmatrix} x_n \\ y_n \end{pmatrix}, \quad n \geq 1,$$

hence

$$\begin{pmatrix} x_n \\ y_n \end{pmatrix} = A_{(x_1,y_1)}^{n-1} \begin{pmatrix} x_1 \\ y_1 \end{pmatrix}.$$

The characteristic equation of matrix $A_{(x_1,y_1)}$ is

$$\lambda^2 - 2x_1\lambda + 1 = 0,$$

with $\lambda_{1,2} = x_1 \pm \sqrt{x_1^2 - 1} = x_1 \pm y_1\sqrt{D}$. From Theorem 6.8.1) we get

$$A_{(x_1,y_1)}^{n-1} = \lambda_1^{n-1}B + \lambda_2^{n-1}C,$$

where

$$B = \frac{1}{\lambda_1 - \lambda_2}(A_{(x_1,y_1)} - \lambda_2 I_2) = \frac{1}{2y_1\sqrt{D}} \begin{pmatrix} x_1 - \lambda_2 & Dy_1 \\ y_1 & x_1 - \lambda_2 \end{pmatrix}$$

$$C = \frac{1}{\lambda_1 - \lambda_2}(A_{(x_1,y_1)} - \lambda_1 I_2) = -\frac{1}{2y_1\sqrt{D}} \begin{pmatrix} x_1 - \lambda_1 & Dy_1 \\ y_1 & x_1 - \lambda_1 \end{pmatrix}.$$

It follows

$$\begin{cases} x_n = \dfrac{1}{2}\left[(x_1 + y_1\sqrt{D})^n + (x_1 - y_1\sqrt{D})^n\right] \\ \\ y_n = \dfrac{1}{2\sqrt{D}}\left[(x_1 + y_1\sqrt{D})^n - (x_1 - y_1\sqrt{D})^n\right], \quad n \geq 1. \end{cases} \tag{6.44}$$

It is clear that formula (6.44) can be extended for $n = 0$ and in this case we find the trivial solution $(x_0, y_0) = (1, 0)$.

Now, let us prove that in this way we generated all solutions in positive integers. Let $(x, y) \in S_D$ be a solution in positive integers. Without loss of generality we can assume that (x, y) is minimal (i.e., x is minimal) solution which has no form (6.44). Then, from the multiplication principle, it follows that $A_{(x,y)}A_{(x_1,y_1)}^{-1}$ generates the solution (x', y'), where

$$\begin{cases} x' = x_1x - Dy_1y \\ y' = -y_1x + x_1y \end{cases}$$

with $x' < x$ and $y' < y$. We get that (x', y') must be of the form (6.44), i.e., $A_{(x,y)}A_{(x_1,y_1)}^{-1} = A_{(x_1,y_1)}^k$ for some integer $k > 0$. Hence, $A_{(x,y)} = A_{(x_1,y_1)}^{k+1}$, i.e., (x, y) is of the form (6.44) with $n = k$.

6.2.3.4 Solutions of the Equation $ax^2 - by^2 = 1$

Here we discuss the positive integers solutions of the equation

$$ax^2 - by^2 = 1. \tag{6.45}$$

Clearly, if ab is a perfect square, then this equation has no integral solutions.

If ab is not a perfect square, then we define the *Pell's resolvent* of (6.45) by

$$u^2 - abv^2 = 1. \tag{6.46}$$

For a pair (x, y) of positive integers we consider the matrix

$$B_{(x,y)} = \begin{pmatrix} x & by \\ y & ax \end{pmatrix}.$$

Let $S_{a,b}$ be the set of all solutions in positive integers to the equation (6.45). Then $(x, y) \in S_{a,b}$ if and only if $\det B_{(x,y)} = 1$. The following relation holds:

$$B_{(x,y)}A_{(u,v)} = B_{(xu+byv,\, axv+yu)},$$

where $A_{(u,v)}$ denotes the matrix for the solution (u, v) of Pell's resolvent (6.46). Taking determinants in the above relation, from (6.45) we deduce that if $(x, y) \in S_{a,b}$ and $(u, v) \in S_{ab}$, then $(xu + byv,\ axv + yu) \in S_{a,b}$, i.e., the product $B_{(x,y)}A_{(u,v)}$ gives the solution $(xu + byv,\ axv + yu)$ of (6.45).

Assume that (6.45) has an integer solution, and let (x_0, y_0) be the minimal solution (i.e., x_0 is minimal) in positive integers. Then all solutions (x_n, y_n) in positive integers are generated by

$$B_{(x_n,y_n)} = B_{(x_0,y_0)}A^n_{(u_1,v_1)}, \quad n \geq 0. \tag{6.47}$$

Indeed, from multiplication principle all pairs (x_n, y_n) generated by (6.47) are solutions of (6.45). Conversely, if $(x, y) \in S_{a,b}$ is a solution of (6.45), then

$$B_{(x,y)}B^{-1}_{(x_0,y_0)} = \begin{pmatrix} x & by \\ y & ax \end{pmatrix}\begin{pmatrix} ax_0 & -by_0 \\ -y_0 & x_0 \end{pmatrix} = \begin{pmatrix} axx_0 - byy_0 & b(xy_0 - yx_0) \\ a(yx_0 - xy_0) & axx_0 - byy_0 \end{pmatrix}.$$

Passing to determinants, we get

$$(axx_0 - byy_0)^2 - ab(xy_0 - yx_0)^2 = 1,$$

i.e., $B_{(x,y)}B^{-1}_{(x_0,y_0)}$ generates a solution to Pell's resolvent (6.46), i.e., we have

$$B_{(x,y)}B^{-1}_{(x_0,y_0)} = A^k_{(u_1,v_1)},$$

for some positive integer k. It follows that if equation (6.45) is solvable, then all its solutions in positive integers are given (x_n, y_n) in (6.47), i.e.

$$\begin{cases} x_n = x_0 u_n + b y_0 v_n \\ y_n = y_0 u_n + a x_0 v_n, \quad n \geq 0, \end{cases} \tag{6.48}$$

where $(u_n, v_n)_{n\geq 0}$ is the general solution to Pell's resolvent.

The special case $b = 1$ gives the *negative Pell's equation*

$$ax^2 - y^2 = 1. \tag{6.49}$$

If this equation is solvable, then all its solutions in positive integers are generated by $B_{(x_0,y_0)} A^n_{(u_1,v_1)}$, $n \geq 0$, where (x_0, y_0) is the minimal solution of (6.49) and $B_{(x_0,y_0)}$ is the matrix

$$B_{(x_0,y_0)} = \begin{pmatrix} x_0 & y_0 \\ y_0 & a x_0 \end{pmatrix}.$$

Here (u_1, v_1) is the fundamental solution to Pell's resolvent $u^2 - av^2 = 1$. The recurrence system describing these solutions is

$$\begin{cases} x_n = x_0 u_n + y_0 v_n \\ y_n = y_0 u_n + a x_0 v_n, \quad n \geq 0. \end{cases}$$

Example 6.10 Find all solutions in positive integers to the Pell's equation

$$x^2 - 2y^2 = 1.$$

Solution The fundamental solution of this equation is $(x_1, y_1) = (3, 2)$, while the matrix generating all solutions is

$$A_{(3,2)} = \begin{pmatrix} 3 & 4 \\ 2 & 3 \end{pmatrix}.$$

The solutions $(x_n, y_n)_{n\geq 0}$ are given by $A^n_{(3,2)}$, i.e., we have

$$\begin{cases} x_n = \frac{1}{2}\left[(3 + 2\sqrt{2})^n + (3 - 2\sqrt{2})^n\right] \\ y_n = \frac{1}{2\sqrt{2}}\left[(3 + 2\sqrt{2})^n - (3 - 2\sqrt{2})^n\right], \ n \geq 0. \end{cases}$$

6.2.4 Sequences Interpolating Geometric Inequalities

Here we present applications given in [20], where interpolating sequences are constructed to generalize many classical triangle inequalities due to Euler, Mitrinović, Weitzenböck, Pólya-Szegő, or Chen.

For the real numbers $a \le b$, a sequence $(u_n)_{n\ge 0}$ is called *increasing interpolating sequence* for $a \le b$, if it is increasing and $a = u_0 \le u_1 \le \ldots \le u_n \le \ldots \le b$ and $u_n \to b$. Similarly, a sequence $(v_n)_{n\ge 0}$ is a *decreasing interpolating sequence* for $a \le b$, if $a \le \ldots \le v_n \le \ldots \le v_1 \le v_0 = b$ and $v_n \to a$.

Let ABC be a triangle having angles A, B, C (measured in radians), length-sides a, b, c, circumradius R, inradius r, semiperimeter s, and area K.

For the fixed nonnegative real numbers x, y, z with $x + y + z = 1$, we define recursively the sequences $(A_n)_{n\ge 0}, (B_n)_{n\ge 0}, (C_n)_{n\ge 0}$ by

$$A_{n+1} = xA_n + yB_n + zC_n$$
$$B_{n+1} = zA_n + xB_n + yC_n$$
$$C_{n+1} = yA_n + zB_n + xC_n,$$

where $A_0 = A$, $B_0 = B$, $C_0 = C$, $n = 0, 1, \ldots$. For convenience, we denote the triangle ABC by T_0, and by T_n the triangle $A_nB_nC_n$, having the length-sides, circumradius, inradius, semiperimeter, and area denoted by $a_n, b_n, c_n, R_n, r_n, s_n, K_n$, respectively. Note that $A_n, B_n, C_n > 0$ and $A_n + B_n + C_n = \pi$, while the triangles T_n, $n = 0, 1, \ldots$, can be shown to have the same circumcenter.

The first result is contained in the following theorem.

Theorem 6.9 *With the above notations, if at most one of x, y, z is equal to 0, then the sequences $(A_n)_{n\ge 0}, (B_n)_{n\ge 0}, (C_n)_{n\ge 0}$ are convergent and*

$$\lim_{n\to\infty} A_n = \lim_{n\to\infty} B_n = \lim_{n\to\infty} C_n = \frac{\pi}{3}. \tag{6.50}$$

Proof It is easy to see that the following matrix relation holds:

$$\begin{pmatrix} A_n \\ B_n \\ C_n \end{pmatrix} = U^n \begin{pmatrix} A \\ B \\ C \end{pmatrix}, \tag{6.51}$$

where U is the circulant matrix given by

$$U = \begin{pmatrix} x & y & z \\ z & x & y \\ y & z & x \end{pmatrix}. \tag{6.52}$$

A simple induction argument shows that

$$U^n = \begin{pmatrix} x_n & y_n & z_n \\ z_n & x_n & y_n \\ y_n & z_n & x_n \end{pmatrix}, \tag{6.53}$$

where the sequences $(x_n)_{n\geq 1}, (y_n)_{n\geq 1}, (z_n)_{n\geq 1}$ verify the recursive relations $x_{n+1} = xx_n + yy_n + zz_n$, $y_{n+1} = zx_n + xy_n + yz_n$, $z_{n+1} = yx_n + zy_n + xz_n$, $x_1 = x$, $y_1 = y$, $z_1 = z$, $n = 1, 2 \ldots$. By summation $x_{n+1} + y_{n+1} + z_{n+1} = x_n + y_n + z_n$, hence the sequence $(x_n + y_n + z_n)_{n\geq 1}$ is equal to 1.

On the other hand, the characteristic polynomial of the matrix U is

$$f_U(t) = (t - x - y - z)(t^2 + (y + z - 2x)t + x^2 + y^2 + z^2 - xy - yz - zx). \tag{6.54}$$

The hypothesis $x + y + z = 1$ implies that the roots of the polynomial f_U are $t_1 = 1, t_2 = \alpha, t_3 = \bar{\alpha}$, where $\alpha \in \mathbb{C} \setminus \mathbb{R}$ and $|\alpha| < 1$. It follows that we have

$$U = P \begin{pmatrix} 1 & 0 & 0 \\ 0 & \alpha & 0 \\ 0 & 0 & \bar{\alpha} \end{pmatrix} P^{-1}, \tag{6.55}$$

for some nonsingular matrix P. Therefore, we obtain

$$U^n = P \begin{pmatrix} 1 & 0 & 0 \\ 0 & \alpha^n & 0 \\ 0 & 0 & \bar{\alpha}^n \end{pmatrix} P^{-1}, \tag{6.56}$$

and we get

$$x_n = a + b\alpha^n + c\bar{\alpha}^n,\ y_n = a' + b'\alpha^n + c'\bar{\alpha}^n,\ z_n = a'' + b''\alpha^n + c''\bar{\alpha}^n,\ n = 1, 2, \ldots.$$

for some real values $a, b, c, a', b', c', a'', b'', c''$ determined by the initial conditions for $(x_n)_{n\geq 1}, (y_n)_{n\geq 1}, (z_n)_{n\geq 1}$. Because $\lim_{n\to\infty} \alpha^n = \lim_{n\to\infty} \bar{\alpha}^n = 0$, from the above formulae it follows that the sequences $(x_n)_{n\geq 1}, (y_n)_{n\geq 1}, (z_n)_{n\geq 1}$ are convergent and

$$\lim_{n\to\infty} x_n = a,\ \lim_{n\to\infty} y_n = a',\ \lim_{n\to\infty} z_n = a''.$$

From $x_n + y_n + z_n = 1, n = 1, 2, \ldots$, we obtain $a + a' + a'' = 1$. On the other hand, the relation (2.7) shows that the eigenvalues of the matrix U^n are $1, \alpha^n, \bar{\alpha}^n$, that is the characteristic polynomial f_{U^n} of the matrix U^n is

$$\begin{aligned} f_{U^n}(t) =& (t - x_n - y_n - z_n) \cdot \\ & \left[t^2 + (y_n + x_n - 2x_n)t + x_n^2 + y_n^2 + z_n^2 - x_n y_n - y_n z_n - z_n x_n\right], \end{aligned}$$

and by Vieta's relations we have $x_n^2 + y_n^2 + z_n^2 - x_n y_n - y_n z_n - z_n x_n = |\alpha|^{2n}$. When $n \to \infty$, we obtain the relation $a^2 + (a')^2 + (a'')^2 - aa' - a'a'' - a''a = 0$, i.e., $(a-a')^2 + (a'-a'')^2 + (a''-a)^2 = 0$. Therefore $a = a' = a'' = \frac{1}{3}$, and the desired result follows from the relation (2). □

We illustrate the above general iterative process by a special geometric case. Recall that if P is a point in the plane of the triangle ABC, the circumcevian triangle of P with respect to ABC is the triangle defined by the intersections of the cevians AP, BP, CP with the circumcircle of ABC. We consider $A_1B_1C_1$ to be the circumcevian triangle of the incenter I of ABC, i.e., the circumcircle mid-arc triangle of ABC. In this case we have

$$A_1 = \frac{1}{2}(B + C), \quad B_1 = \frac{1}{2}(C + A), \quad C_1 = \frac{1}{2}(A + B),$$

that is in the general iterative process we have $x = 0$, $y = \frac{1}{2}$, $z = \frac{1}{2}$. On the other hand, because $A + B + C = \pi$, we have

$$A_1 = \frac{1}{2}(\pi - A), \quad B_1 = \frac{1}{2}(\pi - B), \quad C_1 = \frac{1}{2}(\pi - C).$$

Define recursively the sequence of triangles $\mathcal{T}_n$ as follows: $\mathcal{T}_{n+1}$ is the circumcircle mid-arc triangle with respect to the incenter of $\mathcal{T}_n$, and $\mathcal{T}_0$ is the triangle ABC. The angles of triangles $\mathcal{T}_n$ are given by the relations $A_{n+1} = \frac{1}{2}(\pi - A_n)$, $B_{n+1} = \frac{1}{2}(\pi - B_n)$, $C_{n+1} = \frac{1}{2}(\pi - C_n)$, where $A_0 = A$, $B_0 = B$, $C_0 = C$. Solving these recurrence equations we get

$$A_n = \left(-\frac{1}{2}\right)^n A + \frac{\pi}{3}\left[1 - \left(-\frac{1}{2}\right)^n\right],$$

$$B_n = \left(-\frac{1}{2}\right)^n B + \frac{\pi}{3}\left[1 - \left(-\frac{1}{2}\right)^n\right],$$

$$C_n = \left(-\frac{1}{2}\right)^n C + \frac{\pi}{3}\left[1 - \left(-\frac{1}{2}\right)^n\right],$$

and the conclusion in Theorem 6.9 is verified.

As direct consequences, we provide the following results.

Proposition 6.4 (Theorem 3.2 in [20]) *With the above notations the sequence of inradii $(r_n)_{n\geq 0}$ is increasing and we have*

$$\lim_{n\to\infty} r_n = \frac{R}{2}. \tag{6.57}$$

Proof Using the known formula $\frac{r}{R} = 4\sin\frac{A}{2}\sin\frac{B}{2}\sin\frac{C}{2}$ we have to prove the following inequality:

$$\sin\frac{A_{n+1}}{2}\sin\frac{B_{n+1}}{2}\sin\frac{C_{n+1}}{2} \geq \sin\frac{A_n}{2}\sin\frac{B_n}{2}\sin\frac{C_n}{2}. \tag{6.58}$$

Denote $\frac{A_n}{2} = u$, $\frac{B_n}{2} = v$, $\frac{C_n}{2} = t$ and the inequality (6.57) is equivalent to

$$\sin(xu+yv+zt)\sin(zu+xv+yt)\sin(yu+zv+xt) \geq \sin u \sin v \sin t. \tag{6.59}$$

To prove the inequality (6.59), let us consider the function $f : (0, \pi) \to \mathbb{R}$, defined by $f(s) = \ln\sin s$, which is concave on the interval $(0, \pi)$. Applying the Jensen's inequality we get the inequalities

$$f(xu + yv + zt) \geq xf(u) + yf(v) + zf(t),$$

$$f(zu + xv + yt) \geq zf(u) + xf(v) + yf(t),$$

$$f(yu + zv + xt) \geq yf(u) + zf(v) + xf(t).$$

Summing these inequalities and since $x + y + z = 1$, the inequality (6.59) follows. From $\frac{r_n}{R} = 4\sin\frac{A_n}{2}\sin\frac{B_n}{2}\sin\frac{C_n}{2}$ we obtain the limit (6.57). □

Corollary 6.2 *With the above notations, the sequence of inradii* $(r_n)_{n\geq 0}$ *is an increasing interpolating sequence for Euler's inequality, i.e., we have*

$$r = r_0 \leq r_1 \leq \ldots \leq r_n \leq \ldots \leq \frac{R}{2}. \tag{6.60}$$

Another interesting application is an interpolating sequence for the classical inequality $s \leq \frac{3\sqrt{3}}{2}R$ of Mitrinović. This is involved in the following refinement to Euler's famous $R \geq 2r$ inequality

$$3\sqrt{3}r \leq s \leq \frac{3\sqrt{3}}{2}R.$$

The general Theorem 6.9 is here used to construct an increasing interpolating sequence for the Mitrinović's inequality, and a decreasing interpolating sequence for its counterpart.

Proposition 6.5 (Theorem 4.3 in [20])

1° *The sequence of semiperimeters* $(s_n)_{n\geq 0}$ *is increasing and we have*

$$\lim_{n\to\infty} s_n = \frac{3\sqrt{3}}{2}R. \tag{6.61}$$

2° *The sequence* $(\frac{s_n}{r_n})_{n\geq 0}$ *is decreasing and*

$$\lim_{n\to\infty}\frac{s_n}{r_n}=3\sqrt{3}. \tag{6.62}$$

Proof

1° The function $g:(0,\pi)\to\mathbb{R}$, $g(u)=\sin u$ is concave on $(0,\pi)$ and $s_n=R(\sin A_n+\sin B_n+\sin C_n)$. By Jensen's inequality we obtain $\sin A_{n+1}=\sin(xA_n+yB_n+zC_n)\geq x\sin A_n+y\sin B_n+z\sin C_n$. Similarly, we get the following two inequalities: $\sin B_{n+1}\geq z\sin A_n+x\sin B_n+y\sin C_n$, $\sin C_{n+1}\geq y\sin A_n+z\sin B_n+x\sin C_n$. Summing up these inequalities, it follows that $s_n\leq s_{n+1}$. The relation $\lim\limits_{n\to\infty}s_n=\frac{3\sqrt{3}}{2}R$ follows by Theorem 6.9.

2° From the relation $\cot\frac{A_n}{2}=\frac{s_n-a_n}{r_n}$ and the other two, we obtain

$$\cot\frac{A_n}{2}+\cot\frac{B_n}{2}+\cot\frac{C_n}{2}=\frac{3s_n-2s_n}{r_n}=\frac{s_n}{r_n}.$$

Because the function $h(u)=\cot u$ is convex on the interval $(0,\frac{\pi}{2})$, with a similar argument as in the proof of part 1°, it follows

$$\cot\frac{A_{n+1}}{2}+\cot\frac{B_{n+1}}{2}+\cot\frac{C_{n+1}}{2}\leq\cot\frac{A_n}{2}+\cot\frac{B_n}{2}+\cot\frac{C_n}{2},$$

that is $\frac{s_{n+1}}{r_{n+1}}\leq\frac{s_n}{r_n}$. The limit $\lim\limits_{n\to\infty}\frac{s_n}{r_n}=3\sqrt{3}$ follows from Theorem 6.9.

□

Corollary 6.3

1° *The sequence of semiperimeters* $(s_n)_{n\geq 0}$ *is an increasing interpolating sequence for the Mitrinović's inequality, i.e., we have*

$$s=s_0\leq s_1\leq\ldots\leq s_n\leq\ldots\leq\frac{3\sqrt{3}}{2}R. \tag{6.63}$$

2° *The sequence* $(\frac{s_n}{r_n})_{n\geq 0}$ *is a decreasing interpolating sequence for the counterpart of Mitrinović's inequality, i.e., we have*

$$3\sqrt{3}\leq\ldots\leq\frac{s_n}{r_n}\leq\ldots\leq\frac{s_1}{r_1}\leq\frac{s_0}{r_0}=\frac{s}{r}. \tag{6.64}$$

6.3 The Periodicity of Complex LRS

This section is dedicated to the study of geometric patterns produced by complex linear recurrent sequences of arbitrary order. The context is general, allowing for arbitrary initial conditions and recurrence coefficients. Throughout, orbits are visualized in the complex plane to illustrate the theoretical results.

Under certain conditions, linear recurrent sequences can be periodic. Sufficient conditions for periodicity are formulated in [141] for generalized Lucas sequences over an associative ring with identity, or in [162] for arbitrary sequences over algebraic number fields. An extensive list of periodic recurrent sequences is detailed in [62, Chapter 3], with an emphasis on sequences defined over finite fields.

In this section necessary and sufficient conditions for the periodicity of the sequence $(x_n)_{n\geq 0}$ are established, when the roots of the characteristic polynomial (6.10) are all distinct, all equal, or when the distinct roots have arbitrary multiplicities. Without loss of generality, it is assumed that recurrence order cannot be reduced, therefore $a_m \neq 0$.

6.3.1 Distinct Roots

Here we give sufficient and necessary conditions for the periodicity of $(x_n)_{n\geq 0}$, when the characteristic polynomial (6.10) has distinct roots.

Theorem 6.10 (sufficient condition for periodicity) *Let $z_1, \ldots, z_m$ be distinct kth roots of unity ($m \leq k$) and let the polynomial P be defined as*

$$P(x) = (x - z_1)(x - z_2)\cdots(x - z_m), \quad x \in \mathbb{C}. \tag{6.65}$$

The recurrent sequence $(x_n)_{n\geq 0}$ generated by the characteristic polynomial (6.65) and arbitrary initial conditions

$$x_{i-1} = \alpha_i \in \mathbb{C}, \quad i = 1, \ldots, m, \tag{6.66}$$

is periodic.

Proof The sequence $(x_n)_{n\geq 0}$ having the characteristic polynomial (6.65) satisfies the linear recurrence equation

$$x_n = a_1 x_{n-1} + a_2 x_{n-2} + \cdots + a_m x_{n-m}, \quad m \leq n \in \mathbb{N}, \tag{6.67}$$

with the coefficients $a_1, \ldots, a_m$ given by $a_i = (-1)^{i-1} S_i(z_1, \ldots, z_m)$, where $S_i(z_1, \ldots, z_m)$ represents the symmetric sum of products having i (unordered) factors chosen from $z_1, \ldots, z_m$.

From Proposition 6.1, the sequences

$$f_1(n) = z_1^n,\ f_2(n) = z_2^n, \ldots, f_m(n) = z_m^n,$$

form a basis in the vector space V of solutions of the recurrence relation (6.67), therefore the nth term of the sequence can be written as the linear combination

$$x_n = A_1 z_1^n + A_2 z_2^n + \cdots + A_m z_m^n. \tag{6.68}$$

The coefficients $A_1, \ldots, A_m$ can be obtained from (6.68) and the initial conditions (6.66) by solving the system of linear equations

$$\begin{cases} \alpha_1 & = A_1 + A_2 + \cdots + A_m \\ \alpha_2 & = A_1 z_1 + A_2 z_2 + \cdots + A z_m \\ \cdots & \\ \alpha_m & = A_1 z_1^{m-1} + A_2 z_2^{m-1} + \cdots + A z_m^{m-1}. \end{cases}$$

As $z_1, \ldots, z_m$ are kth roots of unity, $z_i^n = z_i^{n+k}$, $i = 1, \ldots, m$, hence $x_n = x_{n+k}$, $n \in \mathbb{N}$. This shows that the sequence $(x_n)_{n \geq 0}$ is periodic and its period divides k. The period is given by the formula $[\text{ord}(z_1), \text{ord}(z_2), \ldots, \text{ord}(z_m)]$, where $\text{ord}(z_i)$ denotes the order of the root z_i and $[b_1, \ldots, b_m]$ denotes the least common multiple (lcm) for a given set of integers $b_1, \ldots, b_m$.

A procedure that allows the direct computation of sequence terms using elementary matrices and operations is detailed below. The initial condition (6.66) can be written in matrix form as

$$\begin{pmatrix} 1 & 1 & \cdots & 1 \\ z_1 & z_2 & \cdots & z_m \\ z_1^2 & z_2^2 & \cdots & z_m^2 \\ \vdots & \vdots & & \vdots \\ z_1^{m-1} & z_2^{m-1} & \cdots & z_m^{m-1} \end{pmatrix} \begin{bmatrix} A_1 \\ A_2 \\ \vdots \\ A_m \end{bmatrix} = V_{m,m}(z_1, \ldots, z_m) \begin{bmatrix} A_1 \\ A_2 \\ \vdots \\ A_m \end{bmatrix} = \begin{bmatrix} \alpha_1 \\ \alpha_2 \\ \vdots \\ \alpha_m \end{bmatrix}, \tag{6.69}$$

where $V_{n,m}(z_1, \ldots, z_m)$ denotes the Vandermonde matrix defined as

$$V_{n,m}(z_1, \ldots, z_m) = \begin{pmatrix} 1 & 1 & \cdots & 1 \\ z_1 & z_2 & \cdots & z_m \\ \vdots & \vdots & & \vdots \\ z_1^{n-1} & z_2^{n-1} & \cdots & z_m^{n-1} \end{pmatrix}. \tag{6.70}$$

As the numbers $z_1, \ldots, z_m$ are all distinct, the determinant of the square Vandermonde matrix given in (6.70) yields [86]

$$\det\Big(V_{m,m}(z_1,\ldots,z_m)\Big) = \prod_{1\le i<j\le m}(z_j - z_i) \neq 0.$$

By Cramer's rule [127], the unique solution of (6.69) is given by

$$A_i = \frac{\det\Big(V^i_{m,m}(z_1,\ldots,z_m,\alpha_1,\ldots,\alpha_m)\Big)}{\det\Big(V_{m,m}(z_1,\ldots,z_m)\Big)}, \quad i = 1,\ldots,m, \tag{6.71}$$

where matrix $V^i_{m,m}$ is obtained from $V_{m,m}(z_1,\ldots,z_m)$ by replacing the ith column with $(\alpha_1,\ldots,\alpha_m)^T$. The first N terms of $(x_n)_{n\ge 0}$ are given by

$$V_{N,m}(z_1,\ldots,z_m)\begin{bmatrix} A_1 \\ A_2 \\ \vdots \\ A_m \end{bmatrix} = \begin{bmatrix} x_0 \\ x_1 \\ \vdots \\ x_{N-1} \end{bmatrix}. \tag{6.72}$$

Figure 6.1 illustrates the periodic orbits of sequence $(x_n)_{n\ge 0}$ obtained from the recurrence formula (6.67) (diamonds), or direct formula (6.72) (circles) when selecting (a) $m = 3$ and (b) $m = 5$ distinct roots respectively, from the seventh roots of unity. A numerical method for the direct computation of sequence terms with a given index set $I = \{i_1,\ldots,i_N\}$ is given in the Appendix. □

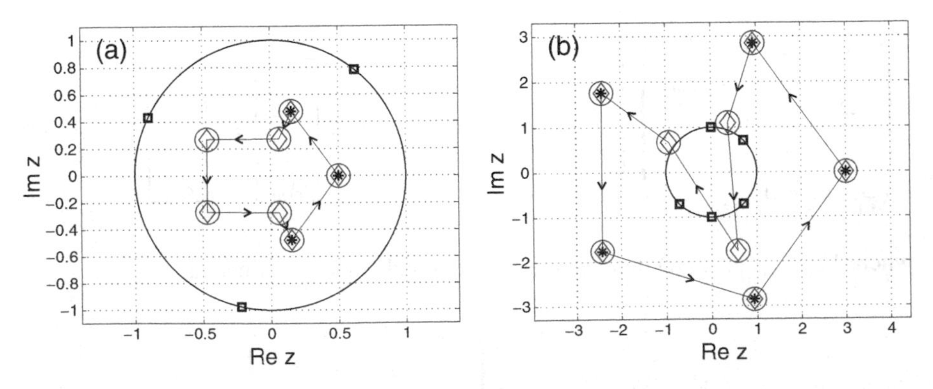

Fig. 6.1 First 15 terms of $(x_n)_{n\ge 0}$ computed from the recurrence relation (6.67) (diamonds) and direct formula (6.72) (circles), for **(a)** $m = 3$, $z_j = e^{\frac{2\pi i}{7}(2j-1)}$, and $\alpha_j = .5e^{\frac{2\pi i}{5}(j+3)}$, $j = 1, 2, 3$; **(b)** $m = 5$, $z_j = e^{\frac{2\pi i}{7}j}$ for $j = 1, 2, 5, 6, 7$ and $\alpha_j = 3e^{\frac{2\pi i}{5}(j+1)}$, $j = 1,\ldots,5$. Also shown, initial conditions (stars), generators (squares), unit circle S (solid line), orbit direction (arrows)

Theorem 6.11 (necessary condition for periodicity) *Let us assume that the roots $z_1, \ldots, z_m$ of (6.65) are all distinct. The recurrent sequence $(x_n)_{n\geq 0}$ having the characteristic polynomial (6.65) and initial conditions $\alpha_1, \ldots, \alpha_m$ is periodic only if there exist $k \in \mathbb{N}$ positive such that*

$$A_i(z_i^k - 1) = 0, \quad i = 1, \ldots, m, \tag{6.73}$$

where $A_1, \ldots, A_m$ are computed from formula (6.71).

Proof Let us assume that the sequence $(x_n)_{n\geq 0}$ is periodic and let $k \in \mathbb{N}$ be the period. Under this assumption, the periodicity can be written as

$$x_n = x_{n+k}, \quad \forall n \in \mathbb{N}. \tag{6.74}$$

For distinct roots $z_1, \ldots, z_m$ the formula (6.68) of the general term x_n and (6.74) give

$$\begin{aligned} x_n &= A_1 z_1^n + A_2 z_2^n + \cdots + A_m z_m^n \\ &= A_1 z_1^{n+k} + A_2 z_2^{n+k} + \cdots + A_m z_m^{n+k} = x_{n+k}, \quad \forall n \in \mathbb{N}, \end{aligned}$$

or equivalently

$$x_{n+k} - x_n = A_1(z_1^k - 1)z_1^n + A_2(z_2^k - 1)z_2^n + \cdots + A_m(z_m^k - 1)z_m^n = 0, \tag{6.75}$$

valid for all $n \in \mathbb{N}$. Writing (6.75) for $n = 0, \ldots, m-1$ one obtains the system

$$V_{m,m}(z_1, \ldots, z_m) \begin{bmatrix} A_1(z_1^k - 1) \\ A_2(z_2^k - 1) \\ \vdots \\ A_m(z_m^k - 1) \end{bmatrix} = \begin{bmatrix} 0 \\ 0 \\ \vdots \\ 0 \end{bmatrix}.$$

As $\det V_{m,m}(z_1, \ldots, z_m) \neq 0$, the system has the unique solution

$$A_i(z_i^k - 1) = 0, \quad i = 1, \ldots, m,$$

which confirms (6.73). For each $i = 1, \ldots, m$ there are two possibilities: either z_i is a kth root of unity, or it doesn't appear in the formula of the general term (6.68) (in this case A_i is zero). The only nondegenerate solution of (6.75) (produced when each generator has a nonzero contribution) satisfies

$$z_1^k = z_2^k = \cdots = z_m^k = 1,$$

in which case $z_1, \ldots, z_m$ are distinct kth roots of unity. □

6.3.2 Equal Roots

Here we present sufficient and necessary conditions for the periodicity of $(x_n)_{n\geq 0}$, when the m roots of the characteristic polynomial (6.10) are equal.

Theorem 6.12 (sufficient condition for periodicity) *Let z be a kth root of unity, m a natural number, and let the polynomial P be defined as*

$$P(x) = (x - z)^m, \quad x \in \mathbb{C}. \tag{6.76}$$

The recurrent sequence $(x_n)_{n\geq 0}$ generated by the characteristic polynomial (6.76) and initial conditions (6.66) is periodic when

$$A_2 = A_3 = \cdots = A_m = 0,$$

where $A_1, \ldots, A_m$ are the coefficients of $(x_n)_{n\geq 0}$ in the basis defined in Proposition 6.2. The sequence $(x_n)_{n\geq 0}$ is divergent otherwise.

Proof Similar to Theorem 6.10, the sequence $(x_n)_{n\geq 0}$ satisfies the linear recurrence equation (6.67), for the coefficients $a_1, \ldots, a_m$ given by

$$a_i = (-1)^{i-1}\binom{m}{i}z^i, \quad i = 1, \ldots, m. \tag{6.77}$$

From Proposition 6.2, the sequences

$$f_1(n) = z^n,\ f_2(n) = nz_2^n, \ldots, f_m(n) = n^{m-1}z^n$$

form a basis in the vector space V of solutions of the recurrence relation generated by the characteristic polynomial (6.76), hence the nth term of the sequence can be written as

$$x_n = A_1 z^n + nA_2 z^n + \cdots + n^{m-1}A_m z^n. \tag{6.78}$$

For $A_i = 0$, $i = 2, \ldots, m$, we have $x_n = A_1 z^n$, so the sequence $(x_n)_{n\geq 0}$ is periodic. Whenever there is $i \geq 2$ such that $A_i \neq 0$, the behavior of x_n is dictated by the divergent coefficient $n^i A_i$ of z^n, hence $(x_n)_{n\geq 0}$ diverges.

This property is illustrated in Figure 6.2, where it is shown that the sequence can either be (a) periodic or (b) divergent. Some asymptotic properties of divergent sequences generated by roots of unity are presented in Section 4.1.

Coefficients $A_1, \ldots, A_m$ are obtained by (6.78) and the initial conditions

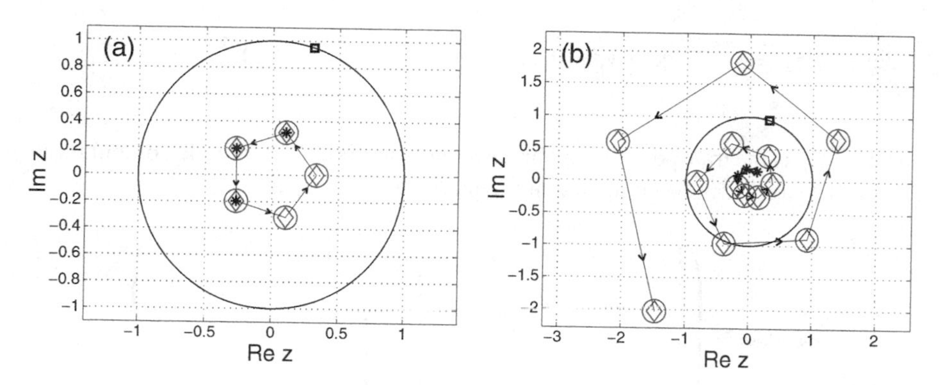

Fig. 6.2 First 16 terms of sequence $(x_n)_{n\geq 0}$ computed from the recurrence formula (6.67) (diamonds) and direct formula (6.82) (circles) for $m = 3$, $z = e^{\frac{2\pi i}{5}}$ and initial conditions (**a**) $\alpha_j = e^{\frac{2\pi i}{5}j}/3$, $j = 1, 2, 3$; (**b**) $\alpha_j = e^{\frac{2\pi i}{7}j}/3$, $j = 1, 2, 3$. Also plotted are initial conditions (stars), generators (squares), and unit circle S (solid line). Orbit's direction is indicated by arrows

$$\begin{cases} \alpha_1 & = A_1 \\ \alpha_2 & = A_1 + A_2 + \cdots + A_m \\ \cdots & \\ \alpha_m & = A_1 z^{m-1} + (m-1)A_2 z^{m-1} + \cdots + (m-1)^{m-1} A z^{m-1}. \end{cases}$$

A procedure that allows the direct computation of sequence terms using elementary matrices and operations is detailed below. The initial condition can be written in matrix form as

$$\begin{pmatrix} 1 & 0 & \cdots & 0 \\ z & z & \cdots & z \\ z^2 & 2z^2 & \cdots & 2^{m-1}z^2 \\ \vdots & \vdots & & \vdots \\ z^{m-1} & (m-1)z^{m-1} & \cdots & (m-1)^{m-1}z^{m-1} \end{pmatrix} \begin{bmatrix} A_1 \\ A_2 \\ \vdots \\ A_m \end{bmatrix} = \mathcal{V}_{m,m}(z) \begin{bmatrix} A_1 \\ A_2 \\ \vdots \\ A_m \end{bmatrix} = \begin{bmatrix} \alpha_1 \\ \alpha_2 \\ \vdots \\ \alpha_m \end{bmatrix}, \tag{6.79}$$

where the matrix $\mathcal{V}_{n,m}(z)$ is defined by

$$\mathcal{V}_{n,m}(z) = \begin{pmatrix} 1 & 0 & \cdots & 0 \\ z & z & \cdots & z \\ z^2 & 2z^2 & \cdots & 2^{m-1}z^2 \\ \vdots & \vdots & & \vdots \\ z^{n-1} & (n-1)z^{n-1} & \cdots & (n-1)^{m-1}z^{n-1} \end{pmatrix}. \tag{6.80}$$

Collecting the powers of z, the determinant of (6.80) for $n = m$ is

$$\det \mathcal{V}_{n,n}(z) = z^{\frac{n(n-1)}{2}} \det\left(V_{n,n}^T(0,\ldots,n-1)\right) = z^{\frac{n(n-1)}{2}} \prod_{0\le i<j\le n-1} (j-i) \neq 0,$$

where $V_{n,n}^T$ is the transpose of the Vandermonde matrix (6.70). Simple computations show that (6.80) can be computed numerically as

$$\mathcal{V}_{n,m}(z) = \begin{pmatrix} 1 & 0 & \cdots & 0 \\ 1 & 1 & \cdots & 1 \\ 1 & 2 & \cdots & 2^{m-1} \\ \vdots & \vdots & & \vdots \\ 1 & n-1 & \cdots & (n-1)^{m-1} \end{pmatrix} .* \begin{pmatrix} 1 & 1 & \cdots & 1 \\ z & z & \cdots & z \\ z^2 & z^2 & \cdots & z^2 \\ \vdots & \vdots & & \vdots \\ z^{n-1} & z^{n-1} & \cdots & z^{n-1} \end{pmatrix},$$

where $.*$ denotes the element-by-element product of matrices from Matlab®, and expressed in terms of Vandermonde matrices as follows:

$$\mathcal{V}_{n,m}(z) = V_{n,m}(z,\ldots,z) .* V_{m,n}^T(0,1,\ldots,n-1).$$

By Cramer's rule and $\det \mathcal{V}_{m,m} \neq 0$, the unique solution of (6.79) is

$$A_i = \frac{\det\left(\mathcal{V}_{m,m}^i(z,\alpha_1,\ldots,\alpha_m)\right)}{\det \mathcal{V}_{m,m}(z)}, \quad i = 1,\ldots,m, \tag{6.81}$$

where $\mathcal{V}_{m,m}^i$ is the matrix obtained by replacing the ith column of $\mathcal{V}_{m,m}(z)$ with $\mathbf{a}^T$.

The first N terms of the sequence $(x_n)_{n\ge0}$ are given by

$$\mathcal{V}_{N,m}(z)\begin{bmatrix} A_1 \\ A_2 \\ \vdots \\ A_m \end{bmatrix} = \begin{bmatrix} x_0 \\ x_1 \\ \vdots \\ x_{N-1} \end{bmatrix}. \tag{6.82}$$

The terms of the sequence $(x_n)_{n\ge0}$ can therefore be obtained in two alternative ways: either from the recurrence relation (6.67) using the coefficients given by (6.77), or directly from formula (6.82). □

Theorem 6.13 (necessary condition for periodicity) *The recurrent sequence $(x_n)_{n\ge0}$ having the characteristic polynomial (6.76) and initial conditions (6.66) is only periodic when*

$$A_1(z^k - 1) = 0, \quad A_2 = A_3 = \cdots = A_m = 0, \tag{6.83}$$

where $A_1,\ldots,A_m$ are computed from formula (6.81).

Proof Let us assume that the sequence $(x_n)_{n\geq 0}$ is periodic and let $k \in \mathbb{N}$ be the period. The periodicity condition and formula (6.78) for the general term of $(x_n)_{n\geq 0}$ give

$$\begin{aligned} x_n &= (A_1 + nA_2 + \cdots + n^{m-1}A_m)z^n \\ &= (A_1 + (n+k)A_2 + \cdots + (n+k)^{m-1}A_m)z^{n+k} = x_{n+k}, \quad \forall n \in \mathbb{N}. \end{aligned}$$

Writing the periodicity condition as $x_{n+k} - x_n = 0$, one obtains

$$\left[A_1\left(z^k - 1\right) + \cdots + A_m\left(z^k(n+k)^{m-1} - n^{m-1}\right)\right]z^n = 0, \quad \forall n \in \mathbb{N}.$$

For a fixed value of $z \in \mathbb{C}$, this is a polynomial of degree $m-1$ in n

$$Q(n) = B_0 + B_1 n + \cdots + B_{m-1}n^{m-1} = 0, \quad \forall n \in \mathbb{N}.$$

As $Q(n)$ has infinitely many zeros, this has to be a null polynomial, therefore

$$\begin{cases} B_0 &= A_1\left(z^k - 1\right) + A_2 k z^k + A_3 k^2 z^k + \cdots + A_m k^{m-1} z^k = 0, \\ B_1 &= A_2\left(z^k - 1\right) + A_3\binom{2}{1}k z^k + A_4\binom{3}{1}k^2 z^k + \cdots + A_m\binom{m-1}{1}k^{m-2}z^k = 0, \\ \ldots & \\ B_j &= A_{j+1}\left(z^k - 1\right) + A_{j+2}\binom{j+1}{j}k z^k + \cdots + A_m\binom{m-1}{j}k^{m-1-j}z^k = 0, \\ \ldots & \\ B_{m-1} &= A_m\left(z^k - 1\right) = 0. \end{cases}$$

The above linear system in $A_1, A_2, \ldots, A_m$ can be written in matrix form as

$$\begin{pmatrix} z^k - 1 & kz^k & k^2 z^k & \cdots & k^{m-1}z^k \\ 0 & z^k - 1 & 2kz^k & \cdots & (m-1)k^{m-2}z^k \\ 0 & 0 & z^k - 1 & \cdots & \binom{m-1}{2}k^{m-3}z^k \\ \vdots & \vdots & \vdots & & \vdots \\ 0 & 0 & 0 & \cdots & z^k - 1 \end{pmatrix} \begin{bmatrix} A_1 \\ A_2 \\ \vdots \\ A_m \end{bmatrix} = \begin{bmatrix} 0 \\ 0 \\ \vdots \\ 0 \end{bmatrix}, \tag{6.84}$$

and two distinct situations emerge.

1. If $z^k - 1 \neq 0$ the determinant of the homogeneous system (6.84) is not zero, hence the system has the unique solution $A_1 = \cdots = A_m = 0$.
2. When $z^k = 1$ the above system is reduced to

$$\begin{pmatrix} k & k^2 & k^3 & \cdots & k^{m-1} \\ 0 & 2k & 3k^2 & \cdots & (m-1)k^{m-2} \\ 0 & 0 & 3k & \cdots & \binom{m-1}{2}k^{m-3} \\ \vdots & \vdots & \vdots & & \vdots \\ 0 & 0 & 0 & \cdots & \binom{m-1}{m-2}k \end{pmatrix} \begin{bmatrix} A_2 \\ A_3 \\ \vdots \\ A_m \end{bmatrix} = \begin{bmatrix} 0 \\ 0 \\ \vdots \\ 0 \end{bmatrix}.$$

As the determinant of the matrix is $(m-1)!k^m \neq 0$, the above homogeneous system has the unique solution $A_2 = \cdots = A_m = 0$, while A_1 can be arbitrary.

Both situations are captured in formula (6.83).

□

6.3.3 Distinct Roots with Arbitrary Multiplicities

Here sufficient and necessary conditions for the periodicity of the recurrence sequence $(x_n)_{n\geq 0}$ are presented in the general case, when sequences have a characteristic polynomial with roots of different multiplicities. a

Theorem 6.14 (sufficient condition for periodicity) *Let* $2 \leq m \leq k$ *and* $d_1, \ldots, d_m$ *be positive integers,* $z_1, \ldots, z_m$ *distinct* k*th roots of unity and the polynomial*

$$P(x) = (x - z_1)^{d_1}(x - z_2)^{d_2} \cdots (x - z_m)^{d_m}, \quad x \in \mathbb{C}. \tag{6.85}$$

The LRS $(x_n)_{n\geq 0}$ *having the characteristic polynomial (6.85) of degree* $d = d_1 + \cdots + d_m$ *and initial conditions* $x_{i-1} = \alpha_i$, $i = 1, \ldots, d$ *is periodic when*

$$A_{ij} = 0, \quad 1 \leq i \leq m, \quad 2 \leq j \leq d_i, \tag{6.86}$$

where A_{ij} $(1 \leq i \leq m,\ 1 \leq j \leq d_i)$ *represent the coefficients of* $(x_n)_{n\geq 0}$ *in the basis defined in Theorem 6.3. The sequence* $(x_n)_{n\geq 0}$ *is divergent otherwise.*

Proof The sequence $(x_n)_{n\geq 0}$ satisfies the linear recurrence relation of order d below

$$x_n = a_1 x_{n-1} + a_2 x_{n-2} + \cdots + a_d x_{n-d}, \quad d \leq n \in \mathbb{N}, \tag{6.87}$$

whose coefficients are given by $a_i = (-1)^{i-1} S_i(Z)$, $i = 1, \ldots, d$ where $S_i(Z)$ is the symmetric sum of products having i factors chosen from the multiset

$$Z = \{\underbrace{z_1, \ldots, z_1}_{d_1}, \underbrace{z_2, \ldots, z_2}_{d_2}, \ldots, \underbrace{z_m, \ldots, z_m}_{d_m}\}.$$

From Proposition 6.3, the sequences

$$f_{ij}(n) = n^{j-1} z_i^n, \quad 1 \leq i \leq m, \quad 1 \leq j \leq d_i, \tag{6.88}$$

form a basis in the vector space V of solutions of the recurrence relation (6.87), therefore the nth term of the sequence can be written as

$$x_n = \sum_{i=1}^{m} \left(A_{i1} + nA_{i2} + \cdots + n^{d_i-1} A_{id_i} \right) z_i^n. \tag{6.89}$$

When the condition (6.86) is fulfilled, formula (6.89) of x_n reduces to

$$x_n = A_{11} z_1^n + A_{21} z_2^n + \cdots + A_{m1} z_m^n,$$

which is similar to (6.68) discussed in Theorem 6.10, which is periodic for distinct $z_1, \ldots, z_m$. As illustrated in Figure 6.3a, the sequence may be periodic even when $d_i \geq 2$, $i = 1, \ldots, m$. When any of the coefficients A_{ij}, $1 \leq i \leq m$, $2 \leq j \leq d_i$ does not vanish, the polynomial (6.89) in n is not constant, therefore the sequence $(x_n)_{n\geq 0}$ diverges as depicted in Figure 6.3b. A detailed explanation of this behavior is presented in Theorem 6.15.

A procedure allowing the computation of sequence terms using elementary matrices is detailed below. For $m, n \in \mathbb{N}$, $z_1, \ldots, z_m = (z_1, \ldots, z_m)$, and $\mathbf{d} = (d_1, \ldots, d_m)$ one can define the matrix

$$\mathcal{W}_{n+1,d}(z_1, \ldots, z_m, \mathbf{d}) = \left(\begin{array}{cccc|c|cccc} 1 & 0 & \cdots & 0 & \cdots & 1 & 0 & \cdots & 0 \\ z_1 & z_1 & \cdots & z_1 & \cdots & z_m & z_m & \cdots & z_m \\ z_1^2 & 2z_1^2 & \cdots & 2^{d_1-1} z_1^2 & \cdots & z_m^2 & 2z_m^2 & \cdots & 2^{d_m-1} z_m^2 \\ \vdots & \vdots & & \vdots & \cdots & \vdots & \vdots & & \vdots \\ z_1^n & nz_1^n & \cdots & n^{d_1-1} z_1^n & \cdots & z_m^n & nz_m^n & \cdots & n^{d_m-1} z_m^n \end{array} \right).$$

The above matrix can be constructed using the matrices $\mathcal{V}_{n,d_i}(z_i)$, $i = 1, \ldots, m$ defined by formula (6.80), as

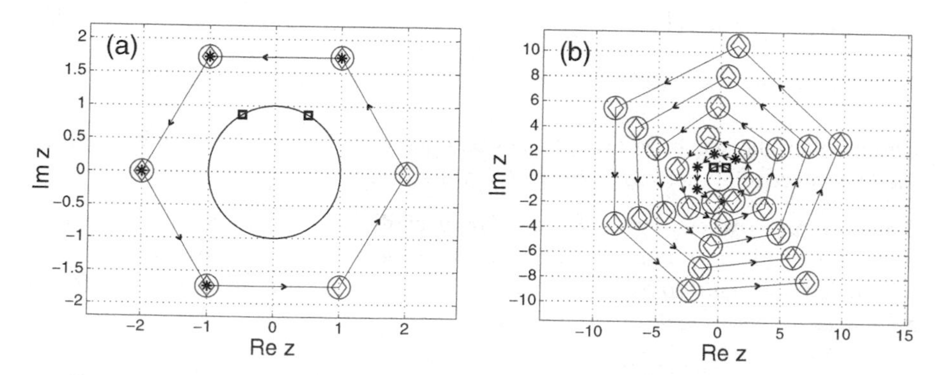

Fig. 6.3 First 31 terms of sequence $(x_n)_{n\geq 0}$ computed from the recurrence relation (6.87) (diamonds) and direct formula (6.92) (circles) for $z_1 = e^{\frac{2\pi i}{6}}$, $z_2 = e^{\frac{4\pi i}{6}}$, $d_1 = d_2 = 2$ and initial conditions (**a**) $\alpha_j = 2e^{\frac{2\pi i}{6} j}$, $j = 1, \ldots, 4$; (**b**) $\alpha_j = 2e^{\frac{2\pi i}{7} j}$, $j = 1, \ldots, 4$. Also plotted are initial conditions (stars), generators (squares), and unit circle S (solid line). Arrows indicate the increase of sequence index

$$\mathcal{W}_{n,d}(z_1,\ldots,z_m,\mathbf{d}) = \left(\mathcal{V}_{n,d_1}(z_1)\middle|\cdots\middle|\mathcal{V}_{n,d_m}(z_m)\right).$$

The columns of the above matrix represent the first $n+1$ terms of each member of basis (6.11) so they are linearly independent, therefore for $d = n+1$ we have $\det \mathcal{W}_{d,d} \neq 0$.

The coefficients A_{ij}, $1 \le i \le m$, $1 \le j \le d_i$ are computed from the initial condition $x_{i-1} = \alpha_i$, $i = 1,\ldots,d$, written with the help of formula (6.89) as the linear system

$$\mathcal{W}_{d,d}(z_1,\ldots,z_m,\mathbf{d})\left(A_{11}\ \cdots\ A_{1d_1}\middle|\cdots\middle|A_{m1}\ \cdots\ A_{md_m}\right)^T = (\alpha_1,\ldots,\alpha_d)^T. \tag{6.90}$$

Using Cramer's rule and that $\det \mathcal{W}_{d,d} \neq 0$, the above system has the unique solution

$$A_{ij} = \frac{\det\left(\mathcal{W}^l_{d,d}(z_1,\ldots,z_m,\mathbf{d},\alpha_1,\ldots,\alpha_m)\right)}{\det\left(\mathcal{W}_{d,d}(z_1,\ldots,z_m,\mathbf{d})\right)}, \quad i = 1,\ldots,m, \tag{6.91}$$

where $\mathcal{W}^l_{d,d}(z_1,\ldots,z_m,\mathbf{d},\alpha_1,\ldots,\alpha_m)$ is the matrix obtained by replacing the lth column of $\mathcal{W}_{d,d}(z_1,\ldots,z_m,\mathbf{d})$ by $(\alpha_1,\ldots,\alpha_m)^T$, with l computed from the formula $l = d_1 + \cdots + d_{i-1} + j$.

The first N terms of the sequence $(x_n)_{n\ge 0}$ can be obtained by

$$\mathcal{W}_{N,d}(z_1,\ldots,z_m,\mathbf{d})\left(A_{11}\ \cdots\ A_{1d_1}\middle|\cdots\middle|A_{m1}\ \cdots\ A_{md_m}\right)^{\mathbf{T}} = \begin{bmatrix} x_0 \\ x_1 \\ \vdots \\ x_{N-1} \end{bmatrix}. \tag{6.92}$$

The terms of the sequence $(x_n)_{n\ge 0}$ can be obtained in two alternative ways: from the recurrence relation (6.87), or directly from (6.92). □

Theorem 6.15 (necessary condition for periodicity) *Let* $2 \le m \le k$ *and* $d_1,\ldots,d_m$ *be natural numbers. The recurrent sequence* $(x_n)_{n\ge 0}$ *having the characteristic polynomial (6.85), degree* $d = d_1 + \cdots + d_m$ *and the initial conditions* $x_{i-1} = \alpha_i$, $i = 1,\ldots,d$, *is periodic only if*

$$\begin{aligned} A_{i1}(z_i^k - 1) &= 0, \quad i = 1,\ldots,m, \\ A_{ij} &= 0, \quad j = 2,\ldots,d_i, \end{aligned} \tag{6.93}$$

where the coefficients A_{ij} *are computed from (6.91), for* $1 \le i \le m$, $1 \le j \le d_i$.

Proof Assume that the sequence has periodicity $k \in \mathbb{N}$. The formula (6.89) of the general term for the sequence $(x_n)_{n\geq 0}$ can be combined with the periodicity condition $x_{n+k} - x_n = 0$ to give

$$\sum_{i=1}^{m} \left[A_{i1} \left(z_i^k - 1 \right) + \cdots + A_{id_i} \left(z_i^k (n+k)^{d_i - 1} - n^{d_i - 1} \right) \right] z_i^n = 0, \qquad (6.94)$$

which holds for $n \in \mathbb{N}$. By Proposition 6.3, the sequences $(z_1^n)_{n\geq 0}, \ldots, (z_m^n)_{n\geq 0}$ are linearly independent, hence their coefficients in (6.94) are zero, i.e.,

$$A_{i1} \left(z_i^k - 1 \right) + \cdots + A_{id_i} \left(z_i^k (n+k)^{d_i - 1} - n^{d_i - 1} \right) = 0, \qquad (6.95)$$

for $i = 1, \ldots, m$ and $n \in \mathbb{N}$. As it was already illustrated for the case of a single root with arbitrary multiplicity presented in Theorem 6.13, for each value $i = 1, \ldots, m$ the only solutions of (6.95) are

1. $z_i^k \neq 1$ when $A_{i1} = A_{i2} = \cdots = A_{id_i} = 0$,
2. $z_i^k = 1$ when A_{i1} is arbitrary and $A_{i2} = \cdots = A_{id_i} = 0$.

This confirms that (6.93) holds. The only nondegenerate solution of (6.94) (each root appears explicitly) satisfies

$$z_1^k = z_2^k = \cdots = z_m^k = 1,$$

in which case $z_1, \ldots, z_m$ are distinct kth roots of unity. □

The periodicity conditions presented in this section can be used to analyze and classify the geometries of self-repeating orbits, as well as to enumerate the sequences having a finite orbit.

6.4 The Geometry and Enumeration of Periodic Patterns

In this section we examine the geometry and number of periodic orbits. First, we explore geometric bounds, then we examine the length of periodic orbits. Finally, we count the periodic orbits, extending the results in Chapter 5.

6.4.1 Geometric Bounds of Periodic Orbits

Here we derive outer bounds for periodic generalized Horadam sequences.

Theorem 6.16 *Let $2 \leq m \leq k$ and $d_1, \ldots, d_m \in \mathbb{N}$. Let $(x_n)_{n\geq 0}$ (6.87) be a complex LRS. Assume that the roots $z_1, \ldots, z_m$ of the polynomial P defined in (6.85)*

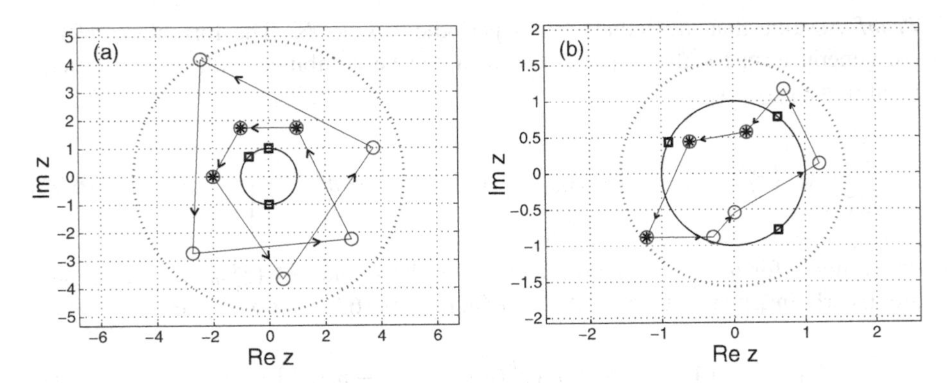

Fig. 6.4 Orbit of sequence $(x_n)_{n\geq 0}$ computed from formula (6.89) (circles) for $m = 3$ simple roots (**a**) $z_1 = e^{2\pi i \frac{2}{8}}$, $z_2 = e^{2\pi i \frac{3}{8}}$, $z_3 = e^{2\pi i \frac{6}{8}}$ and initial conditions $a_j = 2e^{\frac{2\pi i}{6} j}$, $j = 2, 3, 6$; (**b**) $z_1 = e^{\frac{2\pi i}{7}}$, $z_2 = e^{\frac{6\pi i}{7}}$, $z_3 = e^{\frac{12\pi i}{7}}$ and $\alpha_1 = 3/5e^{\frac{2\pi i}{5}}$, $\alpha_2 = 3/4e^{\frac{4\pi i}{5}}$, $\alpha_3 = 3/2e^{\frac{6\pi i}{5}}$. Also plotted, initial conditions $a_1, \ldots, a_m$ (stars), generators $z_1, \ldots, z_m$ (squares), unit circle S (solid line), and circle $S(0,\ |A_{11}| + |A_{21}| + \cdots + |A_{m1}|)$ (dotted line)

are distinct. If $(x_n)_{n\geq 0}$ is periodic, then all sequence terms are located inside the disk of radius $|A_{11}| + |A_{21}| + \cdots + |A_{m1}|$, with A_{j1}, $j = 1, \ldots, m$ given by (6.90).

Proof Any complex numbers u, v verify the triangle inequality [168, p.18]

$$\big|\,|u| - |v|\,\big| \leq |u + v| \leq |u| + |v|. \tag{6.96}$$

In general, for any complex numbers $x_1, \ldots, x_m$ we have

$$|x_1 + \cdots + x_m| \leq |x_1| + \cdots + |x_m|. \tag{6.97}$$

By Theorem 6.15, the periodic solutions can be represented as

$$x_n = A_{11} z_1^n + A_{21} z_2^n + \cdots + A_{m1} z_m^n. \tag{6.98}$$

Using (6.97), (6.98), and $|z_1| = |z_2| = \cdots = |z_m| = 1$ one obtains

$$|x_n| \leq |A_{11} z_1^n| + |A_{21} z_2^n| + \cdots + |A_{m1} z_m^n| = |A_{11}| + |A_{21}| + \cdots + |A_{m1}|.$$

We illustrate this result in Figure 6.4. □

Remark 6.6 For $m = 2$ one can use the left-hand side of (6.96) to obtain a lower bound for the orbit, as illustrated in [27, Theorem 4.1].

Remark 6.7 For $m \geq 3$ the periodic orbit does not generally have an inner bound. Various inequalities for the left-hand side of (6.96) that only involve $|x_1|, \cdots, |x_m|$ are presented in the monograph of Dragomir [61, Chapter 3], but under restrictive assumptions on $|x_1|, \cdots, |x_m|$. The main reason for this can be identified even for $m = 3$, where the sum $x_1 + x_2 + x_3$ can vanish while the radii $|x_1|$, $|x_2|$ and $|x_3|$ cannot be related by an identity.

When $x_1, \ldots, x_m$ are real the following result holds.

Remark 6.8 Let $m \geq 2$ and $x_1, \ldots, x_m$ be real numbers (or complex but aligned). Then the following inequality holds:

$$\min | \pm |x_1| \pm |x_2| \pm \cdots \pm |x_m| | \leq |x_1 + \cdots + x_m| \leq |x_1| + \cdots + |x_m|.$$

6.4.2 The Geometric Structure of Periodic Orbits

Here we investigate the geometric structure of generalized Horadam periodic sequences producing nondegenerated orbits. This is the case when the characteristic polynomial (6.2) has m simple roots (generators), which are roots of unity $z_j = e^{2\pi i p_j/k_j}$ with $j = 1, \ldots, m$. The formula of the general term is

$$x_n = A_1 z_1^n + A_2 z_2^n + \cdots + A_m z_m^n, \tag{6.99}$$

and the arbitrary initial conditions $\alpha_1, \alpha_2, \ldots, \alpha_m$ are such that the coefficients $A_1, \ldots, A_m$ are all nonzero.

Similar results to those presented in Section 2.3 can be formulated. The main result of that section, Theorem 5.11, can be generalized as follows.

Theorem 6.17 *Let $m \geq 2$ and $k_1, k_2, \ldots, k_m$ be natural numbers and consider the distinct primitive roots of unity $z_j = e^{2\pi i p_j/k_j}$, $j = 1, \ldots, m$. Denote the least common multiple of numbers $k_1, \ldots, k_m$ by*

$$K = lcm(k_1, k_2, \ldots, k_m).$$

The orbit of the sequence $(x_n)_{n \geq 0}$ is then a K-gon, whose nodes can be divided into K/k_j regular k_j-gons representing a multipartite graph.

Proof The proof follows the same lines as in Theorem 5.11. One may choose $j = m$ and prove that the orbit is divided into K/k_m regular k_m-gons. The orbit is partitioned into k_m subsets of equal size, obtained from each other by a rotation. The property is illustrated in Figures 5.10 and 5.11 for $m = 2$. □

6.4.3 "Masked" Periodicity

As seen for Horadam sequences in Chapter 5, under certain conditions, the period of the sequence is not completely determined by the generators.

In this section we present the phenomenon of masked periodicity for generalized Horadam sequences, based on the periodicity conditions formulated by Bagdasar and Larcombe in [30]. Findings are illustrated by an example involving third-order recurrent sequences, and have been obtained by Bagdasar and Larcombe [29].

Theorem 6.18 *Let $m \geq 2$ and the distinct primitive roots of unity $z_j = e^{2\pi i p_j/k_j}$, $j = 1, \dots, m$. The recurrence sequence $(x_n)_{n\geq 0}$ generated by the characteristic polynomial (6.65), and the arbitrary initial values $x_j = a_{j+1}$, $j = 1, \dots, m$ is periodic. If the sets I and J satisfy*

$$I \cap J = \emptyset, \quad I \cup J = \{1, \dots, m\},$$

where the solution $A_1, A_2, \dots, A_m$ of the system (6.3.1) *has the property*

$$A_i = 0, \text{ for } \quad i \in I,$$
$$A_j \neq 0, \text{ for } \quad j \in J,$$

then the period of the sequence is the lcm of the numbers k_j with $j \in J$.

Proof For all $i \in I$, the terms corresponding to z_i^n do not feature explicitly in the general formula (6.68). The general term is a linear combination of nonvanishing individual terms of periods k_j, $j \in J$, hence the conclusion. □

We illustrate this result by a third-order recurrent sequence ($m = 3$).

Example 6.11 Consider the distinct primitive roots of unity $z_1 = e^{2\pi i p_1/k_1}$, $z_2 = e^{2\pi i p_2/k_2}$, and $z_3 = e^{2\pi i p_3/k_3}$ where $p_1, p_2, p_3, k_1, k_2, k_3$ are positive integers, and the polynomial $P(x)$

$$P(x) = (x - z_1)(x - z_2)(x - z_3), \quad x \in \mathbb{C}. \tag{6.100}$$

The recurrence sequence $(x_n)_{n\geq 0}$ generated by the characteristic polynomial (6.100), and the arbitrary initial values $x_0 = a_1$, $x_1 = a_2$, $x_2 = a_3$ is periodic. Moreover, if A_1, A_2, and A_3 are the solutions of the system

$$\begin{cases} a_1 &= A_1 + A_2 + A_3, \\ a_2 &= A_1 z_1 + A_2 z_2 + A_3 z_3, \\ a_3 &= A_1 z_1^2 + A_2 z_2^2 + A_3 z_3^2, \end{cases} \tag{6.101}$$

the following periods are possible

1° $A_1A_2A_3 \neq 0$: the period is $\mathrm{lcm}(k_1, k_2, k_3)$;
2° $A_i = 0$, $A_jA_l \neq 0$: the period is $\mathrm{lcm}(k_j, k_l)$ ($i, j, l \in \{1, 2, 3\}$ are distinct);
3° $A_i \neq 0$, $A_j = A_l = 0$: the period is k_i (with $i, j, l \in \{1, 2, 3\}$ distinct);
4° $A_1 = A_2 = A_3 = 0$: the period is 1 (constant sequence).

In addition to the theoretical result, here we also formulate the specific relations between generators and initial conditions, required for the various outcomes. The formula for the general term of sequence $(x_n)_{n\geq 0}$ in this case can be written as

$$x_n = A_1z_1^n + A_2z_2^n + A_3z_3^n, \tag{6.102}$$

where constants A_1, A_2, and A_3 satisfy the system (6.101). In matrix form this is

$$\begin{pmatrix} 1 & 1 & 1 \\ z_1 & z_2 & z_3 \\ z_1^2 & z_2^2 & z_3^2 \end{pmatrix} \begin{bmatrix} A_1 \\ A_2 \\ A_3 \end{bmatrix} = \begin{bmatrix} a_1 \\ a_2 \\ a_3 \end{bmatrix}. \tag{6.103}$$

The determinant of the 3×3 Vandermonde matrix is

$$\Delta = \det\Big(V_{3,3}(z_1, z_2, z_3)\Big) = \prod_{1\leq i<j\leq 3} (z_j - z_i) = (z_2 - z_1)(z_3 - z_2)(z_3 - z_1) \neq 0,$$

hence the solutions of system (6.103) can be written as

$$\begin{aligned} A_1 &= \frac{(z_3 - z_2)\,[a_1\, z_2z_3 - a_2(z_2 + z_3) + a_3]}{\Delta}, \\ -A_2 &= \frac{(z_3 - z_1)\,[a_1\, z_1z_3 - a_2(z_1 + z_3) + a_3]}{\Delta}, \\ A_3 &= \frac{(z_2 - z_1)\,[a_1\, z_1z_2 - a_2(z_1 + z_2) + a_3]}{\Delta}. \end{aligned} \tag{6.104}$$

1° If A_1, A_2, A_3 are nonzero, they all appear in formula (6.102), hence the period of the sequence is given by $\mathrm{lcm}(k_1, k_2, k_3)$. For $k_1 = 2$, $k_2 = 3$, $k_3 = 5$, and $A_1A_2A_3 \neq 0$ the period is 30, as depicted in Figure 6.5a. For distinct z_1, z_2, z_3 the masked periodicity conditions reduce to

$$\begin{aligned} A_1 = 0 &: \quad a_1\, z_2z_3 - a_2(z_2 + z_3) + a_3 = 0, \\ A_2 = 0 &: \quad a_1\, z_1z_3 - a_2(z_1 + z_3) + a_3 = 0, \\ A_3 = 0 &: \quad a_1\, z_1z_2 - a_2(z_1 + z_2) + a_3 = 0. \end{aligned} \tag{6.105}$$

2° If $a_3 = a_2(z_2 + z_3) - a_1z_2z_3$, then $A_1 = 0$ and the term involving z_1^n does not feature explicitly in formula (6.102), hence the orbit period is $\mathrm{lcm}(k_2, k_3) = 15$

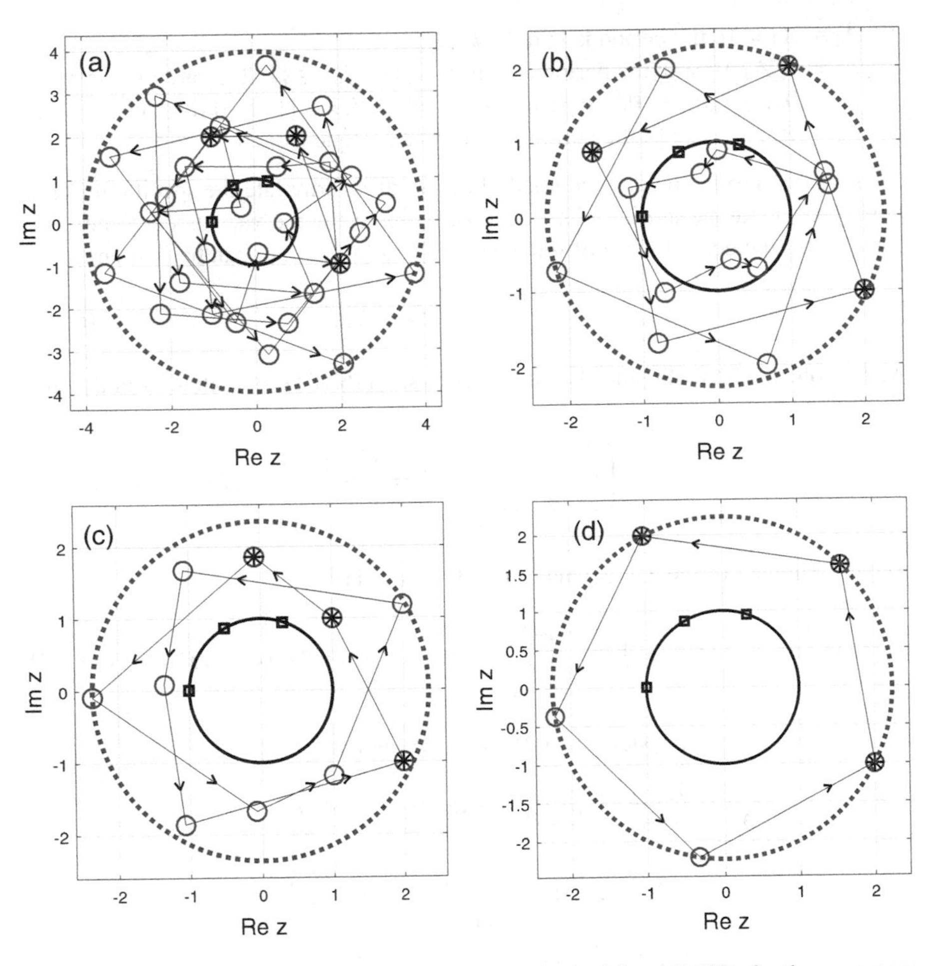

Fig. 6.5 First 30 terms of sequence $(x_n)_{n\geq 0}$ (circles) obtained from (6.102), for the generators $z_1 = e^{2\pi i \frac{1}{2}}$, $z_2 = e^{2\pi i \frac{1}{3}}$ and $z_3 = e^{2\pi i \frac{1}{5}}$ (squares) and initial conditions (stars) for (**a**) $a_1 = 2 - i$, $a_2 = 1 + 2i$, $a_3 = -1 + 2i$ ($A_1 A_2 A_3 \neq 0$); (**b**) $a_1 = 2 - i$, $a_2 = 1 + 2i$, $a_3 = a_2(z_2 + z_3) - a_2 z_2 z_3$ ($A_1 = 0$, $A_2 A_3 \neq 0$); (**c**) $a_1 = 2 - i$, $a_2 = 1 + 2i$, $a_3 = a_2(z_1 + z_3) - a_2 z_1 z_3$ ($A_2 = 0$, $A_1 A_3 \neq 0$); (**d**) $a_1 = 2 - i$, $a_2 = a z_3$, $a_3 = a_2(z_1 + z_3) - a_2 z_1 z_3$ ($A_1 = A_2 = 0$, $A_3 \neq 0$). Arrows indicate the direction of the orbit. Boundaries of circle $U(0, |A_1| + |A_2| + |A_3|)$ (dotted line) with A_1, A_2, A_3 from (6.104) and the unit circle (solid line) are also plotted

as illustrated in Figure 6.5b. A similar example is shown in Figure 6.5c, where $A_2 = 0$ and the period of the sequence is 10.

3° When two of the terms are zero (say for example $A_1 = A_2 = 0$), then the period of the sequence is given by the term z_3^n, hence it is equal to k_3. This case is illustrated in Figure 6.5d. Subtracting the first equation from the second in (6.105), one obtains

$$a_1 z_3(z_2 - z_1) - a_2(z_2 - z_1) = 0,$$

which is equivalent to $a_2 = a_1 z_3$. By substitution in the first line of (6.105) we obtain $a_3 = a_1 z_3^2$.

4° One may also notice that the only possibility for $A_1 = A_2 = A_3 = 0$ is $a_1 = a_2 = a_3 = 0$.

6.4.4 *The Enumeration of Periodic Orbits*

Here we cover generalized Horadam periodic sequences producing nondegenerated orbits. In this case the m distinct generators are all roots of unity $z_j = e^{2\pi i p_j / k_j}$ with $j = 1, \ldots, m$, and the arbitrary initial conditions $\alpha_1, \ldots, \alpha_m$ are such that the coefficients $A_1, \ldots, A_m$ in formula (6.68) are all nonzero.

Denote the order of a root of unity z by $\text{ord}(z)$. The period of the complex generalized Horadam sequence delivered by a generator m-tuple $(z_1, \ldots, z_m)$ is (usually) given by $\text{lcm}(\text{ord}(z_1), \ldots, \text{ord}(z_m))$. Representing the m-tuple $(z_1, \ldots, z_m)$ by the $2m$-tuple $(p_1, k_1, \ldots, p_m, k_m)$, we want to select those producing a sequence having period k. To ensure that the enumeration formula generates all the distinct periodic sequences, we shall assume without loss of generality that $z_1, z_2, \ldots, z_m$ are primitive roots of unity and $k_1 \le k_2 \le \cdots \le k_m$.

The number of distinct sequences having period k can be enumerated from the tuples $(P, K) = (p_1, k_1, p_2, k_2, \ldots, p_m, k_m)$ $(1 \le p_j \le k_j, j = 1, \ldots, m)$ satisfying the conditions

$$H_P^m(k) = \left|\{(P, K) : \gcd(p_j, k_j) = 1,\ j = 1, \ldots, m,\ \text{lcm}(k_1, \ldots, k_m) = k,\ k_1 \le \cdots \le k_m\}\right|. \tag{6.106}$$

Some formulae for this expression are identified, based on the properties of tuples $(k_1, \ldots, k_m)$ satisfying $\text{lcm}(k_1, \ldots, k_m) = k$, and their corresponding generators $z_j = e^{2\pi i p_j / k_j}$ with $j = 1, \ldots, m$.

6.4.5 *A First Formula for $H_P(m; k)$*

We first generate the tuples $(k_1, \ldots, k_m)$ with $\text{lcm}(k_1, \ldots, k_m) = k$, then count the tuples $(p_1, \ldots, p_m)$ such that (P, K) satisfies (6.106).

The first lemma counts the $2m$-tuples (P, K) in (6.106) with $k_1 = \cdots = k_m$.

Lemma 6.3 *If $k_1 = \cdots = k_m$ and $\text{lcm}(k_1, \ldots, k_m) = k$, then $k_1 = \cdots = k_m = k$.*

Proof Clearly, the number of $2m$-tuples (P, K) with (6.106) is given by

$$H'_P(m; k) = \left|\{(p_1, \ldots, p_m) : \gcd(p_j, k) = 1,\ j = 1, \ldots, m,\ p_1 < \cdots < p_m\}\right| = \binom{\varphi(k)}{m},$$

as the distinct values $p_1, \ldots, p_m$ can be chosen in $\varphi(k)$ ways. □

Second lemma counts the $2m$-tuples (P, K) obtained for distinct $k_1, \ldots, k_m$.

Lemma 6.4 *If* $\operatorname{lcm}(k_1, \ldots, k_m) = k$ *and* $k_1, \ldots, k_m$ *are distinct, the number of* $2m$*-tuples* (P, K) *satisfying (6.106) is* $\varphi(k_1) \cdots \varphi(k_m)$.

Proof For distinct $k_1, \ldots, k_m$, any choice of $p_1, \ldots, p_m$ satisfying $1 \le p_j \le k_j$ and $\gcd(p_j, k_j) = 1$ $j = 1, \ldots, m$ generates distinct generators. For each fixed $j = 1, \ldots, m$ there are $\varphi(k_j)$ choices for p_j, giving a total of $\varphi(k_1) \cdots \varphi(k_m)$ configurations. □

Finally, we can draw the main result of this section.

Lemma 6.5 *If the numbers* $k_1, \ldots, k_m$ *satisfying* $\operatorname{lcm}(k_1, \ldots, k_m) = k$ *can be partitioned into* $s > 0$ *sets having* d_j *elements equal to a value* K_s *for* $j = 1, \ldots, s$, *then the* $2m$*-tuples* (P, K) *satisfying (6.106) corresponding to this* m*-tuple is*

$$\binom{\varphi(K_1)}{d_1} \cdots \binom{\varphi(K_s)}{d_s}.$$

Proof If $k_1 = \cdots = k_{d_1} = K_1$, the distinct values $p_1, \ldots, p_{d_1}$ ensuring that $\gcd(p_j, k_j) = 1$, $j = 1, \ldots, d_1$ can be chosen in $\binom{\varphi(K_1)}{d_1}$ ways, as in Lemma 6.4. Repeating this argument for the other subsets one obtains the result. □

Theorem 6.19 *Let us consider the integers* $k, m \ge 2$. *The number of distinct generalized Horadam sequences order* m *and fixed period* k *is equal to*

$$H_P(m; k) = \sum_{s=1}^{m} \Bigg\{ \sum_{\substack{k_1 < \cdots < k_s \\ d_1 + \cdots + d_s = k \\ 1 \le d_i \le k \\ [k_1, \ldots, k_s] = k}} \binom{\varphi(k_1)}{d_1} \cdots \binom{\varphi(k_s)}{d_s} \Bigg\}.$$

Proof The proof follows by Lemmas 6.3, 6.4, and 6.5. □

6.4.5.1 An Example for $m = 3$

Example: $m = 3$ Find all the tuples $(p_1, k_1, p_2, k_2, p_3, k_3)$ satisfying the properties $\gcd(p_1, k_1) = \gcd(p_2, k_2) = \gcd(p_3, k_3) = 1$ and $\operatorname{lcm}(k_1, k_2, k_3) = k$.

Four distinct cases are possible, so $H_P(3; k) = H_1 + H_2 + H_3 + H_4$.

1. $k_1 = k_2 = k_3$: $H_1 = \frac{\varphi(k)(\varphi(k)-1)(\varphi(k)-2)}{3!} = \binom{\varphi(k)}{3}$;
2. $k_1 = k_2 < k_3$: $H_2 = \frac{\varphi(k_1)(\varphi(k_1)-1)}{2!}\varphi(k_3) = \binom{\varphi(k_1)}{2}\varphi(k_3)$;
3. $k_1 < k_2 = k_3$: $H_3 = \varphi(k_1)\frac{\varphi(k_3)(\varphi(k_3)-1)}{2!} = \varphi(k_1)\binom{\varphi(k_3)}{2}$;
4. $k_1 < k_2 < k_3$: $H_4 = \varphi(k_1)\varphi(k_2)\varphi(k_3)$.

To find all triplets (k_1, k_2, k_3) for $k = p_1^{m_1} p_2^{m_2} \cdots p_n^{m_n}$ one may define

$$\text{LCM }(2,k) = |\{(k_1,k_2) : \text{lcm}(k_1,k_2) = k\}| = \prod_{j=1}^{n}[(m_j+1)^2 - m_j^2];$$
$$\text{LCM }(3,k) = |\{(k_1,k_2,k_3) : \text{lcm}(k_1,k_2,k_3) = k\}| = \prod_{j=1}^{n}[(m_j+1)^3 - m_j^3];$$
$$\text{LCM}^{\leq}(3,k) = |\{(k_1,k_2,k_3) : \text{lcm}(k_1,k_2,k_3) fa = k, 1 \leq k_1 \leq k_2 \leq k_3 \leq k\}|.$$

One can prove that (see the details in Section 5.2)

$$\text{LCM}^{\leq}(3,k) = \frac{[\text{LCM}(3,k)-1] + 3\,[\text{LCM}(2,k)-1]}{6} + 1.$$

Example 6.12 (prime numbers) When k is a prime we have $\varphi(k) = k - 1$. For this number we just have two divisor pairs

$$(k_1, k_2, k_3) \in \{(1, k, k), (k, k, k)\},$$

with multiplicities $\varphi(1)\binom{\varphi(k)}{2}$ and $\binom{\varphi(k)}{3} = \frac{k(k-1)(k-2)}{6}$, giving the formula

$$H_P(k) = \frac{(k-1)(k-2)}{2} + \frac{(k-1)(k-2)(k-3)}{6} = \frac{k(k-1)(k-2)}{6} = \binom{k}{3}. \tag{6.107}$$

When $k = 11$, formula (6.107) gives a total of $11 \cdot 10 \cdot 9/2 = 165$ distinct solutions, while for $k = 7$ there is a total of $7 \cdot 6 \cdot 5/2 = 35$ distinct solutions. Explicitly, for $k = 5$ there are 10 solutions given by the fraction pairs

$$\left(\frac{p_1}{k_1}, \frac{p_2}{k_2}, \frac{p_3}{k_3}\right) \in \left\{\left(\frac{1}{1}, \frac{1}{5}, \frac{2}{5}\right), \left(\frac{1}{1}, \frac{1}{5}, \frac{3}{5}\right), \left(\frac{1}{1}, \frac{1}{5}, \frac{4}{5}\right), \left(\frac{1}{1}, \frac{2}{5}, \frac{3}{5}\right), \left(\frac{1}{1}, \frac{2}{5}, \frac{4}{5}\right),\right.$$
$$\left.\left(\frac{1}{1}, \frac{3}{5}, \frac{4}{5}\right), \left(\frac{1}{5}, \frac{2}{5}, \frac{3}{5}\right), \left(\frac{1}{5}, \frac{2}{5}, \frac{4}{5}\right), \left(\frac{1}{5}, \frac{3}{5}, \frac{4}{5}\right), \left(\frac{2}{5}, \frac{3}{5}, \frac{4}{5}\right)\right\}.$$

Example 6.13 (powers of a prime) When $k = p^s$ with p a prime number and $s \geq 2$ we have $\varphi(k) = \varphi(p^s) = p^s(1 - 1/p) = p^s - p^{s-1}$. For this number we have the divisor pairs $(k_1, k_2, k_3) \in \{(p^{s_1}, p^{s_2}, k) : 0 \leq s_1 \leq s_2 \leq s\}$.

As discussed before, each of the triplets may fall into exactly one of the four categories below, with multiplicities

1. $0 \leq s_1 < s_2 < s$: $\varphi(p^{s_1})\varphi(p^{s_2})\varphi(p^s)$;
2. $0 \leq s_1 = s_2 < s$: $\binom{\varphi(p^{s_1})}{2}\varphi(p^s)$;
3. $0 \leq s_1 < s_2 = s$: $\varphi(p^{s_1})\binom{\varphi(p^s)}{2}$;
4. $s_1 = s_2 = s$: $\binom{\varphi(p^s)}{3}$.

Since $\sum_{1\le i<j\le n} x_i x_j = \left[\left(\sum_{i=1}^n x_i\right)^2 - \sum_{i=1}^n x_i^2\right]/2$, the sum of the terms in (1) and (2) give $[p^{2(s-1)} - p^{s-1}]/2$, where we used that $\sum_{i=1}^s \varphi(p^{s_1}) = \varphi(p^s)$. By the same property, the term in (3) is $\varphi(p^{s-1})\binom{\varphi(p^{s_1})}{2}$, hence

$$H_P(3;k) = \frac{\varphi(k)}{6}\left[\varphi(k)^2 + 3(k-1)(k/p-1) - 1\right].$$

For example, when $k = 9 = 3^2$ we have $H_P(k) = \frac{6}{6}[36 + 3\cdot 8\cdot 2 - 1] = 83$, while for $k = 4$ one obtains $H_P(k) = \frac{2}{6}[4 + 3\cdot 3 - 1]$ with the distinct solutions

$$\left(\frac{p_1}{k_1}, \frac{p_2}{k_2}, \frac{p_3}{k_3}\right) \in \left\{\left(\frac{1}{1}, \frac{1}{2}, \frac{1}{4}\right), \left(\frac{1}{1}, \frac{1}{2}, \frac{3}{4}\right), \left(\frac{1}{1}, \frac{1}{4}, \frac{3}{4}\right), \left(\frac{1}{2}, \frac{1}{4}, \frac{3}{4}\right)\right\}.$$

Example 6.14 (product of two primes) For p, q primes, the formula gives

$$H_P(3;k) = \frac{\varphi(k)}{6}\left[k\,(k + (p+q) - 4) + (p+q-1)^2 - 1\right].$$

So far, we have obtained the values

$$k = p: \quad H_P(3;k) = \frac{k(k-1)(k-2)}{6} = \binom{k}{3}.$$

Examples : $H_P(3;2) = 0,\ H_P(3;3) = 1,\ H_P(3;7) = 35,\ H_P(3;11) = 165$

$$k = p^\alpha: \quad H_P(3;k) = \frac{\varphi(k)}{6}\left[\varphi(k)^2 + 3(k-1)(k/p-1) - 1\right]$$

Examples : $H_P(3;4) = 4,\ H_P(3;8) = 52,\ H_P(3;9) = 83; \ldots$

$$k = pq: \quad H_P(3;k) = \frac{\varphi(k)}{6}\left[k\,(k + (p+q) - 4) + (p+q-1)^2 - 1\right].$$

Examples : $H_P(3;6) = 19,\ H_P(3;10) = 110, \ldots.$

As a number sequence $H_P(3;k)$ in $k \ge 1$ starts with the values

$$0, 0, 1, 4, 9, 19, 35, 52, 83, 110, 165, \ldots.$$

To calculate $H_P(m;k)$, one needs to count the m-tuples with the same lcm

$$\mathrm{LCM}^{\le}(m,k) = |\{(k_1,\ldots,k_m) : \mathrm{lcm}(k_1,\ldots,k_m) = k,\ 1 \le k_1 \le \cdots \le k_m \le k\}|\,.$$

6.5 Orbits Generated by Roots of Unity

This section extends the result regarding periodic sequences presented in Theorem 6.14, for sequences generated by repeated roots of unity.

When the nonzero solutions of (6.2) are kth roots of unity, the asymptotic behavior of subsequences $(x_{Nk+j})_{N\geq 0}$, $j = 0, \ldots, k-1$ can be examined. The subsequences converge when the roots' multiplicities are all less than one and diverge otherwise. When divergent, the terms of the subsequences can be collinear if the maximum root multiplicity is two, and approach an asymptote for higher multiplicities.

Theorem 6.20 *Let $2 \leq m \leq k$ and $d_1, \ldots, d_m$ be natural numbers, $z_1, \ldots, z_m$ distinct kth roots of unity and the sequence $(x_n)_{n\geq 0}$ be generated by the characteristic polynomial (6.85). Defining the number*

$$d^* = \max\{j : A_{ij} \neq 0, i \in \{1, \ldots, m\}\}, \tag{6.108}$$

for the coefficients A_{ij} ($1 \leq i \leq m$, $1 \leq j \leq d_i$) given in (6.89), one obtains the following properties of the sequence $(x_n)_{n\geq 0}$ and its subsequences $(x_{Nk+j})_{N\geq 0}$

1° *For $d^* \leq 1$ the sequence $(x_n)_{n\geq 0}$ is periodic.*
2° *For $d^* \geq 2$ the sequence $(x_n)_{n\geq 0}$ is divergent.*
3° *For $d^* \leq 2$ the terms of $(x_{Nk+j})_{N\geq 0}$ are collinear (when periodic, $d^* = 1$).*
4° *For $d^* \geq 3$ the terms of $(x_{Nk+j})_{N\geq 0}$ converge asymptotically to straight lines.*

Proof

1° This follows from Theorem 6.10.
2° Denote by $I_{d^*} \subset \{1, \ldots, m\}$ the index set for which the maximum d^* is attained in (6.108). By Theorem 6.3, the sequences $(z_1^n)_{n\geq 0}, \ldots, (z_m^n)_{n\geq 0}$ are linearly independent, so $(z_i^n)_{n\geq 0}$, $i \in I_{d^*}$ are linearly independent. The term x_n given by (6.89) contains a monomial of degree $d^* - 1$ in n, written as

$$\sum_{i\in I_{d^*}} A_{id^*} n^{d^*-1} z_i^n = n^{d^*-1} \sum_{i\in I_{d^*}} A_{id^*} z_i^n.$$

We show that the coefficients of n^{d^*-1} in x_n cannot be all zero if $d^* \geq 2$.

As $z_1, \ldots, z_m$ are kth roots of unity, the following relation holds:

$$\sum_{i\in I_{d^*}} A_{id^*} z_i^n = \sum_{i\in I_{d^*}} A_{id^*} z_i^{n+k},$$

therefore the above expression can only have one of the values

$$\sum_{i \in I_{d^*}} A_{id^*}, \sum_{i \in I_{d^*}} A_{id^*} z_i, \ldots, \sum_{i \in I_{d^*}} A_{id^*} z_i^{k-1}.$$

If these terms zero, then $\sum_{i \in I_{d^*}} A_{id^*} z_i^n = 0$, and the linear independence of $(z_i^n)_{n \geq 0}$ $(i \in I_{d^*})$ gives $A_{id^*} = 0$ $(i \in I_{d^*})$, in contradiction with (6.108) of d^*. Thus, there is an index $j \in \{0, \ldots, k-1\}$ for which $\sum_{i \in I_{d^*}} A_{id^*} z_i^j \neq 0$, hence $n^{d^*-1} \sum_{i \in I_{d^*}} A_{id^*} z_i^{Nk+j} = n^{d^*-1} \sum_{i \in I_{d^*}} A_{id^*} z_i^j$, is divergent. Since this is the leading term of x_{Nk+j} n, $(x_n)_{n \geq 0}$ diverges.

3° For each $j \in \{0, \ldots, k-1\}$, $N \in \mathbb{N}$ we have $z_i^{Nk+j} = z_i^j$, hence (6.89) gives

$$x_{Nk+j} = \left(A_{11} + (Nk+j)A_{12} + \cdots + (Nk+j)^{d_1-1} A_{1d_1}\right) z_1^j + \cdots +$$
$$\left(A_{m1} + (Nk+j)A_{m2} + \cdots + (Nk+j)^{d_m-1} A_{md_m}\right) z_m^j. \tag{6.109}$$

For a fixed j, (6.109) is a polynomial of degree $d^* - 1$ in N, written as

$$x_{Nk+j} = B_1 + B_2(Nk+j) + \cdots + (Nk+j)^{d^*-1} B_{d^*}, \tag{6.110}$$

where d^* was defined in (6.108). For $d^* = 2$ we obtain

$$x_{Nk+j} - x_j = [B_1 + B_2(Nk+j)] - [B_1 + B_2 j] = NkB_2,$$

whose argument is independent of N. The terms of $(x_{Nk+j})_{N \geq 0}$ are then collinear for each $j \in \{0, \ldots, k-1\}$. For $d^* = 1$ the condition $x_{Nk+j} - x_j = 0$ confirms the periodicity. The property is shown in Figure 6.6a.

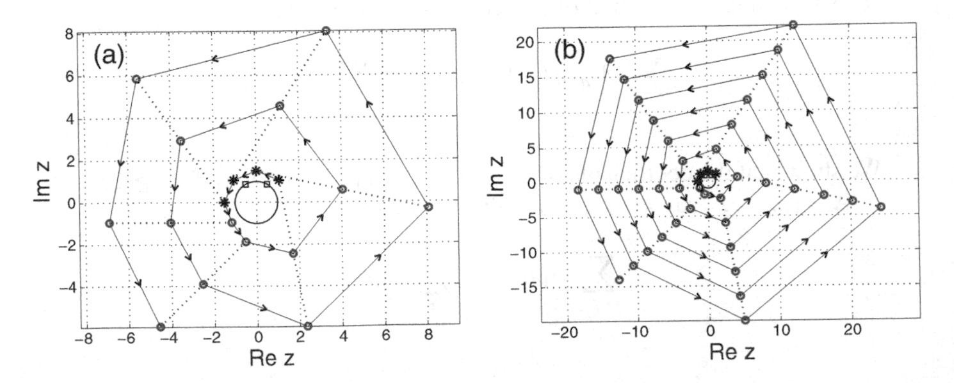

Fig. 6.6 First (**a**) $N = 18$ terms; (**b**) $N = 42$ terms of $(x_n)_{n \geq 0}$ computed from the direct formula (6.92) (circles) for generators $z_j = e^{\frac{2\pi i}{6} j}$, $j = 1, 2$, multiplicities $d^* = d_1 = d_2 = 2$, and initial conditions $\alpha_j = 2e^{\frac{2\pi i}{8} j}$, $j = 1, \ldots, 4$. Also plotted, initial conditions (stars), generators (squares), and unit circle S (solid line). Arrows indicate the increase of n

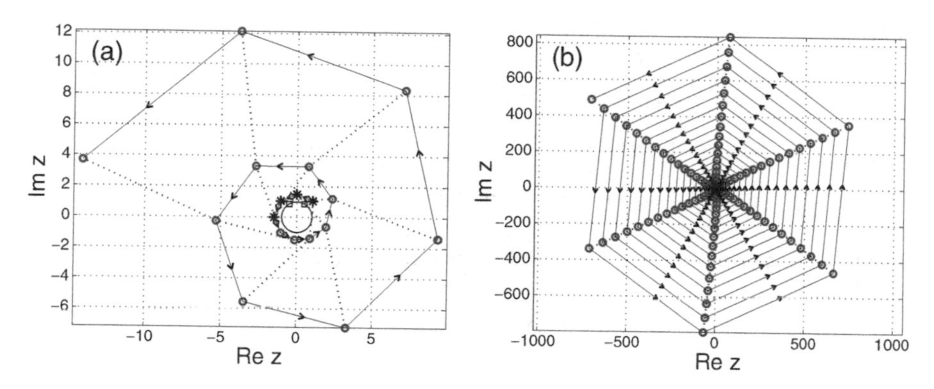

Fig. 6.7 First (**a**) $N = 18$ terms; (**b**) $N = 120$ terms of sequence $(x_n)_{n\geq 0}$ computed from the direct formula (6.92) (circles) for generators $z_j = e^{\frac{2\pi i}{6}j}$, $j = 1, 2$, multiplicities $d^* = d_1 = 3$, $d_2 = 1$, and initial conditions $\alpha_j = 2e^{\frac{2\pi i}{8}j}$, $j = 1, \ldots, 4$. Also plotted are initial conditions (stars), generators (squares), and unit circle S (solid line). Arrows indicate increase in n

4° For $(x_{Nk+j})_{N\geq 0}$ when $d^* \geq 3$, the argument of the term (6.110) verifies

$$\arg(x_{Nk+j}) = \arg\left(B_1 + B_2(Nk+j) + \cdots + (Nk+j)^{d^*-1}B_{d^*}\right)$$
$$= \arg\left[\frac{1}{(Nk+j)^{d^*-1}}\left(B_1 + \cdots + (Nk+j)^{d^*-1}B_{d^*}\right)\right].$$

In the limit $N \to \infty$, the argument of x_{Nk+j} satisfies

$$\lim_{N\to\infty} \arg(x_{Nk+j}) = \lim_{N\to\infty} \arg\left[\frac{1}{(Nk+j)^{d^*-1}}\left(B_1 + \cdots + B_{d^*}(Nk+j)^{d^*-1}\right)\right]$$
$$= \arg\left[\lim_{N\to\infty} \frac{1}{(Nk+j)^{d^*-1}}\left(B_1 + \cdots + B_{d^*}(Nk+j)^{d^*-1}\right)\right]$$
$$= \arg\left(B_{d^*}\right). \tag{6.111}$$

Hence, the terms of the subsequence $(x_{Nk+j})_{N\geq 0}$ align asymptotically to the line of argument $\arg(B_{d^*})$, defined for every value of $j \in \{0, \ldots, k-1\}$. The asymptotic behavior of subsequences is sketched in Figure 6.7. One can notice that in Figure 6.7a the subsequence terms are not collinear, but converge asymptotically to a constant value, as suggested by formula (6.111). □

Example 6.15 Let $(x_n)_{n\geq 1}$ be a sequence satisfying $x_{n+1} = x_n + 2n + 1$ for $n = 0, 1, \ldots$. Show that

$$x_{n+3} = 3x_{n+2} - 3x_{n+1} + x_n, \quad n = 0, 1, \ldots.$$

Solution We have

$$x_{n+3} - x_{n+2} = 2(n+2) + 1 = 2n + 5$$
$$x_{n+2} - x_{n+1} = 2(n+1) + 1 = 2n + 3$$
$$x_{n+1} - x_n = 2n + 1.$$

The relation $x_{n+3} = 3x_{n+2} - 3x_{n+1} + x_n$ is equivalent to

$$(x_{n+3} - x_{n+2}) + (x_{n+1} - x_n) = 2(x_{n+2} - x_{n+1}).$$

6.6 Orbits of Complex General Order LRS

In this section we examine various non-periodic LRS patterns, not generally produced by roots of unity. Similar to the configurations identified for Horadam sequences in Chapter 5, we recover numerous types of stable, convergent or divergent patterns in the complex plane. This long-term behavior is interesting also from an application's perspective, as they can be linked to random-number generation, as seen in Section 5.7.

We shall focus on nondegenerate orbits produced by distinct generators, where the general term formula is given by $x_n = A_1 z_1^n + A_2 z_2^n + \cdots + A_m z_m^n$, and the arbitrary initial conditions $a_1, a_2, \ldots, a_m$ are such that the coefficients $A_1, \ldots, A_m$ are all nonzero.

The $m \geq 2$ distinct generators are here denoted by

$$z_1 = r_1 e^{2\pi i x_1}, \ z_2 = r_2 e^{2\pi i x_2}, \cdots, \ z_m = r_m e^{2\pi i x_m}, \tag{6.112}$$

where $r_1, \ldots, r_m, x_1, \ldots, x_m \in \mathbb{R}$. We may assume $0 \leq r_1 \leq r_2 \leq \cdots \leq r_m$. The generalized Horadam patterns produced by formula (6.99) are

1. Stable for $r_1 = r_2 = \cdots = r_m = 1$;
2. Quasi-convergent for $0 \leq r_1 \leq r_2 \leq \cdots \leq r_m = 1$;
3. Convergent for $0 \leq r_1 \leq r_2 \leq \cdots \leq r_m < 1$;
4. Divergent for $r_m \geq 1$.

The geometric patterns obtained in each case are presented below.

6.6.1 *Stable Orbits:* $r_1 = r_2 = \cdots = r_m = 1$

Here we present the orbits obtained for distinct generators located on the unit circle. We focus on the case $m = 3$, where for initial conditions $x_0 = a$, $x_1 = b$ and $x_2 = c$ $(a, b, c \in \mathbb{C})$, the general term formula is

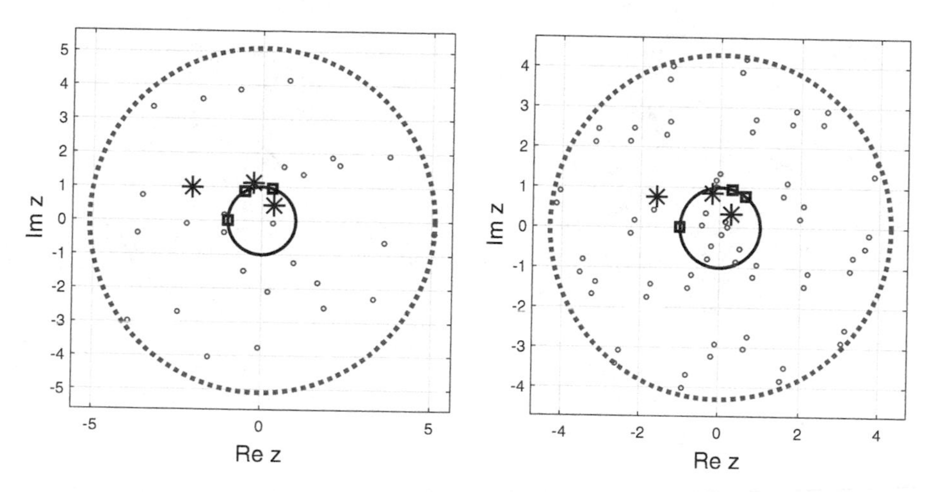

Fig. 6.8 Orbit of $(x_n)_{n\geq 0}$ obtained by (6.113) for $r_1 = r_2 = r_3 = 1$. $(a1)$ $x_1 = \frac{1}{3}$, $x_2 = \frac{1}{2}$, $x_3 = \frac{1}{5}$ (30 points); $(a2)$ $x_1 = \frac{1}{2}$, $x_2 = \frac{1}{7}$, $x_3 = \frac{1}{5}$ (70 points). Seeds x_0, x_1, x_2 (stars), generators z_1, z_2, z_3 (squares), $U(0, 1)$ (solid line), $U(0, |A_1| + |A_2| + |A_3|)$ (dotted line)

$$x_n = A_1 z_1^n + A_2 z_2^n + A_3 z_3^n, \tag{6.113}$$

where the constants A_1, A_2, A_3 can be recovered from the initial conditions.

The patterns recovered in this scenario are finite sets (periodic), or sets dense within certain 1D curves, or unions of 2D annuli and disks.

As illustrated in Section 6.4, sequence orbits are in this case located inside the disk of radius $|A_1| + |A_2| + |A_3|$, which is represented in the diagrams.

The patterns are linked to whether the terms of the set $\{1, x_1, x_2, x_3\}$ are linearly dependent (or independent) over $\mathbb{Q}$, i.e., there exist $p_0, p_1, p_2, p_3 \in \mathbb{Q}$ (not all zero) such that $p_0 + p_1 x_1 + p_2 x_2 + p_3 x_3 = 0$.

(a) Stable periodic finite orbits

When $x_1, x_2, x_3 \in \mathbb{Q}$ the orbit is finite (Figure 6.8). Indeed, when $x_1 = p_1/k_1$, $x_2 = p_2/k_2$, $x_3 = p_3/k_3$ are irreducible, one has $z_1^{k_1} = z_2^{k_2} = z_3^{k_3} = 1$. For most A_1, A_2, A_3, the terms $(x_n)_{n\geq 0}$ repeat with periodicity $\mathrm{lcm}(k_1, k_2, k_3)$.

(b) Orbits dense in unions of circles

Horadam orbits may also be dense within unions of circles, as in Figure 6.9.

(b1) If $x_1, x_2 \in \mathbb{Q}$ and $x_3 \in \mathbb{R} \setminus \mathbb{Q}$, then the orbit is dense in a union of $\mathrm{lcm}(k_1, k_2)$ circles;

(b2) If $x_1 \in \mathbb{R}\setminus\mathbb{Q}$, $x_2 - x_1, x_3 - x_1 \in \mathbb{Q}$, then the orbit is dense in a set of concentric circles.

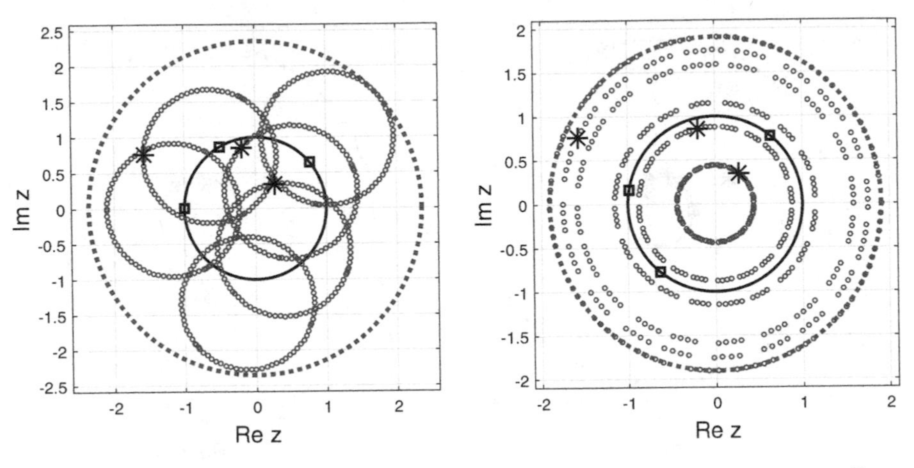

Fig. 6.9 Orbit of $(x_n)_{n\geq 0}$, given by (6.113) for $r_1 = r_2 = r_3 = 1$. $(b1)$ $x_1 = 1/3$, $x_2 = \sqrt{5}/20$, $x_3 = 1/2$; $(b2)$ $x_1 = \pi$, $x_2 = \pi + 1/3$, $x_3 = \pi + 1/3$. Seeds x_0, x_1, x_2 (stars), generators z_1, z_2, z_3 (squares), $U(0, 1)$ (solid line), $U(0, |A_1| + |A_2| + |A_3|)$ (dotted line)

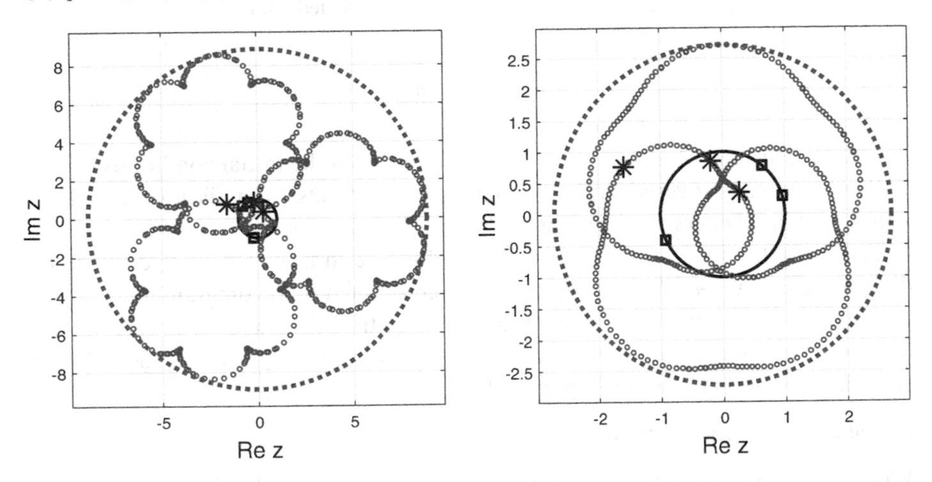

Fig. 6.10 Orbits of $(x_n)_{n\geq 0}$ dense within 1D curves, given by (6.113) for $r_1 = r_2 = r_3 = 1$. $(c1)$ $x_1 = \frac{1}{3}$, $x_2 = e$, $x_3 = \frac{e}{7}$; $(c2)$ $x_1 = \pi$, $x_2 = \frac{1\pi}{3}$, $x_3 = 4\pi$. Seeds x_0, x_1, x_2 (stars), generators z_1, z_2, z_3 (squares), $U(0, 1)$ (solid line), $U(0, |A_1| + |A_2| + |A_3|)$ (dotted line)

(c) Stable Orbits dense within complex curves

Orbits may be dense within unions of rotated curves as in Figure 6.10, or within more complex 1D curves.

(c1) If $x_1 \in \mathbb{Q}$, $x_2 \in \mathbb{R} \setminus \mathbb{Q}$ and $x_3/x_2 \in \mathbb{Q}$, then the orbit's closure is a union of k_1 rotated copies of a curve.

(c2) If $x_1 \in \mathbb{R} \setminus \mathbb{Q}$ but $x_2/x_1, x_3/x_1 \in \mathbb{Q}$, then the orbit's closure is a curve of type $f(z) = az + bz^p + cz^q$.

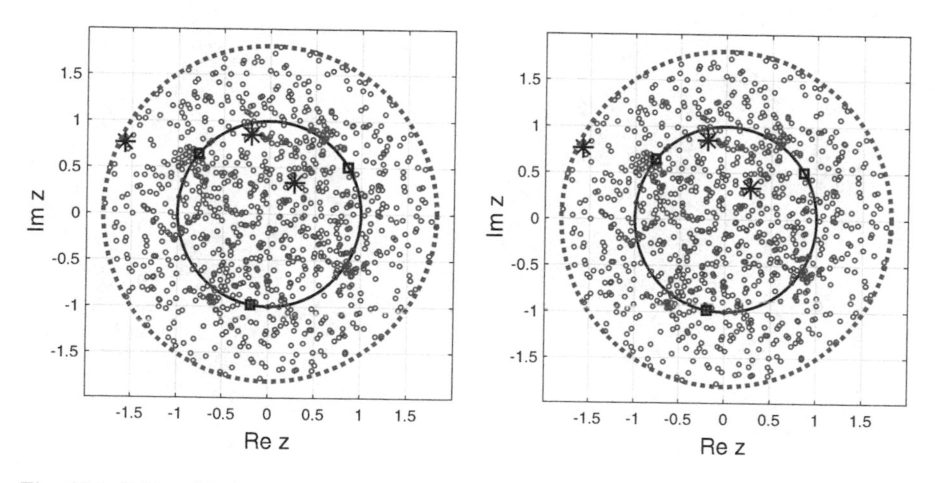

Fig. 6.11 Orbits of $(x_n)_{n\geq 0}$ dense within 2D regions, given by (6.113) for $r_1 = r_2 = r_3 = 1$. $(d1)$ $x_1 = e$, $x_2 = e^2$, $x_3 = e^3$; $(d2)$ $x_1 = \pi$, $x_2 = \pi^2$, $x_3 = \pi^3$. Seeds x_0, x_1, x_2 (stars), generators z_1, z_2, z_3 (squares), $U(0, 1)$ (solid line), $U(0, |A_1| + |A_2| + |A_3|)$ (dotted line)

(d) Stable orbits dense within a disk

If $1, x_1, x_2, x_3$ are linearly independent over $\mathbb{Q}$, then the orbit is dense within the disk of radius $|A_1| + |A_2| + |A_3|$ centered in the origin. This usually happens when we combine square roots, e or π. This case is illustrated in Figure 6.11.

(e) Stable Orbits dense within a 2D Region: unions of annuli

If $x_1, x_2 \in \mathbb{R} \setminus \mathbb{Q}$ with $1, x_1, x_2$ linearly independent over $\mathbb{Q}$ and $x_3 \in \mathbb{Q}$, then the orbit's closure is a collection of k_1 annuli rotated around the origin. The situation is depicted in Figure 6.12 for 3 and 4 annuli respectively.

(f) Stable Orbits dense within a 2D region: "buns"

If $x_1, x_2, x_3 \in \mathbb{R} \setminus \mathbb{Q}$ with $1, x_1, x_3$ linearly dependent over $\mathbb{Q}$ (e.g., $x_3/x_1 \in \mathbb{Q}$), then the orbit's closure is obtained by moving a circle along a closed 1D curve. This case is illustrated in Figure 6.13 for 3 and 5 lobes respectively.

6.6.2 *Quasi-Convergent Orbits:* $0 < r_1 < r_2 = r_3 = 1$

(a) Finite attractor set

When $x_2, x_3 \in \mathbb{Q}$ the orbit has $\text{lcm}(k_2, k_3)$ attractor points. For $x_1 \in \mathbb{Q}$, k_1 rays converge to each attractor, while for $x_1 \in \mathbb{R} \setminus \mathbb{Q}$ we have spirals (Figure 6.14).

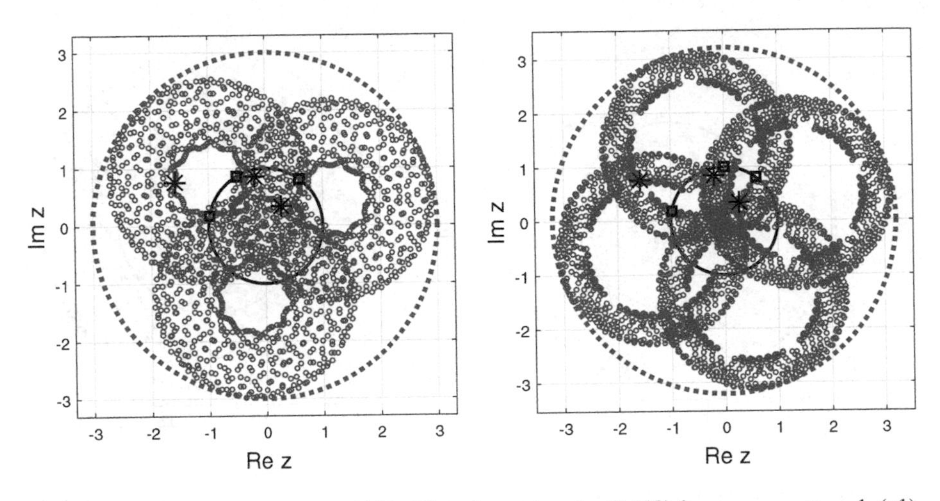

Fig. 6.12 Orbits of $(x_n)_{n\geq 0}$ dense within 2D regions, given by (6.113) for $r_1 = r_2 = r_3 = 1$. $(e1)$ $x_1 = \frac{\sqrt{2}}{3}$, $x_2 = \frac{\sqrt{5}}{15}$, $x_3 = \frac{1}{4}$; $(e2)$ $x_1 = \frac{\sqrt{2}}{3}$, $x_2 = \frac{\sqrt{5}}{15}$, $x_3 = \frac{1}{3}$. Seeds x_0, x_1, x_2 (stars), generators z_1, z_2, z_3 (squares), $U(0, 1)$ (solid line), $U(0, ||A_1| + |A_2| + |A_3|)$ (dotted line)

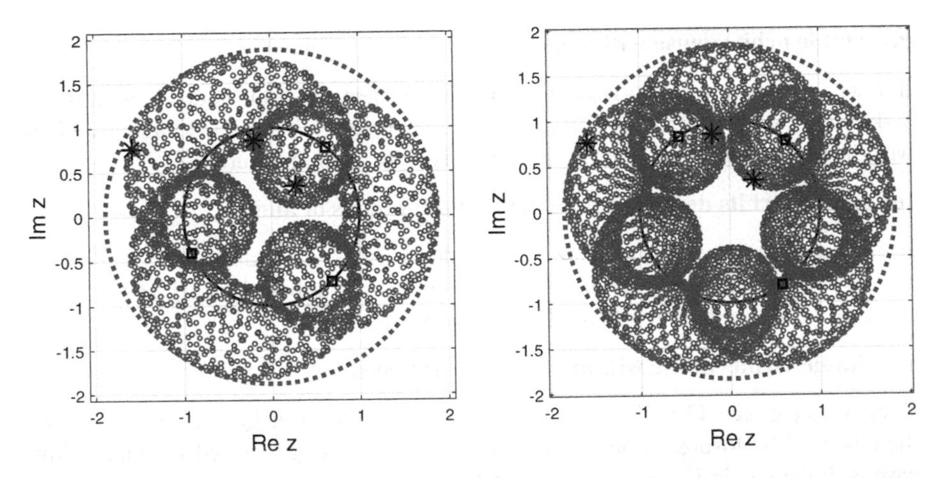

Fig. 6.13 Orbits of $(x_n)_{n\geq 0}$ dense in 2D regions, obtained from (6.113) for $r_1 = r_2 = r_3 = 1$. $(f1)$ $x_1 = \pi$, $x_2 = \pi^2$, $x_3 = 3\pi$; $(f2)$ $x_1 = \pi$, $x_2 = \pi^3/90$, $x_3 = 6\pi$. Seeds x_0, x_1, x_2 (stars), generators z_1, z_2, z_3 (squares), $U(0, 1)$ (solid line), $U(0, |A_1| + |A_2| + |A_3|)$ (dotted line)

(b) 1D attractor set: circles

When $x_2 \in \mathbb{R} \setminus \mathbb{Q}$ the orbit's closure consists of circles. If $x_2 - x_3 \in \mathbb{Q}$ these are concentric, while for $x_3 \in \mathbb{Q}$ we have k_3 rotated circles (Figure 6.15).

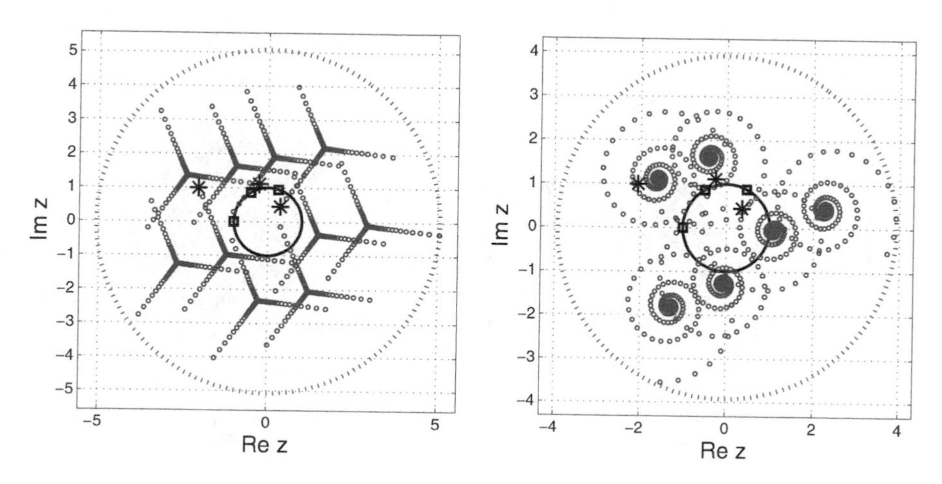

Fig. 6.14 Orbit of $(x_n)_{n\geq 0}$ obtained by (6.113) for $r_1 = .995$, $r_2 = r_3 = 1$. $(a1)$ $x_1 = \frac{1}{3}$, $x_2 = \frac{1}{2}$, $x_3 = \frac{1}{5}$; $(a2)$ $x_1 = \frac{\sqrt{3}}{10}$, $x_2 = \frac{1}{3}$, $x_3 = \frac{1}{2}$. Seeds x_0, x_1, x_2 (stars), generators z_1, z_2, z_3 (squares), $U(0, 1)$ (solid line), $U(0,\ |A_1| + |A_2| + |A_3|)$ (dotted line)

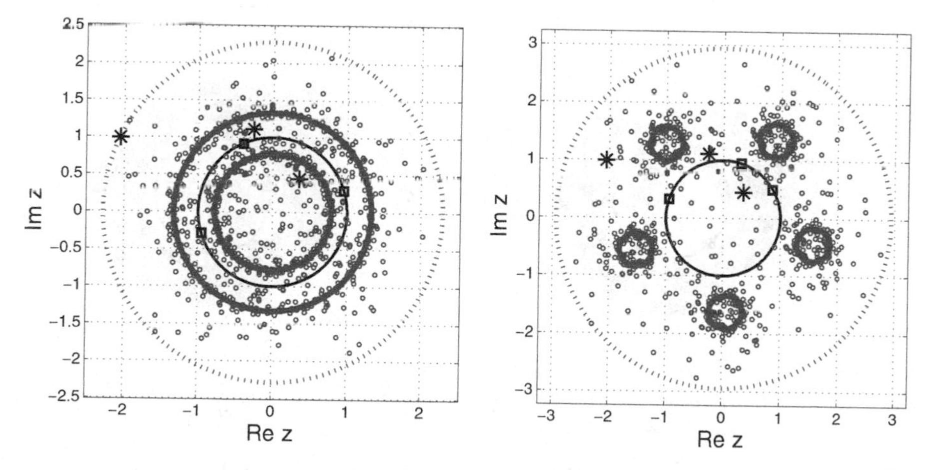

Fig. 6.15 Orbit of $(x_n)_{n\geq 0}$ obtained by (6.113) for $r_1 = .995$, $r_2 = r_3 = 1$. $(b1)$ $x_1 = \frac{\pi}{10}$, $x_2 = \frac{\pi}{3}$, $x_3 = \frac{\pi}{3} + \frac{1}{2}$; $(b2)$ $x_1 = \frac{1}{12}$, $x_2 = \frac{\sqrt{5}}{10}$, $x_3 = \frac{1}{5}$. Seeds x_0, x_1, x_2 (stars), generators z_1, z_2, z_3 (squares), $U(0, 1)$ (solid line), $U(0,\ |A_1| + |A_2| + |A_3|)$ (dotted line)

(c) 1D attractor set: curves

When $x_2 \in \mathbb{R} \setminus \mathbb{Q}$ and $x_3/x_2 = q \in \mathbb{Q}$, the orbit is dense within the graph of the function $f : S \to \mathbb{C}$ defined by $f(z) = A_2 z + A_3 z^q$, as shown in Figure 6.16. The details of this statement are explained in Theorem 5.17(c2).

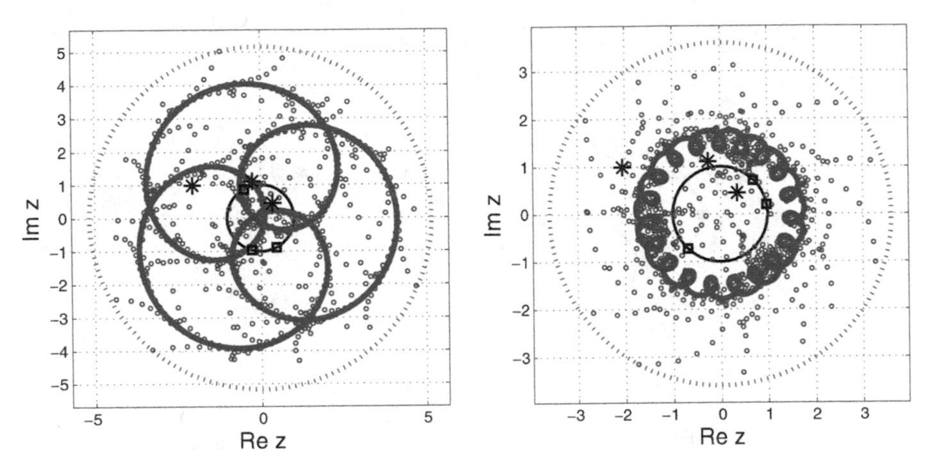

Fig. 6.16 Orbit of $(x_n)_{n\geq 0}$ obtained by (6.113) for $r_1 = .995$, $r_2 = r_3 = 1$. $(c1)$ $x_1 = \frac{1}{3}$, $x_2 = \frac{\sqrt{2}}{2}$, $x_3 = 4x_2$; $(c2)$ $x_1 = \frac{\pi}{25}$, $x_2 = \frac{x_1}{4}$, $x_3 = 20x_2$. Seeds x_0, x_1, x_2 (stars), generators z_1, z_2, z_3 (squares), $U(0, 1)$ (solid line), $U(0, |A_1| + |A_2| + |A_3|)$ (dotted line)

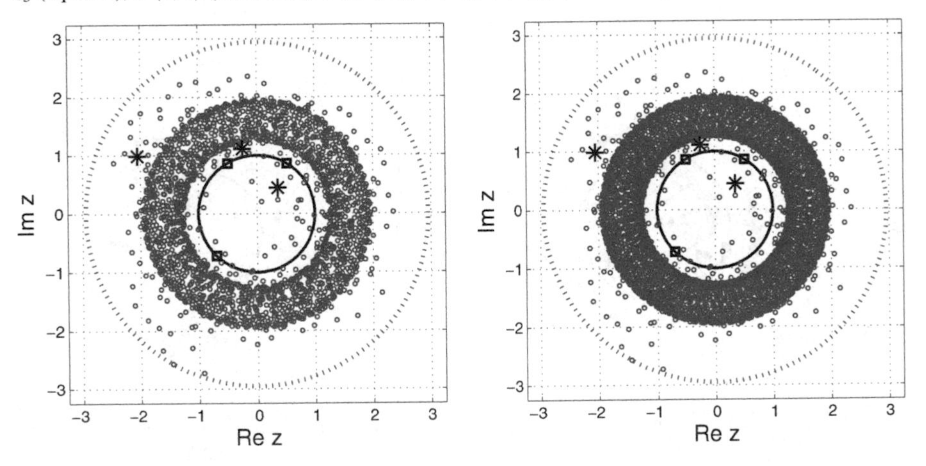

Fig. 6.17 Orbit of $(x_n)_{n\geq 0}$ obtained by (6.113) for $r_1 = .995$, $r_2 = r_3 = 1$ obtained for $x_1 = \frac{1}{3}$, $x_2 = \frac{\exp(.5)}{10}$, $x_3 = \frac{\pi}{5}$, and $(d1)$ 2000 terms; $(d2)$ 4000 terms. Seeds x_0, x_1, x_2 (stars), generators z_1, z_2, z_3 (squares), $U(0, 1)$ (solid), $U(0, |A_1| + |A_2| + |A_3|)$ (dotted)

(d) 2D attractor set: annulus

When $1, x_2, x_3$ are linearly independent over $\mathbb{Q}$, the orbit's closure is dense within an annulus. The convergence property is illustrated in Figure 6.17, when 2000 (d1) or 4000 (d2) sequence terms are evaluated, respectively.

6.6.3 *Convergent Orbits:* $0 \leq r_1 \leq r_2 \leq \cdots \leq r_m < 1$

Here the origin is the unique attractor point, as shown in Figure 6.18.

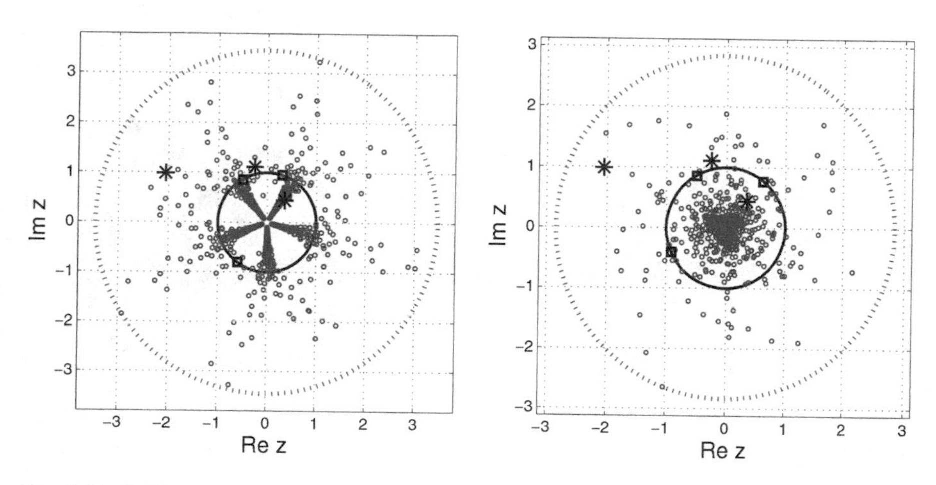

Fig. 6.18 Orbit of $(x_n)_{n\geq 0}$ obtained by (6.113) for $r_1 = .99$, $r_2 = .995$ and $r_3 = .997$. $(a1)$ $x_1 = \frac{1}{2}, x_2 = \frac{1}{3}, x_3 = \frac{1}{4}$; $(a2)$ $x_1 = \frac{1}{3}, x_2 = \pi, x_3 = 4\pi$. Seeds x_0, x_1, x_2 (stars), generators z_1, z_2, z_3 (squares), $U(0, 1)$ (solid), $U(0, |A_1| + |A_2| + |A_3|)$ (dotted)

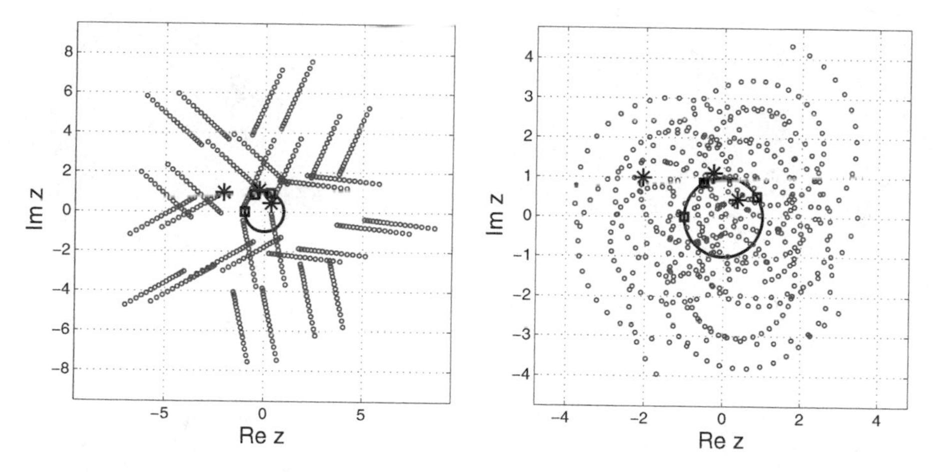

Fig. 6.19 Orbit of $(x_n)_{n\geq 0}$ obtained by (6.113) for $r_1 = r_2 = 1$ and $r_3 = 1.002$. $(a1)$ $x_1 = \frac{1}{3}$, $x_2 = \frac{1}{2}, x_3 = \frac{1}{5}$; $(a2)$ $x_1 = \frac{1}{3}, x_2 = \frac{1}{2}, x_3 = e^3$. Seeds x_0, x_1, x_2 (stars), generators z_1, z_2, z_3 (squares), $U(0, 1)$ (solid line), $U(0, |A_1| + |A_2| + |A_3|)$ (dotted line)

6.6.4 Divergent Orbits: $r_m > 1$

(a) Divergent rays/spirals The orbit diverges along rays for $x_1, x_2, x_3 \in \mathbb{Q}$, and spirals emerge when $x_3 \in \mathbb{R} \setminus \mathbb{Q}$ (Figure 6.19).

(b) General divergent patterns Some other examples of divergent orbits are shown below in Figure 6.20.

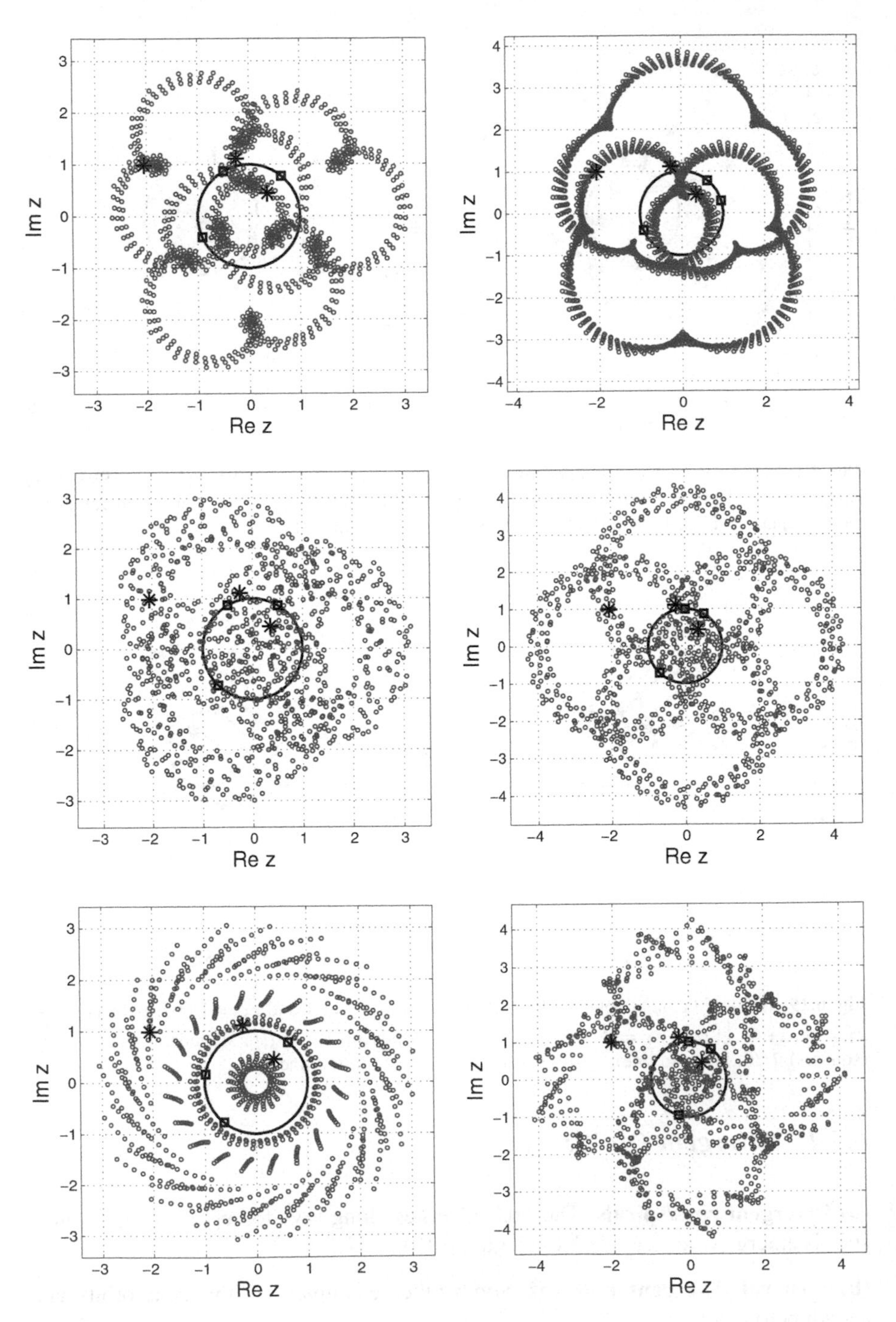

Fig. 6.20 Divergent orbits of $(x_n)_{n\geq 0}$ obtained by (6.113) for $r_1 = r_2 = 1$ and $r_3 = 1.002$. Seeds x_0, x_1, x_2 (stars), generators z_1, z_2, z_3 (squares), $U(0, 1)$ (solid line)

These examples can be linked to those presented for stable orbits. The perturbed patterns are not distinguishable for larger values of n.

6.7 Connection to Finite Differences

This section briefly presents some basic theory on finite differences and their relation to recurrence sequences, largely based on [39, 124].

Definition 6.1 The *difference operator* Δ is defined on the set of sequences of complex numbers. Given a sequence $s = (s_n)_{n\geq 0}$, its difference is the new sequence $\Delta s = (s_{n+1} - s_n)_{n\geq 0}$. We define the nth term of Δs by Δs_n. Clearly, the operator Δ is linear and can be composed with itself. We shall denote $\Delta^2 s = \Delta(\Delta s)$ and $\Delta^{k+1} s = \Delta(\Delta^k s)$.

Remark 6.9 One can easily notice that if $s_n = n^3$ $(n \geq 0)$, one has

$$\begin{aligned}\Delta s_n &= 3n^2 + 3n + 1,\\ \Delta^2 s_n &= 6n + 6,\\ \Delta^3 s_n &= 6.\end{aligned}$$

The following general formulae hold:

$$s_n = n^k, \quad \Delta^k s_n = k!, \tag{6.114}$$

$$s_n = a^n, \quad \Delta s_n = a^{n+1} - a^n = (a-1)s_n. \tag{6.115}$$

A *difference equation* is an equation of the form

$$a_k(n)\Delta^k s_n + a_{k-1}(n)\Delta^{k-1} s_n + \cdots + a_1(n)\Delta s_n + a_0(n) s_n = t_n, \tag{6.116}$$

or written in operator form

$$\left(a_k\Delta^k + a_{k-1}\Delta^{k-1} + \cdots + a_1\Delta + a_0 I\right) s_n = t_n,$$

where a_i are arbitrary functions of n and $(t_n)_{n\geq 0}$ is an arbitrary sequence. Most practical calculations are performed for $a_k = 1$, a_i constants $0 \leq i < k$, and t_n polynomial or exponential functions of n.

Example 6.16 The Fibonacci sequence is usually defined by the recurrence relation $F_{n+1} = F_n + F_{n-1}$, $n = 1, 2, \ldots$, but can also be defined by the difference equation $\Delta^2 F_n + \Delta F_n - F_n = 0$, $n = 1, 2, \ldots$.

6.7.1 Solving Difference Equations

When solving a difference equation, one can adopt a language similar to that commonly used for ordinary differential equations.

First, one has to find the homogeneous solution $(h_n)_{n\geq 0}$ satisfying the associated homogeneous equation

$$\left(a_k\Delta^k + a_{k-1}\Delta^{k-1} + \cdots + a_1\Delta + a_0 I\right) s_n = 0, \tag{6.117}$$

then a particular solution $(p_n)_{n\geq 0}$, of the initial problem. By linearity, any difference of particular solutions will be a solution of the homogeneous problem (6.117), hence for each solution $(s_n)_{n\geq 0}$ there will be a pair $(p_n)_{n\geq 0}$, $(h_n)_{n\geq 0}$, such that $s_n = h_n + p_n$, for all $n \geq 0$.

Since the powers of Δ commute, we can factor the operator polynomial to obtain expressions of the form

$$\Delta^k + a_{k-1}\Delta^{k-1} + \cdots + a_1\Delta + a_0 I = (\Delta - \alpha_1 I)(\Delta - \alpha_2 I)\cdots(\Delta - \alpha_k I),$$

where α_j, $j = 1, \ldots, k$, are the roots of the equation

$$a_k x^k + a_{k-1}x^{k-1} + \cdots + a_1 x + a_0 = 0. \tag{6.118}$$

Note that the factors of the form $(\Delta - \alpha I)$ commute with each other, hence we can solve $\left(\Delta - \alpha_j I\right) s_n = 0$, which is equivalent to $\Delta s_n = \alpha_j s_n$. From the equation (6.115), we infer that a sequence $(h_n^j)_{n\geq 0}$ defined by $h_n^j = C_j(\alpha_j + 1)^n$ is a solution. Clearly, one has

$$\begin{aligned}&(\Delta - \alpha_1 I)(\Delta - \alpha_2 I)\cdots(\Delta - \alpha_k I)\, h_n^j = \\ &= (\Delta - \alpha_1 I)\cdots\left(\Delta - \alpha_{j-1} I\right)\left(\Delta - \alpha_{j+1} I\right)\cdots(\Delta - \alpha_k I)\left[\left(\Delta - \alpha_j I\right) h_n^j\right] = 0,\end{aligned}$$

hence each expression $C_j(\alpha_j + 1)^n$ is a solution of the equation (6.117), and by the linearity of the solution, any linear combination of such solutions is also a solution. When all the roots α_j, $j = 1, \ldots, k$, of the equation (6.118) are distinct, then the general form of the solution is given by

$$s_n = C_1(\alpha_1 + 1)^n + C_2(\alpha_2 + 1)^n + \cdots + C_k(\alpha_k + 1)^n.$$

When a root α has multiplicity m, one can easily check that all the terms

$$(\alpha + 1)^n, n(\alpha + 1)^n, \ldots, n^m(\alpha + 1)^n$$

represent solutions of the homogeneous problem.

Example 6.17 Find the formula for the general term of the Lucas sequence using the method of finite differences.

Solution One can check that the finite difference formula is

$$\Delta^2 L_n + \Delta L_n - L_n = 0, \tag{6.119}$$

with the starting values $L_0 = 2$, $L_1 = 1$. Since this is a homogeneous equation, it is sufficient to solve the homogeneous problem. The polynomial associated with the equation (6.119) is $x^2 + x - 1 = 0$, with the roots

$$\alpha_1 = -\frac{1}{2} + \frac{\sqrt{5}}{2}, \quad \alpha_2 = -\frac{1}{2} - \frac{\sqrt{5}}{2}.$$

Since these are both distinct roots, one obtains

$$L_n = c_1(\alpha_1 + 1)^n + c_2(\alpha_2 + 1)^n = c_1\left(\frac{1}{2} + \frac{\sqrt{5}}{2}\right)^n + c_2\left(\frac{1}{2} - \frac{\sqrt{5}}{2}\right)^n.$$

From the initial conditions, one gets $c_1 = c_2 = 1$ and the well-known formula

$$L_n = \left(\frac{1}{2} + \frac{\sqrt{5}}{2}\right)^n + \left(\frac{1}{2} - \frac{\sqrt{5}}{2}\right)^n.$$

6.7.2 *Finding the Difference Equation for a Sequence*

If a finite difference equation is given, then by writing the difference operators explicitly, one can obtain a recurrence relation by direct calculations. Conversely, finding the difference equation counterpart for a given recurrent sequence can be a complicated problem.

We illustrate this transformation for linear recurrent sequences.

Assume that the recurrence relation has the form

$$b_k s_{n+k} + b_{k-1} s_{n+k-1} + \cdots + b_1 s_{n+1} + b_0 s_n = t_n.$$

This equation can be written as a difference equation by using the identity

$$s_{n+1} = s_n + (s_{n+1} - s_n) = s_n + \Delta s_n = (I + \Delta)s_n.$$

Applying this identify multiple times, one can show that

$$s_{n+j} = (I + \Delta)^j s_n, \quad j > 0,$$

and one can obtain the finite difference equivalent of the original recurrence by using this expression for $s_{n+1}, s_{n+2}, \ldots, s_{n+k}$.

Chapter 7
Recurrences in Olympiad Training

In this chapter we present a thematic collection of 123 olympiad training problems, involving recurrences in various contexts. We have chosen 123 as it represents the 10th Lucas number L_{10}. The problems relate to topics such as linear recurrence sequences of first, second, and higher orders, classical sequences, homographic recurrent sequences, systems of recurrent sequences, and combinatorics.

We indicate the known author(s), the collection, year, or the Olympiad type and country, for the problems which featured in a certain competition.

The sources of the problems are indicated by the following abbreviations: AMM (American Mathematical Monthly), MR (Mathematical Reflections), CM (Crux Mathematicorum), GM (Gazeta Matematică), ME (Mathematical Excalibur), SAMC (Saudi Arabia Mathematics Competitions), as well as the IMO (International Mathematics Olympiad).

7.1 First-Order Recurrent Sequences

Problem 1 (Germany, 1995) Let a and b be positive integers and consider the sequence $(x_n)_{n\geq 0}$ defined by $x_0 = 1$ and

$$x_{n+1} = ax_n + b^{n+1}, \quad n = 0, 1, \ldots .$$

Prove that for any choice of a and b, the sequence $(x_n)_{n\geq 0}$ contains infinitely many composite numbers.

Problem 2 Let m and n be integers greater than 1 such that

$$\gcd(m, n - 1) = \gcd(m, n) = 1.$$

D. Andrica, O. Bagdasar, *Recurrent Sequences*, Problem Books in Mathematics,
https://doi.org/10.1007/978-3-030-51502-7_7

Prove that the first $m-1$ terms of the sequence $n_1, n_2, \ldots$, where $n_1 = mn + 1$ and $n_{k+1} = n \cdot n_k + 1$, $k \geq 1$, cannot all be primes.

Problem 3 Let $a_1 = 1$, $a_{n+1} = a_n + \lfloor \sqrt{a_n} \rfloor$, $n = 1, 2, \ldots$. Show that a_n is a perfect square if and only if n is of the form $2^k + k - 2$.

Problem 4 (IMO Shortlist, 1971) Let $T_k = k - 1$ for $k = 1, 2, 3, 4$ and

$$T_{2k-1} = T_{2k-2} + 2^{k-2}, \quad T_{2k} = T_{2k-5} + 2^k, \quad k = 3, 4, \ldots.$$

Show that for all n,

$$1 + T_{2n-1} = \left\lfloor \frac{12}{7} 2^{n-1} \right\rfloor \text{ and } 1 + T_{2n} = \left\lfloor \frac{17}{7} 2^{n-1} \right\rfloor.$$

Problem 5 Define the sequences $(x_n)_{n\geq 0}$ and $(y_n)_{n\geq 0}$ by $x_0 = \alpha$, $y_0 = \beta$ and $x_{n+1} = 2y_n + 1$, $y_{n+1} = 2x_n - 3$, $n = 0, 1, \ldots$. Find x_n and y_n.

Problem 6 Let $p \geq 3$ be a prime and let $(x_n)_{n\geq 0}$ be the sequence defined by

$$x_{n+1} = 2x_n + 1, \quad n = 0, 1, \ldots, \quad x_0 = p.$$

Prove that for any integer $k \geq 0$, the set $A_k = \{x_k, x_{k+1}, \ldots, x_{k+p-1}\}$ contains at least one composite integer.

Problem 7 (Marius Cavachi, S303, MR-2014) Let $(a_n)_{n\geq 1}$ be the sequence defined by $a_1 = 1$ and $a_{n+1} = \frac{1}{2}\left(a_n + \frac{n}{a_n}\right)$, for $n \geq 1$. Find $\lfloor a_{2014} \rfloor$.

Problem 8 (Albert Stadler, U367, MR-2016) Let $(a_n)_{n\geq 1}$ be the sequence of real numbers satisfying $a_1 = 4$ and $3a_{n+1} = (a_n + 1)^3 - 5$, $n \geq 1$. Prove that a_n is a positive integer for all n, and evaluate

$$\sum_{n=1}^{\infty} \frac{a_n - 1}{a_n^2 + a_n + 1}.$$

Problem 9 (Arkady Alt, U326, MR-2015) Let $(a_n)_{n\geq 0}$ be a sequence with $a_0 > 1$ and $3a_{n+1} = a_n^3 + 2$, for $n \geq 1$. Find

$$\sum_{n=0}^{\infty} \frac{a_n + 2}{a_n^2 + a_n + 1}.$$

Problem 10 (Khakimboy Egamberganov, U327, MR-2015) Let $(a_n)_{n\geq 0}$ be a sequence of real numbers with $a_0 = 1$ and

$$a_{n+1} = \frac{a_n}{na_n + a_n^2 + 1}.$$

Find the limit $\lim\limits_{n\to\infty} n^3 a_n$.

Problem 11 (Titu Andreescu, U333, MR-2015) Evaluate

$$\prod_{n=0}^{\infty}\left(1 - \frac{2^{2^n}}{2^{2^{n+1}}+1}\right).$$

Problem 12 (Marius Cavachi, U272, MR-2013) Let $a > 0$ be a real number and let $(a_n)_{n\geq 0}$ be the sequence defined by $a_0 = \sqrt{a}$, $a_{n+1} = \sqrt{a_n + a}$, for $n \geq 0$. Prove that there are infinitely many irrational numbers among the terms of the sequence.

Problem 13 (Ivan Borsenco, O285, MR-2013) Consider the integer sequence $(a_n)_{n\geq 0}$ given by $a_1 = 1$ and $a_{n+1} = 2^n(2^{a_n} - 1)$, for $n \geq 1$. Prove that $n!$ divides a_n.

Problem 14 (Iosif Pinelis, 11837, AMM-2015) Let $a_0 = 1$ and let $a_{n+1} = a_n + e^{-a_n}$ for $n \geq 0$. Let $b_n = a_n - \log n$. For $n \geq 0$, show that $0 < b_{n+1} < b_n$; also show that $\lim\limits_{n\to\infty} b_n = 0$.

Problem 15 (Titu Andreescu, U64, MR-2007) Let x be a real number. Define the sequence $(x_n)_{n\geq 1}$ recursively by $x_1 = 1$ and $x_{n+1} = x^n + nx_n$ for $n \geq 1$. Prove that

$$\prod_{n=1}^{\infty}\left(1 - \frac{x^n}{x_{n+1}}\right) = e^{-x}.$$

Problem 16 Let I be an interval and consider a function $f : I \to I$. Define the sequence $(a_n)_{n\geq 0}$ by $a_{n+1} = f(a_n)$, for $n \geq 0$ and $a_0 \in I$. Prove that

1. If f is increasing, then $(a_n)_{n\geq 0}$ is monotonic;
2. If f is decreasing, then the sequences $(a_{2n})_{n\geq 0}$ and $(a_{2n+1})_{n\geq 0}$ are monotonic, having different monotonicities.

Problem 17 (Michel Bataille, 4282, CM-2017) Find $\lim\limits_{n\to\infty} u_n$ where the sequence $(u_n)_{n\geq 0}$ is defined by $u_0 = 1$ and the recursion

$$u_{n+1} = \frac{1}{2}\left(u_n + \sqrt{u_n^2 + \frac{u_n}{4^n}}\right),$$

for every nonnegative integer n.

Problem 18 (Mihály Bencze, 4264, CM-2017) Let $x_1 = 4$ and $x_{n+1} = \lfloor \sqrt[3]{2} x_n \rfloor$, where $\lfloor \cdot \rfloor$ denotes the integer part. Determine the largest positive integer n for which x_n, x_{n+1}, x_{n+2} form an arithmetic progression.

Problem 19 (Marcel Chiriţă, 3942, CM-2016) Consider a sequence $(x_n)_{n\geq 1}$ with $x_1 = 1$ and $x_{n+1} = \frac{1}{n+1}\left(x_n + \frac{1}{x_n}\right)$. Find $\lim\limits_{n\to\infty} \sqrt{n}x_n$.

Problem 20 (Laurenţiu Panaitopol, S10, MR-2006) Let $(a_n)_{n\geq 1}$ be a sequence of positive numbers such as $a_{n+1} = a_n^2 - 2$ for all $n \geq 1$. Show that we have $a_n \geq 2$, for all $n \geq 1$.

7.2 Second-Order Recurrent Sequences

Problem 1 (Titu Andreescu, [4]) Find the term a_{2020} for the sequence $(a_n)_{n\geq 1}$ defined by $a_1 = 1$ and

$$a_{n+1} = 2a_n + \sqrt{3a_n^2 - 2}, \quad n = 1, 2, \ldots .$$

Problem 2 Let a and b be positive real numbers. Find the general term of the sequence $(x_n)_{n\geq 0}$ defined by $x_0 = 0$ and

$$x_{n+1} = x_n + a + \sqrt{b^2 + 4ax_n}, \quad n = 0, 1, \ldots .$$

Problem 3 (IMO Shortlist, 1983) Let k be a positive integer. The sequence $(a_n)_{n\geq 0}$ is defined by $a_0 = 0$ and

$$a_{n+1} = (a_n + 1)k + (k+1)a_n + 2\sqrt{k(k+1)a_n(a_n+1)}, \quad n = 0, 1, \ldots .$$

Prove that all a_n are positive integers.

Problem 4 (Iberoamerican Olympiad, 1987) Let $m, n, r > 0$ be integers with

$$1 + m + n\sqrt{3} = (2 + \sqrt{3})^{2r-1}.$$

Prove that m is a perfect square.

Problem 5 (Kürschak Competition, 1988) Let k be a positive integer. Define the sequence $(x_n)_{n\geq 1}$ by $x_1 = k$ and

$$x_{n+1} = kx_n + \sqrt{(k^2 - 1)(x_n^2 - 1)}, \quad n = 1, 2, \ldots .$$

Prove that all x_n are positive integers.

Problem 6 Let the sequence $(a_n)_{n\geq 1}$ given by $a_1 = 1, a_2 = 2, a_3 = 24$, and

$$a_n = \frac{6a_{n-1}^2 a_{n-3} - 8a_{n-1}a_{n-2}^2}{a_{n-2}a_{n-3}}, \quad n = 4, 5, \ldots.$$

Show that for all n, a_n is an integer divisible by n.

Problem 7 (Balkan Olympiad, 1986) Let a, b, c be positive real numbers. The sequence $(a_n)_{n\geq 1}$ is defined by $a_1 = a, a_2 = b$ and

$$a_{n+1} = \frac{a_n^2 + c}{a_{n-1}}, \quad n = 2, 3, \ldots.$$

Prove that the terms of the sequence are all positive integers if and only if a, b and $\dfrac{a^2 + b^2 + c}{ab}$ are positive integers.

Problem 8 (IMO Shortlist, 1984) Let c be a positive integer. The sequence $(f_n)_{n\geq 1}$ is defined as follows:

$$f_1 = 1, \quad f_2 = c, \quad f_{n+1} = 2f_n - f_{n-1} + 2, \quad n = 2, 3, \ldots.$$

Show that for each $k \in \mathbb{N}$ there exists $r \in \mathbb{N}$ such that $f_k f_{k+1} = f_r$.

Problem 9 (IMO Shortlist, 1988) An integer sequence is defined by

$$a_n = 2a_{n-1} + a_{n-2} \quad (n > 1), a_0 = 0, \quad a_1 = 1.$$

Show that 2^k divides a_n if and only if 2^k divides n.

Problem 10 (IMO Shortlist, 1988) The integer sequence $(a_n)_{n\geq 1}$ is defined by $a_1 = 2, a_2 = 7$ and

$$-\frac{1}{2} < a_{n+1} - \frac{a_n^2}{a_{n-1}} \leq \frac{1}{2}, \quad n = 2, 3, \ldots.$$

Show that a_n is odd for all $n \geq 2$.

Problem 11 Let the integer sequence $(x_n)_{n\geq 0}$ be defined by $x_0 = x_1 = 0$ and

$$x_{n+2} = 4^{n+2}x_{n+1} - 16^{n+1}x_n + n \cdot 2^{n^2}, \quad n = 0, 1, \ldots.$$

Show that the numbers x_{1989}, x_{1990} and x_{1991} are divisible by 13.

Problem 12 (IMO Longlist, 1986) Let $(a_n)_{n\geq 1}$ be the integer sequence defined by $a_1 = a_2 = 1$,

$$a_{n+2} = 7a_{n+1} - a_n - 2, \quad n \geq 1.$$

Show that a_n is a perfect square for every n.

Problem 13 (Dorin Andrica and Grigore Călugăreanu, U422, MR-2017) Let a, b be complex numbers and let $(a_n)_{n\geq 0}$ be the sequence defined by $a_0 = 2$ and $a_1 = a$ and

$$a_n = aa_{n-1} + ba_{n-2}, \quad n \geq 2.$$

Write a_n as a polynomial in a and b.

Problem 14 (Titu Andreescu, U297, MR-2014) Let $a_0 = 0$, $a_1 = 2$, and $a_{n+1} = \sqrt{2 - \frac{a_{n-1}}{a_n}}$ for $n \geq 1$. Find $\lim_{n\to\infty} 2^n a_n$.

Problem 15 (Răzvan Gelca, O312, MR-2014) Find all increasing bijections $f : (0, \infty) \to (0, \infty)$ satisfying

$$f(f(x)) - 3f(x) + 2x = 0,$$

and for which there exists $x_0 > 0$ such that $f(x_0) = 2x_0$.

Problem 16 (Albert Stadler, O271, MR-2013) Consider the sequence $(a_n)_{n\geq 0}$ given by $a_0 = 0$, $a_1 = 2$ and $a_{n+2} = 6a_{n+1} - a_n$, for $n \geq 0$. Let $f(n)$ be the highest power of 2 that divides n. Prove that $f(a_n) = f(2n)$ for all $n \geq 0$.

Problem 17 (Problem 49, ME-1997) Let $(u_n)_{n\geq 1}$ be a sequence of integers which satisfies $u_1 = 29$, $u_2 = 45$, and $u_{n+2} = u_{n+1}^2 - u_n$ for $n = 1, 2, \ldots$. Show that 1996 divides infinitely many terms of this sequence.

Problem 18 (Moscow Mathematical Olympiad, 1963) Let $(u_n)_{n\geq 1}$ be a sequence satisfying $a_1 = a_2 = 1$ and $a_n = (a_{n-1}^2 + 2)/a_{n-2}$ for $n \geq 3$. Show that a_n is an integer for $n \geq 3$.

Problem 19 (Ivan Borsenco, J66, MR-2007) Let $a_0 = a_1 = 1$ and $a_{n+1} = 2a_n - a_{n-1} + 2$ for $n \geq 1$. Prove that $a_{n^2+1} = a_{n+1}a_n$ for all $n \geq 0$.

Problem 20 (Titu Andreescu, U46, MR-2007) Let k be a positive integer and

$$a_n = \left\lfloor (k + \sqrt{k^2+1})^n + \left(\frac{1}{2}\right)^n \right\rfloor, \quad n \geq 0.$$

Prove that $\sum_{n=1}^{\infty} \frac{1}{a_{n-1}a_{n+1}} = \frac{1}{8k^2}$.

Problem 21 (Martin Lukarevski, 4117, CM-2017) Find the general term x_n of the sequence $(x_n)_{n\geq 0}$ given recursively by $x_0 = 0$, $x_1 = 1$, and

$$x_{n+1} = x_n\sqrt{x_{n-1}^2 + 1} + x_{n-1}\sqrt{x_n^2 + 1}, \quad n \geq 1.$$

Problem 22 (Dorin Andrica, [4]) Let α and β be nonnegative integers such that $\alpha^2 + 4\beta$ is not a perfect square. Define the sequence $(x_n)_{n\geq 0}$ by

$$x_{n+2} = \alpha x_{n+1} + \beta x_n, \quad n \geq 0,$$

with x_1, x_2 positive integers. Prove that there is no integer $n_0 \geq 1$ such that

$$x_{n_0}^2 = x_{n_0-1}x_{n_0+1}.$$

Problem 23 (Dorin Andrica, [4]) Let x_1, x_2, α, β be real numbers and let the sequence $(x_n)_{n\geq 0}$ be given by

$$x_{n+2} = \alpha x_{n+1} + \beta x_n, \quad n \geq 1.$$

If $x_m^2 \neq x_{m-1}x_{m+1}$ for all integers $m > 1$, prove that there exist real numbers λ_1, λ_2 such that for all $n > 2$ we have

$$\lambda_1 = \frac{x_n^2 - x_{n+1}x_{n-1}}{x_{n-1}^2 - x_n x_{n-2}}, \quad \lambda_2 = \frac{x_n x_{n-1} - x_{n+1}x_{n-2}}{x_{n-1}^2 - x_n x_{n-2}}.$$

7.3 Classical Recurrent Sequences

Problem 1 (Ireland, 1999) Show there is a positive integer in the Fibonacci sequence that is divisible by 1000.

Problem 2 (Ireland, 1996) Prove that

(a) The statement "$F_{n+k} - F_n$ is divisible by 10 for all positive integers n" is true if $k = 60$ and false for any positive integer $k < 60$;
(b) The statement "$F_{n+t} - F_n$ is divisible by 100 for all positive integers n" is true if $t = 300$ and false for any positive integer $t < 300$.

Problem 3 (Dorin Andrica) Let $u_n = F_n^2$, $n = 0, 1, \ldots$, where F_n are the Fibonacci numbers. Prove that $(u_n)_{n\geq 0}$ satisfies a linear recurrence relation of order 3.

Problem 4 (Dorin Andrica) Consider the sequence $v_n = F_n^3$, $n = 0, 1, \ldots$, where F_n are the Fibonacci numbers. Prove that the sequence $(v_n)_{n\geq 0}$ satisfies a linear recurrence relation of order 4.

Problem 5 (Dorin Andrica) Consider the sequence $w_n = F_n F_{n+1}$, $n = 0, 1, \ldots$, where F_n are the Fibonacci numbers. Find a recurrence relation satisfied by the sequence $(w_n)_{n\geq 0}$.

Problem 6 (Ángel Plaza, U460, MR-2018) Let L_k be the kth Lucas numbers. Prove that

$$\sum_{k=1}^{\infty} \tan^{-1}\left(\frac{L_{k+1}}{L_k L_{k+2}}\right) \tan^{-1}\left(\frac{1}{L_{k+1}}\right) = \frac{\pi}{4} \tan^{-1}\left(\frac{1}{3}\right).$$

Problem 7 (Tarit Goswami, U447, MR-2018) Let F_n be the nth Fibonacci numbers. Prove that

$$\sum_{k=1}^{n} \binom{n}{k} F_p^k F_{p-1}^{n-k} F_k = F_{pn}.$$

Problem 8 (Roberto Bosch Cabrera, J254, MR-2013) Solve the equation

$$F_{a_1} + F_{a_2} + \cdots + F_{a_k} = F_{a_1+a_2+\cdots+a_k},$$

where F_i is the ith Fibonacci number.

Problem 9 (Cornel Ioan Vălean, 11910, AMM-2016) Let G_k be the reciprocal of the kth Fibonacci number, for example, $G_4 = 1/3$ and $G_5 = 1/5$. Find

$$\sum_{n=1}^{\infty} \left(\arctan G_{4n-3} + \arctan G_{4n-2} + \arctan G_{4n-1} - \arctan G_{4n}\right).$$

Problem 10 (Hideyuki Ohtsuka, 11978, AMM-2017) Let F_n be the nth Fibonacci number, with $F_0 = 0$, $F_1 = 1$, and $F_n = F_{n-1} + F_{n-2}$, for $n \geq 2$. Find

$$\sum_{n=0}^{\infty} \frac{(-1)^n}{\cosh F_n \cdot \cosh F_{n+3}}.$$

Problem 11 (Mircea Merca, 11736, AMM-2013) For $n \geq 1$, let f be the symmetric polynomial in variables $x_1, \ldots, x_n$ given by

$$f(x_1, \ldots, x_n) = \sum_{k=0}^{n-1} (-1)^{k+1} e_k(x_1 + x_1^2, x_2 + x_2^2, \ldots, x_n + x_n^2),$$

where e_k is the kth elementary polynomial in n variables. For example, when $n = 6$, e_2 has 15 terms, each a product of two distinct variables. Also, let ξ be a primitive nth root of unity. Prove that

$$f(1, \xi, \xi^2, \ldots, \xi^{n-1}) = L_n - L_0,$$

where L_k is the kth Lucas number ($L_0 = 2$, $L_1 = 1$ and $L_k = L_{k-1} + L_{k-2}$ for $k \geq 2$).

Problem 12 (Oliver Knill, 11716, AMM-2013) Let $\alpha = (\sqrt{5} - 1)/2$. Let p_n and q_n be the numerator and denominator of the nth continued fraction convergent to α, denoted by p_n/q_n, where F_n is the nth Fibonacci number and $q_n = p_{n+1}$. Show that

$$\sqrt{5}\left(\alpha - \frac{p_n}{q_n}\right) = \sum_{k=0}^{\infty} \frac{(-1)^{(n+1)(k+1)} C_k}{q_n^{2k+2} 5^k},$$

where C_k denotes the kth Catalan number, given by $C_k = \frac{(2k)!}{k!(k+1)!}$.

Problem 13 (Dorin Andrica, O64, MR-2015) Let F_n be the nth Fibonacci number. Prove that for all $n \geq 4$, $F_n + 1$ is not a prime.

Problem 14 (Gabriel Alexander Reyes, O47, MR-2007) Consider the Fibonacci sequence $F_0 = 1$, $F_1 = 1$, and $F_{n+1} = F_n + F_{n-1}$, for $n \geq 1$. Prove that

$$\sum_{i=0}^{n} \frac{(-1)^{n-i} F_i}{n+1-i} \binom{n}{i} = \begin{cases} \frac{2F_{n+1}}{n+1} & \text{if } n \text{ is odd}, \\ 0 & \text{if } n \text{ is even}. \end{cases}$$

Problem 15 Let $\sigma_n = \sum_{k=1}^{n} F_k^2$, where F_k is the kth Fibonacci number. Find the sum of the series $\sum_{n\geq 0} \frac{(-1)^n}{\sigma_n}$.

Problem 16 Compute $\sum_{n=1}^{\infty} \arctan \frac{1}{F_{2n}}$, where $(F_n)_{n\geq 0}$ is the Fibonacci sequence.

Problem 17 Let $n \geq 1$ be an integer. Show that 2^{n-1} divides the number

$$\sum_{0\leq k<\frac{n}{2}} \binom{n}{2k+1} 5^k. \tag{7.1}$$

Problem 18 (Michel Battaile, 3924, CM-2014) Let $(F_k)_{k\geq 0}$ be the Fibonacci sequence defined by $F_0 = 0$, $F_1 = 1$, and $F_{k+1} = F_k + F_{k-1}$, for $k \geq 1$. If m and n are positive integers with m odd and n not a multiple of 3, prove that $5F_m^2 - 3$ divides $5F_{mn}^2 + 3(-1)^n$.

Problem 19 (SAMC, 2014) A positive integer is called Fib-unique if the way to represent it as a sum of several distinct Fibonacci numbers is unique. For example, 13 is not Fib-unique, as $13 = 13 = 8+5 = 8+3+2$. Find all Fib-unique numbers.

Problem 20 (SAMC, 2014) Consider the function

$$f(x) = (x - F_2)(x - F_3) \cdots (x - F_{3031}),$$

where $(F_n)_{n \geq 0}$ is the Fibonacci sequence, defined by $F_0 = 0$, $F_1 = 1$ and the recurrence relation $F_{n+2} = F_{n+1} + F_n$, for $n \geq 0$. Suppose that on the range (F_1, F_{3031}) the function $| f(x) |$ takes on the maximum value at $x = x_0$. Prove that $x_0 > 2^{2018}$.

Problem 21 (Titu Andreescu, IMO Shortlist, 1983) Let $(F_n)_{n \geq 0}$ be the Fibonacci sequence, given by $F_1 = 1$, $F_2 = 1$ and $F_{n+2} = F_{n+1} + F_n$, for $n \geq 1$. Define P to be the polynomial of degree 990 which satisfies $P(k) = F_k$ for $k = 992, 993, \ldots, 1982$. Prove that $P(1983) = F_{1983} - 1$.

7.4 Higher Order Recurrence Relations

Problem 1 (Titu Andreescu, [4]) Let $(a_n)_{n \geq 0}$ be the sequence defined by $a_0 = 0$, $a_1 = 1$, and

$$a_{n+1} - 3a_n + a_{n-1} = 2(-1)^n, \quad n = 1, 2, \ldots.$$

Prove that a_n is a perfect square for all $n \geq 0$.

Problem 2 Define the sequence $(x_n)_{n \geq 1}$ by $x_0 = 0$, $x_1 = 1$, $x_2 = 1$ and

$$x_{n+3} = 2x_{n+2} + 2x_{n+1} - x_n, \quad n = 0, 1, \ldots.$$

Prove that x_n is a perfect square.

Problem 3 Define the sequence $(a_n)_{n \geq 0}$ by $a_0 = 3$, $a_1 = 1$, $a_2 = 9$ and

$$a_{n+3} = a_{n+2} + 4a_{n+1} - 4a_n, \quad n = 0, 1, \ldots.$$

Find a_n.

Problem 4 Consider all the sequences $(x_n)_{n \geq 0}$ satisfying

$$x_{n+3} = x_{n+2} + x_{n+1} + x_n, \quad n = 0, 1, \ldots.$$

Prove that there are real numbers α and t with $1.34 < \alpha < 1.37$, $2.1 < t < 2.2$ and a, b, c, satisfying the relation

$$x_n = a\alpha^{2n} + (b \cos nt + c \sin nt)\alpha^{-n}.$$

Problem 5 A sequence $(a_n)_{n\geq 1}$ is given by $a_1 = 1, a_2 = 12, a_3 = 20$, and

$$a_{n+3} = 2a_{n+2} + 2a_{n+1} - a_n, \quad n = 1, 2, \ldots.$$

Prove that for every integer $n \geq 1$, the number $1 + 4a_n a_{n+1}$ is a perfect square.

Problem 6 (Dorin Andrica, [4]) A sequence $(a_n)_{n\geq 1}$ is defined by $a_0 = 0, a_1 = 1$, $a_2 = 2, a_3 = 6$ and

$$a_{n+4} = 2a_{n+3} + a_{n+2} - 2a_{n+1} - a_n, \quad n = 1, 2, \ldots.$$

Prove that n divides a_n for all $n \geq 1$.

Problem 7 (Vlad Matei, O334. MR-2015) Consider the sequence $(a_n)_{n\geq 0}$ defined by $a_n = \left\lfloor (\sqrt[3]{65} - 4)^{-n} \right\rfloor$, for $n \geq 0$. Prove that $a_n = 2, 3 \pmod{15}$.

Problem 8 (Dorin Andrica, O346, MR-2015) Consider the sequence $(a_n)_{n\geq 0}$ defined by $a_0 = 0, a_1 = 1, a_0 = 1, a_3 = 6$ and

$$a_{n+4} = 2a_{n+3} + a_{n+2} - 2a_{n+1} - a_n, \quad n \geq 0.$$

Prove that n^2 divides a_n for infinitely many positive integers.

Problem 9 (Bakir Farhi, 11864, AMM-2015) Let p be a prime number and let $(u_n)_{n\geq 0}$ be the sequence defined by $u_n = n$, for $0 \leq n \leq p-1$ and $u_n = pu_{n+1-p} + u_{n-p}$, for $n \geq p$. Prove that for each positive integer n, the greatest power of p dividing u_n is the same as the greatest power of p dividing n.

7.5 Systems of Recurrence Relations

Problem 1 Find the general terms of $(x_n)_{n\geq 0}$, $(y_n)_{n\geq 0}$ if

$$\begin{cases} x_{n+1} = x_n + 2y_n \\ y_{n+1} = -2x_n + 5y_n, \quad n \geq 0, \end{cases}$$

and $x_0 = 1$, $y_0 = 2$.

Problem 2 Solve in positive integers the equation

$$6x^2 - 5y^2 = 1.$$

Problem 3 (Application 3), [7, p. 170]) Find all positive integers n such that $n+1$ and $3n + 1$ are simultaneously perfect squares.

Problem 4 (IMO Shortlist, 1980) Let A and E be a pair of opposite vertices of a regular octagon $AA_1A_2A_3EA_3'A_2'A_1'$. A frog starts jumping at vertex A. From any vertex of the octagon except E, it may jump to either of the two adjacent vertices. When it reaches vertex E, the frog stops and stays there. Let a_n be the number of distinct paths of exactly n jumps ending at E. Derive and solve a recursion for a_n.

Problem 5 (Neculai Stanciu and Titu Zvonaru, S323, MR-2015) Solve in positive integers the equation

$$x + y + (x - y)^2 = xy.$$

Problem 6 (Ivan Borsenco, U325, MR-2015) Let $A_1B_1C_1$ be a triangle with circumradius R_1. For each $n \geq 1$, the incircle of triangle $A_nB_nC_n$ is tangent to its sides at the points $A_{n+1}B_{n+1}C_{n+1}$. The circumradius of triangle $A_{n+1}B_{n+1}C_{n+1}$, which is also the inradius of triangle $A_nB_nC_n$, is R_{n+1}. Find the limit $\lim\limits_{n\to\infty} \frac{R_{n+1}}{R_n}$.

Problem 7 (Dorin Andrica and Mihai Piticari, Romanian TST, 2013) For a positive integer n we consider the expression

$$(\sqrt[3]{2} - 1)^n = a_n + b_n\sqrt[3]{2} + c_n\sqrt[3]{4},$$

where $a_n, b_n, c_n \in \mathbb{Z}$. Show that $c_{80} \neq 0$.

Problem 8 (Jean-Charles Mathieux, S38, MR-2007) Prove that for each positive integer n, there is a positive integer m such that

$$(1 + \sqrt{2})^n = \sqrt{m} + \sqrt{m + 1}.$$

Problem 9 (Gauss formula) Let $a > b > 0$ and the function

$$G(a, b) = \int_0^{\frac{\pi}{2}} \frac{\mathrm{d}x}{\sqrt{a^2 \cos^2 x + b^2 \sin^2 x}}.$$

Define the sequences $(a_n)_{n\geq 0}$ and $(b_n)_{n\geq 0}$ by the recurrence relations

$$a_n = \frac{a_{n-1} + b_{n-1}}{2}, \quad b_n = \sqrt{a_{n-1}b_{n-1}}, \quad a_0 = a, \quad b_0 = b.$$

(a) Prove that the sequences $(a_n)_{n\geq 0}$ and $(b_n)_{n\geq 0}$ are convergent to a common limit $\lim\limits_{n\to\infty} a_n = \lim\limits_{n\to\infty} a_n = \mu(a, b)$.

(b) Show that

$$G(a,b) = \int_0^{\frac{\pi}{2}} \frac{dx}{\sqrt{a_n^2 \cos^2 x + b_n^2 \sin^2 x}}.$$

(c) $G(a,b) = \frac{\pi}{2\mu(a,b)}$.

Problem 10 (Dorin Marghidaru and Leonard Giugiuc, 4264, CM-2017) Let $(a_n)_{n\geq 0}$ and $(b_n)_{n\geq 0}$ be two sequences such that $a_0, b_0 > 0$ and

$$a_{n+1} = a_n + \frac{1}{2b_n}, \quad b_{n+1} = b_n + \frac{1}{2a_n}, \quad n \geq 0.$$

Prove that

$$\max\{a_{2017}, b_{2017}\} > 44.$$

Problem 11 (SAMC, 2015) Let k be a positive integer. Prove that there exist integers x, y neither of which is divisible by 7, such that $x^2 + 6y^2 = 7^k$.

Problem 12 (SAMC, 2015) Arrange the numbers 1, 2, 3, 4 around a circle in this order. One starts at 1, and with every step moves to an adjacent number on either side. How many ways can you move such that the sum of the numbers visited in the path (including the starting number) is equal to 21?

Problem 13 (Dorin Andrica, O97, GM-1979) Let the sequences $(u_n)_{n\geq 0}, (v_n)_{n\geq 0}$ be defined by $u_1 = 3$, $v_1 = 2$ and

$$\begin{cases} u_{n+1} = 3u_n + 4v_n \\ v_{n+1} = 2u_n + 3v_n, \quad n \geq 1. \end{cases}$$

Define $x_n = u_n + v_n$, $y_n = u_n + 2v_n$, $n \geq 1$. Prove that $y_n = \lfloor x_n\sqrt{2} \rfloor$ for $n \geq 1$.

7.6 Homographic Recurrent Sequences

Problem 1 Let $(x_n)_{n\geq 0}$ be the sequence defined by $x_0 = 2$ and

$$x_{n+1} = \frac{2}{1 + x_n}, \quad n = 0, 1, \ldots.$$

Find x_{2009}.

Problem 2 Define the sequence $(x_n)_{n\geq 0}$ by

$$x_{n+1}x_n + x_{n+1} = x_n - 1, \quad n = 0, 1, \ldots.$$

Given that for all n we have $x_n \notin \{-1, 0, 1\}$ and $x_{2009} = 40$, find x_n.

Problem 3 (Austria, 1979) Define the sequence $(x_n)_{n\geq 0}$ by $x_0 = 1979$ and

$$x_{n+1} = \frac{1979(1 + x_n)}{1979 + x_n}, \quad n = 0, 1, \ldots.$$

Find x_n and $\lim_{n\to\infty} x_n$.

Problem 4 The sequence $(a_n)_{n\geq 0}$ is defined by $x_0 = a$ and

$$a_{n+1} = \frac{a_n + \sqrt{3}}{1 - \sqrt{3}a_n}, \quad n = 0, 1, \ldots.$$

Find a_{2009}.

Problem 5 Let $(x_n)_{n\geq 0}$ be the sequence defined by $x_0 = a$ and

$$x_{n+1} = \frac{x_n + \sqrt{2} - 1}{1 - (\sqrt{2} - 1)x_n}, \quad n = 0, 1, \ldots.$$

1. Prove that $(x_n)_{n\geq 0}$ is periodic and find its period.
2. Find x_{2009}.

Problem 6 Let $a, b > 0$ be real numbers such that $a^2 > b$. Define the sequence $(x_n)_{n\geq 0}$ by $x_0 = \alpha > 0$ and

$$x_{n+1} = \frac{ax_n + b}{x_n + a}, \quad n = 0, 1, \ldots.$$

Prove that $(x_n)_{n\geq 0}$ is convergent and find $\lim_{n\to\infty} x_n$.

Problem 7 Find all sequences $(a_n)_{n\geq 1}$ satisfying the following properties:

1. for all positive integers n, a_n is an integer;
2. $a_{n+2} = \dfrac{na_n + 1}{a_n + n}$, $n = 1, 2, \ldots.$

7.7 Complex Recurrent Sequences

Problem 1 (Ovidiu Bagdasar) Define the sequence $(w_n)_{n\geq 0}$ by

$$2w_{n+2} = \left(1 + (2 + \sqrt{3})i\right) w_{n+1} + \left(\sqrt{3} - i\right) w_n, \quad w_0 = 1, w_1 = i, \quad n = 0, 1, \ldots.$$

1. Find the formula of w_n;
2. Prove that the sequence is periodic and find the period;
3. Discuss the behavior of the sequence $(w_n)_{n\geq 0}$, if this satisfies the same recurrence relation, but with the starting values $w_0 = 1$, $w_1 = 2i$.

Problem 2 (Ovidiu Bagdasar) Define the sequence $(w_n)_{n\geq 0}$ by the recurrence relation

$$2w_{n+2} = \left(1 + (2+\sqrt{3})i\right) w_{n+1} + \left(\sqrt{3} - i\right) w_n, \quad w_0 = 1, w_1 = i, \quad n = 0, 1, \ldots.$$

1. Find the formula of w_n;
2. Prove that the sequence is periodic and find the period;
3. Discuss the behavior of the sequence $(w_n)_{n\geq 0}$, if this satisfies the same recurrence relation, but with the starting values $w_0 = 1$, $w_1 = 2i$.

Problem 3 (Ovidiu Bagdasar) Let the sequence $(w_n)_{n\geq 0}$ be defined by $w_{n+2} = pw_{n+1} + qw_n$ with $p, q \in \mathbb{C}$, and $w_0 = 1$, $w_1 = i$.

1. Find three distinct sequences having period 15.
2. How many such sequences exist?

Problem 4 (Ovidiu Bagdasar) Consider the sequence $(w_n)_{n\geq 0}$ defined by $w_{n+2} = pw_{n+1} + qw_n$ with $p, q \in \mathbb{C}$, and $w_0 = 1$, $w_1 = i$. Find the number of periodic sequences $(w_n)_{n\geq 0}$, whose period is the cube of a given prime number.

Problem 5 (Ovidiu Bagdasar) Let $(w_n)_{n\geq 0}$ be the sequence defined by the initial conditions $w_0 = a$, $w_1 = b$, and the recurrence equation

$$w_{n+2} = e^{2\pi i \sqrt[3]{2}} \left(e^{2\pi i \sqrt[3]{2}} + 1\right) w_{n+1} - e^{2\pi i \sqrt[3]{2}\left(\sqrt[3]{2}+1\right)} w_n, \quad n = 0, 1, \ldots.$$

1. Show that the closure of the set $\{w_n : n \in \mathbb{N}\}$ is an annulus.
2. Find a, b for which the orbit of $(w_n)_{n\geq 0}$ fills the annulus $U(0; 1, 2)$.

Problem 6 (Ovidiu Bagdasar) Let $(w_n)_{n\geq 0}$ be a recurrent sequence which has the initial values $w_0 = 1$, $w_1 = i$, $w_2 = 1 + i$.

1. Show that if $(w_n)_{n\geq 0}$ satisfies the recurrence relation

$$w_{n+3} = (1+i)w_{n+2} - (1+i)w_{n+1} + iw_n,$$

then it is periodic, and find its period.
2. Find the complex number q for which the sequence $(w_n)_{n\geq 0}$ defined by

$$w_{n+3} = (q^2 + q - 1)w_{n+2} + (q + q^2 - q^3)w_{n+1} - q^3 w_n,$$

is periodic. What are the possible values of the period?

Problem 7 (Ovidiu Bagdasar) Let $(w_n)_{n\geq 0}$ be a sequence defined by $w_0 = 1$, $w_1 = i$, $w_2 = 1 + i$, and $w_{n+3} = pw_{n+2} + qw_{n+1} + rw_n$, where $p, q, r \in \mathbb{C}$.

1. Find p, q, r such that $(w_n)_{n\geq 0}$ is periodic of period 2019;
2. How many such sequences of period 2019 exist, if the characteristic polynomial has distinct roots?

Problem 8 (Dorin Andrica) Let $(z_n)_{n\geq 1}$ be the complex sequence defined by $z_{n+1} = z_n^2 - z_n + 1$, $n = 1, 2, \ldots$, where $z_n^2 - z_n + 1 \neq 0$ and $z_1 \neq 0, 1$. Prove that there are two points O_1 and O_2 in the complex plane such that for any $n \geq 1$, the points of coordinates z_{n+1} and $\frac{1}{z_1} + \cdots + \frac{1}{z_n}$ are located on circles with centers O_1 and O_2 and radius 1.

7.8 Recurrent Sequences in Combinatorics

Problem 1 Prove the identity

$$\sum_{k=0}^{n} \binom{n+k}{k} \frac{1}{2^k} = 2^n.$$

Problem 2 Prove the identity

$$\sum_{k=0}^{n} (-1)^k \binom{2n-k}{k} = \begin{cases} 1 & \text{if } n = 3p; \\ 0 & \text{if } n = 3p+1, \\ -1 & \text{if } n = 3p+2. \end{cases} \quad p \in \mathbb{N}.$$

.

Problem 3 Determine the number of functions $f : \{1, 2, , \ldots, n\} \to \{1, 2, 3, 4, 5\}$ satisfying the property $\mid f(k+1) - f(k) \mid \geq 3$, for all $k \in 1, \ldots, n-1$.

Problem 4 In how many ways can you pave a $2 \times n$ rectangle with 1×2 tiles?

Problem 5 Prove the identity

$$\frac{1}{p!} \sum_{k=0}^{n} (-1)^{n-k} \binom{n}{k} k^p = \begin{cases} 0 & \text{if } 0 \leq p < n, \\ 1 & \text{if } p = n, \\ \frac{n}{2} & \text{if } p = n+1, \\ \frac{n(3n+1)}{24} & \text{if } p = n+2, \\ \frac{n^2(n+1)}{48} & \text{if } p = n+3, \\ \frac{n(15n^3+30n^2+5n+1)}{1152} & \text{if } p = n+4. \end{cases}$$

Problem 6 (Italian Mathematical Olympiad, 1996) Given an alphabet with three letters a, b, c, find the number of words of n letters which contain an even number of a's.

Problem 7 Find the number of n-words from the alphabet $A = \{0, 1, 2\}$, if any two neighbors can differ by at most 1.

Problem 8 (Romanian Mathematical Olympiad, 1995) Let $A_1, A_2, \ldots, A_n$, be points on a circle. Find the number of possible colorings of these points with p colors, $p \geq 2$, such that any two neighbors have distinct colors.

Problem 9 Define a set S of integers to be *fat* if each of its elements is greater than its cardinal $|S|$. For example, the empty set and $\{5, 7, 91\}$ are fat, but the set $\{3, 5, 10, 14\}$ is not. Let f_n denote the number of fat subsets of $\{1, \ldots, n\}$. Derive a recursion for f_n.

Problem 10 Consider a cube of dimensions $1 \times 1 \times 1$. Let O and A be two of its vertices such that OA is the diagonal of a face of the cube. Which one is larger: the number of paths of length 1386 beginning at O and ending at O, or the number of paths of length 1386 beginning at O and ending at A? (A path of length n on the cube is a sequence of $n + 1$ vertices, such that the distance between each two consecutive vertices is 1.)

Problem 11 (Rishub Thaper, O443, MR-2018) Let $f(n)$ be the number of permutations of the set $\{1, 2, \ldots, n\}$, such that no pair of consecutive integers appears in that order, i.e., 2 does not follow 1, 3 does not follow 2, and so on.

1. Prove that $f(n) = (n-1)f(n-1) + (n-2)f(n-2)$.
2. If $[\alpha]$ denotes the nearest integer to a real number α, prove that

$$f(n) = \frac{1}{n}\left[\frac{(n+1)!}{e}\right].$$

Problem 12 (Mircea Merca, 11767, AMM-2014) Prove that

$$\sum \frac{(1 + t_1 + t_2 + \cdots + t_n)!}{(1 + t_1)! t_2! \cdots t_n!} = 2^n - F_n,$$

where F_k is the kth Fibonacci number and the sum is over all nonnegative integer solutions to $t_1 + 2t_2 + \cdots + nt_n = n$.

Problem 13 (David Beckwith, 11754, AMM-2015) When a fair coin is tossed n times, let $P(n)$ be the probability that the lengths of all runs (maximal constant strings) in the resulting sequence are of the same parity as n. Denoting by F_n the nth Fibonacci number, defined by $F_0 = 0$, $F_1 = 1$, and $F_n = F_{n-1} + F_{n-2}$ for $n \geq 2$, prove that

$$P(n) = \begin{cases} \left(\frac{1}{2}\right)^{n/2} & \text{if } n \text{ is even,} \\ \left(\frac{1}{2}\right)^{n-1} F_n & \text{if } n \text{ is odd.} \end{cases}$$

Problem 14 Find the number of ways u_n in which we can climb a ladder with n steps, if we can climb either one, or two steps at a time?

7.9 Miscellaneous

Problem 1 (Josef Tkadlec, O466, MR-2018) Let $n \geq 2$ be an integer. Prove that there exists a set S of $n-1$ real numbers such that whenever $a_1, \ldots, a_n$ are distinct numbers satisfying

$$a_1 + \frac{1}{a_2} = a_2 + \frac{1}{a_3} = \cdots = a_{n-1} + \frac{1}{a_n} = a_n + \frac{1}{a_1},$$

then the common value of all these sums is a number from S.

Problem 2 (Titu Andreescu, S334, MR-2015) Let $(a_n)_{n\geq 0}$ be a sequence of real numbers with $a_0 \geq 0$ and $a_{n+1} = a_0 \cdots a_n + 4$ for $n \geq 0$. Prove that

$$a_n - \sqrt[4]{(a_{n+1}+1)(a_n^2+1) - 4} = 1, \quad n \geq 1.$$

Problem 3 (Ángel Plaza and Sergio Falcón, 11920, AMM-2016) For a positive integer k, let $\langle F_k \rangle$ be the sequence defined by the initial condition $F_{k,0} = 0$, $F_{k,1} = 1$, and the recurrence relation $F_{k,n+1} = kF_{k,n} + F_{k,n-1}$. Find a closed form for

$$\sum_{i=0}^{n} \binom{2n+1}{i} F_{k,2n+1-i}.$$

Problem 4 (Problem 32, ME-1997) Let $a_0 = 1996$ and $a_{n+1} = \frac{a_n^2}{a_n+1}$ for $n = 0, 1, 2, \ldots$. Prove that $\lfloor a_n \rfloor = 1996 - n$ for $n = 0, 1, 2, \ldots, 999$, where $\lfloor x \rfloor$ is the greatest integer less than or equal to x.

Problem 5 (Problem 21, ME-1997) Show that if a polynomial P satisfies $P(2x^2 - 1) = \frac{P(x)^2}{2}$, then it must be constant.

Problem 6 (Titu Andreescu, J55, MR-2007) Consider the sequence defined by $a_0 = 1$ and $a_{n+1} = a_0 \cdots a_n + 4$, $n \geq 0$. Prove that $a_n - \sqrt{a_{n+1}} = 2$, for $n \geq 1$.

Problem 7 (APMO, 2014) A sequence of real numbers $(a_n)_{n\geq 0}$ is said to be good if the following three conditions hold:

1. The value of a_0 is a positive integer;
2. For each nonnegative integer i we have $a_{i+1} = 2a_i + 1$ or $a_{i+1} = \frac{a_i}{a_i+2}$;
3. There exists a positive integer k such that $a_k = 2014$.

Find the smallest positive integer n such that there exists a good sequence $(a_n)_{n\geq 0}$ of real numbers with the property that $a_n = 2014$.

Problem 8 (Dorin Andrica) Define $x_n = 2^{2^{n-1}} + 1$ for all $n \geq 1$. Prove that

1. $x_n = x_1 x_2 \cdots x_{n-1} + 2$, $n \geq 1$;
2. $(x_k, x_l) = 1$, for distinct $k, l \in \mathbb{N}$;
3. x_n ends in 7 for all $n \geq 3$.

Chapter 8
Solutions to Proposed Problems

In this chapter we present detailed solutions (sometimes two or three) for the proposed problems in Chapter 7.

8.1 First-Order Recurrent Sequences

Problem 1 (Germany, 1995) *Let a and b be positive integers and consider the sequence* $(x_n)_{n\geq 0}$ *defined by* $x_0 = 1$ *and*

$$x_{n+1} = ax_n + b^{n+1}, \quad n = 0, 1, \ldots.$$

Prove that for any choice of a and b, the sequence $(x_n)_{n\geq 0}$ *contains infinitely many composite numbers.*

Solution The formula (2.5) in the case $a = 1$ gives

$$x_n = 1 + b + \cdots + b^n = \frac{b^{n+1} - 1}{b - 1}, \quad n \geq 0.$$

If n is odd, $n = 2k - 1$, then

$$x_n = (b^k + 1)\frac{b^k - 1}{b - 1} = (b^k + 1)x_k, \quad k \geq 0,$$

and the conclusion follows.

Let $a \neq 1$. Assume to the contrary that x_n is composite for only finitely many n. Take N larger than all such n, so that x_m is prime for all $n > N$. Choose such a

D. Andrica, O. Bagdasar, *Recurrent Sequences*, Problem Books in Mathematics,
https://doi.org/10.1007/978-3-030-51502-7_8

prime $x_m = p$ not dividing $a - 1$ (this excludes only finitely many candidates). Let t be such that $t(1-a) \equiv b^{n+1} \pmod{p}$. Then

$$x_{n+1} - t = ax_n + b^{n+1} - t \equiv ax_n + b^{n+1} - b^{n+1} - ta \pmod{p},$$

hence $x_{n+1} - t \equiv a(x_n - t) \pmod{p}$. In particular,

$$x_{m+p-1} = t + (x_{m+p-1} - t) \equiv t + a^{p-1}(x_m - t) \equiv (1 - a^{p-1})t \equiv 0 \pmod{p}.$$

However, x_{m+p-1} is a prime greater than p, yielding a contradiction. Hence infinitely many of the x_n are composite.

Problem 2 *Let m and n be integers greater than 1 such that*

$$\gcd(m, n-1) = \gcd(m, n) = 1.$$

Prove that the first $m-1$ terms of the sequence $n_1, n_2, \ldots$, where $n_1 = mn + 1$ and $n_{k+1} = n \cdot n_k + 1$, $k \geq 1$, cannot all be primes.

Solution It is easy to show that

$$n_k = n^k m + n^{k-1} + \cdots + n + 1 = n^k m + \frac{n^k - 1}{n - 1},$$

for every positive integer k. Hence

$$n_{\varphi(m)} = n^{\varphi(m)} \cdot m + \frac{n^{\varphi(m)} - 1}{n - 1}.$$

By Euler's theorem, we have $m \mid (n^{\varphi(m)} - 1)$, and since $\gcd(m, n-1) = 1$, it follows that

$$m \mid \frac{n^{\varphi(m)} - 1}{n - 1}.$$

Consequently, m divides $n_{\varphi(m)}$. Because $\varphi(m) \leq m - 1$, $n_{\varphi(m)}$ is not a prime.

Problem 3 *Let $a_1 = 1$, $a_{n+1} = a_n + \lfloor \sqrt{a_n} \rfloor$, $n = 1, 2, \ldots$. Show that a_n is a perfect square if and only if n is of the form $2^k + k - 2$.*

Solution 1. Let $n_k = 2^k + k - 2$. We prove by induction first that $a_{n_k} = (2^{k-1})^2$ and second that $n_{k-1} < m < n_k$ implies that a_m is not a perfect square. The assertion holds for $k = 1$. By induction on i one can prove that the relations

$$a_{n_k+2i} = (2^{k-1} + 1 - i)^2 + 2k,$$

$$a_{n_k+2i+1} = (2^{k-1} + i)^2 + 2^{k-1} - i,$$

hold for $0 \leq i \leq 2^{k-1}$. In particular, if $i = 2^{k-1}$, we obtain $a_{n_{k+1}} = (2^k)^2$.

The other values are not perfect squares, since they are located between two consecutive squares. □

Problem 4 (IMO Shortlist, 1971) *Let* $T_k = k - 1$ *for* $k = 1, 2, 3, 4$ *and*

$$T_{2k-1} = T_{2k-2} + 2^{k-2}, \quad T_{2k} = T_{2k-5} + 2^k, \quad k = 3, 4, \ldots.$$

Show that for all n,

$$1 + T_{2n-1} = \left\lfloor \frac{12}{7} 2^{n-1} \right\rfloor \text{ and } 1 + T_{2n} = \left\lfloor \frac{17}{7} 2^{n-1} \right\rfloor.$$

Solution We use induction. Since $T_1 = 0$, $T_2 = 1$, $T_3 = 2$, $T_4 = 3$, $T_5 = 5$, and $T_6 = 8$, the statement is true for $n = 1, 2, 3$. Suppose that both formulae from the problem hold for some $n \geq 3$. In this case

$$T_{2n+1} = 1 + T_{2n} + 2^{n-1} = \left\lfloor \frac{17}{7} 2^{n-1} + 2^{n-1} \right\rfloor = \left\lfloor \frac{12}{7} 2^n \right\rfloor,$$

$$T_{2n+2} = 1 + T_{2n-3} + 2^{n+1} = \left\lfloor \frac{12}{7} 2^{n-2} + 2^{n+1} \right\rfloor = \left\lfloor \frac{17}{7} 2^n \right\rfloor.$$

This shows that the formulae also hold for $n + 1$, which ends the proof.

Problem 5 *Define the sequences* $(x_n)_{n \geq 0}$ *and* $(y_n)_{n \geq 0}$ *by* $x_0 = \alpha$, $y_0 = \beta$, *and* $x_{n+1} = 2y_n + 1$, $y_{n+1} = 2x_n - 3$, $n = 0, 1, \ldots$. *Find* x_n *and* y_n.

Solution Define the sequences $(u_n)_{n \geq 0}$ and $(v_n)_{n \geq 0}$ by $u_n = x_n + y_n$ and $v_n = x_n - y_n$, $n = 0, 1, \ldots$. From the recursion relations we obtain

$$x_{n+1} + y_{n+1} = 2(x_n + y_n) - 2, \quad n = 0, 1, \ldots,$$

hence

$$u_{n+1} = 2u_n - 2, \quad n = 0, 1, \ldots, \quad u_0 = x_0 + y_0 = \alpha + \beta.$$

By applying formula (2.2) for $a = 2$ and $b = -2$, we obtain

$$u_n = 2^n(\alpha + \beta - 2) + 2, \quad n = 0, 1, \ldots.$$

From the recursive relations we obtain

$$x_{n+1} - y_{n+1} = -2(x_n - y_n) + 4, \quad n = 0, 1, \ldots,$$

hence

$$v_{n+1} = -2v_n + 4, \quad n = 0, 1, \ldots, \quad v_0 = x_0 - y_0 = \alpha - \beta.$$

By applying formula (2.2) for $a = -2$ and $b = 4$, we obtain

$$v_n = (-2)^n \left(\alpha - \beta - \frac{4}{3}\right) + \frac{4}{3}, \quad n = 0, 1, \ldots.$$

From the system $x_n + y_n = u_n$ and $x_n - y_n = v_n$, we get

$$x_n = \frac{1}{2}\left[2^n(\alpha + \beta - 2) + (-2)^n \left(\alpha - \beta - \frac{4}{3}\right) + \frac{10}{3}\right]$$

$$y_n = \frac{1}{2}\left[2^n(\alpha + \beta - 2) - (-2)^n \left(\alpha - \beta - \frac{4}{3}\right) + \frac{2}{3}\right], \quad n = 0, 1, \ldots.$$

Problem 6 *Let $p \geq 3$ be a prime and let $(x_n)_{n \geq 0}$ be the sequence defined by*

$$x_{n+1} = 2x_n + 1, \quad n = 0, 1, \ldots, \quad x_0 = p.$$

Prove that for any integer $k \geq 0$, the set $A_k = \{x_k, x_{k+1}, \ldots, x_{k+p-1}\}$ contains at least one composite integer.

Solution Using formula (2.2) for $\alpha = p$, $a = 2$, $b = 1$, we get

$$x_n = 2^n(p + 1) - 1, \quad n = 0, 1, 2, \ldots.$$

Take $n \in \{k, k+1, \ldots, k+p-1\}$ such that $p - 1 | n$, that is $n = m(p - 1)$ for some positive integer m. Then

$$x_n = 2^n p + 2^n - 1 = 2^n p + (2^m)^{p-1} - 1 \equiv 0 \pmod{p},$$

since according to Fermat's little Theorem we have $(2^m)^{p-1} \equiv 1 \pmod{p}$.

Problem 7 (Marius Cavachi, S303, MR-2014) *Let $(a_n)_{n \geq 1}$ be the sequence defined by $a_1 = 1$ and $a_{n+1} = \frac{1}{2}\left(a_n + \frac{n}{a_n}\right)$, for $n \geq 1$. Find $\lfloor a_{2014} \rfloor$.*

Solution By the AM-GM inequality we deduce that for every $n \geq 1$ we have

$$a_{n+1} = \frac{1}{2}\left(a_n + \frac{n}{a_n}\right) \geq \sqrt{n},$$

hence

$$a_{n+1} \leq \frac{1}{2}\left(a_n + \frac{n}{\sqrt{n-1}}\right).$$

By induction, it is easy to show that

$$2^n a_{n+1} \leq a_1 + 2^{n-1} b_n + \cdots + 2b_3 + b_2,$$

where $b_n = \frac{n}{\sqrt{n-1}}$, and also notice that

$$b_{n+1} > b_n \leftrightarrow n^2 > n+1.$$

This shows that $(b_n)_{n\geq 2}$ is an increasing sequence and that

$$a_{n+1} \leq \frac{1+(2^{n-1}+\cdots+1)b_n}{2^n} = \frac{1+(2^n-1)b_n}{2^n} < \frac{1}{2^n} + b_n.$$

We conclude that

$$\sqrt{n} \leq a_{n+1} \leq \frac{1}{2^n} + \frac{n}{\sqrt{n-1}}.$$

Applying these results for $n = 2013$, one obtains $\lfloor a_{2014} \rfloor = 44$.

Problem 8 (Albert Stadler, U367, MR-2016) *Let $(a_n)_{n\geq 1}$ be the sequence of real numbers satisfying the properties $a_1 = 4$ and $3a_{n+1} = (a_n+1)^3 - 5$, $n \geq 1$. Prove that a_n is a positive integer for all n, and evaluate*

$$\sum_{n=1}^{\infty} \frac{a_n - 1}{a_n^2 + a_n + 1}.$$

Solution Denoting $a_n = 3b_n + 1$, we obtain a new sequence with $b_1 = 1$ and

$$b_{n+1} = 3b_n(b_n+1)^2 + b_n,$$

therefore since b_1 is a positive integer, by induction, it easily follows that a_n is also an integer. Moreover,

$$\frac{a_n - 1}{a_n^2 + a_n + 1} = \frac{b_n}{3b_n^2 + 3b_n + 1} = \frac{b_n(b_n+1)}{b_{n+1}+1} = \frac{b_{n+1} - b_n}{3(b_n+1)(b_{n+1}+1)}$$
$$= \frac{1}{3(b_n+1)} - \frac{1}{3(b_{n+1}+1)},$$

which gives

$$\sum_{n=1}^{N} \frac{a_n - 1}{a_n^2 + a_n + 1} = \frac{1}{3(b_1+1)} - \frac{1}{3(b_{N+1}+1)} = \frac{1}{6} - \frac{1}{3(b_{n+1}+1)}.$$

Since b_N converges to infinity, we deduce that

$$\sum_{n=1}^{\infty} \frac{a_n - 1}{a_n^2 + a_n + 1} = \frac{1}{6}.$$

Problem 9 (Arkady Alt, U326, MR-2015) *Let $(a_n)_{n\geq 0}$ be a sequence with $a_0 > 1$ and $3a_{n+1} = a_n^3 + 2$, for $n \geq 1$. Find*

$$\sum_{n=0}^{\infty} \frac{a_n + 2}{a_n^2 + a_n + 1}.$$

Solution By the identity $3a_{n+1} = a_n^3 + 2$, we have $a_n^3 - 1 = 3(a_{n+1} - 1)$, hence

$$\frac{a_n + 2}{a_n^2 + a_n + 1} = \frac{(a_n - 1)(a_n + 2)}{a_n^3 - 1} = \frac{a_n^2 + a_n - 2}{3(a_{n+1} - 1}.$$

One can write

$$\begin{aligned} S = \sum_{n=0}^{\infty} \frac{a_n + 2}{a_n^2 + a_n + 1} &= \sum_{n=0}^{\infty} \frac{a_n^2 + a_n - 2}{3(a_{n+1} - 1)} \\ = \sum_{n=0}^{\infty} \frac{(a_n - 1)(a_n^2 + a_n - 2)}{3(a_n - 1)(a_{n+1} - 1)} &= \sum_{n=0}^{\infty} \frac{a_{n+1} - a_n}{(a_n - 1)(a_{n+1} - 1)} \\ = \sum_{n=0}^{\infty} \left(\frac{1}{a_n - 1} - \frac{1}{a_{n+1} - 1} \right) &= \frac{1}{(a_0 - 1)}. \end{aligned}$$

Problem 10 (Khakimboy Egamberganov, U327, MR-2015) *Let $(a_n)_{n\geq 0}$ be a sequence of real numbers with $a_0 = 1$ and*

$$a_{n+1} = \frac{a_n}{na_n + a_n^2 + 1}.$$

Find the limit $\lim\limits_{n\to\infty} n^3 a_n$.

Solution Note that $a_{n+1} = \frac{a_n}{n^2 a_n}$, hence by the comparison test, $\sum_{n=0}^{\infty} a_n$ converges to a value $S \in \mathbb{R}$. One can rewrite the initial relation as

$$\begin{aligned} \frac{1}{a_1} &= 0^2 + a_0 + \frac{1}{a_0} \\ \frac{1}{a_2} &= 1^2 + a_1 + \frac{1}{a_1} \\ &\vdots \\ \frac{1}{a_n} &= (n-1)^2 + a_{n-1} + \frac{1}{a_{n-1}}. \end{aligned}$$

By summation we obtain

$$\frac{1}{a_n} = \frac{(n-1)n(2n-1)}{6} + \sum_{k=0}^{n-1} a_k + 1.$$

Dividing by n^3 and taking limits, notice first that

$$\lim_{n\to\infty} \frac{\sum_{k=0}^{n-1} a_k + 1}{n^3} = \lim_{n\to\infty} \frac{S+1}{n^3} = 0,$$

therefore

$$\lim_{n\to\infty} \frac{1}{n^3 a_n} = \lim_{n\to\infty} \frac{(n-1)n(2n-1)}{6n^3} + \lim_{n\to\infty} \frac{\sum_{k=0}^{n-1} a_k + 1}{n^3} = \frac{1}{3}.$$

Finally, we conclude that $\lim_{n\to\infty} n^3 a_n = 3$.

Problem 11 (Titu Andreescu, U333, MR-2015) *Evaluate*

$$\prod_{n=0}^{\infty} \left(1 - \frac{2^{2^n}}{2^{2^{n+1}} + 1}\right).$$

Solution Denoting $x_n = 2^{2^n}$, one obtains $x_{n+1} = x_n^2$ and

$$\begin{aligned}
\prod_{n=0}^{\infty} \left(1 - \frac{2^{2^n}}{2^{2^{n+1}} + 1}\right) &= \prod_{n=0}^{\infty} \left(1 - \frac{1/x_n}{1/x_n^2 + 1}\right) \\
&= \prod_{n=0}^{\infty} \frac{x_n^2 - x_n + 1}{1 + x_n^2} \cdot \frac{1 + x_n}{1 + x_n} \cdot \frac{1 - x_n}{1 - x_n} \cdot \frac{1 - x_n^3}{1 - x_n^3} \\
&= \prod_{n=0}^{\infty} \frac{(1 - x_n^6)(1 - x_n)}{(1 - x_n^4)(1 - x_n^3)} \\
&= \prod_{n=0}^{\infty} \frac{1 - x_{n+1}^3}{1 - x_n^3} \cdot \prod_{n=0}^{\infty} \frac{1 - x_n}{1 - x_{n+2}} \\
&= \frac{(1 - x_0)(1 - x_1)}{1 - x_0^3} = \frac{(1 - 1/2)(1 - 1/4)}{1 - 1/8} = \frac{3}{7}.
\end{aligned}$$

Problem 12 (Marius Cavachi, U272, MR-2013) *Let a be a positive real number and let $(a_n)_{n\geq 0}$ be the sequence defined by $a_0 = \sqrt{a}$, $a_{n+1} = \sqrt{a_n + a}$, for $n \geq 0$. Show that the sequence has infinitely many irrational terms.*

Solution Assuming the result is false, either all sequence terms are rational, or there is a last rational term a_N, i.e., $a_{N+1}, a_{N+2}, \dots$, are all rational. Notice that if two consecutive terms are rational, then $a = a_{N+2}^2 - a_{N+1}$ must be rational. From the relation $a_{N+1}^2 = a_N + a$, we deduce that a_N is also rational, which suggests that all terms in the sequence would be rational.

Assume there is a real number $a > 0$, such that the sequence $(a_n)_{n\geq 0}$ has only rational terms. One can show that $a_1 = \sqrt{a+\sqrt{a}} > \sqrt{a} = a_0$, and then also in general, if $a_n > a_{n-1}$, we have $a_{n+1} = \sqrt{a+\sqrt{a_n}} > \sqrt{a+a_{n-1}} = a_n$, hence the sequence is strictly increasing.

Denote by $a_0 = \sqrt{a} = \frac{u}{v}$ with u, v coprime. We prove that we can define the increasing integer sequence $(u_n)_{n\geq 0}$, given by $u_0 = u$ and $u_{n+1} = \sqrt{u^2 + u_n v}$. Notice that if $a_n = \frac{u_n}{v}$, then $a_{n+1} = \frac{u_{n+1}}{v}$. Also, since a_{n+1} is rational, if u_n is an integer, then u_{n+1} is rational and at the same time, the root of an integer, hence it is an integer too. Since u_0 is an integer, it is clear that the integer sequence $u_n = va_n$ is strictly increasing, hence unbounded.

One can find therefore a value of n for which

$$a_n = vu_n > \frac{1+\sqrt{1+4a}}{2}.$$

From this we deduce that

$$a < \frac{(2a_n^2-1)^2-1}{4} = a_n^2 - a_n, \quad a_{n+1} = \sqrt{a+a_n} < \sqrt{a_n^2} = a_n.$$

Since $(a_n)_{n\geq 0}$ was increasing, this is a contradiction, hence the assumption that $(a_n)_{n\geq 0}$ had a finite number of irrational terms was false.

Problem 13 (Ivan Borsenco, O285, MR-2013) *Consider the integer sequence $(a_n)_{n\geq 0}$ given by $a_1 = 1$ and $a_{n+1} = 2^n(2^{a_n} - 1)$, $n \geq 1$. Prove that $n!$ divides a_n.*

Solution Note first that given a prime p and an integer $n \geq p$, the multiplicity with which p divides $n!$ is at most $n - p + 1$. Indeed, this multiplicity is given by the exact formula

$$\sum_{k=1}^{\infty} \left\lfloor \frac{n}{p^k} \right\rfloor < n \sum_{k=1}^{\infty} \frac{1}{p^k} = \frac{n}{p-1}.$$

Clearly, it is sufficient to show that $\frac{n}{p-1} \leq n - p + 1$, which is equivalent to $(n - p)(p-2) \geq 1$. This is clearly true, except for $p = 2$ or $n = p$. If $p = 2$, note that the multiplicity with which 2 divides $n!$ is less than n, that is at most $n - 1 = n - p + 1$, in agreement with the stated result. On the other hand, if $n = p$, the multiplicity with which p divides $n! = p!$ is $1 = n - p + 1$, hence the result also holds in this case. This confirms that the result is always true for any prime p and integer $n \geq p$.

By the previous result, it suffices to show that for any prime p and any integer $n \geq p$, p divides a_n with multiplicity at least $n - p + 1$. This is trivially true for $p = 2$, as for any $n \geq 2$, $2^{n-1} = 2^{n-2+1}$ divides a_2, hence $2^{a_2} \equiv$ (mod 3), hence 3^{3-3+1} divides a_3, whereas if the result is true for n, then a_n is a multiple of $2 \cdot 3^{n-p+1}$, so by Euler-Fermat's theorem, $3^{(n+1)-p+1}$ divides 2^{a_n}, or it divides a_{n+1}. The proof of this result is done by induction over the set of odd primes.

Given any odd prime $p > 3$, assume that the result is true for all primes less than p, or since $p - 1$ is a product of primes less than p. By the induction hypothesis and using the previous arguments, $(p-1)!$ divides a_{p-1}, or in particular $p - 1$ divides a_{p-1}, hence by Euler-Fermat's theorem, $2^{a_{p-1}-1}$ is divisible by p, or for $n = p$, $a_n = a_p$ is divisible by $p = p^{n-p+1}$. For any $n \geq p$, if p^{n-p+1} divides p, and since $p - 1 < n$, then $p - 1$ divides $n!$, which in turn divides a_n, it follows that $(p-1)p^{n-p+1}$ divides a_n, or by Euler-Fermat's theorem, $p^{(n+1)-p+1}$ divides $2^{a_n} - 1$, hence it divides a_{n+1}.

Problem 14 (Iosif Pinelis, 11837, AMM-2015) *Let $a_0 = 1$ and let $a_{n+1} = a_n + e^{-a_n}$ for $n \geq 0$. Let $b_n = a_n - \log n$. For $n \geq 0$, show that $0 < b_{n+1} < b_n$; also show that $\lim_{n\to\infty} b_n = 0$.*

Solution The sequence $(a_n)_{n\geq 0}$ satisfies $a_{n+1} = f(a_n)$, where $f(t) = t + e^{-t}$. Clearly, function f is increasing and positive on the interval $[0, \infty)$. For any positive integer n, we have the inequalities

$$\log\left(1 + \frac{1}{n}\right) < \frac{1}{n}, \quad \log\left(1 - \frac{1}{n+1}\right) < -\frac{1}{n+1},$$

hence

$$\log(n+1) < \log n + \frac{1}{n}, \tag{8.1}$$

and

$$\frac{1}{n+1} < \log\left(1 + \frac{1}{n}\right) < \frac{1}{n}. \tag{8.2}$$

We will prove by induction that $a_n > \log(n+1)$ for all $n \geq 1$. Clearly, we have $a_0 = 1 > \log 2$. Next, if $a_n > \log(n+1)$, then

$$a_{n+1} = f(a_n) > f(\log(n+1)) = \log(n+1) + \frac{1}{n+1},$$

hence by (8.1), $a_{n+1} > \log(n+2)$. Notice that since $a_n > \log(n+1) > \log n$, it follows that $b_n > 0$, for any $n \geq 1$. Also, because $a_{n+1} - a_n = e^{-a_n}$, we have

$$a_{n+1} - a_n < e^{-\log(n+1)} = \frac{1}{n+1}, \quad n \geq 1.$$

By (8.2) we obtain $a_{n+1} - a_n < \log(n+1) - \log n$, hence $b_{n+1} < b_n$.

Clearly, a_n diverges to infinity, so we have

$$\lim_{n\to\infty} \frac{e^{-a_{n+1}} - e^{-a_n}}{(n+1)-n} = \lim_{n\to\infty} \frac{e^{e^{-a_n}} - 1}{e^{-a_n}} = \lim_{h\to 0} \frac{e^h - 1}{h} = 1.$$

By the Stolz-Cesàro lemma, it follows that $\lim_{n\to\infty} e^{a_n}/n = 1 = \lim_{n\to\infty} e^{b_n}$, hence $\lim_{n\to\infty} b_n = 0$.

Problem 15 (Titu Andreescu, U64, MR-2007) *Let x be a real number. Define the sequence $(x_n)_{n\geq 1}$ by $x_1 = 1$ and $x_{n+1} = x^n + nx_n$ for $n \geq 1$. Prove that*

$$\prod_{n=1}^{\infty} \left(1 - \frac{x^n}{x_{n+1}}\right) = e^{-x}.$$

Solution The condition $x_{n+1} = x^n + nx_n$ gives $\frac{x_{n+1}}{n!} = \frac{x^n}{n!} + \frac{x_n}{(n-1)!}$, for $n \geq 1$. Then

$$\sum_{k=1}^{n} \frac{x^k}{k!} = \sum_{k=1}^{n} \left(\frac{x_{k+1}}{k!} - \frac{x_k}{(k-1)!}\right) = \frac{x_{n+1}}{n!} - \frac{x_1}{0!} = \frac{x_{n+1}}{n!} - 1,$$

yielding

$$\frac{x_{n+1}}{n!} = \sum_{k=0}^{n} \frac{x^k}{k!}.$$

It follows that

$$\prod_{k=1}^{n} \left(1 - \frac{x^k}{x_{k+1}}\right) = \prod_{k=1}^{n} \left(\frac{x_{k+1} - x^k}{x_{k+1}}\right) = \prod_{k=1}^{n} \frac{kx_k}{x_{k+1}} = n! \prod_{k=1}^{n} \frac{x_k}{x_{k+1}} = \frac{n!}{x_{n+1}}.$$

Finally

$$\prod_{n=1}^{\infty} \left(1 - \frac{x^n}{x_{n+1}}\right) = \lim_{n\to\infty} \frac{n!}{x_{n+1}} = \frac{1}{\lim_{n\to\infty} \sum_{k=0}^{n} \frac{x^k}{k!}} = \frac{1}{e^x} = e^{-x}.$$

Problem 16 *Let I be an interval and consider a function $f : I \to I$. Define the sequence $(a_n)_{n\geq 0}$ by the relation $a_{n+1} = f(a_n)$, for $n \geq 0$ and $a_0 \in I$. Prove that*

1. *If f is increasing, then $(a_n)_{n\geq 0}$ is monotonic;*
2. *If f is decreasing, then the sequences $(a_{2n})_{n\geq 0}$ and $(a_{2n+1})_{n\geq 0}$ are monotonic, having different monotonicities.*

Solution

1. If $a_0 \le a_1$, then $f(a_0) \le f(a_1)$, hence $a_1 \le a_2$. By induction we show that $a_n \le a_{n+1}$ for $n \ge 0$. If $a_0 \ge a_1$, then the sequence is decreasing.
2. We have

$$a_{2n+1} = f(a_{2n}) = (f \circ f)(a_{2n-1}), \quad n \ge 0$$
$$a_{2n} = f(a_{2n-1}) = (f \circ f)(a_{2n-2}), \quad n \ge 0.$$

Since $g = f \circ f$ is increasing, by 1) it follows that $(a_{2n})_{n\ge0}$ and $(a_{2n+1})_{n\ge0}$ are monotonic. Assuming that the sequence $(a_{2n})_{n\ge0}$ is increasing, by the relation $a_{2n} \le a_{2n+2}$ it follows that $a_{2n+1} = f(a_{2n}) \ge f(a_{2n+2}) = a_{2n+3}$, hence the sequence $(a_{2n+1})_{n\ge0}$ is decreasing. Similarly, if $(a_{2n})_{n\ge0}$ is decreasing, then the sequence $(a_{2n+1})_{n\ge0}$ is increasing.

Problem 17 (Michel Bataille, 4282, CM-2017) *Find* $\lim_{n\to\infty} u_n$ *where the sequence* $(u_n)_{n\ge0}$ *is defined by* $u_0 = 1$ *and the recursion*

$$u_{n+1} = \frac{1}{2}\left(u_n + \sqrt{u_n^2 + \frac{u_n}{4^n}}\right),$$

for every nonnegative integer n.

Solution For each nonnegative integer n, let $u_n = \frac{1}{4v_n^2}$, with $v_n > 0$. We have $v_0 = 1$ and v_n satisfies the recurrence relation

$$v_{n+1}^2 = \frac{\sqrt{1+v_n^2} - 1}{2}.$$

Let $v_n = \sinh x_n$ for some $x_n \ge 0$. Then

$$(\sinh x_{n+1})^2 = \frac{\cosh x_n - 1}{2} = \left(\sinh \frac{x_n}{2}\right)^2,$$

therefore $x_{n+1} = x_n/2$ for each $n \ge 0$. It follows that

$$x_n = \frac{x_0}{2^n} = \frac{\sinh^{-1} 1}{2^n} = \frac{\ln(1+\sqrt{2})}{2^n}.$$

It follows that

$$u_n = \frac{1}{4^n\left[\sinh\left(2^{-n}\ln(1+\sqrt{2})\right)\right]^2}$$

$$= 4^{-n} \left(\frac{2^{-n} \ln(1+\sqrt{2})}{\sinh\left(2^{-n}\ln(1+\sqrt{2})\right)} \right)^2 \left(\frac{2^n}{\ln(1+\sqrt{2})} \right)^2.$$

Letting n tend to infinity yields the desired limit $(\ln(1+\sqrt{2}))^{-2} \sim 1.287$.

Problem 18 (Mihály Bencze, 4264, CM-2017) *Let $x_1 = 4$ and $x_{n+1} = \lfloor \sqrt[3]{2} x_n \rfloor$, where $\lfloor \cdot \rfloor$ denotes the integer part. Determine the largest positive integer n for which x_n, x_{n+1}, x_{n+2} for an arithmetic progression.*

Solution We show that the largest such number is $n^* = 7$. By direct computations, one obtains $x_7 = 12, x_8 = 15, x_9 = 18, x_{10} = 22$. This confirms that x_7, x_8, x_9 are in arithmetic progression, hence $n^* \geq 7$, while also, $n^* \neq 8$.

Since x_n is increasing, for any $n \geq 9$ we have $x_n \geq 18 = x_9$. Note that

$$x_{n+1} = \lfloor \sqrt[3]{2} x_n \rfloor = x_n + \left\lfloor (\sqrt[3]{2} - 1) x_n \right\rfloor = x_n + y_n,$$

where $y_n \geq \lfloor 0.25 x_n \rfloor \geq 4$. Since $4(\sqrt[3]{2} - 1) > 1$, we have

$$\begin{aligned} x_{n+2} &= \lfloor \sqrt[3]{2} x_{n+1} \rfloor = x_{n+1} + \left\lfloor (\sqrt[3]{2} - 1) x_{n+1} \right\rfloor = x_{n+1} + \left\lfloor (\sqrt[3]{2} - 1)(x_n + y_n) \right\rfloor \\ &\geq x_{n+1} + \left\lfloor (\sqrt[3]{2} - 1)(x_n + 4) \right\rfloor \geq x_{n+1} + \left\lfloor (\sqrt[3]{2} - 1) x_n + 1 \right\rfloor \\ &= x_{n+1} + \left\lfloor (\sqrt[3]{2} - 1) x_n \right\rfloor + 1 = x_{n+1} + y_n + 1. \end{aligned}$$

Hence

$$x_{n+2} - x_{n+1} \geq y_n + 1 = \left\lfloor (\sqrt[3]{2} - 1) x_n \right\rfloor + 1 = \lfloor \sqrt[3]{2} x_n \rfloor - x_n + 1 > x_{n+1} - x_n,$$

so x_n, x_{n+1}, x_{n+2} cannot be in arithmetic progression.

Problem 19 (Marcel Chiriţă, 3942, CM-2016) *Consider a sequence $(x_n)_{n \geq 1}$ with $x_1 = 1$ and $x_{n+1} = \frac{1}{n+1}\left(x_n + \frac{1}{x_n}\right)$. Find $\lim\limits_{n \to \infty} \sqrt{n} x_n$.*

Solution We prove by induction that we have

$$\frac{1}{n} \leq x_n^2 \leq \frac{1}{n-1}, \quad n \geq 2.$$

This clearly holds for $n = 2$, since $x_2 = 1$. Assume that this holds for some n. Since the function $f(t) = t + \frac{1}{t}$ is decreasing on $(0, 1)$, we have

$$f\left(\frac{1}{n-1}\right) \leq f\left(x_n^2\right) \leq f\left(\frac{1}{n}\right). \tag{8.3}$$

Clearly, it follows that

$$\frac{f\left(\frac{1}{n-1}\right)+2}{(n+1)^2} \leq x_{n+1}^2 = \frac{f(x_n^2)+2}{(n+1)^2} \leq \frac{f\left(\frac{1}{n}\right)+2}{(n+1)^2},$$

from where we deduce

$$\frac{\frac{1}{n-1}+n+1}{(n+1)^2} \leq x_{n+1}^2 \leq \frac{\frac{1}{n}+n+1}{(n+1)^2}.$$

We obtain

$$\frac{1}{n+1} \leq \frac{n^2}{n^2-1} \cdot \frac{1}{n+1} \leq x_{n+1}^2 \leq \frac{(n+1)^2}{n(n+1)^2} = \frac{1}{n},$$

which completes the induction step.

Since $x_n \geq 0$ for all n, we deduce that

$$1 \leq x_n\sqrt{n} \leq \sqrt{\frac{n}{n-1}},$$

from where we conclude that $\lim\limits_{n\to\infty} \sqrt{n}x_n = 1$.

Problem 20 (Laurenţiu Panaitopol, S10, MR-2006) *Let $(a_n)_{n\geq 1}$ be a sequence of positive numbers such as $a_{n+1} = a_n^2 - 2$ for all $n \geq 1$. Show that we have $a_n \geq 2$, for all $n \geq 1$.*

Solution Define the sequence $(b_n)_{n\geq 1}$ as follows:

$$b_1 = \sqrt{2}, \quad b_{n+1} = \sqrt{2+b_n}, \quad n \geq 1.$$

Observe that $a_n > b_m$ is equivalent to $a_n^2 - 2 > b_m^2 - 1$, or

$$a_{n+1} > b_{m-1}.$$

Since $a_{n+1} > 0$, it follows that $a_n > \sqrt{2}$, hence $a_n > b_1$. We deduce that

$$a_{n-1} > b_2, \quad a_{n-2} > b_3, \quad \ldots, \quad a_1 > b_n, \quad n \geq 1.$$

It is easy to show inductively that

$$b_n = 2\cos\frac{\pi}{2^{n+1}},$$

therefore

$$\lim_{n\to\infty} b_n = 2.$$

It follows that $a_1 \geq 2$, and we obtain inductively that $a_n \geq 2$, for all n.

8.2 Second-Order Recurrent Sequences

Problem 1 (Titu Andreescu, [4]) *Find a_{2020} for the sequence $(a_n)_{n\geq 1}$ defined by $a_1 = 1$ and*

$$a_{n+1} = 2a_n + \sqrt{3a_n^2 - 2}, \quad n = 1, 2, \ldots.$$

Solution It is clear that $a_1 < a_2 < \cdots$, hence we can write the recurrence relation in the equivalent form

$$a_{n+1}^2 - 4a_n a_{n+1} + a_n^2 + 2 = 0, \quad n = 1, 2, \ldots.$$

Replacing n by $n-1$ we obtain

$$a_n^2 - 4a_{n-1}a_n + a_{n-1}^2 + 2 = 0, \quad n = 2, 3, \ldots.$$

It follows that a_{n+1} and a_{n-1} are the roots of the quadratic equation

$$t^2 - 4a_n t + a_n^2 + 2 = 0,$$

hence

$$a_{n+1} + a_{n-1} = 4a_n.$$

From here one obtains $a_1 = 1$, $a_2 = 3$ and $a_{n+1} = 4a_n - a_{n-1}$, $n = 2, 3, \ldots$. The characteristic equation is $t^2 - 4t + 1 = 0$ with the roots $t_1 = 2 + \sqrt{3}$ and $t_2 = 2 - \sqrt{3}$. By Binet-type formula the general term is given by

$$a_n = c_1 t_1^n + c_2 t_2^n, \quad n = 1, 2, \ldots,$$

where c_1 and c_2 are given by the system

$$\begin{cases} c_1 t_1 + c_2 t_2 = 1 \\ c_1 t_1^2 + c_2 t_2^2 = 3. \end{cases}$$

Simple calculations show that $c_1 = \frac{1}{6}(3 - \sqrt{3})$ and $c_2 = \frac{1}{6}(3 + \sqrt{3})$, hence

$$a_n = \frac{1}{6}\left[(3 - \sqrt{3})(2 + \sqrt{3})^n + (3 + \sqrt{3})(2 - \sqrt{3})^n\right], \quad n = 1, 2, \ldots.$$

Problem 2 *Let a and b be positive real numbers. Find the general term of the sequence $(x_n)_{n \geq 0}$ defined by $x_0 = 0$ and*

$$x_{n+1} = x_n + a + \sqrt{b^2 + 4ax_n}, \quad n = 0, 1, \ldots.$$

Solution We have

$$b^2 + 4ax_{n+1} + b^2 + 4a(x_n + a + \sqrt{b^2 + 4ax_n})$$
$$= 4a^2 + 4a\sqrt{b^2 + 4ax_n} + (b^2 + 4ax_n) = (2a + \sqrt{b^2 + 4ax_n})^2,$$

therefore

$$\sqrt{b^2 + 4ax_{n+1}} = 2a + \sqrt{b^2 + 4ax_n}, \quad n = 0, 1, \ldots. \tag{8.4}$$

Using the initial recursive relation we obtain

$$\sqrt{b^2 + 4ax_{n+1}} = x_{n+1} - x_n + a, \quad n = 0, 1, \ldots. \tag{8.5}$$

Replacing n by $n - 1$ in (8.5) we obtain

$$\sqrt{b^2 + 4ax_n} = x_n - x_n + a, \quad n = 1, 2, \ldots. \tag{8.6}$$

Using the initial recurrence relation again it follows that

$$x_{n+1} - x_n - a = x_n - x_{n-1} + a,$$

that is

$$x_{n+1} - 2x_n + x_{n-1} = 2a, \quad n = 1, 2, \ldots. \tag{8.7}$$

Letting $y_n = x_n - x_{n-1}$, $n = 1, 2, \ldots$, the relation (8.7) is equivalent to

$$y_{n+1} - y_n = 2a, \quad n = 0, 1, \ldots. \tag{8.8}$$

From (8.8) one clearly obtains $y_n = 2an + (a + b)$, $n = 0, 1, \ldots$, which finally results in $x_n = an^2 + bn$, $n = 0, 1, \ldots$.

Problem 3 (IMO Shortlist, 1983) *Let k be a positive integer. The sequence $(a_n)_{n\geq 0}$ is defined by $a_0 = 0$ and*

$$a_{n+1} = (a_n + 1)k + (k+1)a_n + 2\sqrt{k(k+1)a_n(a_n+1)}, \quad n = 0, 1, \ldots.$$

Prove that all a_n are positive integers.

Solution Clearly, each a_n is positive and

$$\sqrt{a_{n+1}} = \sqrt{a_n}\sqrt{k+1} + \sqrt{a_n+1}\sqrt{k}.$$

Notice that

$$\sqrt{a_{n+1}+1} = \sqrt{k+1}\sqrt{a_n+1} + \sqrt{k}\sqrt{a_n}.$$

Therefore, one obtains

$$\begin{aligned}&(\sqrt{k+1} - \sqrt{k})(\sqrt{a_n+1} - \sqrt{a_n})\\ =&(\sqrt{k+1}\sqrt{a_n+1} + \sqrt{k}\sqrt{a_n}) - (\sqrt{a_n}\sqrt{k+1} + \sqrt{a_n+1}\sqrt{k})\\ =&\sqrt{a_{n+1}+1} - \sqrt{a_{n+1}}.\end{aligned}$$

By induction, one obtains

$$\sqrt{a_n+1} - \sqrt{a_n} = (\sqrt{k+1} - \sqrt{k})^n.$$

Similarly

$$\sqrt{a_n+1} + \sqrt{a_n} = (\sqrt{k+1} + \sqrt{k})^n.$$

It follows that

$$\sqrt{a_n} = \frac{1}{2}\left[(\sqrt{k+1} + \sqrt{k})^n - (\sqrt{k+1} - \sqrt{k})^n\right],$$

from which the conclusion follows.

Problem 4 (Iberoamerican Olympiad, 1987) *Let $m, n, r > 0$ be integers with*

$$1 + m + n\sqrt{3} = (2 + \sqrt{3})^{2r-1}.$$

Prove that m is a perfect square.

Solution Taking the conjugate we get

$$1 + m - n\sqrt{3} = (2 - \sqrt{3})^{2r-1},$$

hence

$$1 + m = \frac{1}{2}\left[(2+\sqrt{3})^{2r-1} + (2-\sqrt{3})^{2r-1}\right].$$

Consider the sequence $(x_n)_{n\geq 1}$ defined by $x_1 = 2$, $x_2 = 5$, and

$$x_{n+2} = 4x_{n+1} - x_n, \quad n = 0, 1, \ldots.$$

The characteristic equation is $t^2 - 4t - 1 = 0$, having the roots $t_1 = 2 + \sqrt{3}$ and $t_2 = 2 - \sqrt{3}$. By the Binet-type formula, the general term is given by

$$x_n = c_1(2+\sqrt{3})^n + c_2(2-\sqrt{3})^n, \quad n = 1, 2, \ldots,$$

where c_1 and c_2 are given by the system

$$\begin{cases} c_1(2+\sqrt{3}) + c_2(2-\sqrt{3}) = 2 \\ c_1(2+\sqrt{3})^2 + c_2(2-\sqrt{3})^2 = 5. \end{cases}$$

We obtain the general formula

$$x_n = \frac{1}{1+\sqrt{3}}(2+\sqrt{3})^n + \frac{1}{1-\sqrt{3}}(2-\sqrt{3})^n, \quad n = 1, 2, \ldots.$$

Using this formula, one may notice that

$$x_r^2 = \frac{(2+\sqrt{3})^{2r}}{4+2\sqrt{3}} + \frac{(2-\sqrt{3})^{2r}}{4-2\sqrt{3}} - 1 = \frac{1}{2}\left[(2+\sqrt{3})^{2r-1} + (2-\sqrt{3})^{2r-1}\right] - 1 = m,$$

and the conclusion follows for all n, since x_n is an integer.

Problem 5 (Kürschak Competition, 1988) *Let k be a positive integer. Define the sequence $(x_n)_{n\geq 1}$ by $x_1 = k$ and*

$$x_{n+1} = kx_n + \sqrt{(k^2-1)(x_n^2-1)}, \quad n = 1, 2, \ldots.$$

Prove that all x_n are positive integers.

Solution Proceed as in some problems above. One can prove that the following recurrence relation is satisfied

$$x_{n+1} = 2kx_n - x_{n-1}, \quad n = 2, 3, \ldots,$$

where $x_1 = k$ and $x_2 = 2k^2 - 1$.

Problem 6 *Let the sequence $(a_n)_{n\geq 1}$ given by $a_1 = 1$, $a_2 = 2$, $a_3 = 24$ and*

$$a_n = \frac{6a_{n-1}^2 a_{n-3} - 8a_{n-1}a_{n-2}^2}{a_{n-2}a_{n-3}}, \quad n = 4, 5, \ldots.$$

Show that for all n, a_n is an integer divisible by n.

Solution We have

$$\frac{a_n}{a_{n-1}} = 6\frac{a_{n-1}}{a_{n-2}} - 8\frac{a_{n-2}}{a_{n-3}}.$$

Denoting $b_n = a_n/a_{n-1}$, one obtains $b_2 = 2$ and $b_3 = 12$, while the general term of the sequence is given by $b_n = 2^{n-1}(2^{n-1} - 1)$. Simple computations show that

$$a_n = 2^{\frac{n(n-1)}{2}} \prod_{i=1}^{n} (2^i - 1).$$

To prove that a_n is a multiple of n, let $n = 2^k m$, where m is odd.

Since $k \leq n \leq n(n-1)/2$, and there exists $i \leq m - 1$ such that m is a divisor of $2^i - 1$ (for example, $i = \varphi(m)$, where φ denotes Euler's function). This confirms that a_n is a multiple of n.

Problem 7 (Balkan Olympiad, 1986) *Let a, b, c be positive real numbers. The sequence $(a_n)_{n\geq 1}$ is defined by $a_1 = a$, $a_2 = b$ and*

$$a_{n+1} = \frac{a_n^2 + c}{a_{n-1}}, \quad n = 2, 3, \ldots.$$

Prove that the terms of the sequence are all positive integers if and only if a, b and $\dfrac{a^2 + b^2 + c}{ab}$ are positive integers.

Solution Clearly, all the sequence terms are positive numbers. Writing the recurrence relation as

$$a_{n+1}a_{n-1} = a_n^2 + c,$$

and replacing n by $n - 1$, one obtains

$$a_n a_{n-2} = a_{n-1}^2 + c.$$

By subtracting the two equations we get

$$a_{n-1}(a_{n+1} + a_{n-1}) = a_n(a_n + a_{n-2}).$$

Therefore

$$\frac{a_{n+1}+a_{n-1}}{a_n}=\frac{a_n+a_{n-2}}{a_{n-1}}, \quad n=3,4,\ldots,$$

from where we deduce that the sequence $b_n = \frac{a_{n+1}+a_{n-1}}{a_n}$ is constant, which has the value $b_n = k$, for all $n \geq 2$. It follows that the sequence $(a_n)_{n\geq 1}$ satisfies the recurrence equation

$$a_{n+1}=ka_n-a_{n-1}, \quad n=2,3,\ldots,$$

and since $a_3 = \frac{b^2+c}{a} = kb-a$, we deduce that

$$k=\frac{a^2+b^2+c}{ab}.$$

Since a, b, k are positive integers, it follows by induction that a_n is a positive integer for all $n \geq 1$.

Conversely, suppose that a_n is a positive integer for all $n \geq 1$. Then a and b are positive integers, and $k = \frac{a_3+a}{b}$ is a rational number. Let $k = \frac{p}{q}$, where p and q a relatively prime positive integers. We want to show that $q = 1$. Suppose that $q > 1$. By the recurrence relation we obtain

$$q(a_{n+1}+a_{n-1})=pa_n,$$

and hence q divides a_n for all $n \geq 2$. We prove by induction on s that q^s divides a_n for all $n \geq s+1$. This is true for $s = 1$. Suppose that q^{s-1} divides a_n for all $n \geq s$. We have

$$a_{n+2}=\frac{p}{q}a_{n+1}-a_n,$$

which is equivalent to

$$\frac{a_{n+2}}{q^{s-1}}=p\frac{a_{n+1}}{q^s}-\frac{a_n}{q^{s-1}}.$$

If $n \geq s$, then q^{s-1} divides a_n and a_{n+2}, hence q^s divides a_{n+1}. It follows that q^s divides a_n for all $n \geq s+1$. Finally, we have

$$a_{s+2}=\frac{a_{s+1}^2+c}{a_s},$$

which implies that c is divisible by $q^{2(s-1)}$, for all $s \geq 1$. Because $c > 0$, this is a contradiction.

Problem 8 (IMO Shortlist, 1984) *Let c be a positive integer. The sequence $(f_n)_{n\geq 1}$ is defined as follows:*

$$f_1 = 1, \quad f_2 = c, \quad f_{n+1} = 2f_n - f_{n-1} + 2, \quad n = 2, 3, \ldots.$$

Show that for each $k \in \mathbb{N}$ there exists $r \in \mathbb{N}$ such that $f_k f_{k+1} = f_r$.

Solution From the given recurrence relation we infer that $f_{n+1} - f_n = f_n - f_{n-1} + 2$. Consequently, $f_{n+1} - f_n = f_2 - f_1 + 2(n-1) = c - 1 + 2(n-1)$. Summing up for $n = 1, 2, \ldots, k-1$ yields the explicit formula

$$f_k = f_1 + (k-1)(c-1) + (k-1)(k-2) = k^2 - bk - b,$$

where $b = c - 4$. Simple computations show that

$$f_k f_{k+1} = k^4 + 2(b+1)k^3 + (b^2 + b + 1)k^2 + (b^2 + b)k - b.$$

We are looking for an r for which the last expression equals f_r. Considering $r = k^2 + pk + q$ we determine by direct calculations that $p = b + 1$, $q = -b$, and $r = k^2 + (b+1)k - b = f_k + k$. Hence, it results that $f_k f_{k+1} = f_{f_k + k}$.

Problem 9 (IMO Shortlist, 1988) *An integer sequence is defined by*

$$a_n = 2a_{n-1} + a_{n-2} \quad (n > 1), \quad a_0 = 0, \quad a_1 = 1.$$

Show that 2^k divides a_n if and only if 2^k divides n.

Solution Using a Binet-type formula, one can easily check that the solution of the recurrence equation is

$$a_n = \frac{1}{2\sqrt{2}}\left[(1+\sqrt{2})^n - (1-\sqrt{2})^n\right] = \binom{n}{1} + 2\binom{n}{3} + 2^2\binom{n}{5} + \cdots.$$

Let $n = 2^k m$, with m odd. Then for $p > 0$ the summand

$$2^p\binom{n}{2p+1} = 2^{k+p} m \frac{(n-1)\cdots(n-2p)}{(2p+1)!} = 2^{k+p}\frac{m}{2p+1}\binom{n-1}{2p},$$

which is divisible by 2^{k+p}, since the term $2p + 1$ at the denominator is odd. It follows that

$$a_n = n + \sum_{p>0} 2^p \binom{n}{2p+1} = 2^k m + 2^{k+1} N,$$

for some integer N, so that a_n is exactly divisible by 2^k.

Problem 10 (IMO Shortlist, 1988) *The integer sequence $(a_n)_{n\geq 1}$ is defined by $a_1 = 2$, $a_2 = 7$ and*

$$-\frac{1}{2} < a_{n+1} - \frac{a_n^2}{a_{n-1}} \leq \frac{1}{2}, \quad n = 2, 3, \ldots.$$

Show that a_n is odd for all $n \geq 2$.

Solution The interval $\left(-\frac{1}{2} + \frac{a_n^2}{a_{n-1}}, \frac{1}{2} + \frac{a_n^2}{a_{n-1}}\right]$ has length 1, so it contains exactly one integer. Therefore, the sequence a_n is uniquely defined by the conditions. We have $a_1 = 2$, $a_2 = 7$, $a_3 = 25$, $a_4 = 89$, $a_5 = 317$, ... It seems that these terms satisfy the recurrence relation $a_n = 3a_{n-1} + 2a_{n-2}$ for $n = 3, 4, 5$. We show that the sequence b_n defined by $b_1 = 2$, $b_2 = 7$ and the recurrence relation $b_n = 3b_{n-1} + 2b_{n-2}$ satisfies the same inequality property, from which we will deduce that $a_n = b_n$ for $n = 1, 2, \ldots$. Clearly, for $n > 2$, we have

$$b_{n+1}b_{n-1} - b_n^2 = (3b_n + 2b_{n-1})b_{n-1} - b_n(3b_{n-1} + 2b_{n-2}) = -2(b_n b_{n-2} - b_{n-1}^2).$$

This gives $b_{n+1}b_{n-1} - b_n^2 = (-2)^{n-2}$ for all $n \geq 2$. But then

$$\left| b_{n+1} - \frac{b_n^2}{b_{n-1}} \right| = \frac{2^{n-2}}{b_{n-1}} \leq \frac{1}{2},$$

since it is easily shown that $b_n > 2^{n-1}$ for all n. Now it is obvious that all the terms $a_n = b_n$ are odd for $n > 1$.

Problem 11 *Consider the integer sequence $(x_n)_{n\geq 0}$ is defined by $x_0 = x_1 = 0$ and*

$$x_{n+2} = 4^{n+2}x_{n+1} - 16^{n+1}x_n + n \cdot 2^{n^2}, \quad n = 0, 1, \ldots.$$

Show that the numbers x_{1989}, x_{1990} and x_{1991} are divisible by 13.

Solution Denoting $x_n = y_n \cdot 2^{n^2}$, the recurrence relation transforms into

$$y_{n+2} - 2y_{n+1} + y_n = n \cdot 16^{-n-1}, \quad n = 0, 1, \ldots.$$

By summing this equation for all indices from 0 to $n - 1$ we get

$$y_{n+1} - y_n = \frac{1}{15^2}\left(1 - (15n + 1)16^{-n}\right).$$

By summing up again from 0 to $n - 1$ we obtain

$$y_n = \frac{1}{15^3}\left(15n - 32(15n + 2)16^{-n+1}\right),$$

hence

$$x_n = \frac{1}{15^3}\left(15n + 2 + (15n - 32)16^{n-1}\right)2^{(n-2)^2}.$$

We can now look at the values of $f(n)$ module 13, where $f(n) = 15n + 2 + (15n - 32)16^{n-1}$. First, notice that

$$f(n) \equiv 2n + 2 + (2n - 6)3^{n-1}.$$

We get $1989 = 3 \cdot 663 = 13 \cdot 153$, while for a positive integer k one has

$$3^{3k} = (26+1)^k \equiv 1(\mod 13)$$
$$3^{3k+1} = 3(26+1)^k \equiv 3(\mod 13)$$
$$3^{3k+2} = 9(26+1)^k \equiv 9(\mod 13).$$

It follows that

$$f(1989) \equiv 2 + (-6) \cdot 9 \equiv 0(\mod 13)$$
$$f(1990) \equiv 2 + 2 + (2 - 6) \cdot 1 \equiv 0(\mod 13)$$
$$f(1991) \equiv 4 + 2 + (-2) \cdot 3 \equiv 0(\mod 13).$$

Problem 12 (IMO Longlist, 1986) *Let $(a_n)_{n\geq 1}$ be the integer sequence defined by $a_1 = a_2 = 1$,*

$$a_{n+2} = 7a_{n+1} - a_n - 2, \quad n \geq 1.$$

Show that a_n is a perfect square for every n.

Solution The first few terms of the sequence are $a_3 = 4$, $a_4 = 5$, $a_5 = 169$, $a_6 = 1156$, which confirms that for $n = 1, 2, \ldots, 6$ we have $a_n = b_n^2$, where b_n satisfies the relations $b_1 = b_2 = 1$, $b_{n+2} = 3b_{n+1} - b_n$ for $n \geq 1$. We will also prove that the sequence b_n satisfies $b_{n+2}^2 = 7b_{n+1}^2 - b_n^2 - 2$, for $n \geq 1$, which will show that $a_n = b_n$. Indeed, replacing b_{n+2} in the recurrence equation, one obtains

$$7b_{n+1}^2 - b_n^2 - 2 = b_{n+2}^2 = 9b_{n+1}^2 - 6b_nb_{n+1} + b_n^2,$$

or equivalently

$$b_{n+1}^2 - 3b_nb_{n+1} + b_n^2 + 1,$$

which can be easily checked by induction, or by finding the terms b_n directly, through a Binet-type formula.

Problem 13 (Dorin Andrica and Grigore Călugăreanu, U422, MR-2017) *Let a, b be complex numbers and let $(a_n)_{n\geq 0}$ be the sequence defined by $a_0 = 2$ and $a_1 = a$ and*

$$a_n = aa_{n-1} + ba_{n-2}, \quad n \geq 2.$$

Write a_n as a polynomial in a and b.

Solution Using standard techniques for the solution of recursive equations, we find that

$$a_n = \left(\frac{a + \sqrt{a^2+b}}{2}\right)^n + \left(\frac{a - \sqrt{a^2+b}}{2}\right)^n.$$

Using Newton's binomial formula twice we find that

$$\begin{aligned} a_n &= 2\sum_{u=0}^{\lfloor \frac{n}{2} \rfloor} \binom{n}{2u} \frac{a^{n-2u}(a^2+4b)^u}{2^n} \\ &= 2\sum_{u=0}^{\lfloor \frac{n}{2} \rfloor} \binom{n}{2u} \frac{a^{n-2u}}{2^n} \sum_{v=0}^{n} \binom{u}{v} a^{2u-2v} 4^v b^v \\ &= \sum_{u=0}^{\lfloor \frac{n}{2} \rfloor} c_{n,v}\, a^{n-2v} b^v, \end{aligned}$$

where

$$c_{n,v} = \frac{1}{2^{n-2v-1}} \sum_{u=v}^{\lfloor \frac{n}{2} \rfloor} \binom{n}{2u}\binom{u}{v}.$$

This confirms that a_n can be expressed as a polynomial in a and b.

Problem 14 (Titu Andreescu, U297, MR-2014) *Let $a_0 = 0$, $a_1 = 2$ and the sequence defined recursively by $a_{n+1} = \sqrt{2 - \frac{a_{n-1}}{a_n}}$ for $n \geq 1$. Find $\lim_{n\to\infty} 2^n a_n$.*

Solution One can first show by induction on n that

$$a_n = 2\sin\frac{\pi}{2^n}.$$

Clearly, we have $2\sin\frac{\pi}{2^0} = 2\sin 0 = 0 = a_0$, and $2\sin\frac{\pi}{2^1} = 2\sin\frac{\pi}{2} = 2 = a_1$. Assuming that the identity holds for $k = 2, \ldots, n$, we obtain

$$a_{n+1} = \sqrt{2 - \frac{a_{n-1}}{a_n}} = \sqrt{2 - \frac{2\sin\frac{\pi}{2^{n-1}}}{2\sin\frac{\pi}{2^n}}}$$
$$= \sqrt{2 - 2\cos\frac{\pi}{2^n}} = \sqrt{4\sin^2\frac{\pi}{2^{n+1}}} = 2\sin\frac{\pi}{2^{n+1}},$$

which confirms the induction hypothesis. Finally

$$\lim_{n\to\infty} 2^n a_n = \lim_{n\to\infty} 2^{n+1}\sin\frac{\pi}{2^n} = \lim_{n\to\infty}\left(2\pi\cdot\frac{\sin\frac{\pi}{2^n}}{\frac{\pi}{2^n}}\right) = 2\pi.$$

Problem 15 (Răzvan Gelca, O312, MR-2014) *Find all the increasing bijections $f:(0,\infty)\to(0,\infty)$ satisfying*

$$f(f(x)) - 3f(x) + 2x = 0,$$

and for which there exists $x_0 > 0$ such that $f(x_0) = 2x_0$.

Solution Suppose that there is $\xi > 0$ such that $f(\xi) \neq 2\xi$. Then we have $f(f(\xi)) - 3f(\xi) + 2\xi = 0$ and

$$f^{n+1}(\xi)) - 3f^n(\xi) + 2f^{n-1}(\xi) = 0,$$

where $f^n(\xi) = f(f\dots f(f(x)))\dots)$ is the composition of f by itself n times.

Denoting by $a_n = f^n(\xi)$, we define a positive sequence $(a_n)_{n\in\mathbb{Z}}$, which satisfies the recurrence equation $a_{n+2} = 3a_{n+1} - 2a_n$ for $n \geq 0$, having the characteristic equation $t^2 - 3t + 2$, with roots $t_1 = 1$, $t_2 = 2$, and the explicit solution

$$a_n = f^n(\xi) = A + B\cdot 2^n, \quad n\in\mathbb{Z}, \tag{8.9}$$

where A and B are constants. Here the index is an integer, as the function f is bijective, hence has an inverse f^{-1}, and f^{-n} is also defined.

Clearly, $B > 0$, as otherwise the sequence a_n would be constant for $B = 0$ (hence not bijective), or could diverge to $-\infty$ in the limit $n\to\infty$ for $B < 0$ (hence out of range).

We now prove that $A = 0$. Indeed, if $A < 0$, then there is a value n_0 such that $f^{n_0} = A + B\cdot 2^{n_0} \leq 0$, for example, if $n_0 \to -\infty$, hence $A \geq 0$.

Assume that $A > 0$. By the condition there is $x_0 > 0$ such that $f(x_0) = 2x_0$. In fact, one can easily check by (8.9), that

$$f^n(x_0) = 2f(f^{n-1}(x_0)) = \cdots = 2^n x_0, \quad n\in\mathbb{Z},$$

hence there are infinitely many values for which $f(x) = 2x$. It also follows that there exists $x_0 > 0$ such that $f(x_0) = 2x_0$ and also $x_0 < 2A$. If $x_0 \leq A$, then there exists $n_0 \in \mathbb{Z}$ such that $x_0 < A + B\cdot 2^{n_0} < x_0 + A$. If $A < x_0 < 2A$, then

$$\left(\log_2 x_0 - \log_2 B\right) - \left(\log_2(x_0 - A) - \log_2 B\right) = \log_2 \frac{x_0}{x_0 - A} > 1,$$

and there exists $n_0 \in \mathbb{Z}$ such that $\log_2(x_0 - A) - \log_2 B < n_0 < \log_2 x_0 - \log_2 B$. Hence there is $n_0 \in \mathbb{Z}$ such that

$$x_0 < A + B \cdot 2^{n_0} < x_0 + A.$$

Denote $\alpha = A + B \cdot 2^{n_0}$. Since f is an increasing bijection, we obtain $2x_0 < f(x_0) < f(\alpha)A + B \cdot 2^{n_0+1} = 2\alpha - A$, hence $x_0 + \frac{A}{2} < \alpha$. It follows that

$$2^m x_0 = f^m(x_0) < f^m(\alpha) = 2^m \alpha - (2^m - 1) \cdot A, \quad m \in \mathbb{Z}.$$

Taking $m \to \infty$, we obtain

$$\alpha > \lim_{m\to\infty} \left(x_0 + \frac{2^m - 1}{2^m} \cdot A\right),$$

hence $\alpha \geq x_0 + A$, a contradiction. It then follows that $A = 0$ and $f(\xi) = 2\xi$, which is also false. This shows that the only function with the desired property is $f(x) = 2x$, for all $x > 0$.

Problem 16 (Albert Stadler, O271, MR-2013) *Consider the sequence $(a_n)_{n\geq 0}$ given by $a_0 = 0$, $a_1 = 2$, and $a_{n+2} = 6a_{n+1} - a_n$, for $n \geq 0$. Let $f(n)$ be the highest power of 2 that divides n. Prove that $f(a_n) = f(2n)$ for all $n \geq 0$.*

Solution The recurrence relation $a_{n+2} = 6a_{n+1} - a_n$ has the characteristic equation $t^2 - 6t + 1 = 0$, having the roots $t_{1,2} = 3 \pm 2\sqrt{2}$, and the general solution

$$a_n = A(3 + 2\sqrt{2})^n + B(3 - 2\sqrt{2})^n,$$

where constants A and B are calculated from the initial conditions, giving the exact formula

$$a_n = \frac{1}{2\sqrt{2}}\left[(3 + 2\sqrt{2})^n - (3 - 2\sqrt{2})^n\right].$$

Since $3 + 2\sqrt{2} = (1 + \sqrt{2})^2$ and $3 - 2\sqrt{2} = (1 - \sqrt{2})^2$, one obtains

$$a_n = \frac{(1 + \sqrt{2})^{2n} - (1 - \sqrt{2})^{2n}}{(1 + \sqrt{2}) - (1 - \sqrt{2})}.$$

This last form can be seen as Pell numbers of even values, that is $a_n = P_{2n}$.

From the first values of the function $f(n)$, we get $f(0) = 0$, $f(1) = 0$, $f(2) = 1$, $f(3) = 0$, $f(4) = 2$, $f(5) = 0$, and so on. The first few terms are given below:

$$0, 0, 1, 0, 2, 0, 1, 0, 3, 0, 1, 0, 2, 0, 1, 0, 4, 0, 1, 0, 2, 0, \ldots.$$

This recovers the OEIS sequence $A007814$ and confirms that $f(2n) = f(P_{2n})$.

Problem 17 (Problem 49, ME-1997) *Let $(u_n)_{n\geq 1}$ be a sequence of integers which satisfies the relations $u_1 = 29$, $u_2 = 45$, and $u_{n+2} = u_{n+1}^2 - u_n$ for $n = 1, 2, \ldots$. Show that 1996 divides infinitely many terms of this sequence.*

Solution Let U_n be the reminder of u_n when it is divided by 1996. Consider the sequence of pairs (U_n, U_{n+1}). Clearly, there are at most 1996^2 pairs. Let $(U_p, U_{p+1}) = (U_q, U_{q+1})$ be the first repetition with $p < q$. If $p > 1$, then the recurrence relation implies $(U_{p-1}, U_p) = (U_{q-1}, U_q)$, resulting in an earlier repetition. This shows that $p = 1$ and the sequence of pairs (U_n, U_{n+1}) is periodic with period $q - 1$. Since $u_3 = 1996$, we have $0 = U_3 = U_{3+k(k-1)}$ and so 1996 divides $U_{3+k(k-1)}$ for every positive integer k.

Problem 18 (Moscow Mathematical Olympiad, 1963) *Let $(u_n)_{n\geq 1}$ be a sequence satisfying $a_1 = a_2 = 1$ and $a_n = (a_{n-1}^2 + 2)/a_{n-2}$ for $n \geq 3$. Show that a_n is an integer for $n \geq 3$.*

Solution Since $a_1 = a_2 = 1$ and $a_n a_{n-2} = a_{n-1}^2 + 2$ for all integers $n \geq 3$, we have $a_n \neq 0$ and

$$a_n a_{n-2} - a_{n-1}^2 = 2 = a_{n+1} a_{n-1} - a_n^2, \quad n \geq 3.$$

From this we can deduce that

$$\frac{a_{n+1} + a_{n-1}}{a_n} = \frac{a_n + a_{n-2}}{a_{n-1}} = \cdots = \frac{a_3 + a_1}{a_2} = 4.$$

It follows that $a_n = 4a_{n-1} - a_{n-2}$ for $n \geq 3$. This shows that a_n is an odd integer for all $n \geq 1$.

Problem 19 (Ivan Borsenco, J66, MR-2007) *Let $a_0 = a_1 = 1$ and $a_{n+1} = 2a_n - a_{n-1} + 2$ for $n \geq 1$. Prove that $a_{n^2+1} = a_{n+1} a_n$ for all $n \geq 0$.*

First Solution Writing $a_{n+1} = 2a_n - a_{n-1} + 2$ for $n = 1, \ldots, m$, we have

$$\begin{aligned} a_2 + a_0 &= 2a_1 + 2 \\ a_3 + a_1 &= 2a_2 + 2 \\ \vdots &= \vdots \\ a_{m+1} + a_{m-1} &= 2a_m + 2. \end{aligned}$$

Summing up, one obtains

$$a_0 + a_1 + 2(a_2 + \cdots + a_{m-1}) + a_m + a_{m+1} = 2(a_1 + \cdots + a_m) + 2m.$$

From here we deduce that $a_{m+1} = a_m + m$, hence $a_m = m^2 - m + 1$. One can now easily check that

$$a_{n^2+1} = (n^2+1)^2 - (n^2+1) + 1 = (n^2+n+1)(n^2-n+1) = a_{n+1}a_n.$$

Second Solution Clearly, $a_{n+1} - a_n - 2n = a_n - a_{n-1} - 2(n-1)$, $n \geq 1$. It follows that the sequence $a_{n+1} - a_n - 2n = c$, where c is a constant. Also, we can deduce that $a_n = (n-1)n + cn + b$, for $n \geq 0$. From the initial conditions, we obtain $c = 0$ and $b = 1$, therefore $a_n = n^2 - n + 1$. The property can now be checked by elementary computations.

Third Solution The characteristic equation associated with the difference equation $a_{n+1} = 2a_n - a_{n-1} + 2$ is $z^2 - 2z + 1 = 0$, having 1 as a double root. For this reason, the complementary solution associated with it is $c_1 + c_2 n$, for some constants c_1 and c_2. A particular solution for the nonhomogeneous difference equation takes the form λn^2 for some constant λ. Substituting back into the equation, we obtain

$$\lambda(n+1)^2 = 2\lambda n^2 - \lambda(n-1)^2 + 2,$$

which gives $\lambda = 1$, and the general solution

$$a_n = c_1 + c_2 n + n^2.$$

To satisfy the initial conditions, we get $c_1 = 1$ and $c_2 = -1$, which gives

$$a_n = 1 - n + n^2.$$

Simple calculations confirm the desired formula.

Problem 20 (Titu Andreescu, U46, MR-2007) *Let k be a positive integer and let*

$$a_n = \left\lfloor (k + \sqrt{k^2+1})^n + \left(\frac{1}{2}\right)^n \right\rfloor, \quad n \geq 0.$$

Prove that $\sum_{n=1}^{\infty} \frac{1}{a_{n-1}a_{n+1}} = \frac{1}{8k^2}$.

Solution Consider the sequence

$$\begin{cases} b_0 = 2 \\ b_1 = 2k \\ b_n = 2k(b_{n-1}) + b_{n-2}. \end{cases}$$

From the Binet-type formula, the general term is given by

$$b_n = (k + \sqrt{k^2+1})^n + (k - \sqrt{k^2+1})^n.$$

The inequalities

$$b_n \le (k + \sqrt{k^2+1})^n + \left(\frac{1}{2}\right)^n < b_n + 1,$$

hold for all $n \ge 1$, we conclude that

$$a_n \le \left\lfloor (k + \sqrt{k^2+1})^n + \left(\frac{1}{2}\right)^n \right\rfloor = b_n, \quad n \ge 1.$$

It follows that

$$\begin{aligned}\sum_{n=1}^{\infty} \frac{1}{a_{n-1}a_{n+1}} &= \sum_{n=1}^{\infty} \frac{1}{b_{n-1}b_{n+1}} \\ &= \sum_{n=1}^{\infty} \left(\frac{1}{2k}\right)\left(\frac{1}{b_{n-1}b_n} - \frac{1}{b_n b_{n+1}}\right) \\ &= \frac{1}{2k}\sum_{n=1}^{\infty}\left(\frac{1}{b_{n-1}b_n} - \frac{1}{b_n b_{n+1}}\right) = \frac{1}{8k^2},\end{aligned}$$

which confirms the desired result.

Problem 21 (Martin Lukarevski, 4117, CM-2017) *The sequence $(x_n)_{n\ge 0}$ is given recursively by $x_0 = 0$, $x_1 = 1$, and*

$$x_{n+1} = x_n\sqrt{x_{n-1}^2+1} + x_{n-1}\sqrt{x_n^2+1}, \quad n \ge 1.$$

Find x_n.

Solution Clearly, $x_n \ge 0$ for all $n \ge 0$.

The function $f : [0,\infty) \to [0,\infty)$ defined by

$$f(x) = \sinh(x) = \frac{e^x - e^{-x}}{2},$$

is bijective. Hence, for every nonnegative integer n there is $\theta_n \in [0,\infty)$ such that $x_n = \sinh(\theta_n)$. From the given values for x_0 and x_1, one obtains $\theta_0 = 0$ and $\theta_1 = \ln(1+\sqrt{2})$. Recall that

$$\begin{aligned}\cosh^2(x)^2 &= 1 + \sinh^2(x) \\ \sinh(x+y) &= \sinh(x)\cosh(y) + \cosh(x)\sinh(y).\end{aligned}$$

Replacing $x_n = \sinh(\theta_n)$ into the original recurrence relation, we obtain

$$\sinh(\theta_{n+1}) = \sinh(\theta_n)\cosh(\theta_{n-1}) + \sinh(\theta_{n-1})\cosh(\theta_n) = \sinh(\theta_n + \theta_{n-1}).$$

Since sinh is bijective on $[0, \infty)$, it follows that $\theta_{n+1} = \theta_n + \theta_{n-1}$ for $n \geq 1$, hence $\theta_n = \theta_1 F_n$, where F_n is the nth Fibonacci number.

It follows that

$$x_n = \sinh(\ln(1+\sqrt{2})F_n) = \frac{e^{\ln(1+\sqrt{2})F_n} - e^{-\ln(1+\sqrt{2})F_n}}{2}$$
$$= \frac{(1+\sqrt{2})^{F_n} - (1-\sqrt{2})^{F_n}}{2}.$$

Problem 22 (Dorin Andrica, [4]) *Let α and β be nonnegative integers such that $\alpha^2 + 4\beta$ is not a perfect square. Define the sequence $(x_n)_{n\geq 0}$ by*

$$x_{n+2} = \alpha x_{n+1} + \beta x_n, \quad n \geq 0,$$

where x_1 and x_2 are positive integers.
Prove that there is no positive integer n_0 such that

$$x_{n_0}^2 = x_{n_0-1}x_{n_0+1}.$$

First Solution Note that all sequence terms are positive integers. Assume that there is a positive integer n_0 for which

$$x_{n_0}^2 = x_{n_0-1}x_{n_0+1}.$$

It follows that

$$\frac{x_{n_0+1}}{x_{n_0}} = \frac{x_{n_0}}{x_{n_0-1}} = t,$$

where t is rational. Since $x_{n_0+1} = \alpha x_{n_0} + \beta x_{n_0-1}$, we have

$$\frac{x_{n_0+1}}{x_{n_0}} = \alpha + \beta\frac{x_{n_0-1}}{x_{n_0}} = \alpha + \beta\left(\frac{x_{n_0}}{x_{n_0-1}}\right)^{-1},$$

or

$$t^2 - \alpha t - \beta = 0.$$

This equation has no rational roots, since the discriminant $\Delta = \alpha^2 + 4\beta$ is not a perfect square. This is a contradiction, hence the problem is solved.

Second Solution From the condition of the problem we obtain

$$\alpha x_n = x_{n+1} - \beta x_{n-1}, \quad n \geq 1.$$

Assume that there is a positive integer n_0 for which

$$x_{n_0}^2 = x_{n_0-1} x_{n_0+1}.$$

From the two relations above we deduce that

$$\left(\alpha^2 + 4\beta\right) x_{n_0}^2 = \left(x_{n_0+1} + \beta x_{n_0-1}\right)^2,$$

which is false, since $\alpha^2 + 4\beta$ is not a square. This ends the proof.

Problem 23 (Dorin Andrica, [4]) *Let x_1, x_2, α, β be real numbers and let the sequence $(x_n)_{n\geq 0}$ be given by*

$$x_{n+2} = \alpha x_{n+1} + \beta x_n, \quad n \geq 1.$$

If $x_m^2 \neq x_{m-1} x_{m+1}$ for all $m > 1$, prove that there are real numbers λ_1, λ_2 such that

$$\lambda_1 = \frac{x_n^2 - x_{n+1} x_{n-1}}{x_{n-1}^2 - x_n x_{n-2}}, \qquad \lambda_2 = \frac{x_n x_{n-1} - x_{n+1} x_{n-2}}{x_{n-1}^2 - x_n x_{n-2}},$$

for all $n > 2$.

Solution For an arbitrary integer we have

$$\begin{aligned} x_{n+1} &= \alpha x_n + \beta x_{n-1}, \\ x_n &= \alpha x_{n-1} + \beta x_{n-2}, \quad n \geq 2. \end{aligned}$$

This can be seen as a system of linear equations in the variables α and β, having the solutions

$$\alpha = \frac{x_n x_{n-1} - x_{n+1} x_{n-2}}{x_{n-1}^2 - x_n x_{n-2}}$$

$$-\beta = \frac{x_n^2 - x_{n+1} x_{n-1}}{x_{n-1}^2 - x_n x_{n-2}}.$$

Since α and β are constant, the problem is solved.

8.3 Classical Recurrent Sequences

Problem 1 (Ireland, 1999) *Show there is a positive integer in the Fibonacci sequence that is divisible by* 1000.

Solution In fact, for any natural number n, there exist infinitely many positive Fibonacci numbers divisible by n.

Consider the ordered pairs of consecutive Fibonacci numbers (F_0, F_1), (F_1, F_2), ..., taken modulo n. Because the Fibonacci sequence is infinite and there are only n^2 possible ordered pairs of integers modulo n, two such pairs (F_j, F_{j+1}) must be congruent

$$F_i \equiv F_{i+m} \text{ and } F_{i+1} \equiv F_{i+m+1} \pmod{n} \text{ for some } i \text{ and } m.$$

If $i \geq 1$, then

$$F_{i-1} \equiv F_{i+1} - F_i \equiv F_{i+m+1} - F_{i+m} \equiv F_{i+m-1} \pmod{n}.$$

Likewise

$$F_{i+2} \equiv F_{i+1} + F_i \equiv F_{i+m+1} + F_{i+m} \equiv F_{i+2+m} \pmod{n}.$$

Continuing similarly, we have

$$F_j \equiv F_{j+m} \pmod{n} \text{ for all } j \geq 0.$$

In particular

$$0 = F_0 \equiv F_m \equiv F_{2m} \equiv \dots \pmod{n},$$

so the numbers $F_m, F_{2m}, \dots$, are positive Fibonacci numbers divisible by n. Applying for $n = 1000$, we are done.

Problem 2 (Ireland, 1996) *Prove that*

(a) *The statement "$F_{n+k} - F_n$ is divisible by* 10 *for all positive integers n" is true if $k = 60$ and false for any positive integer $k < 60$;*
(b) *The statement "$F_{n+t} - F_n$ is divisible by* 100 *for all positive integers n" is true if $t = 300$ and false for any positive integer $t < 300$.*

First Solution By direct computations, the Fibonacci sequence has period 3 modulo 3, and period 20 modulo 5 (compute terms until the initial terms 0 and 1 repeat, at which time the entire sequence repeats), yielding (a). As for (b), one computes that the period modulo 4 is 6.

The period mod 25 turns out to be 100, which is awfully many terms to compute by hand, but knowing that the period must be a multiple of 20 helps, and verifying the recursion $F_{n+8} = 7F_{n+4} - F_n$ shows that the period divides 100; finally, an explicit computation shows that the period is not 20.

Second Solution By Binet's formula and the binomial theorem we have

$$2^{n-1}F_n = \sum_{k=0}^{n/2} 5^k \binom{n}{2k+1}.$$

Therefore

$$F_n \equiv 3^{n-1}\left(n + 5\binom{n}{3}\right) \pmod{25}.$$

Modulo 25, 3^{n-1} has period 20, n has period 25, and $5\binom{n}{3}$ has period 5. Therefore F_n clearly has period dividing 100. The period cannot divide 50, since this formula gives

$$F_{n+50} \equiv 3^{10}F_n \equiv -F_n \pmod{25},$$

and the period cannot divide 20 since it gives

$$F_{n+20} \equiv F_n + 3^{n-1} \cdot 20 \pmod{25}.$$

Problem 3 *Let $u_n = F_n^2$, $n = 0, 1, \ldots$, where F_n are the Fibonacci numbers. Prove that the sequence $(u_n)_{n\geq 0}$ satisfies a linear recurrence relation of order 3.*

Solution By Binet's formula, we have

$$F_n = \frac{1}{\sqrt{5}}(a^n - b^n),$$

where $a = \frac{1+\sqrt{5}}{2}$, $b = \frac{1+\sqrt{5}}{2}$. We have

$$\begin{aligned} u_n = F_n^2 &= \frac{1}{5}(a^n - b^n)^2 = \frac{1}{5}(a^{2n} - 2a^n b^n + b^{2n}) \\ &= \frac{1}{5}\left[(a^2)^n - 2(-1)^n + (b^2)^n\right], \quad n = 0, 1, \ldots, \end{aligned}$$

since $ab = -1$. It becomes clear that the sequence $(u_n)_{n\geq 0}$ satisfies a third-order linear recurrence relation whose characteristic equation is a cubic with the roots $t_1 = -1$, $t_2 = a^2 = \frac{3+\sqrt{5}}{2}$, $t_3 = b^2 = \frac{3-\sqrt{5}}{2}$, and has the explicit form $t^3 - 2t^2 - 2t + 1 = 0$. Finally, $(u_n)_{n\geq 0}$ is given by $u_0 = 0$, $u_1 = 1$, $u_2 = 1$, satisfying the recurrence relation

$$S = \sum_{k=1}^{\infty} \tan^{-1}\left(\frac{L_{k+1}}{L_k L_{k+2}}\right) \tan^{-1}\left(\frac{1}{L_{k+1}}\right)$$

$$= \sum_{k=1}^{\infty} \left[\tan^{-1}\left(\frac{1}{L_k}\right) \tan^{-1}\left(\frac{1}{L_{k+1}}\right) - \tan^{-1}\left(\frac{1}{L_{k+1}}\right) \tan^{-1}\left(\frac{1}{L_{k+2}}\right)\right]$$

$$= \tan^{-1}\left(\frac{1}{L_1}\right) \tan^{-1}\left(\frac{1}{L_2}\right) = \frac{\pi}{4} \tan^{-1}\left(\frac{1}{3}\right).$$

Problem 7 (Tarit Goswami, U447, MR-2018) *Let F_n be the nth Fibonacci numbers. Prove that*

$$\sum_{k=1}^{n} \binom{n}{k} F_p^k F_{p-1}^{n-k} F_k = F_{pn}.$$

Solution Let $a = \frac{1+\sqrt{5}}{2}$. For a positive integer m, the following identity holds:

$$a^m = F_{m-1} + aF_m.$$

It follows that

$$aF_{pn} + F_{pn-1} = a^{pn} = (a^p)^n = (aF_p + F_{p-1})^n$$

$$= \sum_{k=1}^{n} \binom{n}{k} (aF_p)^k F_{p-1}^{n-k} = \sum_{k=1}^{n} \binom{n}{k} F_p^k F_{p-1}^{n-k} (aF_k + F_{k-1})$$

$$= a\left[\sum_{k=1}^{n} \binom{n}{k} F_p^k F_{p-1}^{n-k} F_k\right] + \sum_{k=1}^{n} \binom{n}{k} F_p^k F_{p-1}^{n-k} F_{k-1}.$$

Finally, we obtain that

$$F_{pn} = \sum_{k=1}^{n} \binom{n}{k} F_p^k F_{p-1}^{n-k} F_k.$$

Problem 8 (Roberto Bosch Cabrera, J254, MR-2013) *Solve the equation*

$$F_{a_1} + F_{a_2} + \cdots + F_{a_k} = F_{a_1+a_2+\cdots+a_k},$$

where F_i is the ith Fibonacci number.

Solution If $k = 1$ the equation has infinitely many solutions. Let $k \geq 2$ and suppose that $1 \leq a_1 \leq a_2 \leq \cdots \leq a_k$.

For $k > 5$, notice that we have no solution. Indeed, one always has $a_1 + a_2 + \cdots + a_k \geq k$. If $a_k = 1$, then $a_1 = a_2 = \cdots = a_k = 1$, but $F_k > k$, for $k > 5$. If $a_k \geq 2$, then we have two situations.

If $a_{k-1} = 1$, then $a_1 = \cdots = a_{k-1} = 1$ and

$$\begin{aligned} F_{a_1+a_2+\cdots+a_k} &= F_{a_1+a_2+\cdots+a_k-1} + F_{a_1+a_2+\cdots+a_k-2} \\ &> F_{a_k} + F_{a_1+a_2+\cdots+a_{k-1}} = F_{a_k} + F_{k-1} \\ &\geq F_{a_k} + k - 1 \\ &= F_{a_k} + F_{a_{k-1}} + \cdots + F_{a_1}. \end{aligned}$$

If $a_{k-1} \geq 2$, then

$$\begin{aligned} F_{a_1+a_2+\cdots+a_k} &> F_{a_k} + F_{a_1+a_2+\cdots+a_{k-1}} \\ &\geq F_{a_k} + F_{a_{k-1}} + F_{a_1+a_2+\cdots+a_{k-2}} \\ &\vdots \\ &\geq F_{a_k} + F_{a_{k-1}} + \cdots + F_{a_1} - 1. \end{aligned}$$

This shows that the solutions must satisfy $k \leq 5$.

Case 1. $k = 2$ If $a_1 + a_2 \leq 4$, one can see that $(1, 2)$ and $(1, 3)$ are solutions. Now assume that $a_1 + a_2 \geq 5$. If $a_1 = 1$, then $a_2 \geq 4$, and

$$F_{a_1+a_2} = F_{a_1+a_2-1} + F_{a_1+a_2-2} = F_{a_2} + F_{a_1+a_2-1} > F_{a_2} + F_{a_1}.$$

If $a_1 > 1$, then $a_2 \geq 2$, and

$$F_{a_1+a_2} = F_{a_1+a_2-1} + F_{a_1+a_2-2} > F_{a_2} + F_{a_1},$$

hence the equation has no solution if $a_1 + a_2 \geq 5$.

Case 2. $k = 3$ There are no solutions when $a_3 = 1$. If $a_3 = 2$, one can check the solution $(1, 1, 2)$. It is easy to see that there are no solutions for $a_1 = a_2 = 1$ and $a_3 > 2$. We now prove that there are also no solutions for $2 \leq a_2 \leq a_3$. Indeed, in this case

$$F_{a_1+a_2+a_3} = F_{a_1+a_2+a_3-1} + F_{a_1+a_2+a_3-2} > F_{a_3} + F_{a_1+a_2} \geq F_{a_3} + F_{a_2} + F_{a_1}.$$

Case 3. $k = 4$ There are no solutions when $a_4 = 1$. If $a_3 = 1$ there are also no solutions, while for $a_3 \geq 2$, similar to the previous case we deduce that there are no solutions, since

$$F_{a_1+a_2+a_3+a_4} => F_{a_4} + F_{a_1+a_2+a_3} \geq F_{a_4} + F_{a_3} + F_{a_2} + F_{a_1}.$$

Case 4. $k = 5$ One obtains the solution $(1, 1, 1, 1, 1)$. There are no solutions for $a_4 = 1$, and also no solutions if $a_4 \geq 2$, since

$$F_{a_1+a_2+a_3+a_4+a_5} => F_{a_5} + F_{a_1+a_2+a_3+a_4} \geq F_{a_5} + F_{a_4} + F_{a_3} + F_{a_2} + F_{a_1}.$$

Finally, the only solutions are $(1, 2)$, $(1, 3)$, $(1, 1, 2)$, and $(1, 1, 1, 1, 1)$.

Problem 9 (Cornel Ioan Vălean, 11910, AMM-2016) *Let G_k be the reciprocal of the kth Fibonacci number, for example, $G_4 = 1/3$ and $G_5 = 1/5$. Find*

$$\sum_{n=1}^{\infty} \left(\arctan G_{4n-3} + \arctan G_{4n-2} + \arctan G_{4n-1} - \arctan G_{4n}\right).$$

Solution Recall Catalan's identity for Fibonacci numbers which states that $F_n^2 - F_{n-1}F_{n+1} = (-1)^{n+1}$, and d'Ocagne's identity $F_nF_{n+1} + F_nF_{n-1} = F_{2n}$.

By Catalan's identity we have

$$\begin{aligned}\arctan \frac{1}{F_{2n}} - \arctan \frac{1}{F_{2n+2}} &= \arctan \frac{F_{2n+2} - F_{2n}}{F_{2n}F_{2n+2} + 1} \\ &= \arctan \frac{F_{2n+1}}{F_{2n+1}^2} = \arctan \frac{1}{F_{2n+1}},\end{aligned}$$

from which we can deduce that

$$\begin{aligned}&\arctan \frac{1}{F_{4n-3}} + \arctan \frac{1}{F_{4n-1}} \\ &= \arctan \frac{1}{F_{4n-4}} - \arctan \frac{1}{F_{4n-2}} + \arctan \frac{1}{F_{4n-2}} - \arctan \frac{1}{F_{4n}} \\ &= \arctan \frac{1}{F_{4n-4}} - \arctan \frac{1}{F_{4n}}.\end{aligned}$$

This equation holds for all positive integers $n \geq 1$, since in the limit we may consider $\arctan 1/0 = \pi/2$. Notice also that by d'Ocagne's identity we have

$$\begin{aligned}\arctan \frac{F_{n-1}}{F_n} - \arctan \frac{F_n}{F_{n+1}} &= \arctan \frac{F_{n+1}F_{n-1} - F_n^2}{F_nF_{n+1} + F_nF_{n-1}} \\ &= \arctan \frac{(-1)^n}{F_{2n}} = (-1)^n \arctan \frac{1}{F_{2n}}.\end{aligned}$$

By these identities it follows that

$$\sum_{n=1}^{\infty}\left(\arctan\frac{1}{F_{4n-3}}+\arctan\frac{1}{F_{4n-2}}+\arctan\frac{1}{F_{4n-1}}-\arctan\frac{1}{F_{4n}}\right)$$

$$=\sum_{n=1}^{\infty}\left(\arctan\frac{1}{F_{4n-4}}-\arctan\frac{1}{F_{4n}}\right)+\sum_{n=1}^{\infty}\left(\arctan\frac{1}{F_{4n-2}}-\arctan\frac{1}{F_{4n}}\right)$$

$$=\frac{\pi}{2}+\sum_{n=1}^{\infty}(-1)^n\arctan\frac{1}{F_{2n}}=\frac{\pi}{2}-\sum_{n=1}^{\infty}\left(\arctan\frac{F_{n-1}}{F_n}-\arctan\frac{F_n}{F_{n+1}}\right)$$

$$=\frac{\pi}{2}+\lim_{n\to\infty}\arctan\frac{F_n}{F_{n+1}}=\frac{\pi}{2}+\arctan\frac{1}{\varphi},$$

where $\varphi=\frac{1+\sqrt{5}}{2}$.

Problem 10 (Hideyuki Ohtsuka, 11978, AMM-2017) *Let F_n be the nth Fibonacci number, with $F_0=0$, $F_1=1$, and $F_n=F_{n-1}+F_{n-2}$, for $n\geq 2$. Find*

$$\sum_{n=0}^{\infty}\frac{(-1)^n}{\cosh F_n\cosh F_{n+3}}.$$

Solution We use the following identities for cosh and Fibonacci numbers

$$\cosh(\alpha+\beta)+\cosh(\alpha-\beta)=2\cosh\alpha\cosh\beta$$

$$F_{n+3}=F_{n+2}+F_{n+2},\quad F_n=F_{n+2}-F_{n+1},$$

to obtain

$$\cosh F_n+\cosh F_{n+3}=2\cosh F_{n+1}\cosh F_{n+2}.$$

Noting that

$$\lim_{n\to\infty}\frac{1}{\cosh F_n\cosh F_{n+3}}=0,$$

one can rewrite the sum as a telescopic series, as follows:

$$\sum_{n=0}^{\infty}\frac{(-1)^n}{\cosh F_n\cosh F_{n+3}}=\sum_{n=0}^{\infty}\frac{(-1)^n}{\cosh F_n+\cosh F_{n+3}}\left(\frac{1}{\cosh F_n}+\frac{1}{\cosh F_{n+3}}\right)$$

$$=\sum_{n=0}^{\infty}\frac{(-1)^n}{2\cosh F_{n+1}\cosh F_{n+2}}\left(\frac{1}{\cosh F_n}+\frac{1}{\cosh F_{n+3}}\right)$$

$$= \frac{1}{2\cosh F_0 \cosh F_1 \cosh F_2} + \sum_{n=0}^{\infty} \frac{(-1)^n + (-1)^{n+1}}{2\cosh F_{n+1} \cosh F_{n+2} \cosh F_{n+3}}$$

$$= \frac{1}{2\cosh^2 1} = \frac{2e^2}{e^4 + 2e^2 + 1}.$$

Note: Fibonacci numbers can actually be replaced with any sequence satisfying the same recurrence relation.

Problem 11 (Mircea Merca, 11736, AMM-2013) *For $n \geq 1$, let f be the symmetric polynomial in variables $x_1, \dots, x_n$ given by*

$$f(x_1, \dots, x_n) = \sum_{k=0}^{n-1} (-1)^{k+1} e_k(x_1 + x_1^2, x_2 + x_2^2, \dots, x_n + x_n^2),$$

where e_k is the kth elementary polynomial in n variables. For example, when $n = 6$, e_2 has 15 terms, each a product of two distinct variables. Also, let ξ be a primitive nth root of unity. Prove that

$$f(1, \xi, \xi^2, \dots, \xi^{n-1}) = L_n - L_0,$$

where L_k is the kth Lucas number.

Solution Define the function

$$g(t_1, \dots, t_n) = \sum_{k=0}^{n-1} (-1)^{k+1} e_k(t_1, t_2, \dots, t_n) = (-1)^n \prod_{k=1}^{n} t_k - \prod_{k=1}^{n} (1 - t_k),$$

and consider

$$p(x, y) = \prod_{k=0}^{n-1} (x - \xi^k y) = x^n - y^n.$$

Let $\alpha = \frac{1+\sqrt{5}}{2}$ and $\beta = \frac{1-\sqrt{5}}{2}$, so that $1 - x - x^2 = (1 - \alpha x)(1 - \beta x)$, where $\alpha\beta = -1$, and $L_n = \alpha^n + \beta^n$. We compute

$$f(1, \xi, \xi^2, \dots, \xi^{n-1}) = g(1 + 1, \xi + \xi^2, \dots, \xi^{n-1} + \xi^{2(n-1)})$$

$$= (-1)^n \prod_{k=0}^{n-1} (\xi^k + \xi^{2k}) - \prod_{k=0}^{n-1} (1 - \xi^k - \xi^{2k})$$

$$= \prod_{k=0}^{n-1} (-\xi)^k \prod_{k=0}^{n-1} (1 + \xi^k) - \prod_{k=0}^{n-1} (1 - \alpha\xi^k) \prod_{k=0}^{n-1} (1 - \beta\xi^k)$$

$$
\begin{aligned}
&= p(0,1)p(1,-1) - p(1,\alpha)p(1,\beta) \\
&= (-1)[1-(-1)^n] - (1-\alpha^n)(1-\beta^n) \\
&= [-1+(-1)^n] - [1-\alpha^n-\beta^n+(\alpha\beta)^n] \\
&= \alpha^n + \beta^n - 2 = L_n - L_0.
\end{aligned}
$$

Problem 12 (Oliver Knill, 11716, AMM-2013) *Let $\alpha = (\sqrt{5}-1)/2$. Let p_n and q_n be the numerator and denominator of the nth continued fraction convergent to α, denoted by p_n/q_n, where F_n is the nth Fibonacci number and $q_n = p_{n+1}$. Show that*

$$
\sqrt{5}\left(\alpha - \frac{p_n}{q_n}\right) = \sum_{k=0}^{\infty} \frac{(-1)^{(n+1)(k+1)}C_k}{q_n^{2k+2}5^k},
$$

where C_k denotes the kth Catalan number, given by $C_k = \frac{(2k)!}{k!(k+1)!}$.

Solution We shall use the notations $p_n = F_{n-1}$ and $q_n = p_{n+1} = F_n$. Consider the generating function for the Catalan numbers (see Example 4.12)

$$
\sum_{k=0}^{\infty} C_k x^k = \frac{1-\sqrt{1-4x}}{2x}.
$$

By the ratio test, the power series has the radius of convergence $1/4$. Setting $x = (-1)^{n+1}/(5F_n^2)$ in the generating function, we obtain

$$
\sum_{k=0}^{\infty} \frac{(-1)^{(n+1)k}C_k}{5^k F_n^{2k}} = \frac{\sqrt{5}F_n}{(-1)^{n+1}} \cdot \frac{\sqrt{5}F_n - \sqrt{5F_n^2 - 4(-1)^{n+1}}}{2}.
$$

It follows that

$$
\begin{aligned}
\sum_{k=0}^{\infty} \frac{(-1)^{(n+1)(k+1)}C_k}{5^k q_n^{2k+2}} &= \frac{(-1)^{n+1}}{5F_n^2} \sum_{k=0}^{\infty} \frac{(-1)^{(n+1)k}C_k}{5^k F_n^{2k}} \\
&= \frac{\sqrt{5}}{F_n} \cdot \frac{\sqrt{5}F_n - \sqrt{5F_n^2 + 4(-1)^n}}{2}.
\end{aligned}
$$

By Catalan's identity for Fibonacci numbers $F_n^2 + (-1)^n = F_{n-1}F_{n+1}$, we can rewrite the terms under the square root as

$$
\begin{aligned}
5F_n^2 + 4(-1)^n &= F_n^2 + 4(F_n^2 + (-1)^n) = (F_{n+1} - F_{n-1})^2 + 4F_{n-1}F_{n+1} \\
&= (F_{n+1} + F_{n-1})^2 = (F_n + 2F_{n-1})^2,
\end{aligned}
$$

and obtain

$$\sum_{k=0}^{\infty}\frac{(-1)^{(n+1)(k+1)}C_k}{5^k q_n^{2k+2}}=\frac{\sqrt{5}}{F_n}\cdot\frac{\sqrt{5}F_n-(F_n+2F_{n-1})}{2}=\sqrt{5}\left(\frac{\sqrt{5}-1}{2}-\frac{F_{n-1}}{F_n}\right),$$

which ends the proof.

Problem 13 (Dorin Andrica, O64, MR-2015) *Let F_n be the nth Fibonacci number. Prove that for all $n \geq 4$, $F_n + 1$ is not a prime.*

Solution Let us first recall that

$$F_n^2 = F_{n-1}F_{n+1} + (-1)^{n+1} \qquad \text{(Cassini)}$$

$$F_n^4 = F_{n-2}F_{n-1}F_{n+1}F_{n+2} + 1 \qquad \text{(Gelin-Cesàro).}$$

By this identity, one obtains the factorization

$$(F_n^2+1)(F_n-1)(F_n+1) = F_{n-2}F_{n-1}F_{n+1}F_{n+2}.$$

Then if $F_n + 1 = p$ with p a prime, we get $p \mid F_{n+2}$.

This follows from $F_{n-2}, F_{n-1} < p$, while we have $F_{n+1} < 2F_n = 2p$ and $\gcd(F_n, F_{n+1}) = 1$. Notice that $2F_n < F_{n+2} < 3F_n$, hence $2p < F_{n+2} < 3p$, for $n \geq 4$, a contradiction. The conclusion follows.

Problem 14 (Gabriel Alexander Reyes, O47, MR-2007) *Consider the Fibonacci sequence $F_0 = 0$, $F_1 = 1$, and $F_{n+1} = F_n + F_{n-1}$, for $n \geq 1$. Prove that*

$$\sum_{i=0}^{n}\frac{(-1)^{n-i}F_i}{n+1-i}\binom{n}{i}=\begin{cases}\frac{2F_{n+1}}{n+1} & \text{if } n \text{ is odd,}\\ 0 & \text{if } n \text{ is even.}\end{cases}$$

Solution Notice that

$$\frac{1}{n+1-i}\binom{n}{i}=\frac{1}{n+1}\binom{n+1}{i},$$

therefore

$$\sum_{i=0}^{n}\frac{(-1)^{n-i}F_i}{n+1-i}\binom{n}{i}=\sum_{i=0}^{n}\frac{(-1)^{n-i}F_i}{n+1}\binom{n+1}{i}=\frac{1}{n+1}\sum_{i=0}^{n}(-1)^{n-i}F_i\binom{n+1}{i}.$$

We just have to prove that

$$\sum_{i=0}^{n}(-1)^{n-i}F_i\binom{n+1}{i}=\begin{cases}2F_{n+1} & \text{if } n \text{ is odd,}\\ 0 & \text{if } n \text{ is even.}\end{cases}$$

To this end, one can use the classical formula $F_i = \frac{\varphi^i-(1-\varphi)^i}{\sqrt{5}}$, where $\varphi = \frac{1+\sqrt{5}}{2}$. We have

$$\sum_{i=0}^{n}(-1)^{n-i}F_i\binom{n+1}{i} = \sum_{i=0}^{n}(-1)^{n-i}\frac{\varphi^i-(1-\varphi)^i}{\sqrt{5}}\binom{n+1}{i}$$
$$= \frac{1}{\sqrt{5}}\left[\sum_{i=0}^{n}\binom{n+1}{i}(-1)^{n-i}\varphi^i - \sum_{i=0}^{n}\binom{n+1}{i}(-1)^{n-i}(1-\varphi)^i\right]$$
$$= \frac{1}{\sqrt{5}}\left[-(-1+\varphi)^{n+1}+\varphi^{n+1}-\left(-(-1+(1-\varphi))^{n+1}+(1-\varphi)^{n+1}\right)\right]$$
$$= \frac{1}{\sqrt{5}}\left[-(-1)^{n+1}(1-\varphi)^{n+1}+\varphi^{n+1}+(-1)^{n+1}\varphi^{n+1}-(1-\varphi)^{n+1}\right]$$
$$= \left[1+(-1)^{n+1}\right]F_{n+1}.$$

This is clearly 0 if n is even and $2F_{n+1}$ if n is odd.

Problem 15 *Let $\sigma_n = \sum_{k=1}^{n}F_k^2$, where F_k is the kth Fibonacci number with $F_0 = 0$ and $F_1 = 1$. Find the sum of the series $\sum_{n\geq 1}\frac{(-1)^{n+1}}{\sigma_n}$.*

Solution The proof requires the general formula of the Fibonacci numbers, and the Cassini relation

$$F_n = \frac{1}{\sqrt{5}}\left[\left(\frac{1+\sqrt{5}}{2}\right)^n - \left(\frac{1-\sqrt{5}}{2}\right)^n\right]$$
$$F_n^2 = F_{n-1}F_{n+1}+(-1)^{n+1}, \quad n \geq 0.$$

By the definition of F_k, it follows that

$$F_{k+1}F_k = F_k^2 + F_{k-1}F_k, \quad k \geq 0.$$

Summing the above relations for $k = 1, \ldots, n$ we obtain

$$\sigma_n = F_{n+1}F_n, \quad n \geq 0.$$

From the above formula and the Cassini relation, we have

$$S_n = \sum_{k=1}^{n}\frac{(-1)^{k+1}}{\sigma_k} = \sum_{k=1}^{n}\frac{(-1)^{k+1}}{F_kF_{k+1}} = 1 - \sum_{k=2}^{n}\frac{F_{k-1}F_{k+1}-F_k^2}{F_kF_{k+1}}$$
$$= 1 - \sum_{k=2}^{n}\left(\frac{F_{k-1}}{F_k} - \frac{F_k}{F_{k+1}}\right) = \frac{F_n}{F_{n+1}}.$$

From the exact formula, the sum of the series is given by

$$\sum_{n\geq 1}\frac{(-1)^{n+1}}{\sigma_n}=\lim_{n\to\infty}\frac{F_n}{F_{n+1}}=\frac{2}{1+\sqrt{5}}.$$

Problem 16 *Find* $\sum_{n=1}^{\infty}\arctan\frac{1}{F_{2n+1}}$, *where* $(F_n)_{n\geq 0}$ *is the Fibonacci sequence.*

Solution From the Cassini relation $F_n^2=F_{n-1}F_{n+1}+(-1)^{n+1}$ we deduce

$$\arctan\frac{1}{F_{2n}}-\arctan\frac{1}{F_{2n+2}}=\arctan\frac{F_{2n+2}-F_{2n}}{F_{2n}F_{2n+2}+1}$$
$$=\arctan\frac{F_{2n+1}}{F_{2n+1}^2}=\arctan\frac{1}{F_{2n+1}}.$$

Summing these relations from $n=1$ we get

$$\sum_{k=1}^{n}\arctan\frac{1}{F_{2k+1}}=\arctan\frac{1}{F_2}-\arctan\frac{1}{F_{2n+2}}.$$

By taking limits it follows that

$$\sum_{n=1}^{\infty}\arctan\frac{1}{F_{2n+1}}=\arctan\frac{1}{F_2}=\arctan 1=\frac{\pi}{4}.$$

Problem 17 *Let $n\geq 1$ be an integer. Show that 2^{n-1} divides the number*

$$\sum_{0\leq k<\frac{n}{2}}\binom{n}{2k+1}5^k. \tag{8.10}$$

Solution Using the general formula of the Fibonacci sequence terms

$$F_n=\frac{1}{\sqrt{5}}\left[\left(\frac{1+\sqrt{5}}{2}\right)^n-\left(\frac{1-\sqrt{5}}{2}\right)^n\right],$$

and Newton's binomial formula, we have

$$F_n=\frac{1}{2^{n-1}}\left[\binom{n}{1}+\binom{n}{3}5+\cdots+\binom{n}{t}5^{\frac{t-1}{2}}\right],\quad t=2\left\lfloor\frac{n-1}{2}\right\rfloor+1.$$

We obtain

$$2^{n-1}F_n = \sum_{0 \le k < \frac{n}{2}} \binom{n}{2k+1} 5^k,$$

and the problem is solved.

Problem 18 (Michel Battaile, 3924, CM-2014) *Let $(F_k)_{k\ge 0}$ be the Fibonacci sequence defined by $F_0 = 0$, $F_1 = 1$, and $F_{k+1} = F_k + F_{k-1}$, for $k \ge 1$. If m and n are positive integers with m odd and n not a multiple of 3, prove that $5F_m^2 - 3$ divides $5F_{mn}^2 + 3(-1)^n$.*

First Solution Let $\alpha = \frac{1+\sqrt{5}}{2}$ and $\beta = \frac{1-\sqrt{5}}{2}$. Denoting by $(L_n)_{n\ge 0}$ the Lucas sequence, the Binet-type formula gives $F_n = \frac{\alpha^n - \beta^n}{\alpha - \beta}$ and $L_n = \alpha^n + \beta^n$ for all $n \ge 0$. Let m and k be nonnegative integers with m odd. Since $\alpha\beta = -1$, we obtain

$$5F_k^2 + 2(-1)^k = 5\left(\frac{\alpha^k - \beta^k}{\alpha - \beta}\right)^2 + 2(-1)^k = \alpha^{2k} + \beta^{2k} = L_{2k}, \tag{8.11}$$

and furthermore

$$L_{2m}L_{2m(k+1)} = L_{2mk} + L_{2m(k+2)}. \tag{8.12}$$

Let $P(n)$ denote the assertion that for every odd natural number m

$$L_{2mn} = \begin{cases} (-1)^{n+1} \pmod{L_{2m} - 1} & \text{if } n \not\equiv 0 \pmod 3, \\ 2(-1)^n \pmod{L_{2m} - 1} & \text{if } n \equiv 0 \pmod 3). \end{cases}$$

We show by induction on n that $P(n)$ is valid for $n \ge 0$. Since we clearly have $L_0 = 2 \equiv 2(\text{mod } L_{2m} - 1)$ and $L_{2m} \equiv 1(\text{mod } L_{2m} - 1)$, $P(0)$ and $P(1)$ hold.

Assume that $P(k)$ holds for $k = 2, \ldots, n-1$. We have three cases.

Case 1 $n \equiv 0(\text{mod } 3)$. By (8.12) and the induction hypothesis we have

$$\begin{aligned} L_{2mn} = L_{2m}L_{2m(n-1)} - L_{2m(n-2)} &\equiv 1 \cdot (-1)^n - (-1)^{n-1} \\ &\equiv 2(-1)^n (\text{mod } L_{2m} - 1). \end{aligned}$$

Case 2 $n \equiv 1(\text{mod } 3)$. Here

$$L_{2mn} \equiv 1 \cdot 2(-1)^{n-1} - (-1)^{n-1} \equiv (-1)^{n+1} (\text{mod } L_{2m} - 1).$$

Case 3 $n \equiv 2(\text{mod } 3)$. Here

$$L_{2mn} \equiv 1 \cdot (-1)^n - 2(-1)^{n-2} \equiv (-1)^{n+1} (\text{mod } L_{2m} - 1).$$

This confirms that $P(n)$ is valid in all three cases.

Let m and n be positive integers with m odd and n not a multiple of 3. By (8.11) we have $5F_m^2 - 3 = L_{2m} - 1$ and $5F_{mn}^2 + 3(-1)^n = L_{2mn} - (-1)^{n+1}$. From $P(n)$, we conclude that $5F_m^2 - 3$ divides $5F_{mn}^2 + 3(-1)^n$.

Second Solution For m, n positive integers with m odd, $n \not\equiv 0 \pmod 3$, let

$$Q_{mn} = \frac{5F_{mn}^2 + 3(-1)^n}{5F_m^2 - 3}.$$

Let $x = \frac{1+\sqrt{5}}{2}$ and $y = \frac{1-\sqrt{5}}{2}$, such that $F_t = \frac{x^t - y^t}{\sqrt{5}}$. For any odd integer m we have

$$\begin{aligned} 5F_m^2 - 3 &= (x^m - y^m)^2 - 3 = x^{2m} + y^{2m} - 2x^m y^m - 3 \\ &= x^{2m} + y^{2m} - 2(-1)^m - 3 = x^{2m} + y^{2m} - 1 \\ &= x^{2m} + x^m y^m + y^{2m}. \end{aligned}$$

Furthermore

$$\begin{aligned} 5F_{mn}^2 + 3(-1)^n &= (x^{mn} - y^{mn})^2 + 3(-1)^n \\ &= x^{2mn} + y^{2mn} - 2x^{mn} y^{mn} + 3(xy)^{mn} \\ &= x^{2mn} + x^{mn} y^{mn} + y^{2mn}. \end{aligned}$$

One can check that for any positive integer n, the polynomial $x^{2n} + x^n + 1$ is divisible by $x^2 + x + 1$ if and only if $n \not\equiv 0 \pmod 3$. This can be obtained from the Remainder Theorem. Indeed, if ω is a nonreal cubic root of unity, then $(x - \omega)$ and $(x - \omega^2)$ are both factors of $x^{2n} + x^n + 1$. In the same way, the polynomial $x^{2n} + x^n y^n + y^{2n}$ is divisible by $x^2 + xy + y^2$ for $n \not\equiv 0 \pmod 3$, as it has factors $(x - \omega y)$ and $(x - \omega^2 y)$ as a polynomial in x. Consequently, Q_{mn} is a polynomial in x and y with integer coefficients. Observe that both its numerator and denominator are symmetric polynomials in x and y, hence Q_{mn} is also symmetric in x and y. Also, notice that Q_{mn} can be expressed as a polynomial with integer coefficients in terms of $(x + y)$ and xy. Since $x + y = 1$ and $xy = -1$, it follows that Q_{mn} is an integer, as desired.

Also, when $n \equiv 0 \pmod 3$, $5F_m^2 - 3$ divides F_{mn}^2.

Problem 19 *A positive integer is called Fib-unique if the way to represent it as a sum of several distinct Fibonacci numbers is unique. For example, 13 is not Fib-unique, as* $13 = 13 = 8 + 5 = 8 + 3 + 2$. *Find all Fib-unique numbers.*

Solution According to Theorem 2.22, an integer n can be written as

$$n = f_{j_1} + f_{j_2} + \cdots + f_{j_k}, \quad j_1 < j_2 < \cdots < j_k.$$

We need to find all numbers n for which this representation is unique.

First, notice that if $i_1 \geq 3$, then we can replace f_{i_1} by $f_{i_1-1} + f_{i_1-2}$ and get another representation of n. So, if n is Fib-unique, then $i_1 \in \{1, 2\}$.

Second, if there is some t such that $i_{t+1} - i_t \geq 3$, then we can replace $f_{i_{t+1}}$ by $f_{i_{t+1}-1} + f_{i_{t+1}-2}$ and get another representation of n. Hence, another necessary condition for n to be Fin-unique is $i_{t+1} - i_t \leq 2$, for every $0 \leq t \leq k-1$.

Third, if there is t such that $i_{t+1} - i_t = 1$, then we choose the t which is largest. We can replace $f_{i_t} + f_{i_{t+1}}$ by $f_{i_{t+1}+1}$, and obtain another representation for n, hence we have $i_{t+1} - i_t = 2$, for every $0 \leq t \leq k-1$.

So, every Fib-unique number n has one of the forms

1. $n = f_1 + f_3 + \cdots + f_{2k-1} = f_{2k} - 1$;
2. $n = f_2 + f_4 + \cdots + f_{2k} = f_{2k+2} - 1$.

In conclusion, the only Fib-unique numbers are f_1, f_2, and $f_k - 1$ with $k \geq 3$.

Problem 20 *Consider the function*

$$f(x) = (x - F_2)(x - F_3) \cdots (x - F_{3031}),$$

where $(F_n)_{n \geq 0}$ is the Fibonacci sequence, defined by $F_0 = 0$, $F_1 = 1$, and the recurrence relation $F_{n+2} = F_{n+1} + F_n$, for $n \geq 0$. Suppose that on the range (F_1, F_{3031}) the function $\mid f(x) \mid$ takes on the maximum value at $x = x_0$. Prove that $x_0 > 2^{2018}$.

Solution We prove that $x_0 \in (F_{3030}, F_{3031})$ by showing that for $x^* \in (F_2, F_{3030}]$, there is some $x^{**} \in (F_{3030}, F_{3031})$ for which $\mid f(x^{**}) \mid > \mid f(x^*) \mid$.

Indeed, if

$$x^* \in \{F_2, F_3, \ldots, F_{3030}\},$$

then $\mid f(x^*) \mid = 0$, hence the conclusion holds.

Suppose that $x^* \in (F_k, F_{k+1})$ for $2 \leq k \leq 3029$, and denote $m = x^* - x_k$ and let $x^{**} = F_{3031} - m$. The relation $\mid f(x^{**}) \mid > \mid f(x^*) \mid$ can be written as

$$\left| \prod_{i=2}^{3031} (x^{**} - F_i) \right| > \left| \prod_{i=2}^{3031} (x^* - F_i) \right|.$$

Each side has 3030 positive factors, and we pair one on the left with one on the right, in such a way that the value on the left is bigger, with the exception of the case $i = k$, when $\mid x^* - F_k \mid = m = \mid x^{**} - F_k \mid$.

This can be checked easily.

1. If $i = 2, 3, \ldots, k$, we have $x^{**} - F_i > x^* - F_i$.

2. If $2 \le i \le 3031 - k$, then we have $| x^{**} - F_{3031-i} | > | x^* - F_{k+i} |$. Note that we just consider the disjoint ranges, i.e., $k + i \le 3031 - i$. Otherwise, if some ranges overlap, then we remove that part, so

$$\begin{aligned} F_{k+i} - x^* &= (F_{k+1} - x^*) + (F_{k+2} - F_{k+1}) + \cdots + (F_{k+i} - F_{k+i-1}) \\ &< (x^{**} - F_{3030}) + (F_{3030} - F_{3029}) + \cdots + (F_{3031-(i-1)} - F_{3031-i}) \\ &= x^{**} - F_{3031-i}, \end{aligned}$$

which proves the statement. From here it follows that

$$\begin{aligned} x_0 > F_{3030} &= \frac{1}{\sqrt{5}}\left[\left(\frac{1+\sqrt{5}}{2}\right)^{3030} - \left(\frac{1-\sqrt{5}}{2}\right)^{3030}\right] \\ &> \frac{1}{\sqrt{5}}\left(\frac{1+\sqrt{5}}{2}\right)^{3029} > \left(\frac{1+\sqrt{5}}{2}\right)^{3027}. \end{aligned}$$

One can easily check that $\frac{1+\sqrt{5}}{2} > 2^{\frac{2}{3}}$. Substituting into the above inequality, one obtains the desired result.

Problem 21 (Titu Andreescu, IMO Shortlist, 1983) *Let $(F_n)_{n\ge 0}$ be the Fibonacci sequence, given by $F_1 = 1$, $F_2 = 1$ and $F_{n+2} = F_{n+1} + F_n$, for $n \ge 1$. Define P to be the polynomial of degree 990 which satisfies $P(k) = F_k$ for $k = 992, 993, \ldots, 1982$. Prove that $P(1983) = F_{1983} - 1$.*

Solution Denote by P_n the unique polynomial of degree n such that

$$P_n(k) = F_k, \quad k = n+2, n+3, \ldots, 2n+2. \tag{8.13}$$

We will show by induction that $P_n(2n+3) = F_{2n+3} - 1$.

Clearly, $P_0(x) = 1$ and the claim holds for $n = 0$. Suppose it holds for $k = 1, \ldots, n-1$. The polynomial

$$Q(x) = P_n(x+2) - P_n(x+1)$$

has degree at most $n - 1$. In view of (8.13) one has

$$Q(k) = P_n(k+2) - P_n(k-1) = F_{k+2} - F_{k+1} = F_k, \quad k = n+1, n+2, \ldots, 2n.$$

It follows that the polynomials Q and P_{n-1} agree at n distinct points, hence they are identical, i.e., $Q(x) = P_{n-1}(x)$ for all x.

This shows that $P_n(x+2) = P_n(x+1) + P_{n-1}(x)$ for all x. From the inductive hypothesis we have $P_{n-1}(2n+1) = F_{2n+1} - 1$, hence

$$P_n(2n+3) = P_n(2n+2) + P_{n-1}(2n+1) = F_{2n+2} + F_{2n+1} - 1 = F_{2n+3} - 1.$$

The desired conclusion follows by setting $n = 990$.

8.4 Higher Order Recurrent Sequences

Problem 1 (Titu Andreescu, [4]) *Let $(a_n)_{n\geq 0}$ be the sequence defined by $a_0 = 0$, $a_1 = 1$, and*

$$a_{n+1} - 3a_n + a_{n-1} = 2(-1)^n, \quad n = 1, 2, \ldots.$$

Prove that a_n is a perfect square for all $n \geq 0$.

Solution Note that $a_2 = 1$, $a_3 = 4$, $a_4 = 9$, $a_5 = 25$, so $a_0 = F_0^2$, $a_1 = F_1^2$, $a_2 = F_2^2$, $a_3 = F_3^2$, $a_4 = F_4^2$, $a_5 = F_5^2$, where $(F_n)_{n\geq 0}$ is the Fibonacci sequence.

We induct on n to prove that $a_n = F_n^2$ for all $n \geq 0$. Assume that $a_k = F_k^2$ for all $k \leq n$. Hence

$$a_n = F_n^2, \quad a_{n-1} = F_{n-1}^2, \quad a_{n-2} = F_{n-2}^2. \tag{8.14}$$

From the given relation we obtain

$$a_{n+1} - 3a_n + a_{n-1} = 2(-1)^n,$$

and

$$a_n - 3a_{n-1} + a_{n-2} = 2(-1)^{n-1}, \quad n \geq 2.$$

Summing up these equalities yields

$$a_{n+1} - 2a_n - 2a_{n-1} + a_{n-2} = 0, \quad n \geq 2. \tag{8.15}$$

Using the relations (8.14) and (8.15) we obtain

$$\begin{aligned} a_{n+1} &= 2F_n^2 + 2F_{n-1}^2 - F_{n-2}^2 = (F_n + F_{n-1})^2 + (F_n - F_{n-1})^2 - F_{n-2}^2 \\ &= F_{n+1}^2 + F_{n-2}^2 - F_{n-2}^2 = F_{n+1}^2. \end{aligned}$$

Problem 2 *Define the sequence $(x_n)_{n\geq 1}$ by $x_0 = 0$, $x_1 = 1$, $x_2 = 1$, and*

$$x_{n+3} = 2x_{n+2} + 2x_{n+1} - x_n, \quad n = 0, 1, \ldots.$$

Prove that x_n is a perfect square.

First Solution The characteristic equation of $(x_n)_{n\geq 1}$ is

$$t^3 - 2t^2 - 2t - 1 = 0.$$

This is equivalent to

$$(t+1)(t^2 - 3t + 1) = 0,$$

having the distinct roots

$$t_1 = -1, \quad t_2 = \frac{3+\sqrt{5}}{2}, \quad t_3 = \frac{3-\sqrt{5}}{2}.$$

The general form of the solution is given by

$$x_n = c_1(-1)^n + c_2\left(\frac{3+\sqrt{5}}{2}\right)^n + c_3\left(\frac{3-\sqrt{5}}{2}\right)^n, \quad n = 0, 1, \ldots,$$

where the coefficients c_1, c_2, c_3 are obtained from the system

$$\begin{cases} c_1 + c_2 + c_3 = 0 \\ -c_1 + c_2\left(\frac{3+\sqrt{5}}{2}\right) + c_3\left(\frac{3-\sqrt{5}}{2}\right) = 1, \\ c_1 + c_2\left(\frac{3-\sqrt{5}}{2}\right)^2 + c_3\left(\frac{3-\sqrt{5}}{2}\right)^2 = 1. \end{cases}$$

Solving this system one obtains $c_1 = -\frac{2}{5}, c_2 = \frac{1}{5}, c_3 = \frac{1}{5}$. After some calculations, we get

$$\begin{aligned} x_n &= -\frac{2}{5}(-1)^n + \frac{1}{5}\left(\frac{3+\sqrt{5}}{2}\right)^n + \frac{1}{5}\left(\frac{3-\sqrt{5}}{2}\right)^n \\ &= \frac{1}{5}\left[\left(\frac{1+\sqrt{5}}{2}\right)^n - \left(\frac{1-\sqrt{5}}{2}\right)^n\right]^2 = F_n^2 \quad n = 0, 1, \ldots, \end{aligned}$$

where F_n is the nth Fibonacci number.

Second Solution One can also prove by induction that $x_n = F_n^2$, $n = 0, 1, \ldots$. Clearly, we have $x_0 = F_0^2, x_1 = F_1^2, x_2 = F_2^2$. Assume that $x_n = F_n^2, x_{n+1} = F_{n+1}^2$, $x_{n+2} = F_{n+2}^2$. Denoting $a = \frac{1+\sqrt{5}}{2}$, $b = \frac{1+\sqrt{5}}{2}$, we obtain that

$$\begin{aligned} x_{n+3} &= 2F_{n+2}^2 + 2F_{n+1}^2 - F_n^2 \\ &= \frac{2}{5}(a^{n+2} - b^{n+2})^2 + \frac{2}{5}(a^{n+1} - b^{n+1})^2 - \frac{1}{5}(a^n - b^n)^2 \end{aligned}$$

$$= \frac{1}{5}\left[(2a^{2n+4} + 2a^{2n+2} - a^{2n}) + (2b^{2n+4} + 2b^{2n+2} - b^{2n})\right] + \frac{2}{5}(-1)^n$$

$$= \frac{1}{5}\left[(a^{2n+6} + b^{2n+6} - 2(-1)^{n+3}\right]$$

$$= \frac{1}{5}(a^{n+3} - b^{n+3})^2 = F_{n+3}^2.$$

In the proof we have used the relations $2a^{2n+4} + 2a^{2n+2} - a^{2n} = a^{2n+6}$, $2b^{2n+4} + 2b^{2n+2} - a^{2n} = b^{2n+6}$, and $ab = -1$.

Problem 3 *Define the sequence* $(a_n)_{n\geq 0}$ *by* $a_0 = 3$, $a_1 = 1$, $a_2 = 9$, *and*

$$a_{n+3} = a_{n+2} + 4a_{n+1} - 4a_n, \quad n = 0, 1, \ldots.$$

Find a_n.

Solution The characteristic equation is

$$t^3 - t^2 - 4t + 4 = 0 = (t-1)(t^2 - 4),$$

having the distinct roots

$$t_1 = 1, \quad t_2 = 2, \quad t_3 = -2.$$

The general form of the solution is given by

$$a_n = c_1 t_1^n + c_2 t_2^n + c_3 t_3^n = c_1 + c_2 2^n + c_3(-2)^n, \quad n = 0, 1, \ldots,$$

where the coefficients c_1, c_2, c_3 are obtained from the system

$$\begin{cases} c_1 + c_2 + c_3 = 3 \\ c_1 + c_2 + c_3 = 1 \\ c_1 + 4c_2 + 4c_3 = 9, \end{cases}$$

hence $c_1 = c_2 = c_3 = 1$. It follows that

$$a_n = 1 + 2^n + (-2)^n = 1 + \left[1 + (-1)^n\right]2^n, \quad n = 0, 1, \ldots.$$

Problem 4 *Consider all the sequences* $(x_n)_{n\geq 0}$ *satisfying*

$$x_{n+3} = x_{n+2} + x_{n+1} + x_n, \quad n = 0, 1, \ldots.$$

Prove that there are real numbers α *and* t *with* $1.34 < \alpha < 1.37$, $2.1 < t < 2.2$, *and* a, b, c, *satisfying the relation*

$$x_n = a\alpha^{2n} + (b\cos nt + c\sin nt)\alpha^{-n}.$$

Solution The characteristic equation $t^3 - t^2 - t - 1 = 0$ clearly has a real root $y_1 = \alpha^2$, where $1.8 < \alpha^2 < 1.9$, hence $1.34 < \alpha < 1.37$. The other two roots satisfy the quadratic equation

$$y^2 + (\alpha^2 - 1)y + (\alpha^4 - \alpha^2 - 1) = 0.$$

The discriminant of this equation is

$$\Delta = -3\alpha^4 + 2\alpha^2 + 5 = (\alpha^2 + 1)(5 - 3\alpha^2) < 0,$$

hence y_2, y_3 are complex, so

$$y_{2,3} = r(\cos t \pm i\sin t).$$

We have

$$r^2 = y_2y_3 = \alpha^4 - \alpha^2 - 1 = \frac{1}{\alpha^2},$$

that is $r = \frac{1}{\alpha}$. From $y_2 + y_3 = 1 - \alpha^2$, it follows that $2\frac{1}{\alpha}\cos t = 1 - \alpha^2$, hence $\cos t = \frac{\alpha}{2}(1 - \alpha^2)$. Since $1.34 < \alpha < 1.37$, we get $0.536 < -\cos t < 0.6201$, hence there is a value t satisfying $2.13 < t < 2.24$.

Problem 5 *An integer sequence* $(a_n)_{n\geq 1}$ *is given by* $a_1 = 1$, $a_2 = 12$, $a_3 = 20$, *and*

$$a_{n+3} = 2a_{n+2} + 2a_{n+1} - a_n, \quad n = 1, 2, \ldots.$$

Prove that for every positive integer n, the number $1 + 4a_na_{n+1}$ *is a perfect square.*

Solution Define the sequence $(b_n)_{n\geq 1}$ by $b_n = a_{n-2} - a_{n+1} - a_n$, for all $n \geq 1$, and observe that $(b_n)_{n\geq 1}$ and $(a_n)_{n\geq 1}$ satisfy the same recurrence relation.

We want to prove that for all n we have the identity $1 + 4a_na_{n+1} = b_n^2$. Clearly, this holds for $n - 1$, and we assume now that it holds for all index values between 1 and $n - 1$, hence in particular, we have $1 + 4a_{n-1}a_n = b_{n-1}^2$. Indeed, we have

$$\begin{aligned} b_n &= (2a_{n+1} - a_n - a_{n-1}) - a_{n+1} - a_n = a_{n+1} + a_n - a_{n-1} \\ &= (a_{n+1} - a_n - a_{n-1}) + 2a_n = b_{n-1} + 2a_n, \quad n = 0, 1, \ldots. \end{aligned}$$

We obtain

$$\begin{aligned} 1 + 4a_{n-1}a_n &= b_{n-1}^2 = (b_n - 2a_n)^2 = b_n^2 - 4a_nb_n + 4a_n^2 \\ &= b_n^2 - 4a_n(a_{n+1} + a_n - a_{n-1}) + 4a_n^2 \end{aligned}$$

$$= b_n^2 - 4a_n a_{n+1} - 4a_n^2 + 4a_n a_{n-1} + 4a_n^2$$
$$= b_n^2 - 4a_n a_{n+1} + 4a_n a_{n-1}.$$

Subtracting $4a_n a_{n-1}$ we find $1 + 4a_n a_{n+1} = b_n^2$, which ends the proof.

Problem 6 (Dorin Andrica, [4]) *A sequence $(a_n)_{n\geq 1}$ is defined by $a_0 = 0$, $a_1 = 1$, $a_2 = 2$, $a_3 = 6$, and*

$$a_{n+4} = 2a_{n+3} + a_{n+2} - 2a_{n+1} - a_n, \quad n = 1, 2, \ldots.$$

Prove that n divides a_n for all $n \geq 1$.

Solution From the hypothesis it follows that $a_4 = 12$,$a_5 = 25$, $a_6 = 48$. We have $\frac{a_1}{1} = 1$, $\frac{a_2}{2} = 1$ $\frac{a_3}{3} = 2$ $\frac{a_4}{4} = 3$ $\frac{a_5}{5} = 5$ $\frac{a_6}{6} = 8$, which shows that $\frac{a_n}{n} = F_n$ for all $n = 1, 2, 3, 4, 5, 6$, where $(F_n)_{n\geq 1}$ is the Fibonacci sequence. We now prove by induction that $a_n = nF_n$ for all n. Let us assume for a start that $a_k = kF_k$ for $k \leq n + 3$. We have

$$\begin{aligned} a_{n+4} &= 2(n+3)F_{n+3} + (n+2)F_{n+2} - 2(n+1)F_{n+1} - nF_n \\ &= 2(n+3)F_{n+3} + (n+2)F_{n+2} - 2(n+1)F_{n+1} - n(F_{n+2} - F_{n+1}) \\ &= 2(n+3)F_{n+3} + 2F_{n+2} - (n+2)F_{n+1} \\ &= 2(n+3)F_{n+3} + 2F_{n+2} - (n+2)(F_{n+3} - F_{n+2}) \\ &= (n+4)(F_{n+3} + F_{n+2}) = (n+4)F_{n+4}, \end{aligned}$$

which is the desired result.

Problem 7 (Vlad Matei, O334. MR-2015) *Consider the sequence $(a_n)_{n\geq 0}$ defined by $a_n = \left\lfloor (\sqrt[3]{65} - 4)^{-n} \right\rfloor$, for $n \geq 0$. Prove that $a_n = 2, 3 \pmod{15}$.*

Solution Denote

$$r = \sqrt[3]{65}, \quad v = (\sqrt[3]{65} - 4)^{-1} = \sqrt[3]{65}^2 + 4\sqrt[3]{65} + 16 = r^2 + 4r + 16.$$

Note that we also have

$$u^2 = 48r^2 + 193r + 776, \quad u^3 = 2316r^2 + 9312r + 37441,$$

hence u is one of the roots of the equation

$$x^3 - 48x^2 - 12x - 1 = 0.$$

Denoting by v, w the other two, we have

$$v + w = 48 - u = 32 - 4r - r^2 = -12(r-4) - (r-4)^2,$$

and $vw = \frac{1}{u} = r - 4$. Note first that

$$r^3 = 4^3 + 1 = 4^3 + \frac{3 \cdot 4}{48} < \left(4 + \frac{1}{48}\right)^3,$$

or $0 < r - 4 < \frac{1}{48}$, hence $0 > v + w > -1$ and $1 > vw > 0$, or $v, w < 0$. For every positive integer n we have

$$0 <| v^n + w^n |=| v^n | + | w^n |\leq| v | + | w |< 1,$$

or $1 > v^n + w^n > 0$ iff n is even, and $0 > v^n + w^n > -1$ iff n is odd. Note that

$$v^2 + w^2 = (v+w)^2 - 2vw = 1552 - 193r - 48r^2,$$

$$v^3 + w^3 = (v+w)^3 - 3vw(v+w) = 74882 - 9312r - 2316r^2,$$

for

$$u + v + w = 48 \equiv 3 \pmod{15}$$

$$u^2 + v^2 + w^2 = 2328 \equiv 3 \pmod{15}$$

$$u^3 + v^3 + w^3 = 112323 \equiv 3 \pmod{15}.$$

The sequence $(b_n)_{n\geq 1}$ defined by $b_{n+3} = 48b_{n+2} + 12b_{n+1} + b_n$, with initial conditions $b_1 = 48$, $b_2 = 2328$, $b_3 = 112323$, has characteristic equation with roots u, v, w, or $b_n = u^n + v^n + w^n$ for $n \geq 1$. Clearly, all b_n are integers, and $b_n \equiv 3 \pmod{15}$ for all $n \geq 1$, since $b_1 \equiv b_2 \equiv b_3 \equiv 3 \pmod{15}$, and by $b_{n+3} \equiv 3(b_{n+2} - b_{n+1}) + b_n \pmod{15}$. We therefore conclude that for all even integers n, we have

$$a_n = \lfloor u^n \rfloor = u^n + v^n + w^n - 1 \equiv 2 \pmod{15},$$

while for all odd integers n, we have

$$a_n = \lfloor u^n \rfloor = u^n + v^n + w^n \equiv 3 \pmod{15}.$$

Finally, this confirms that $a_n = 2, 3 \pmod{15}$.

Problem 8 (Dorin Andrica, O346, MR-2015) *Consider the sequence* $(a_n)_{n\geq 0}$ *defined by* $a_0 = 0$, $a_1 = 1$, $a_0 = 1$, $a_3 = 6$ *and*

$$a_{n+4} = 2a_{n+3} + a_{n+2} - 2a_{n+1} - a_n, \quad n \geq 0.$$

Prove that n^2 *divides* a_n *for infinitely many positive integers.*

First Solution From the recursive relation it follows that $a_4 = 12, a_5 = 25, a_6 = 48$, hence we have $\frac{a_1}{1} = 1, \frac{a_2}{2} = 1, \frac{a_3}{3} = 2, \frac{a_4}{4} = 3, \frac{a_5}{5} = 5, \frac{a_6}{6} = 8$, that is $\frac{a_n}{n} = F_n$, for all $n = 1, 2, 3, 4, 5, 6$, where $(F_n)_{n\geq 1}$ is the Fibonacci sequence.

We prove by induction that $a_n = nF_n$ for all $n \geq 1$. Indeed, assuming that $a_k = kF_k$ for $k = n, n+1, n+2, n+3$, we have

$$\begin{aligned} a_{n+4} &= 2(n+3)F_{n+3} + (n+2)F_{n+2} - 2(n+1)F_{n+1} - nF_n \\ &= 2(n+3)F_{n+3} + (n+2)F_{n+2} - 2(n+1)F_{n+1} - n(F_{n+2} - F_{n+1}) \\ &= 2(n+3)F_{n+3} + 2F_{n+2} - (n+2)F_{n+1} \\ &= 2(n+3)F_{n+3} + 2F_{n+2} - (n+2)(F_{n+3} - F_{n+2}) \\ &= (n+4)(F_{n+3} + F_{n+2}) = (n+4)F_{n+4}, \end{aligned}$$

as desired.

It suffices to show that n divides F_n for infinitely many positive integers n. Using the well-known Binet formula for the Fibonacci numbers, we have

$$\begin{aligned} F_n &= \frac{1}{\sqrt{5}}\left[\left(\frac{1+\sqrt{5}}{2}\right)^n - \left(\frac{1-\sqrt{5}}{2}\right)^n\right] \\ &= \frac{1}{2^n\sqrt{5}}\left[\sum_{k=0}^{n}\binom{n}{k}(\sqrt{5})^k - \sum_{k=0}^{n}(-1)^k\binom{n}{k}(\sqrt{5})^k\right] \\ &= \frac{1}{2^n\sqrt{5}}\sum_{k=0}^{n}\binom{n}{k}\left[1-(-1)^k\right](\sqrt{5})^k. \end{aligned}$$

From the previous relation it follows

$$F_{5^l} = \frac{1}{2^{5^l-1}}\sum_{k=0}^{\frac{5^l-1}{2}}\binom{5^l}{2k+1}5^k. \tag{8.16}$$

We will prove that each of the first l terms in (1) are divisible by 5^l, that is 5^l divides $\binom{5^l}{2k+1}5^k$ for $k = 0, \cdots, l-1$. Indeed, we have

$$\binom{5^l}{2k+1} = \frac{5^l(5^l-1)\cdots(5^l-2k)}{1\cdot 2\cdots(2k+1)}.$$

Moreover, for every $a < 5^l$, the relation $\exp_5(5^a - a) = \exp_5(a)$, implies $\exp_5((2k)!) = \exp_5(5^l - 1)\cdots(5^l - 2k))$. It follows

$$\exp_5(\binom{5^l}{2k+1}) = l - \exp_5((2k+1)) \geq l - \exp_5(5^k) = l - k,$$

since clearly we have $2k+1 \leq 5^k$. From (1) we obtain that for every positive integer l, 5^l divides F_{5^l}, hence $(5^l)^2$ divides a_{5^l} and we are done.

Second Solution Denote by $(F_n)_{n\geq 0}$ the Fibonacci sequence, which satisfies $F_0 = 0$, $F_1 = 1$ and $F_{n+2} = F_{n+1} + F_n$, $n \geq 0$. One may notice that $a_n = nF_n$ for $n = 0, 1, 2, 3$. We will prove by induction that this statement is valid in general. Assume for now that it is true for $k = 0, 1, \ldots, n$. We have

$$\begin{aligned} a_{n+1} &= 2nF_n + (n-1)F_{n-1} - 2(n-2)F_{n-2} - (n-3)F_{n-3} \\ &= 2nF_n + 2F_{n-1} - (n-1)F_{n-2} \\ &= (n+1)F_n + (n+1)F_{n-1} = (n+1)F_{n+1}, \end{aligned}$$

which confirms that desired relation. Now it is sufficient to prove that $n \mid F_n$ for infinitely many n.

This statement can be proved in various ways. One of them is to use Problem U316, Math. Reflections, **5** (2014), where it was shown that $2^{m+2} | F_{3\cdot 2^m}$, for all $m \geq 1$. Also, for $m \geq 2$, $144 = F_{(12,3\cdot 2^m)} = (F_{12}, F_{3\cdot 2^m})$, so $3|F_{3\cdot 2^m}$. It follows that $n \mid F_n$ for all $n = 3\cdot 2^m$, with $m \geq 2$, which ends the proof.

Problem 9 (Bakir Farhi, 11864, AMM-2015) *Let p be a prime number and let $(u_n)_{n\geq 0}$ be the sequence defined by $u_n = n$, for $0 \leq n \leq p-1$ and $u_n = pu_{n+1-p} + u_{n-p}$, for $n \geq p$. Prove that for each positive integer n, the greatest power of p dividing u_n is the same as the greatest power of p dividing n.*

Solution Let $v_p(n)$ be the greatest integer e such that p^e divides n. For $n \geq 1$, we first prove by induction that $v_p(n!) \leq n - 1$, with strict inequality for $p > 2$ and $n > 1$. This is clearly the case for $n < p$. If $n = ip + j$ such that $i > 0$ and $0 \leq j < p$, using the induction hypothesis we compute

$$v_p((ip+j)!) = v_p(p(2p)\cdots(ip)) = i + v_p(i!) \leq i + i - 1 \leq ip - 1,$$

with strict inequality for $p > 2$.

We now show that $u_{ip+j} = \sum_{k=0}^{i}\binom{i}{k}p^k u_{j+k}$ for $i \geq 0$ and $j \geq 0$, by using induction on i. Clearly, the result holds for $i = 0$. For $i \geq 1$ we compute

$$
\begin{aligned}
u_{ip+j} &= u_{(i-1)p+j} + pu_{(i-1)p+j+1} \\
&= \sum_{k=0}^{i-1} \binom{i-1}{k} p^k u_{j+k} + \sum_{k=0}^{i-1} \binom{i-1}{k} p^{1+k} u_{j+1+k} \\
&= \sum_{k=0}^{i-1} \binom{i-1}{k} p^k u_{j+k} + \sum_{k=1}^{i} \binom{i-1}{k-1} p^k u_{j+k} \\
&= \sum_{k=0}^{i} \binom{i}{k} p^k u_{j+k}.
\end{aligned}
$$

We must prove that $v_p(n) = v_p(u_n)$, for $n \geq 1$. For $1 \leq j < p$ and $i \geq 0$, we have $u_{ip+j} = \sum_{k=1}^{i} \binom{i}{k} p^k u_{j+k} + u_j$. Since $u_j = j$, u_{ip+j} is clearly not divisible by p, and also, $ip + j$ is not divisible by p, hence $v_p(u_{ip+j}) = 0 = v_p(ip + j)$.

Since $u_0 = 0$ and $u_1 = 1$, we have $u_{ip+0} = \sum_{k=0}^{i} \binom{i}{k} p^k u_k = ip + \sum_{k=2}^{i} \binom{i}{k} p^k u_k$. Assuming that $p > 2$, for $k > 1$ we have

$$
v_p\left(\binom{i}{k} p^k u_k\right) \geq v_p(i) - v_p(k!) + k > v_p(i) - k + 1 + k = v_p(ip).
$$

Since the terms in the sum are divisible by higher powers of p, the powers of p dividing u_{ip} are just the powers dividing ip, and we have $v_p(u_{ip}) = v_p(ip)$.

The last case to be covered is $p = 2$. For $k = 2$, we have

$$
v_2\left(\binom{i}{2} 2^2 u_2\right) \geq v_2(i) - 1 + 2 + 1 = v_2(2i) + 1.
$$

For $k > 2$, we use the factors $i(i-1)(i-2)$ in the numerator of $\binom{i}{k}$ (with either $i - 1$ or $i - 2$ being even), to obtain

$$
\begin{aligned}
v_2\left(\binom{i}{2} 2^k u_k\right) &\geq v_2(i) + v_2((i-1)(i-2)) - v_2(k!) + k \\
&\geq v_2(i) + 1 - k + 1 + k = v_2(2i) + 1.
\end{aligned}
$$

Using the same argument as before, we obtain $v_2(u_{2i}) = v_2(2i)$.

8.5 Systems of Recurrence Relations

Problem 1 *Find the general terms of* $(x_n)_{n\geq 0}$, $(y_n)_{n\geq 0}$ *if*

$$\begin{cases} x_{n+1} = x_n + 2y_n \\ y_{n+1} = -2x_n + 5y_n, \quad n \geq 0, \end{cases}$$

and $x_0 = 1$, $y_0 = 2$.

Solution The matrix of coefficients is $A = \begin{pmatrix} 1 & 2 \\ -2 & 5 \end{pmatrix}$ and its characteristic equation is $\lambda^2 - 6\lambda + 9 = 0$ with $\lambda_1 = \lambda_2 = 3$. From Theorem 6.8.2) we get

$$A^n = \lambda_1^n B + n\lambda_1^{n-1} C = 3^n B + n3^{n-1} C,$$

where $B = I_2$ and

$$C = A - \lambda_1 I_2 = A - 3I_2 = \begin{pmatrix} -2 & 2 \\ -2 & 2 \end{pmatrix}.$$

It follows that

$$A^n = 3^n I_2 + n3^{n-1} \begin{pmatrix} -2 & 2 \\ -2 & 2 \end{pmatrix} = \begin{pmatrix} (3-2n)3^{n-1} & 2n3^{n-1} \\ -2n3^{n-1} & (3+2n)3^{n-1} \end{pmatrix}.$$

From (6.38) with $x_0 = 1$ and $y_0 = 2$ we get

$$x_n = (2n+3)3^{n-1} \text{ and } y_n = 2(n+3)3^{n-1}, \quad n \geq 0.$$

Problem 2 *Solve in positive integers the equation*

$$6x^2 - 5y^2 = 1.$$

Solution This equation is solvable and its minimal solution is given by $(x_0, y_0) = (1, 1)$. The Pell's resolvent is $u^2 - 30v^2 = 1$ and its fundamental solution is $(u_1, v_1) = (11, 2)$. All solutions (x_n, y_n) to equation $6x^2 - 5y^2 = 1$ are generated by

$$\begin{pmatrix} 1 & 5 \\ 1 & 6 \end{pmatrix} \begin{pmatrix} 11 & 60 \\ 2 & 11 \end{pmatrix}^n, \quad n \geq 0.$$

By elementary computations using (6.48) and then (6.44), we get

$$\begin{cases} x_n = \dfrac{6+\sqrt{30}}{12}(11+2\sqrt{30})^n + \dfrac{6-\sqrt{30}}{12}(11-2\sqrt{30})^n \\ \\ y_n = \dfrac{5+\sqrt{30}}{12}(11+2\sqrt{30})^n + \dfrac{5-\sqrt{30}}{12}(11-2\sqrt{30})^n, \; n \geq 0. \end{cases}$$

Problem 3 (Application 3, [7, p. 170]) *Find all positive integers n such that $n + 1$ and $3n + 1$ are simultaneously perfect squares.*

Solution If $n + 1 = x^2$ and $3n + 1 = y^2$, then we obtain $3x^2 - y^2 = 2$. This equation is equivalent to Pell's equation $u^2 - 3v^2 = 1$, where

$$u = \frac{1}{2}(3x - y) \text{ and } v = \frac{1}{2}(y - x).$$

The fundamental solution of Pell's equation $u^2 - 3v^2 = 1$ is given by $(u_1, v_1) = (2, 1)$ and matrix generating all solutions (u_k, v_k)

$$A_{(2,1)} = \begin{pmatrix} 2 & 3 \\ 1 & 2 \end{pmatrix}.$$

We find

$$u_k = \frac{1}{2}[(2 + \sqrt{3})^k + (2 - \sqrt{3})^k], \quad v_k = \frac{1}{2\sqrt{3}}[(2 + \sqrt{3})^k - (2 - \sqrt{3})^k],$$

hence

$$n_k = x_k^2 - 1 = (u_k + v_k)^2 - 1 = \frac{1}{6}[(2 + \sqrt{3})^{2k+1} + (2 - \sqrt{3})^{2k+1} - 4], \quad k \geq 0.$$

Problem 4 (IMO Shortlist, 1980) *Let A and E be a pair of opposite vertices of a regular octagon $AA_1A_2A_3EA_3'A_2'A_1'$. A frog starts jumping at vertex A. From any vertex of the octagon except E, it may jump to either of the two adjacent vertices. When it reaches vertex E, the frog stops and stays there. Let a_n be the number of distinct paths of exactly n jumps ending at E. Derive and solve a recursion for a_n.*

Solution Let a_n be the number of paths with n jumps that end at A, b_n the number of paths with n jumps that end at A_1 (and at A_1'), c_n the number of paths with n jumps that end at A_2 (and at A_2'), d_n the number of paths with n jumps that end at A_3 (and at A_3'), and e_n the number of paths with n jumps that end at E. Clearly, for any positive integer k, we have $e_{2k-1} = 0$. The following recurrence relations are easy to derive

$$\begin{aligned} a_n &= 2b_{n-1} \\ b_n &= a_{n-1} + c_{n-1} \\ c_n &= b_{n-1} + d_{n-1} \\ d_n &= c_{n-1} \\ e_n &= 2d_{n-1}. \end{aligned}$$

We want to find $e_n = 2d_{n-1} = 2c_{n-2}$. We have $b_n = 2b_{n-2} + c_{n-1}$ and $c_n = b_{n-1} + c_{n-2}$. By replacing in the next equation, we obtain the recursive relation $c_{n+1} - c_{n-1} = 2c_{n-1} - c_{n-3} + c_{n-1}$, that is

$$c_n = 4c_{n-2} - 2c_{n-4}, \quad n \geq 4.$$

Notice that c_{2k+1} is 0. Denoting $x_k = c_{2k}$, the previous recurrence relation becomes $x_n = 4x_{n-1} - 2x_{n-2}$, with initial values $x_0 = 0$ and $x_1 = 1$, and characteristic roots $2 + \sqrt{2}$ and $2 - \sqrt{2}$. By the Binet-type formula we obtain

$$x_n = \frac{(2+\sqrt{2})^n - (2-\sqrt{2})^n}{2\sqrt{2}}, \quad n \geq 0,$$

which gives the final formula

$$e_{2n} = \frac{(2+\sqrt{2})^{n-1} - (2-\sqrt{2})^{n-1}}{\sqrt{2}}, \quad n \geq 1.$$

Problem 5 (Neculai Stanciu and Titu Zvonaru, S323, MR-2015) *Solve in positive integers the equation*

$$x + y + (x-y)^2 = xy.$$

Solution Denote by $s = x + y$, $d = x - y$, and notice that the proposed equation can be written as

$$s^2 - d^2 = 4xy = 4x + 4y + 4(x-y)^2 = 4s + 4d^2,$$

from where we obtain

$$(s-2)^2 - 5d^2 = 4.$$

This is a Pell-like equation of the form $c^2 - 5d^2 = 4$, where $s - d = c$, where c and d are integers having the same parity. This equation has infinitely many solutions (c_n, d_n), given by the recurrent relations

$$c_{n+2} = 3c_{n+1} - c_n, \quad d_{n+2} = 3d_{n+1} - d_n, \quad n \geq 0,$$

with the initial conditions $(c_0, d_0) = (2, 0)$ and $(c_1, d_1) = (3, 1)$. Therefore, all the solutions are of the general form

$$s_n = \left(\frac{\sqrt{5}+1}{2}\right)^{2n} + \left(\frac{\sqrt{5}-1}{2}\right)^{2n} + 2,$$

$$d_n = \frac{1}{\sqrt{5}}\left(\frac{\sqrt{5}+1}{2}\right)^{2n} - \frac{1}{\sqrt{5}}\left(\frac{\sqrt{5}-1}{2}\right)^{2n},$$

or equivalently

$$x_n = \frac{s_n + d_n}{2} = \frac{1}{\sqrt{5}}\left(\frac{\sqrt{5}+1}{2}\right)^{2n+1} + \frac{1}{\sqrt{5}}\left(\frac{\sqrt{5}-1}{2}\right)^{2n+1} + 1,$$

$$y_n = \frac{s_n - d_n}{2} = \frac{1}{\sqrt{5}}\left(\frac{\sqrt{5}+1}{2}\right)^{2n-1} + \frac{1}{\sqrt{5}}\left(\frac{\sqrt{5}-1}{2}\right)^{2n-1} + 1, \quad n \geq 0.$$

One can check that the solutions (x_n, y_n), $n \geq 0$, satisfy the initial equation, as well as the solutions (y_n, x_n), $n \geq 0$.

Problem 6 (Ivan Borsenco, U325, MR-2015) *Let $A_1B_1C_1$ be a triangle with circumradius R_1. For each $n \geq 1$, the incircle of triangle $A_nB_nC_n$ is tangent to its sides at the points $A_{n+1}B_{n+1}C_{n+1}$. The circumradius of triangle $A_{n+1}B_{n+1}C_{n+1}$, which is also the inradius of triangle $A_nB_nC_n$, is R_{n+1}. Find $\lim\limits_{n\to\infty} \frac{R_{n+1}}{R_n}$.*

Solution Suppose that the triangle $\Delta A_nB_nC_n$ has an incircle of radius r_n, then we have $R_{n+1} = r_n$. The following relation is well known:

$$\frac{r_n}{R_n} = 4\sin\frac{A_n}{2}\sin\frac{B_n}{2}\sin\frac{C_n}{2},$$

and we also have the relations between angles

$$\measuredangle A_{n+1} = \frac{\pi}{2} - \frac{\measuredangle A_n}{2}, \quad \measuredangle B_{n+1} = \frac{\pi}{2} - \frac{\measuredangle B_n}{2}, \quad \measuredangle C_{n+1} = \frac{\pi}{2} - \frac{\measuredangle C_n}{2}.$$

Denoting by I the center of the incircle, we obtain

$$\measuredangle A_n + \measuredangle B_{n+1}IC_{n+1} = \pi, \quad \measuredangle B_{n+1}IC_{n+1} = 2\measuredangle C_{n+1}A_{n+1}B_{n+1}.$$

From this we deduce that

$$\measuredangle A_{n+1} - \frac{\pi}{3} = -\frac{1}{2}\left(\measuredangle A_n - \frac{\pi}{3}\right),$$

hence

$$\measuredangle A_n = \frac{\pi}{3} + \left(\measuredangle A_1 - \frac{\pi}{3}\right)\left(-\frac{1}{2}\right)^{n-1}$$

$$\measuredangle B_n = \frac{\pi}{3} + \left(\measuredangle B_1 - \frac{\pi}{3}\right)\left(-\frac{1}{2}\right)^{n-1}$$

$$\measuredangle C_n = \frac{\pi}{3} + \left(\measuredangle C_1 - \frac{\pi}{3}\right)\left(-\frac{1}{2}\right)^{n-1}.$$

It follows that

$$\lim_{n\to\infty} A_n = \lim_{n\to\infty} B_n = \lim_{n\to\infty} C_n = \frac{\pi}{3},$$

therefore

$$\lim_{n\to\infty} \frac{R_{n+1}}{R_n} = \lim_{n\to\infty} \frac{r_n}{R_n} = \lim_{n\to\infty} 4\sin\frac{A_n}{2}\sin\frac{B_n}{2}\sin\frac{C_n}{2} = 4\left(\sin\frac{\pi}{6}\right)^3 = \frac{1}{2}.$$

Problem 7 (Dorin Andrica and Mihai Piticari, Romanian TST, 2013) *For a positive integer n we consider the expression*

$$(\sqrt[3]{2}-1)^n = a_n + b_n\sqrt[3]{2} + c_n\sqrt[3]{4},$$

where $a_n, b_n, c_n \in \mathbb{Z}$. Show that $c_{80} \neq 0$.

Solution It is known that if $a + b\sqrt[3]{2} + c\sqrt[3]{4} = 0$ with $a, b, c \in \mathbb{Z}$, then we have $a = b = c = 0$ (by Proposition A.2, this holds if 2 is replaced with any prime). Hence, whenever $a + b\sqrt[3]{2} + c\sqrt[3]{4} = a' + b'\sqrt[3]{2} + c'\sqrt[3]{4}$, with $a, b, c, a', b', c' \in \mathbb{Z}$, then $a = a', b = b', c = c'$.

By the relation $(\sqrt[3]{2}-1)^n = a_n + b_n\sqrt[3]{2} + c_n\sqrt[3]{4}$, it results

$$(\sqrt[3]{2}-1)^n(1+\sqrt[3]{2}+\sqrt[3]{4}) = (a_n + b_n\sqrt[3]{2} + c_n\sqrt[3]{4})(1+\sqrt[3]{2}+\sqrt[3]{4}),$$

therefore

$$(\sqrt[3]{2}-1)^{n-1} = (a_n + b_n\sqrt[3]{2} + c_n\sqrt[3]{4})(1+\sqrt[3]{2}+\sqrt[3]{4}).$$

Furthermore

$$a_{n-1} + b_{n-1}\sqrt[3]{2} - c_{n-1}\sqrt[3]{4} = (a_n + b_n\sqrt[3]{2} + c_n\sqrt[3]{4})(1+\sqrt[3]{2}+\sqrt[3]{4}),$$

from where we obtain

$$c_{n-1} = a_n + b_n + c_n.$$

Should we have $c_{80} = 0$, then using the above relation, it follows that $a_{81} + b_{81} + c_{81} = 0$. As a consequence, we obtain

$$\left[1 - 2\binom{81}{3} + 2^2\binom{81}{6} + \cdots - 2^{27}\binom{81}{81}\right]$$
$$+\left[\binom{81}{1} + 2\binom{81}{4} + \cdots - 2^{26}\binom{81}{79}\right]$$
$$+\left[\binom{81}{2} - \binom{81}{5} + \cdots - 2^{26}\binom{81}{80}\right] = 0.$$

However, the numbers $\binom{81}{1}, \binom{81}{2}, \cdots, \binom{81}{80}$ are divisible by 3, hence, from the above relationship, it follows that 3 must divide $2^{27} - 1$. On the other hand, we have $2^{27} - 1 = (3-1)^{27} - 1 \equiv -2(\text{mod}3)$, a contradiction.

Problem 8 (Jean-Charles Mathieux, S38, MR-2007) *Prove that for each positive integer n, there is a positive integer m such that*

$$(1+\sqrt{2})^n = \sqrt{m} + \sqrt{m+1}.$$

First Solution We first show by induction that we can find the sequences $(x_n)_{n\geq 0}$ and $(y_n)_{n\geq 0}$ of positive integers such that

$$(1+\sqrt{2})^n = x_n + y_n\sqrt{2},$$

where $x_1 = y_1 = 1$. It then follows

$$(1+\sqrt{2})^{n+1} = (x_n + y_n\sqrt{2}) \cdot (1+\sqrt{2}) = (x_n + 2y_n) + (x_n + y_n)\sqrt{2},$$

hence

$$x_{n+1} = x_n + 2y_n$$
$$y_{n+1} = x_n + y_n.$$

Also by induction, we can prove that

$$2y_n^2 - x_n^2 = (-1)^{n+1}.$$

This is clearly true for $n = 1$, and one can prove that

$$2y_{n+1}^2 - x_{n+1}^2 = 2(x_n + y_n)^2 - (x_n + 2y_n)^2 = -(2y_n^2 - x_n^2) = (-1)^{n+2}.$$

In our problem, for all odd positive integer n, one may choose

$$m = x_n^2 = 2y_n^2 - 1,$$

and for all even positive integer n

$$m = 2y_n^2 = x_n^2 - 1.$$

An integrated solution can also be obtained, since we can show that

$$x_n = \frac{(1+\sqrt{2})^n + (1-\sqrt{2})^n}{2},$$

$$y_n = \frac{(1+\sqrt{2})^n - (1-\sqrt{2})^n}{2},$$

and for each n we can select m_n given by the formula

$$m_n = \frac{(1+\sqrt{2})^{2n} + (1-\sqrt{2})^{2n}}{2}.$$

Second Solution We only prove the statement for $n = 2j$, the solution for $n = 2j+1$ being analogous. By the binomial theorem we have

$$(1+\sqrt{2})^n = \sum_{k=0}^{n} \binom{n}{k} (\sqrt{2})^k = 1 + \binom{n}{1}\sqrt{2} + \cdots + \binom{n}{n-1}(\sqrt{2})^{n-1} + (\sqrt{2})^n$$

$$= \left[\binom{n}{0} + 2\binom{n}{2} + \cdots + 2^j \binom{n}{2j}\right] + \left[\binom{n}{1} + \cdots + 2^{j-1}\binom{n}{2j-1}\right]\sqrt{2}$$

$$= \sqrt{\left[\binom{n}{0} + 2\binom{n}{2} + \cdots + 2^j \binom{n}{2j}\right]^2} + \sqrt{2\left[\binom{n}{1} + \cdots + 2^{j-1}\binom{n}{2j-1}\right]^2}$$

$$= \sqrt{A+1} + \sqrt{B}.$$

We show that $A = B$. Indeed, notice that

$$(1+\sqrt{2})^n + (1-\sqrt{2})^n = 2\left[\binom{n}{0} + 2\binom{n}{2} + \cdots + 2^j \binom{n}{2j}\right]$$

$$(1+\sqrt{2})^n - (1-\sqrt{2})^n = 2\left[\binom{n}{1} + \cdots + 2^{j-1}\binom{n}{2j-1}\right].$$

The proof is now reduced to some simple algebraic manipulations.

Third Solution Another proof is based on algebraic number theory.

Consider the quadratic field $\mathbb{Q}[\sqrt{2}] = \{a + b\sqrt{2} : a, b \in \mathbb{Q}\}$, with the norm function defined by $N(a+b\sqrt{2}) = a^2 - 2b^2$. We know that $1+\sqrt{2}$ is a unit, because $N(1+\sqrt{2}) = 1 - 2 \cdot 1^2 = -1$. Since the set of units is closed under multiplication, it follows that $(1+\sqrt{2})^n$ is also a unit for every integer $n \geq 1$. If $(1+\sqrt{2})^n = A_n + B_n\sqrt{2}$, for some integers A_n, B_n, then it follows that $N((1+\sqrt{2})^n) = \pm 1$, that is $A_n^2 - 2B_n^2 = \pm 1$.

The proof is now complete, since $(1+\sqrt{2})^n = A_n + B_n\sqrt{2} = \sqrt{A_n^2} + \sqrt{2B_n^2}$, and we have showed that the difference between A_n^2 and $2B_n^2$ is 1.

Problem 9 (Gauss Formula) *Let $a > b > 0$ and the function*

$$G(a,b) = \int_0^{\frac{\pi}{2}} \frac{dx}{\sqrt{a^2\cos^2 x + b^2 \sin^2 x}}.$$

Define the sequences $(a_n)_{n\geq 0}$ and $(b_n)_{n\geq 0}$ by the recurrence relations

$$a_n = \frac{a_{n-1} + b_{n-1}}{2}, \quad b_n = \sqrt{a_{n-1}b_{n-1}}, \quad a_0 = a, \quad b_0 = b.$$

(a) Prove that the sequences $(a_n)_{n\geq 0}$ and $(b_n)_{n\geq 0}$ are convergent to a common limit $\lim_{n\to\infty} a_n = \lim_{n\to\infty} a_n = \mu(a,b)$.

(b) Show that

$$G(a,b) = \int_0^{\frac{\pi}{2}} \frac{dx}{\sqrt{a_n^2\cos^2 x + b_n^2 \sin^2 x}}.$$

(c) $G(a,b) = \frac{\pi}{2\mu(a,b)}$.

Solution

(a) Clearly, $a_1 = \frac{a_0+b_0}{2}$ and $b_1 = \sqrt{a_0 b_0}$, hence

$$b_0 < b_1 < a_1 < a_0.$$

One can show by induction that

$$b_0 < b_1 < \cdots < b_n < a_n < \cdots < a_1 < a_0,$$

therefore $(a_n)_{n\geq 0}$ and $(b_n)_{n\geq 0}$ are bounded and monotonic. Denoting their limits by A and B, taking limits in $a_n = \frac{a_{n-1}+b_{n-1}}{2}$, we obtain $A = B$.

(b) Defining the new variable

$$\sin x = \frac{2a\sin t}{a+b+(a-b)\sin^2 t}, \quad t \in \left[0, \frac{\pi}{2}\right],$$

we obtain by differentiation

$$\cos x \, dx = 2a\frac{a+b-(a-b)\sin^2 t}{\left[a+b+(a-b)\sin^2 t\right]^2}\cos t \, dt.$$

From the initially defined variable we also obtain

$$\cos x = \frac{\sqrt{(a+b)^2 - (a-b)^2 \sin^2 t}}{a+b+(a-b)\sin^2 t} \cos t,$$

from where it follows

$$\mathrm{d}x = 2a \frac{a+b-(a-b)\sin^2 t}{a+b+(a-b)\sin^2 t} \cdot \frac{\mathrm{d}t}{\sqrt{(a+b)^2 - (a-b)^2 \sin^2 t}}.$$

We also have

$$\sqrt{a^2 \cos^2 x + b^2 \sin^2 x} = a \frac{a+b-(a-b)\sin^2 t}{a+b+(a-b)\sin^2 t},$$

from where it yields

$$\frac{\mathrm{d}x}{\sqrt{a^2 \cos^2 x + b^2 \sin^2 x}} = \frac{\mathrm{d}t}{\sqrt{\left(\frac{a+b}{2}\right)^2 \cos^2 t + ab \sin^2 t}}.$$

Since $a_1 = \frac{a+b}{2}$ and $b_1 = \sqrt{ab}$, it follows that

$$G(a,b) = \int_0^{\frac{\pi}{2}} \frac{\mathrm{d}x}{\sqrt{a_1^2 \cos^2 x + b_1^2 \sin^2 x}},$$

and by applying the same argument, one obtains

$$G(a,b) = \int_0^{\frac{\pi}{2}} \frac{\mathrm{d}x}{\sqrt{a_n^2 \cos^2 x + b_n^2 \sin^2 x}}, \quad n \geq 0.$$

(c) One can check that the inequalities below hold:

$$\frac{\pi}{2a_n} \leq G(a,b) \leq \frac{\pi}{2b_n}, \quad n \geq 0.$$

By taking the limit as $n \to \infty$ it follows that $G(a,b) = \frac{\pi}{2\mu(a,b)}$.

Problem 10 (Dorin Marghidaru and Leonard Giugiuc, 4264, CM-2017) *Let $(a_n)_{n\geq 0}$ and $(b_n)_{n\geq 0}$ be two sequences such that $a_0, b_0 > 0$ and*

$$a_{n+1} = a_n + \frac{1}{2b_n}, \quad b_{n+1} = b_n + \frac{1}{2a_n}, \quad n \geq 0.$$

Prove that

$$\max\{a_{2017}, b_{2017}\} > 44.$$

Solution We first show that the quantity $f(a_n, b_n) = \frac{a_n}{b_n}$ is invariant. Indeed

$$f(a_{n+1}, b_{n+1}) = \frac{a_n + \frac{1}{2b_n}}{b_n + \frac{1}{2a_n}} = \frac{a_n}{b_n} = f(a_n, b_n).$$

From here we conclude that

$$\frac{a_n}{b_n} = f(a_n, b_n) = f(a_0, b_0) = \frac{a_0}{b_0},$$

hence

$$a_{n+1} = a_n + \frac{1}{2a_n} \cdot \frac{a_0}{b_0}, \quad b_{n+1} = b_n + \frac{1}{2b_n} \cdot \frac{b_0}{a_0}.$$

Therefore

$$a_{n+1}^2 + b_{n+1}^2 = a_n^2 + b_n^2 + \left(\frac{a_0}{b_0} + \frac{b_0}{a_0}\right) + \left(\frac{1}{4a_n^2} \cdot \frac{a_0^2}{b_0^2} + \frac{1}{4b_n^2} \cdot \frac{b_0^2}{a_0^2}\right) \geq a_n^2 + b_n^2 + 2,$$

since the first bracket is greater than 2 by the AM-GM inequality, while the second one is nonnegative. We deduce that

$$a_{n+1}^2 + b_{n+1}^2 \geq 2(n+1) + a_0^2 + b_0^2 > 2(n+1).$$

Moreover

$$\max\{a_{2017}, b_{2017}\} \geq \frac{\sqrt{a_{2017}^2 + b_{2017}^2}}{\sqrt{2}} > \sqrt{2017} > 44.9.$$

Problem 11 (SAMC, 2015) *Let k be a positive integer. Prove that there exist integers x, y neither of which is divisible by 7, such that $x^2 + 6y^2 = 7^k$.*

Solution Take $x_1 = y_1 = -1$, and consider the relations

$$x_{k+1} = x_k - 6y_k, \quad y_{k+1} = x_k + y_k, \quad k = 1, 2, 3, \ldots.$$

One can check that

$$x_{k+1}^2 + 6x_{k+1}^2 = 7\left(x_k^2 + 6x_k^2\right),$$

and

$$x_k \equiv y_k \equiv (-1)^k \pmod 7.$$

Hence, there exist infinitely many integers x, y with the desired property.

Problem 12 (SAMC, 2015) *Arrange the numbers* 1, 2, 3, 4 *around a circle in this order. One starts at* 1, *and with every step, he moves to an adjacent number on either side. How many ways can he move such that the sum of the numbers he visits in his path (including the starting number) is equal to* 21?

Solution Let a_i, b_i, c_i, d_i be the number of paths that end by 1, 2, 3, 4 respectively and having the sum equal to i. Since all these paths start from 1, it is easy to check that

$$a_1 = 1,\ b_1 = c_1 = d_1 = 0, \qquad a_2 = b_2 = c_2 = d_2 = 0,$$

$$a_3 = 0,\ b_3 = 1, c_3 = d_3 = 0, \qquad a_4 = 1,\ b_4 = c_4 = d_4 = 0.$$

We can visit 1 from 2 or 4, visit 2 from 1 or 3, visit 3 from 2 or 4, and visit 4 from 3 or 1, hence the following relations hold for $n \geq 5$:

$$\begin{cases} a_n = b_{n-1} + d_{n-1} \\ b_n = a_{n-2} + c_{n-2} \\ c_n = b_{n-3} + d_{n-3} \\ d_n = c_{n-4} + a_{n-4}. \end{cases}$$

We have to calculate $a_{21} + b_{21} + c_{21} + d_{21}$.

Denoting by $u_n = a_n + c_n$ and $v_n = b_n + d_n$ for $n \geq 1$, we have

$$\begin{cases} u_n = v_{n-1} + v_{n-3} \\ v_n = u_{n-2} + u_{n-4}. \end{cases}$$

We also have $u_1 = 1, u_2 = 0, u_3 = 0, u_4 = 1$ and $v_1 = 0, v_2 = 0, v_3 = 1, v_4 = 0$. It also follows that $v_n = v_{n-3} + 2v_{n-5} + v_{n-7}$ and $u_n = u_{n-3} + 2u_{n-5} + u_{n-7}$.

Setting $s_n = u_n + v_n$, we obtain

$$s_n = s_{n-3} + 2s_{n-5} + s_{n-7},$$

where $s_1 = 1, s_2 = 0, s_3 = 1, s_4 = 1, s_5 = 1, s_6 = 3, s_7 = 1$.

One may calculate directly that

$$s_8 = 4,\ s_9 = 5,\ s_{10} = 4,\ s_{11} = 11,\ s_{12} = 8,\ s_{13} = 15,\ s_{14} = 22,$$

$$s_{15} = 20,\ s_{16} = 42,\ s_{17} = 42,\ s_{18} = 61,\ s_{19} = 94,\ s_{20} = 97,\ s_{21} = 167.$$

In conclusion, there are 167 paths that have their sum equal to 21.

Problem 13 (Dorin Andrica, O97, GM-1979) *Consider the sequences* $(u_n)_{n\geq 0}$, $(v_n)_{n\geq 0}$ *defined by* $u_1 = 3$, $v_1 = 2$ *and*

$$\begin{cases} u_{n+1} = 3u_n + 4v_n \\ v_{n+1} = 2u_n + 3v_n, \quad n \geq 1. \end{cases}$$

Define $x_n = u_n + v_n$, $y_n = u_n + 2v_n$, $n \geq 1$. *Prove that* $y_n = \lfloor x_n\sqrt{2} \rfloor$ *for all* $n \geq 1$.

Solution We show by induction that

$$u_n^2 - 2v_n^2 = 1, \quad n \geq 1.$$

This is clearly true for $n = 1$. Assuming that the relation holds for n, we have

$$u_{n+1}^2 - 2v_{n+1}^2 = (3u_n + 4v_n)^2 - 2(2u_n + 3v_n)^2 = u_n^2 - 2v_n^2 = 1.$$

We now prove that we also have

$$2x_n^2 - y_n^2 = 1, \quad n \geq 1.$$

Indeed, we have

$$2x_n^2 - y_n^2 = 2(u_n + v_n)^2 - (u_n + 2v_n)^2 = u_n^2 - 2v_n^2 = 1,$$

hence

$$\left(x_n\sqrt{2} + y_n\right)\left(x_n\sqrt{2} - y_n\right) = 1, \quad n \geq 1.$$

Since $x_n\sqrt{2} + y_n > 1$, it follows that

$$0 < x_n\sqrt{2} - y_n < 1, \quad n \geq 1.$$

Therefore, $y_n = \lfloor x_n\sqrt{2} \rfloor$, as claimed.

8.6 Homographic Recurrent Sequences

Problem 1 *Let* $(x_n)_{n\geq 0}$ *be the sequence defined by* $x_0 = 2$ *and*

$$x_{n+1} = \frac{2}{1 + x_n}, \quad n = 0, 1, \ldots.$$

Find x_{2019}.

Solution The homographic function defining the recurrence relation is

$$f(z) = \frac{2}{z+1},$$

of matrix $A_f = \begin{pmatrix} 0 & 2 \\ 1 & 1 \end{pmatrix}$, and characteristic equation $\lambda^2 - \lambda - 2 = 0$, with roots $\lambda_1 = -1$ and $\lambda_2 = 2$. By Cayley's Theorem we have

$$(A_f)^n = B\lambda_1^n + C\lambda_2^n = B(-1)^n + C\,2^n,$$

where

$$B = \frac{1}{\lambda_1 - \lambda_2}(A_f - \lambda_2 I_2) = -\frac{1}{3}\begin{pmatrix} -2 & 2 \\ 1 & -1 \end{pmatrix}$$

$$C = \frac{1}{\lambda_2 - \lambda_1}(A_f - \lambda_1 I_2) = \frac{1}{3}\begin{pmatrix} 1 & 2 \\ 1 & 2 \end{pmatrix}.$$

It follows that

$$(A_f)^n = \begin{pmatrix} \frac{1}{3}(2(-1)^n + 2^n) & -\frac{2}{3}((-1)^n - 2^n) \\ \frac{1}{3}(-(-1)^n + 2^n) & \frac{1}{3}((-1)^n + 2^{n+1}) \end{pmatrix}.$$

For $n = 0, 1, \ldots$, we therefore obtain

$$f^n(2) = \frac{\frac{1}{3}(2(-1)^n + 2^n)\cdot 2 - \frac{2}{3}((-1)^n - 2^n)}{\frac{1}{3}(-(-1)^n + 2^n)\cdot 2 + \frac{1}{3}((-1)^n + 2^{n+1})} = \frac{2(-1)^n + 4\cdot 2^n}{-(-1)^n + 4\cdot 2^n}.$$

Problem 2 *Define the sequence* $(x_n)_{n\geq 0}$ *by*

$$x_{n+1}x_n + x_{n+1} = x_n - 1, \quad n = 0, 1, \ldots.$$

Given that for all n *we have* $x_n \notin \{-1, 0, 1\}$ *and* $x_{2009} = 40$, *find* x_n.

Solution The recurrence relation is equivalent to

$$x_{n+1} = \frac{x_n - 1}{x_n + 1}, \quad n = 0, 1, \ldots,$$

that is the sequence $(x_n)_{n\geq 0}$ is defined by the homographic function

$$f(z) = \frac{z-1}{z+1}.$$

The matrix of this function is

$$A_f = \begin{pmatrix} 1 & -1 \\ 1 & 1 \end{pmatrix} = \sqrt{2} \begin{pmatrix} \cos \frac{\pi}{4} & -\sin \frac{\pi}{4} \\ \sin \frac{\pi}{4} & \cos \frac{\pi}{4} \end{pmatrix}.$$

One can easily prove by induction that

$$(A_f)^n = (\sqrt{2})^n \begin{pmatrix} \cos \frac{n\pi}{4} & -\sin \frac{n\pi}{4} \\ \sin \frac{n\pi}{4} & \cos \frac{n\pi}{4} \end{pmatrix},$$

hence

$$x_n = f^n(x_0) = \frac{\left(\cos \frac{n\pi}{4}\right) x_0 - \sin \frac{n\pi}{4}}{\left(\sin \frac{n\pi}{4}\right) x_0 + \cos \frac{n\pi}{4}}, \quad n = 0, 1, \ldots.$$

Clearly, the sequence $(x_n)_{n\geq 0}$ has period 8, i.e., $x_{n+8} = x_n$, $n = 0, 1, \ldots$. Since $2009 = 251 \cdot 8 + 1$, hence we have

$$x_{2009} = x_1 = \frac{x_0 - 1}{x_0 + 1}.$$

Therefore, $\frac{x_0-1}{x_0+1} = 40$, hence $x_0 = -\frac{41}{39}$. It follows that

$$x_n = \frac{41 \cdot \left(\cos \frac{n\pi}{4}\right) + 39 \cdot \left(\sin \frac{n\pi}{4}\right)}{41 \cdot \left(\sin \frac{n\pi}{4}\right) - 39 \cdot \left(\cos \frac{n\pi}{4}\right)}, \quad n = 0, 1, \ldots.$$

Problem 3 (Austria, 1979) *Define the sequence* $(x_n)_{n\geq 0}$ *by* $x_0 = 1979$ *and*

$$x_{n+1} = \frac{1979(1 + x_n)}{1979 + x_n}, \quad n = 0, 1, \ldots.$$

Find x_n *and* $\lim_{n\to\infty} x_n$.

Solution Let us consider the general case

$$x_{n+1} = \frac{a(1 + x_n)}{a + x_n}, n = 0, 1, \ldots,$$

where $a > 0$ and $x_0 > 0$. The sequence $(x_n)_{n\geq 0}$ is defined by the homographic function

$$f(z) = \frac{az + a}{z + a}.$$

The matrix of this function is

$$A_f = \begin{pmatrix} a & a \\ 1 & a \end{pmatrix},$$

which has the characteristic equation $\lambda^2 - 2a\lambda + a^2 - a = 0$, having the roots $\lambda_1 = a + \sqrt{a}$, $\lambda_2 = a - \sqrt{a}$.

By Cayley's Theorem one can deduce that

$$(A_f)^n = B\lambda_1^n + C\lambda_2^n = B(a + \sqrt{a})^n + C\,(a - \sqrt{a})^n,$$

where

$$B = \frac{1}{\lambda_1 - \lambda_2}(A_f - \lambda_2 I_2) = \frac{1}{2\sqrt{a}} \begin{pmatrix} \sqrt{a} & a \\ 1 & \sqrt{a} \end{pmatrix}$$

$$C = \frac{1}{\lambda_2 - \lambda_1}(A_f - \lambda_1 I_2) = \frac{1}{\lambda_1 - \lambda_2}(A_f - \lambda_2 I_2) = -\frac{1}{2\sqrt{a}} \begin{pmatrix} -\sqrt{a} & a \\ 1 & -\sqrt{a} \end{pmatrix}.$$

Simple computations give

$$(A_f)^n = \frac{1}{2\sqrt{a}} \begin{pmatrix} \sqrt{a}(\lambda_1^n + \lambda_2^n) & a(\lambda_1^n - \lambda_2^n) \\ \lambda_1^n - \lambda_2^n & \sqrt{a}(\lambda_1^n + \lambda_2^n) \end{pmatrix},$$

hence

$$x_n = \frac{\sqrt{a}(\lambda_1^n + \lambda_2^n)x_0 + a(\lambda_1^n - \lambda_2^n)}{(\lambda_1^n - \lambda_2^n)x_0 + \sqrt{a}(\lambda_1^n + \lambda_2^n)}, \qquad n = 0, 1, \ldots.$$

It follows that

$$\lim_{n\to\infty} x_n = \lim_{n\to\infty} \frac{\sqrt{a}\left(1 + \left(\frac{\lambda_2}{\lambda_1}\right)^n\right)x_0 + a\left(1 - \left(\frac{\lambda_2}{\lambda_1}\right)^n\right)}{\left(1 - \left(\frac{\lambda_2}{\lambda_1}\right)^n\right)x_0 + \sqrt{a}\left(1 + \left(\frac{\lambda_2}{\lambda_1}\right)^n\right)} = \frac{\sqrt{a}x_0 + a}{x_0 + \sqrt{a}} = \sqrt{a}.$$

Problem 4 *The sequence $(a_n)_{n\geq 0}$ is defined by $x_0 = a$ and*

$$a_{n+1} = \frac{a_n + \sqrt{3}}{1 - \sqrt{3}a_n}, \qquad n = 0, 1, \ldots.$$

Find a_{2019}.

Solution The sequence $(a_n)_{n\geq 0}$ is defined by the homographic function

$$f(z) = \frac{z + \sqrt{3}}{-\sqrt{3}z + 1}.$$

The matrix is

$$A_f = \begin{pmatrix} 1 & \sqrt{3} \\ -\sqrt{3} & 1 \end{pmatrix} = (-2)\begin{pmatrix} \frac{1}{2} & -\frac{\sqrt{3}}{2} \\ \frac{\sqrt{3}}{2} & -\frac{1}{2} \end{pmatrix} = (-2)\begin{pmatrix} \cos\frac{2\pi}{3} & -\sin\frac{2\pi}{3} \\ \sin\frac{2\pi}{3} & \cos\frac{2\pi}{3} \end{pmatrix}.$$

A simple inductive argument shows that

$$(A_f)^n = (-2)^n \begin{pmatrix} \cos\frac{2n\pi}{3} & -\sin\frac{2n\pi}{3} \\ \sin\frac{2n\pi}{3} & \cos\frac{2n\pi}{3} \end{pmatrix},$$

hence

$$a_n = f^n(a_0) = \frac{\left(\cos\frac{2n\pi}{3}\right)a - \sin\frac{2n\pi}{3}}{\left(\sin\frac{2n\pi}{3}\right)a + \cos\frac{2n\pi}{3}}, \qquad n = 0, 1, \ldots.$$

Clearly, $a_{n+3} = a_n$, $n = 0, 1, \ldots$, i.e., $(a_n)_{n\geq 0}$ is periodic having period 3. Since $2009 = 669 \cdot 3 + 2$, we have

$$a_{2009} = a_2 = \frac{a - \sqrt{3}}{\sqrt{3}a + 1}.$$

Problem 5 *Let $(x_n)_{n\geq 0}$ be the sequence defined by $x_0 = a$ and*

$$x_{n+1} = \frac{x_n + \sqrt{2} - 1}{1 - (\sqrt{2} - 1)x_n}, \qquad n = 0, 1, \ldots.$$

1. *Prove that $(x_n)_{n\geq 0}$ is periodic and find its period.*
2. *Find x_{2009}.*

Solution

1. The sequence is defined by

$$f(z) = \frac{z + \sqrt{2} - 1}{-(\sqrt{2} - 1)z + 1}.$$

Note that

$$1 = \tan\frac{\pi}{4} \tan 2\frac{\pi}{8} = \frac{2\tan\frac{\pi}{8}}{1 - \tan^2\frac{\pi}{8}},$$

hence $\tan\frac{\pi}{8} = \sqrt{2} - 1$. It follows that

$$f(z) = \frac{z + \sqrt{2} - 1}{-(\sqrt{2} - 1)z + 1} = \frac{z + \tan\frac{\pi}{8}}{-(\tan\frac{\pi}{8})z + 1} = \frac{\left(\cos\frac{\pi}{8}\right)z + \sin\frac{\pi}{8}}{-\left(\sin\frac{\pi}{8}\right)z + \cos\frac{\pi}{8}}.$$

The matrix of f is

$$A_f = \begin{pmatrix} \cos\frac{\pi}{8} & \sin\frac{\pi}{8} \\ -\sin\frac{\pi}{8} & \cos\frac{\pi}{8} \end{pmatrix} = \begin{pmatrix} \cos\left(-\frac{\pi}{8}\right) & -\sin\left(-\frac{\pi}{8}\right) \\ \sin\left(-\frac{\pi}{8}\right) & \cos\left(-\frac{\pi}{8}\right) \end{pmatrix}.$$

By an inductive argument we can deduce that

$$(A_f)^n = \begin{pmatrix} \cos\left(-\frac{n\pi}{8}\right) & -\sin\left(-\frac{n\pi}{8}\right) \\ \sin\left(-\frac{n\pi}{8}\right) & \cos\left(-\frac{n\pi}{3}\right) \end{pmatrix} = \begin{pmatrix} \cos\left(\frac{n\pi}{8}\right) & \sin\left(\frac{n\pi}{8}\right) \\ -\sin\left(\frac{n\pi}{8}\right) & \cos\left(\frac{n\pi}{8}\right) \end{pmatrix}.$$

It follows that

$$x_n = f^n(x_0) = \frac{\left(\cos\frac{n\pi}{8}\right)a + \sin\frac{n\pi}{8}}{-\left(\sin\frac{n\pi}{8}\right)a + \cos\frac{n\pi}{8}}, \quad n = 0, 1, \ldots.$$

Clearly, $x_{n+8} = x_n$, $n = 0, 1, \ldots$, i.e., $(x_n)_{n \geq 0}$ is periodic having period 8.

2. Since $2009 = 251 \cdot 8 + 1$, we obtain

$$x_{2009} = x_1 = \frac{a + \sqrt{2} - 1}{1 - (\sqrt{2} - 1)a}.$$

Problem 6 *Let $a, b > 0$ be real numbers such that $a^2 > b$. Define the sequence $(x_n)_{n \geq 0}$ by $x_0 = \alpha > 0$ and*

$$x_{n+1} = \frac{ax_n + b}{x_n + a}, \quad n = 0, 1, \ldots.$$

Prove that $(x_n)_{n \geq 0}$ is convergent and find $\lim_{n \to \infty} x_n$.

Solution The sequence $(x_n)_{n \geq 0}$ is defined by the homographic function

$$f(z) = \frac{az + b}{z + a}.$$

The matrix is

$$A_f = \begin{pmatrix} a & b \\ 1 & a \end{pmatrix},$$

which has the characteristic equation $\lambda^2 - 2a\lambda + a^2 - b = 0$, having the roots $\lambda_1 = a + \sqrt{b}$, $\lambda_2 = a - \sqrt{b}$.

By Cayley's Theorem one can deduce that

$$(A_f)^n = \frac{1}{2\sqrt{b}} \begin{pmatrix} \sqrt{b}(\lambda_1^n + \lambda_2^n) & b(\lambda_1^n - \lambda_2^n) \\ \lambda_1^n - \lambda_2^n & \sqrt{b}(\lambda_1^n + \lambda_2^n) \end{pmatrix},$$

hence

$$x_n = \frac{\sqrt{b}(\lambda_1^n + \lambda_2^n)x_0 + b(\lambda_1^n - \lambda_2^n)}{(\lambda_1^n - \lambda_2^n)x_0 + \sqrt{b}(\lambda_1^n + \lambda_2^n)}, \quad n = 0, 1, \dots,$$

and we can deduce that $\lim_{n\to\infty} x_n = \sqrt{b}$.

Problem 7 *Find all sequences $(a_n)_{n\geq 1}$ satisfying the following properties:*

1. *for all positive integers n, a_n is an integer;*
2. $a_{n+2} = \dfrac{na_n + 1}{a_n + n}$, $n = 1, 2, \dots$.

Solution Not that the sequence $(a_n)_{n\geq 1}$ is not defined by a linear fractional (homographic) transformation. Clearly, $a_1 \neq -1$, and then we have

$$a_3 = \frac{a_1 + 1}{a_1 + 1} = 1, \quad a_5 = \frac{3 + 1}{1 + 3} = 1.$$

Assuming that $a_{2k+1} = 1$, it follows that

$$a_{2k+3} = \frac{(2k + 1)a_{2k+1} + 1}{a_{2k+1} + 2k + 1} = \frac{2k + 2}{2k + 2} = 1,$$

hence $a_{2k+1} = 1$, $k = 1, 2, \dots$.

We also have

$$a_4 = \frac{(2a_2 + 1}{a_2 + 1} = 2 - \frac{3}{a_2 + 2}.$$

Since a_2 and a_4 are integers, it follows that $a_2 + 2 \in \{-3, -1, 1, 3\}$, hence $a_2 \in \{-5, -3, -1, 1\}$ and $a_4 \in \{-1, 1, 3, 5\}$. Computing all possibilities for a_6 we get $a_6 \in \left\{-1, 1, \frac{21}{9}, \frac{13}{7}\right\}$, and since a_6 is also an integer, we have $a_6 \in \{-1, 1\}$.

Observe that $a_{2k+2} = 1$ if and only if $a_{2k} = 1$, $k = 1, 2, \dots$, and $a_{2k+2} = -1$ if and only if $a_{2k} = -1$, $k = 1, 2, \dots$. In conclusion, the sequences $(a_n)_{n\geq 1}$ with the desired property are

$$a_1, 1, 1, 1, \dots,$$

$$a_1, -1, 1, -1, \dots,$$

where $a_1 \in \mathbb{Z} \setminus \{-1\}$.

8.7 Complex Recurrent Sequences

Problem 1 (Ovidiu Bagdasar) *Define the sequence $(w_n)_{n\geq 0}$ by the recurrence relation*

$$w_{n+2} = 2w_{n+1} + 3w_n, \quad w_0 = 1, w_1 = i, \quad n = 0, 1, \ldots.$$

1. *Find the general formula for w_n and compute the first 9 terms.*
2. *Show that $|\Re w_n - \Im w_n)| = 1$ for all $n \geq 1$.*

Solution

1. The characteristic equation $t^2 - 2t - 3 = 0$ has the eigenvalues $t_1 = -1$ and $t_2 = 3$, hence by the Binet-type formula we have

$$w_n = c_1(-1)^n + c_2 3^n,$$

where c_1 and c_2 are obtained from the initial conditions $w_0 = c_1 + c_2 = 1$ and $w_1 = -c_1 + 3c_2 = i$, from where we obtain $c_2 = \frac{3-i}{4}$ and $c_2 = \frac{1+i}{4}$.
Simple computations show that the first 9 terms of the sequence are

$$1, i, 3 + 2i, 6 + 7i, 21 + 20i, 60 + 61i, 183 + 182i, 546 + 547i, 1641 + 1640i.$$

2. Denoting $w_n = a_n + b_n i$, we actually prove that the proposition $P(n)$: $a_n - b_n = (-1)^n$ is true for all $n \geq 1$. From the above calculations, this clearly holds up to $n = 9$. Assume that $P(k)$ holds for $k = 1, \ldots, n$. Clearly,

$$\begin{aligned} a_{n+1} + b_{n+1}i = w_{n+1} &= 2w_n + 3w_{n-1} = 2(a_n + b_n i) + 3(a_{n-1} + b_{n-1}i) \\ &= 2a_n - 3a_{n-1} + (2b_n + 3b_{n-1})i. \end{aligned}$$

By the induction hypothesis, $a_n - b_n = (-1)^n$ and $a_{n-1} - b_{n-1} = (-1)^{n-1}$, therefore

$$\begin{aligned} a_{n+1} - b_{n+1} &= 2(a_n - b_n) + 3(a_{n-1} - b_{n-1}) \\ &= 2[(-1)^n + (-1)^{n-1}] + (-1)^{n-1} = (-1)^{n+1}, \end{aligned}$$

which ends the proof.

An alternative proof can be obtained by deriving the recursions satisfied by the integer sequences $(a_n)_{n\geq 0}$ and $(b_n)_{n\geq 0}$.

Problem 2 (Ovidiu Bagdasar) *Define the sequence* $(w_n)_{n\geq 0}$ *by the recurrence relation*

$$2w_{n+2} = \left(1 + (2+\sqrt{3})i\right) w_{n+1} + \left(\sqrt{3} - i\right) w_n, \quad w_0 = 1, w_1 = i, \quad n = 0, 1, \dots.$$

1. *Find the formula of* w_n*;*
2. *Prove that the sequence is periodic and find the period;*
3. *Discuss the behavior of the sequence* $(w_n)_{n\geq 0}$*, if this satisfies the same recurrence relation, but with the starting values* $w_0 = 1$, $w_1 = 2i$.

Solution

1. The characteristic equation for this problem has the formula

$$2z^2 - \left(1 + (2+\sqrt{3})i\right) z - (\sqrt{3} - i) = 0,$$

which divided by two is proven to produce the factorization

$$(z - z_1)(z - z_2) = 0,$$

where $z_1 = e^{\frac{2\pi}{3}i} = \frac{1}{2} + \frac{\sqrt{3}}{2}i$ and $z_2 = e^{\frac{2\pi}{4}i} = i$.
Applying the Binet-type formula, one obtains

$$w_n = Az_1^n + Bz_2^n,$$

where $A = \frac{z_2 - i}{z_2 - z_1}$ and $B = \frac{i - z_1}{z_2 - z_1}$.
2. Since the sequences $(z_1^n)_{n\geq 0}$ and $(z_2^n)_{n\geq 0}$ have periods 3 and 4 respectively, the sequence w_n is periodic and its period divides $\text{lcm}(3, 4) = 12$. Since $z_2 = i$, $A = 0$, the sequence terms satisfy $w_n = Bz_2^n$, which has period 4. The resulting orbit is

$$\{1, i, -1, -i, 1, i, \dots\}.$$

3. In this case $A = \frac{z_2 - i}{z_2 - z_1} \neq 0$ and $B = \frac{i - z_1}{z_2 - z_1} \neq 0$, and the sequence has period 12. Based on Theorem 5.11 or [33], the orbit can be decomposed into either three squares or four equilateral triangles.

Problem 3 (Ovidiu Bagdasar) *Let the sequence* $(w_n)_{n\geq 0}$ *be defined by the recurrence* $w_{n+2} = pw_{n+1} + qw_n$ *with* $p, q \in \mathbb{C}$*, and* $w_0 = 1$, $w_1 = i$.

1. *Find three distinct sequences having period* 15.
2. *How many such sequences exist?*

Solution Based on Section 5.4 and [28], a second-order linear recurrence relation of the form has the characteristic equation

$$z^2 - pz - q = (z - z_1)(z - z_2) = 0.$$

If $z_1 \neq z_2$, the general term is given by the formula

$$w_n = Az_1^n + Bz_2^n,$$

where $A = \frac{z_2 - i}{z_2 - z_1}$ and $B = \frac{i - z_1}{z_2 - z_1}$.

A sufficient condition for $(w_n)_{n\geq 0}$ to be periodic of period 15 is for z_1 to be a primitive root of order 3, z_2 a primitive root of order 5, and $AB \neq 0$.

One can choose $z_1 = e^{\frac{2\pi j}{3}i}$, $j = 1, 2$ and $z_2 = e^{\frac{2\pi l}{5}i}$, $l = 1, 2, 3, 4$, while in this case AB is clearly not zero.

From Example 5.5, the number of periodic orbits of length $k = pq$ with p and q primes is

$$H_P(k) = (p-1)(q-1)(pq + p + q)/2.$$

For $p = 3$ and $q = 5$ this gives $H_P(15) = 92$.

Problem 4 (Ovidiu Bagdasar) *Consider the sequence $(w_n)_{n\geq 0}$ defined by $w_{n+2} = pw_{n+1} + qw_n$ with $p, q \in \mathbb{C}$, and $w_0 = 1$, $w_1 = i$. Find the number of periodic sequences $(w_n)_{n\geq 0}$, whose period is the cube of a given prime number.*

Solution The sequence has the characteristic equation

$$z^2 - pz - q = 0,$$

whose roots are denoted by z_1 and z_2.

For distinct roots $z_1 \neq z_2$ of (5.2), the general term of Horadam's sequence $(w_n)_{n\geq 0}$ is $w_n = Az_1^n + Bz_2^n$ (5.4), where A and B are given by (5.5). Clearly, in this case we have $AB \neq 0$, hence the orbit is not degenerated.

Periodic orbits are obtained when z_1 and z_2 are distinct roots of unity $z_1 = e^{2\pi i p_1/k_1}$ and $z_2 = e^{2\pi i p_2/k_2}$. The distinct sequences of period k are enumerated from the quadruples (p_1, k_1, p_2, k_2) such that $\gcd(p_1, k_1) = \gcd(p_2, k_2) = 1$, $\text{lcm}(k_1, k_2) = k$, and $k_1 \leq k_2$. Their number is denoted by $H_P(k)$.

By formula (5.34), this number is given by

$$H_P(k) = \sum_{[k_1, k_2] = k,\ k_1 < k_2} \varphi(k_1)\varphi(k_2) + \frac{1}{2}\varphi(k)\left(\varphi(k) - 1\right).$$

If s is prime and $k = s^3$, then $\varphi(k) = s^3(1 - 1/s) = s^3 - s^2$. The divisor pairs (k_1, k_2) in the set $\{(1, k), (s, k), (s^2, k), (k, k)\}$, have multiplicities $\varphi(s^j)\varphi(k)$, with $j = 0, 1, 2$, and $\varphi(k)(\varphi(k) - 1)/2 = (k - k/s)(k - k/s - 1)/2$. We obtain

$$H_P(k) = \left[1 + (s-1) + (s^2 - s^1) + (s^3 - s^2 - 1)/2\right]\varphi(k) = \frac{s^6 - s^4 - s^3 + s^2}{2}.$$

A more general case is presented in Example 5.4.

Problem 5 (Ovidiu Bagdasar) *Let $(w_n)_{n\geq 0}$ be the sequence defined by the initial conditions $w_0 = a$, $w_1 = b$, and the recurrence relation*

$$w_{n+2} = e^{2\pi i\sqrt[3]{2}}\left(e^{2\pi i\sqrt[3]{2}} + 1\right)w_{n+1} - e^{2\pi i\sqrt[3]{2}\left(\sqrt[3]{2}+1\right)}w_n, \quad n = 0, 1, \ldots.$$

1. *Show that the closure of the set $\{w_n : n \in \mathbb{N}\}$ is an annulus.*
2. *Find a, b for which the orbit of $(w_n)_{n\geq 0}$ fills the annulus $U(0; 1, 2)$.*

Solution

1. One can easily check that the characteristic equation of the recurrence can be factorized as

$$(z - z_1)(z - z_2) = 0,$$

where $z_1 = e^{2\pi i\sqrt[3]{2}} = e^{2\pi i x_1}$ and $z_1 = e^{2\pi i x_2}$. By Proposition A.2, the numbers $1, \sqrt[3]{2}, \sqrt[3]{4}$ are linearly independent over $\mathbb{Q}$, hence as seen in Section 5.7.1, the orbit of $(w_n)_{n\geq 0}$ is dense within an annulus. The graph obtained for $a = i$ and $b = 1 + i$ is given in Figure 8.1.

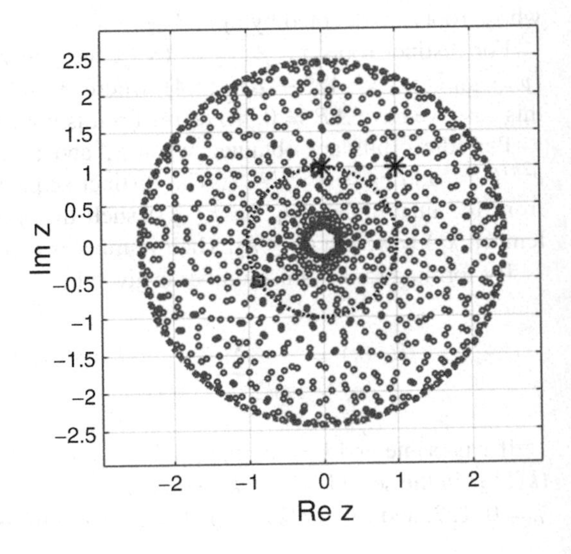

Fig. 8.1 First 1000 terms of the sequence $(w_n)_{n\geq 0}$ in Problem 5 (1) (circles), computed for $a = i$, $b = 1 + i$ (stars), $z_1 = e^{2\pi i\sqrt[3]{2}}$, $z_2 = e^{2\pi i\sqrt[3]{4}}$. Boundaries of $U(0, ||A| - |B||, |A| + |B|)$ (dotted line) with A, B from (5.5) and unit circle (solid line) are also plotted

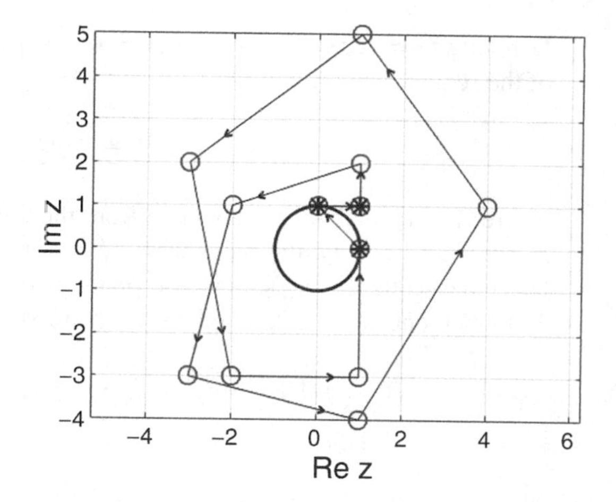

Fig. 8.2 Sequence $(w_n)_{n\geq 0}$ (circles) for Problem 6. Initial conditions $w_0 = 1$, $w_1 = i$, $w_2 = 1 + i$ (stars) and the unit circle are also plotted (solid line). Arrows indicate the orbit direction

2. Given two real numbers $0 < R_1 < R_2$, the condition a and b should satisfy is

$$|az_2 - b| = \frac{R_1 + R_2}{2}|z_2 - z_1|, \quad |az_1 - b| = \frac{R_2 - R_1}{2}|z_2 - z_1|,$$

where a, b are initial conditions and $|z_2 - z_1|$ is the distance between z_1, z_2. In our case, we would have to replace $R_1 = 1$ and $R_2 = 2$.

Problem 6 (Ovidiu Bagdasar) *Let $(w_n)_{n\geq 0}$ be a recurrent sequence which has the initial values $w_0 = 1$, $w_1 = i$, $w_2 = 1 + i$ (Figure 8.2).*

1. *Show that if $(w_n)_{n\geq 0}$ satisfies the recurrence relation*

$$w_{n+3} = (1 + i)w_{n+2} - (1 + i)w_{n+1} + iw_n,$$

then it is periodic, and find its period.

2. *Find the complex number q for which the sequence $(w_n)_{n\geq 0}$ defined by*

$$w_{n+3} = (q^2 + q - 1)w_{n+2} + (q + q^2 - q^3)w_{n+1} - q^3 w_n,$$

is periodic. What are the possible values of the period?

Solution

1. The sequence has the characteristic equation

$$z^3 - (1 + i)z^2 + (1 + i)z - i = 0,$$

which can be factorized as

$$(z - i)(z^2 - z + 1) = (z - e^{2\pi i \frac{1}{4}})(z - e^{2\pi i \frac{1}{6}})(z - e^{2\pi i \frac{5}{6}}) = 0.$$

Denoting $z_1 = i$, $z_2 = e^{2\pi i \frac{1}{6}}$ and $z_3 = e^{2\pi i \frac{5}{6}}$, and using (6.68), the general term of the sequence is given by

$$w_n = A_1 z_1^n + A_2 z_2^n + A_3 z_3^n,$$

where A_1, A_2, and A_3 are obtained from the initial conditions. Since z_1, z_2 and z_3 are roots of unity, the sequences $(z_1^n)_{n\geq 0}$, $(z_2^n)_{n\geq 0}$, and $(z_1^3)_{n\geq 0}$ are periodic, hence the sequence $(w_n)_{n\geq 0}$ is also periodic, of period lcm$(4, 6, 6) = 12$.

In fact, one can check that the sequence terms are:

$$1,\ i,\ 1+i,\ 1+2i,\ -2+i,\ -3-3i,\ 1-4i,\ 4+i,$$
$$1+5i,\ -3+2i,\ -2-3i,\ 1-3i,\ 1,\ i,\ \ldots.$$

2. One can check that the characteristic equation is

$$z^3 - (q^2+q-1)z^2 - (q+q^2-q^3)z + q^3 = 0,$$

which can be factorized as

$$(z+1)(z-q)(z-q^2) = 0.$$

From Section 6.3, the sequence $(w_n)_{n\geq 0}$ can only be periodic if the roots of the characteristic equation are roots of unity. We have three cases.

Case 1. $q = 1$ Here $z = 1$ is a double root and from (6.89), the general term of the sequence is

$$w_n = A_1(-1)^n + (A_2 + nA_3)1^n,$$

where from the initial conditions we obtain

$$A_1 = \frac{1}{2} - \frac{1}{4}i, \quad A_2 = \frac{1}{2} + \frac{1}{4}i, \quad A_3 = \frac{1}{2}i.$$

This clearly shows that $(w_n)_{n\geq 0}$ is not periodic.

Case 2. $q = -1$ Here $z = -1$ is a double root and 1 is a single root. By the same argument, the general solution is

$$w_n = A_1 + (A_2 + nA_3)(-1)^n,$$

where from the initial condition

$$A_1 = \frac{1}{2} - \frac{3}{4}i, \quad A_2 = \frac{1}{2} + \frac{3}{4}i, \quad A_3 = \frac{1}{2}i,$$

hence the sequence $(w_n)_{n\geq 0}$ is not periodic.

Case 3. $q = e^{2\pi i \frac{p}{k}}$, $k \geq 3$, $\gcd(p,k) = 1$ In this case, the roots of the characteristic equation $z_1 = -1$, $z_2 = q$ and $z_3 = q^2$ are distinct, and the general term is given by

$$w_n = A_1(-1)^n + A_2 q^n + A_3 q^{2n}.$$

The sequence is clearly periodic and has the period $\mathrm{lcm}(2,k)$.

Problem 7 (Ovidiu Bagdasar) *Let $(w_n)_{n\geq 0}$ be a sequence defined by $w_0 = 1$, $w_1 = i$, $w_2 = 1+i$ and $w_{n+3} = pw_{n+2} + qw_{n+1} + rw_n$, where $p, q, r \in \mathbb{C}$.*

1. *Find p, q, r such that $(w_n)_{n\geq 0}$ is periodic of period 2019;*
2. *How many such sequences of period 2019 exist, if the characteristic polynomial has distinct roots?*

Solution

1. If the roots z_1, z_2, z_3 of the characteristic equation

$$z^3 - pz^2 - qz - r = 0,$$

are distinct, then the general term of the sequence is given by

$$w_n = A_1 z_1^n + A_2 z_2^n + A_3 z_3^n.$$

Clearly, $2019 = 3 \cdot 673$, and since for the given starting values $A_1 A_2 A_3 \neq 0$, it is sufficient to consider $z_2 = e^{2\pi i \frac{1}{3}}$, $z_2 = e^{2\pi i \frac{2}{3}}$ and $z_3 = e^{2\pi i \frac{1}{673}}$.

2. Following the steps outlined in Section 6.4.5.1, when $k = pq$ is a product of two primes, the number of periodic sequence of length k is given by the explicit formula

$$H_P(3;k) = \frac{\varphi(k)}{6}\left[k\left(k + (p+q) - 4\right) + (p+q-1)^2 - 1\right],$$

while

$$\varphi(k) = pq\left(1 - \frac{1}{p}\right)\left(1 - \frac{1}{q}\right) = (p-1)(q-1).$$

For $p = 3$ and $q = 673$ this gives $H_P(3; 2019) = 1319080672$.

Problem 8 (Dorin Andrica) *Let $(z_n)_{n\geq 1}$ be the complex sequence defined by $z_{n+1} = z_n^2 - z_n + 1$, $n = 1, 2, \ldots$, where $z_n^2 - z_n + 1 \neq 0$ and $z_1 \neq 0, 1$. Prove that there are two points O_1 and O_2 in the complex plane such that for any $n \geq 1$, the points of coordinates z_{n+1} and $\frac{1}{z_1} + \cdots + \frac{1}{z_n}$ are located on circles with centers O_1 and O_2 and radius* 1.

Solution Clearly, $z_n \neq 1$ and $z_n \neq 0$, $n = 1, 2, \ldots$. The recurrence relation is equivalent to $z_{n+1} - 1 = z_n(z_n - 1)$, hence we have

$$\frac{1}{z_{n+1} - 1} = \frac{1}{z_n(z_n - 1)} = \frac{1}{z_n - 1} - \frac{1}{z_n}, \quad n = 1, 2, \ldots.$$

From the relation

$$\frac{1}{z_n} = \frac{1}{z_n - 1} - \frac{1}{z_{n+1} - 1},$$

we obtain by summation that

$$\frac{1}{z_1} + \cdots + \frac{1}{z_n} = \frac{1}{z_1 - 1} - \frac{1}{z_{n+1} - 1},$$

therefore

$$(z_{n+1} - 1)\left[\frac{1}{z_1 - 1} - \left(\frac{1}{z_1} + \cdots + \frac{1}{z_n}\right)\right] = 1.$$

It follows that

$$\left|z_{n+1} - 1\right| \cdot \left|\frac{1}{z_1} + \cdots + \frac{1}{z_n} - \frac{1}{z_1 - 1}\right| = 1.$$

Considering the points $O_1(1)$ and $O_2\left(\frac{1}{z_1 - 1}\right)$, we have $A(z_{n+1}) \in \mathcal{C}(O_1; R_n)$ and $B\left(\frac{1}{z_1} + \cdots + \frac{1}{z_n}\right) \in \mathcal{C}\left(O_1; \frac{1}{R_n}\right)$, which ends the proof.

8.8 Recurrent Sequences in Combinatorics

Problem 1 *Prove the identity*

$$\sum_{k=0}^{n} \binom{n+k}{k} \frac{1}{2^k} = 2^n.$$

Solution Denote by $a_n = \sum_{k=0}^{n} \binom{n+k}{k} \frac{1}{2^k}$. Clearly, $a_1 = 2$ and

$$\begin{aligned}
a_{n+1} &= \sum_{k=0}^{n+1} \binom{n+1+k}{k} \frac{1}{2^k} \\
&= \sum_{k=0}^{n+1} \binom{n+k}{k} \frac{1}{2^k} + \sum_{k=1}^{n+1} \binom{n+k}{k-1} \frac{1}{2^k} \\
&= a_n + \binom{2n+1}{n+1} \frac{1}{2^{n+1}} + \frac{1}{2} \sum_{k=0}^{n+2} \binom{n+1+k-1}{k-1} \frac{1}{2^{k-1}} \\
&- \binom{2n+2}{n+1} \frac{1}{2^{n+2}} = a_n + \frac{1}{2} a_{n+1}.
\end{aligned}$$

This confirms that $a_{n+1} = 2a_n$ for all integers $n \geq 1$, hence $a_n = 2^n$.

Problem 2 *Prove the identity*

$$\sum_{k=0}^{n} (-1)^k \binom{2n-k}{k} = \cos\left(\frac{2\pi n}{3}\right) + \frac{1}{\sqrt{3}} \sin\left(\frac{2\pi n}{3}\right)$$

$$= \begin{cases} 1 & \textit{if}\, n = 3p; \\ 0 & \textit{if}\, n = 3p+1, \quad p \in \mathbb{N}. \\ -1 & \textit{if}\, n = 3p+2. \end{cases}$$

Solution The desired sum can also be written as

$$S_n = \sum_{k \geq 0}^{n} (-1)^k \binom{2n-k}{k} = \binom{2n}{0} - \binom{2n-1}{1} + \binom{2n-2}{2} - \cdots .$$

Using the identity

$$\binom{2n-k}{k} = \binom{2n+1-k}{k} - \binom{2n-k}{k-1},$$

one can easily prove that $S_n = S_{n+1} + 2S_n + S_{n-1}$, hence the numbers S_n satisfy the recurrence relation $S_{n+1} = -(S_n + S_{n-1})$. Since $S_1 = 0$, $S_2 = -1$, and $S_3 = 1$, the desired result can be proved by applying induction on n.

Alternatively, one can solve the recurrence equation explicitly. The characteristic equation is $x^2 + x + 1 = 0$, which has as solutions

$$x_1 = \cos\left(\frac{2\pi}{3}\right) + i \sin\left(\frac{2\pi}{3}\right), \quad x_2 = \cos\left(\frac{4\pi}{3}\right) + i \sin\left(\frac{4\pi}{3}\right).$$

It follows that the general solution of the recurrence equation is given by the formula

$$S_n = A\cos\left(\frac{2\pi n}{3}\right) + B\sin\left(\frac{2\pi n}{3}\right),$$

where the constants A and B are computed from $S_1 = 0$ and $S_2 = -1$, as

$$A\left(-\frac{1}{2}\right) + B\left(\frac{\sqrt{3}}{2}\right) = 0, \quad A\left(-\frac{1}{2}\right) + B\left(-\frac{\sqrt{3}}{2}\right) = 1,$$

hence $A = 1$ and $B = \frac{1}{\sqrt{3}}$. Simple calculations confirm that $S_3 = A = 1$.

Problem 3 *Determine the number of functions $f : \{1, 2, , \ldots, n\} \to \{1, 2, 3, 4, 5\}$ satisfying the property $\mid f(k+1) - f(k) \mid\geq 3$, for all $k \in 1, \ldots, n-1$.*

Solution From the hypothesis, we clearly have $f(k) \neq 3$, for $k = 1, \ldots, n$. Let a_n, b_n, c_n, and c_n, be the number of functions having this property, for which $f(n) = 1, 2, 4$, or 5. The following recurrence relations hold true:

$$\begin{cases} a_{n+1} &= c_n + d_n \\ b_{n+1} &= d_n \\ c_{n+1} &= a_n \\ d_{n+1} &= a_n + b_n. \end{cases}$$

The number we are looking for is $x_n = a_n + b_n + c_n + d_n$. One has $a_2 = 2$ (since $f(2) = 1$ implies $f(1) = 4$ or $f(1) = 5$), $b_2 = 1$ (since $f(2) = 1$ implies $f(1) = 5$), $c_2 = 1$ ($f(2) = 4$ implies $f(1) = 1$), and $d_2 = 2$ (since $f(2) = 5$ implies $f(1) = 1$ or $f(1) = 2$). From here it follows that $a_n = d_n$ and $b_n = c_n$, while using the previous relations one obtains

$$\begin{cases} a_{n+1} &= a_n + b_n \\ b_{n+1} &= a_n, \end{cases}$$

therefore $a_{n+1} = a_n + a_{n-1}$, for $n \geq 2$. We therefore have $x_n = 2(a_n + b_n)$, where $x_2 = 6 = 2 \cdot 3$, and $x_3 = 10 = 2 \cdot 5$. If F_n is the nth Fibonacci number, from $x_2 = 2F_4$, $x_3 = 2F_5$, and $x_{n+1} = x_n + x_{n-1}$, we have $x_n = 2 \cdot F_{n+2}$.

Problem 4 *In how many ways can you pave a $2 \times n$ rectangle with 1×2 tiles?*

Solution Denote the desired number by a_n. The different possible coverings for a rectangle of size $2 \times (n+1)$ is denoted by a_{n+1}. Each of the a_{n+1} coverings may finish with a horizontal, or vertical tile, as in Figure 8.3.

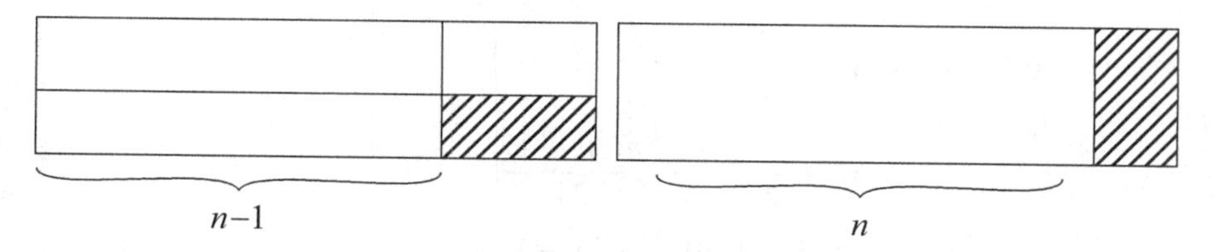

Fig. 8.3 Possible tile configurations

If the last tile is horizontal, then there are actually two such tiles stacked on top of the other, and there are a_{n-1} ways of covering the remaining $2 \times (n-1)$ rectangle. If the covering finishes with a vertical rectangle of size 1×2, then the remaining $2 \times n$ rectangle can be covered in a_n ways. We obtain the recurrence relation

$$a_{n+1} = a_n + a_{n-1}, \quad a_1 = 1, a_2 = 2,$$

hence $a_n = F_{n+1}$, where $(F_n)_{n\geq 0}$ denotes the Fibonacci sequence.

Problem 5 *Prove the identity*

$$\sum_{k=0}^{n}(-1)^{n-k}\binom{n}{k}k^p = \begin{cases} 0 & \text{if } 0 \leq p < n, \\ p! & \text{if } p = n, \\ \frac{n}{2}p! & \text{if } p = n+1, \\ \frac{n(3n+1)}{24}p! & \text{if } p = n+2, \\ \frac{n^2(n+1)}{48}p! & \text{if } p = n+3, \\ \frac{n(15n^3+30n^2+5n+1)}{1152}p! & \text{if } p = n+4. \end{cases}$$

First Solution Consider the function $(e^x - 1)^n$, in which we first compute the binomial and then consider the Taylor series expansion

$$(e^x-1)^n = \sum_{k=0}^{n}(-1)^{n-k}\binom{n}{k}e^{kx} = \sum_{k=0}^{n}(-1)^{n-k}\binom{n}{k}\left(\sum_{j=0}^{\infty}\frac{1}{j!}k^j x^j\right)$$
$$= \sum_{j=0}^{\infty}\frac{1}{j!}\left(\sum_{k=0}^{n}(-1)^{n-k}\binom{n}{k}k^j\right)x^j.$$

For the same function we can expand the exponential function, and then the binomial to obtain

$$
\begin{aligned}
(e^x-1)^n &= \left(x+\frac{1}{2!}x^2+\frac{1}{3!}x^3+\cdots\right)^n \\
&= x^n+\frac{n}{2}x^{n+1}+\frac{n(3n+1)}{24}x^{n+2}+\frac{n^2(n+1)}{48}x^{n+3} \\
&+\frac{n(15n^3+30n^2+5n+1)}{1152}x^{n+4}+\cdots .
\end{aligned}
$$

Identifying the coefficients in the two series, we obtain the desired formulae.

Second Solution [Dorin Andrica] A simple computation with the binomial coefficients shows that the following double recurrence relations hold [10]:

$$s_{n,p+1}=n(s_{n,p}+s_{n-1,p}), \quad p\ge 0.$$

In particular, one recovers the results from the problem.

Indeed, denoting $s_{n,p}=\sum_{k=0}^{n}(-1)^{n-k}\binom{n}{k}k^p$, for any real number p one has

$$
\begin{aligned}
n\left(s_{n,p}+s_{n-1,p}\right) &= n\left(\sum_{k=0}^{n}(-1)^{n-k}\binom{n}{k}k^p+\sum_{k=0}^{n-1}(-1)^{n-k-1}\binom{n-1}{k}k^p\right) \\
&= n\left(n^p+\sum_{k=0}^{n-1}(-1)^{n-k}\left[\binom{n}{k}-\binom{n-1}{k}\right]k^p\right) \\
&= n\left(n^p+\sum_{k=0}^{n-1}(-1)^{n-k}\frac{k}{n}\binom{n}{k}k^p\right) \\
&= \sum_{k=0}^{n}(-1)^{n-k}\binom{n}{k}k^{p+1}=s_{n,p+1}.
\end{aligned}
$$

We have used the identity

$$\binom{n}{k}-\binom{n-1}{k}=\binom{n-1}{k-1}=\frac{k}{n}\binom{n}{k}.$$

Problem 6 (Italian Mathematical Olympiad, 1996) *Given an alphabet with three letters a, b, c, find the number of words of n letters having an even number of a's.*

Solution Let S_n be the number of n letter words with even number of a's, and let T_n be the number of n letter words with odd number of a's. Clearly, one has $S_n+T_n=3^n$. Among the S_n words, there are T_{n-1} words ended in a and $2S_{n-1}$ words ended in b or c, hence $S_n=T_{n-1}+2S_{n-1}$. Similarly, one has $T_n=S_{n-1}+2T_{n-1}$. Subtracting these, we get $S_n-T_n=S_{n-1}-T_{n-1}$, therefore $S_n-T_n=S_1-T_1=2-1=1$. Therefore, $S_n=(3^n+1)/2$.

Problem 7 *Find the number of n-words from the alphabet $A = \{0, 1, 2\}$, if any two neighbors can differ by at most* 1.

Solution Let x_n be the number of n-words satisfying the condition. So $x_1 = 3$ and $x_2 = 7$. Let y_n be the number of n-words satisfying the condition and beginning with 0 (by interchanging 0 and 2, y_n is also the number of n-words satisfying the condition and beginning with 2). Considering a 0, 1, or 2 in front of an n-word, we get $x_{n+1} = 3x_n - 2y_n$ and $y_{n+1} = x_n - y_n$. Solving for y_n in the first equation, then substituting into the second equation, we get $x_{n+2} - 2x_{n+1} - x_n = 0$. For convenience, set $x_0 = x_2 - 2x_1 = 1$. Since $r^2 - 2r - 1$ has the roots $1 \pm \sqrt{2}$ and $x_0 = 1$ and $x_1 = 3$, one obtains that $x_n = \alpha(1+\sqrt{2})^n + \beta(1-\sqrt{2})^n$, where $\alpha = (1+\sqrt{2})/2$ and $\beta = (1-\sqrt{2})/2$. Therefore, $x_n = [(1+\sqrt{2})^{n+1} + (1-\sqrt{2})^{n+1}]/2$.

Problem 8 (Romanian Mathematical Olympiad, 1995) *Let $A_1, A_2, \ldots, A_n$ be points on a circle. Find the number of possible colorings of these points with p colors, $p \geq 2$, such that any two neighbors have distinct colors.*

Solution Let C_n be the answer for n points. We have $C_1 = p$, $C_2 = p(p-1)$, and $C_3 = p(p-1)(p-2)$. For $n+1$ points, if A_1 and A_n have different colors, then $A_1, \ldots, A_n$ can be colored in C_n ways, while A_{n+1} can be colored in $p-2$ ways. If A_1 and A_n have the same color, then $A_1, \ldots, A_n$ can be colored in C_{n-1} ways, while A_{n+1} can be colored in $p-1$ ways. It follows that $C_{n+1} = (p-2)C_n + (p-1)C_{n-1}$, for $n \geq 3$, which can be written as $C_{n+1} + C_n = (p-1)(C_n + C_{n-1})$.

This implies that $C_{n+1} + C_n = (p-1)^{n-2}(C_3 + C_2) = p(p-1)^n$. Then, one can use induction to show that $C_n = (p-1)^n + (-1)^n(p-1)$, for $n > 3$.

Problem 9 *Define a set S of integers to be* fat *if each of its elements is greater than its cardinal $|S|$. For example, the empty set and $\{5, 7, 91\}$ are fat, but $\{3, 5, 10, 14\}$ is not. Let f_n denote the number of fat subsets of $\{1, \ldots, n\}$. Derive a recursive relation for f_n.*

Solution The fat subsets of $\{1, 2, \ldots, n+1\}$ are of two types: subsets that contain $n+1$ and subsets that do not contain $n+1$. There are obviously f_n subsets that do not contain $n+1$ (these are just the fat subsets of $\{1, 2, \ldots, n\}$).

Consider now the fat subsets containing $n+1$. These are in a one-to-one correspondence with the fat subsets of $\{1, 2, \ldots, n-1\}$, by associating with each set $\{x_1, \ldots, x_k, n+1\}$ with $x_1 < x_2 < \cdots < x_k$ the set $\{x_1 - 1, x_2 - 1, \ldots, x_k - 1\}$. It is easy to check that this correspondence is well defined and bijective. Therefore, there are f_{n-1} fat subsets of $\{1, 2, \ldots, n+1\}$ that contain $n+1$.

The recurrence relation we are looking for is

$$f_{n+1} = f_n + f_{n-1}, \quad n \geq 2,$$

where $f_1 = 2$ and $f_2 = 3$. It can be easily checked that $f_n = F_{n+2}$, where F_n denotes the nth term of the Fibonacci sequence.

Problem 10 *Consider a cube of dimensions $1 \times 1 \times 1$. Let O and A be two of its vertices such that OA is the diagonal of a face of the cube. Which one is larger: the*

number of paths of length 1386 *beginning at* O *and ending at* O, *or the number of paths of length* 1386 *beginning at* O *and ending at* A?
(A path of length n *on the cube is a sequence of* $n+1$ *vertices, such that the distance between each two consecutive vertices is* 1.*)*

Solution Let a_n be the number of paths of length n from O to itself, and let b_n be the number of paths of length n from O to A.

First, notice that the number of paths of length 2 from O to itself is 3, while the number of paths of length 2 from A to O is 2.

Consider the $(n-1)$th vertex of a path from O to itself. Clearly, this has to be either O itself, or the other endpoint of a diagonal of a face containing O as the other endpoint, hence we have the relation

$$a_n = 2(b_{n-2} + b_{n-2} + b_{n-2}) + 3a_{n-2}.$$

Similarly, for considering the possibilities concerning the $(n-1)$th vertex of a path from O to A we obtain

$$b_n = 2(a_{n-2} + b_{n-2} + b_{n-2}) + 3b_{n-2}.$$

Taking the difference, one obtains

$$a_n - b_n = a_{n-2} - b_{n-2}.$$

Finally, we deduce that

$$a_{1386} - b_{1386} = a_{1384} - b_{1384} = a_0 - b_0 = 1,$$

hence $a_{1386} > b_{1386}$.

Problem 11 (Rishub Thaper, O443, MR-2018) *Let* $f(n)$ *be the number of permutations of the set* $\{1, 2, \ldots, n\}$, *such that no pair of consecutive integers appears in that order; that is,* 2 *does not follow* 1, 3 *does not follow* 2, *and so on.*

1. *Prove that* $f(n) = (n-1)f(n-1) + (n-2)f(n-2)$.
2. *If* $[\alpha]$ *denotes the nearest integer to a real number* α, *prove that*

$$f(n) = \frac{1}{n}\left[\frac{(n+1)!}{e}\right].$$

Solution

1. Given such a permutation of the integers 1 through n, removing n yields a permutation of the integers 1 through $n-1$ with the same property, unless n is between a pair of increasing consecutive integers. But if n is between k and $k+1$, then removing $k+1$ and decreasing each of the remaining integers greater than $k+1$ by 1 yield a permutation of the integers 1 through $n-2$ with the same

property. Conversely, given a permutation of the integers 1 through $n-1$ with the given property, $n-1$ permutations of the integers 1 through n can be obtained by placing n either before all the integers, or after any integer $1, \dots, n-2$. And given a permutation of the integers 1 through $n-2$ with the given property, $n-2$ permutations of the integers 1 through n can be obtained, one for each $k = 1, \dots, n-2$, as follows: increase each number $k+1, \dots, n-2$ by 1, then insert $n, k+1$ after k. By this process, each permutation with the given property of $1, \dots, n$ is either among $n-1$ associated with a permutation with the given property of $1, \dots, n-1$, or among $n-2$ associated with a permutation having the given property of $1, \dots, n-2$. Thus, one obtains

$$f(n) = (n-1)f(n-1) + (n-2)f(n-2).$$

2. Simple calculations show that

$$\frac{(n+1)!}{e} = (n+1)!\sum_{k=0}^{\infty}\frac{(-1)^k}{k!} = \sum_{k=2}^{n+1}(-1)^k\frac{(n+1)!}{k!} + (n+1)!\sum_{k=n+2}^{\infty}\frac{(-1)^k}{k!},$$

where $\sum_{k=2}^{n+1}(-1)^k\frac{(n+1)!}{k!}$ is an integer, and

$$\left|(n+1)!\sum_{k=n+2}^{\infty}\frac{(-1)^k}{k!}\right| < (n+1)!\sum_{k=n+2}^{\infty}\frac{1}{k!} <$$

$$< (n+1)!\sum_{k=n+2}^{\infty}\frac{1}{(n+1)!(n+2)^{k-n-1}} = \frac{1}{n+1} < 1,$$

so

$$\left[\frac{(n+1)!}{e}\right] = \sum_{k=2}^{n+1}(-1)^k\frac{(n+1)!}{k!}.$$

The values $f(2) = 1$ (permutation $(2,1)$) and $f(3) = 3$ (permutations $(1,3,2)$, $(2,1,3)$, $(3,2,1)$) match the formula for $f(n)$; assume by induction that it holds for $n-1$ and $n-2$ for $n \geq 3$. Then

$$\begin{aligned} f(n) &= (n-1)f(n-1) + (n-2)f(n-2) \\ &= \left[\frac{n!}{e}\right] + \left[\frac{(n-1)!}{e}\right] \\ &= \sum_{k=2}^{n}(-1)^k\frac{n!}{k!} + \sum_{k=2}^{n-1}(-1)^k\frac{(n-1)!}{k!} \end{aligned}$$

$$= (-1)^n + \sum_{k=2}^{n-1}(-1)^k \frac{n! + (n-1)!}{k!}$$

$$= \frac{1}{n}\left((n+1)(-1)^n + (-1)^{n+1} + \sum_{k=2}^{n-1}(-1)^k \frac{(n+1)!}{k!}\right)$$

$$= \frac{1}{n}\sum_{k=2}^{n+1}(-1)^k \frac{(n+1)!}{k!} = \frac{1}{n}\left[\frac{(n+1)!}{e}\right],$$

which completes the induction step.

Problem 12 (Mircea Merca, 11767, AMM-2014) *Prove that*

$$\sum \frac{(1+t_1+t_2+\cdots+t_n)!}{(1+t_1)!t_2!\cdots t_n!} = 2^n - F_n,$$

where F_k is the kth Fibonacci number and the sum is over all nonnegative integer solutions to $t_1 + 2t_2 + \cdots + nt_n = n$.

First Solution View the sum as over all partitions of $n+1$ having at least 1, considering $t_1 + 1$ as the number of copies of 1 and t_j as the number of copies of j, for $2 \le j \le n$. The summand counts the ways to permute parts, so the sum is the number of composites of $n+1$ having at least 1.

The number of compositions of $n+1$ is 2^n, so it suffices to prove that the number a_n of compositions of $n+1$ in which 1 does not feature is F_n. Clearly, this is true for $n = 0$ and $n = 1$. For $n \ge 2$, these compositions have last part 2 or greater than 2. Deleting the last part shows that there are a_{n-2} of the first type, and subtracting 1 from the last part shows that there are a_{n-1} of the second type. It follows that $a_n = a_{n-1} + a_{n-2}$, hence $a_n = F_n$, by induction.

Second Solution Rewrite the sum as

$$\sum \frac{(t_1+t_2+\cdots+t_n)!}{t_1!t_2!\cdots t_n!},$$

summed over all integer solutions to $t_1 + 2t_2 + \cdots + nt_n = n+1$ with $t_1 \ge 1$ and $t_i \ge 0$ for $i \ge 2$. The sum is the coefficient of x^{n+1} in the series

$$f(x) = \sum_{m=0}^{\infty}\left[\left(x + x^2 + \cdots\right)^m - \left(x^2 + x^3 + \cdots\right)^m\right]$$

$$= \sum_{m=0}^{\infty}\left[\left(\frac{x}{1-x}\right)^m - \left(\frac{x^2}{1-x}\right)^m\right]$$

$$= \frac{1-x}{1-2x} - \frac{1-x}{1-x-x^2} = \frac{x(1-x)^2}{(1-2x)(1-x-x^2)}.$$

The coefficient we are looking for is also that of x^n in

$$\frac{f(x)}{x} = \frac{(1-x)^2}{(1-2x)(1-x-x^2)} = \frac{1}{1-2x} - \frac{x}{1-x-x^2}.$$

The coefficient subtracted in the second term is the number of 1, 2-lists with sum $n-1$, well known to be F_n, so the answer is $2^n - F_n$.

Problem 13 (David Beckwith, 11754, AMM-2015) *When a fair coin is tossed n times, let $P(n)$ be the probability that the lengths of all runs (maximal constant strings) in the resulting sequence are of the same parity as n. Denoting by F_n the nth Fibonacci number, prove that*

$$P(n) = \begin{cases} \left(\frac{1}{2}\right)^{n/2} & \text{if } n \text{ is even,} \\ \left(\frac{1}{2}\right)^{n-1} F_n & \text{if } n \text{ is odd.} \end{cases}$$

Solution When n is even, all the runs have even length if and only if tosses $2i-1$ and $2i$ have the same outcome, for all i. Each constraint holds with probability $\frac{1}{2}$, so $P(n) = \left(\frac{1}{2}\right)^{n/2}$.

For any n, let $Q(n)$ be the number of lists of length n whose runs all have odd length. It is sufficient to prove that $Q_n = 2F_n$. Clearly, we have $Q(1) = Q(2) = 2$. Such lists of length n are obtained from a list of length $n-1$ by adding a run of length 1, or from a list of length $n-2$, by extending the last run by length 2. This shows that $Q(n) = Q(n-1) + Q(n-2)$ for $n \geq 2$, which ends the proof.

Problem 14 *Find the number of ways u_n in which we can climb a ladder with n steps, if we can climb either one, or two steps at a time?*

Solution Clearly, u_n corresponds to the number of ways in which n can be written as a sum of 1s and 2s, where the order of numbers matters. One can check that $u_1 = 1$, $u_2 = 2$, $u_3 = 3$, and $u_4 = 5$, since

$$2 = 2 = 1 + 1$$

$$3 = 2 + 1 = 1 + 2 = 1 + 1 + 1$$

$$4 = 2 + 2 = 2 + 1 + 1 = 1 + 2 + 1 = 1 + 1 + 2 = 1 + 1 + 1 + 1.$$

Notice that the first term of such a decomposition is either 1 or 2. There are u_{n-1} such decomposition starting with 1, and u_{n-2} which begin with 2, giving the recurrence relation

$$u_n = u_{n-1} + u_{n-2}.$$

Since the first two terms and the recurrent sequence coincide, we deduce that $u_n = F_n$, where F_n denotes the nth Fibonacci number.

8.9 Miscellaneous

Problem 1 (Josef Tkadlec, O466, MR-2018) *Let $n \geq 2$ be an integer. Prove that there exists a set S of $n-1$ real numbers such that whenever $a_1, \ldots, a_n$ are distinct numbers satisfying*

$$a_1 + \frac{1}{a_2} = a_2 + \frac{1}{a_3} = \cdots = a_{n-1} + \frac{1}{a_n} = a_n + \frac{1}{a_1},$$

then the common value of all these sums is a number from S.

Solution We show that $S = \{2\cos\frac{i\pi}{n} : i = 1, \ldots, n-1\}$. For any nonzero a_1, define $r_0(x) = a_1$ and $r_k(x) = 2x - \frac{1}{r_{k-1}(x)}$, $k \geq 1$. Denote the Chebysev polynomial of the second kind by $U_k(x)$, which is defined by $U_{-1}(x) = 0$, $U_0(x) = 1$, and $U_k(x) = 2xU_{k-1}(x) - U_{k-2}(x)$, $k \geq 1$. We claim that

$$r_k(x) = a_1 + \frac{(2a_1x - a_1^2 - 1)U_{k-1}(x)}{a_1U_{k-1}(x) - U_{k-2}(x)}, \qquad k = 1, 2, \ldots.$$

This is clearly true for $k = 1$, and we use induction to prove that it is generally true. Assume that it is true for a fixed value k. Then

$$\begin{aligned}
r_{k+1}(x) = 2x - \frac{1}{r_k(x)} &= 2x - \frac{1}{a_1 + \frac{(2a_1x - a_1^2 - 1)U_{k-1}(x)}{a_1U_{k-1}(x) - U_{k-2}(x)}} \\
&= 2x - \frac{a_1U_{k-1}(x) - U_{k-2}(x)}{a_1U_k(x) - U_{k-1}(x)} \\
&= a_1 + \frac{(2x - a_1)\left[a_1U_k(x) - U_{k-1}(x)\right] - a_1U_{k-1}(x) + U_{k-2}(x)}{a_1U_k(x) - U_{k-1}(x)} \\
&= a_1 + \frac{(2a_1x - a_1^2 - 1)U_k(x)}{a_1U_k(x) - U_{k-1}(x)}.
\end{aligned}$$

Now, if $a_1, \ldots, a_n$ are distinct numbers satisfying

$$a_1 + \frac{1}{a_2} = a_2 + \frac{1}{a_3} = \cdots = a_{n-1} + \frac{1}{a_n} = a_n + \frac{1}{a_1} = 2x,$$

then $2a_1x - a_1^2 - 1 \neq 0$ and

$$a_n = 2x - \frac{1}{a_1} = r_1(x) \to a_{n-1} = 2x - \frac{1}{a_n} = r_2(x) \to \cdots \to a_1 = 2x - \frac{1}{a_2} = r_n(x).$$

Hence, we can deduce that $U_{n-1}(x) = 0$. Finally, it is well known that $U_{n-1}(x) = 2^{n-1}\prod_{i=1}^{n-1}\left(x - \cos\frac{i\pi}{n}\right)$, completing the proof.

Problem 2 (Titu Andreescu, S334, MR-2015) *Let $(a_n)_{n\geq 0}$ be a sequence of real numbers with $a_0 \geq 0$ and $a_{n+1} = a_0 \cdots a_n + 4$ for $n \geq 0$. Prove that*

$$a_n - \sqrt[4]{(a_{n+1}+1)(a_n^2-1) - 4} = 1, \quad n \geq 1.$$

Solution The recurrence relation can be written as $a_{n+1} = (a_n - 4)a_n + 4$, hence $a_{n+1} + 1 = a_n^2 - 4a_n + 5$. From here, one can write

$$\begin{aligned}(a_{n+1}+1)(a_n^2+1) - 4 &= (a_n^2 - 4a_n + 5)(a_n^2+1) - 4 \\ &= a_n^4 - 4a_n^3 + 6a_n^2 - 4a_n + 1 \\ &= (a_n - 1)^4.\end{aligned}$$

After taking the fourth root and subtracting, we obtain

$$a_n - \sqrt[4]{(a_{n+1}+1)(a_n^2+1) - 4} = 1, \quad n \geq 1.$$

Problem 3 (Ángel Plaza and Sergio Falcón, 11920, AMM-2016) *For a positive integer k, let $\langle F_k\rangle$ be the sequence defined by the initial condition $F_{k,0} = 0$, $F_{k,1} = 1$, and the recurrence relation $F_{k,n+1} = kF_{k,n} + F_{k,n-1}$. Find a closed form for*

$$\sum_{i=0}^{n} \binom{2n+1}{i} F_{k,2n+1-i}.$$

Note: The kth sequence $(F_{k,n})_{n>0}$ is also known as the k-Fibonacci sequence.

Solution Consider the Fibonacci polynomial $f_n(x,s)$ defined by

$$f_n(x,s) = xf_{n-1}(x,s) + sf_n(x,s), \quad f_0(x,s) = 0, \quad f_1(x,s) = 1.$$

Denote the roots of the characteristic equation $z^2 - xz - s = 0$ by

$$\alpha = \frac{x + \sqrt{x^2+4s}}{2}, \quad \beta = \frac{x - \sqrt{x^2+4s}}{2}.$$

Since we have $\alpha^2 = x\alpha + s$ and $\beta^2 = x\beta + s$, we can derive by induction the well-known Binet formula giving $f_n(x, s) = (\alpha^n - \beta^n)/(\alpha - \beta)$.

Since $s = -\alpha\beta$, we have

$$\alpha + \frac{s}{\alpha} = -\left(\beta + \frac{s}{\beta}\right) = \alpha - \beta = x + \sqrt{x^2 + 4s}.$$

By the binomial formula, we obtain

$$\sum_{i=0}^{n} \binom{2n+1}{i} s^i f_{2n+1-2i}(x, s) = \frac{1}{\alpha - \beta} \sum_{i=0}^{n} \binom{2n+1}{i} s^i (\alpha^{2n+1-2i} - \beta^{2n+1-2i})$$

$$= \frac{1}{\alpha - \beta} \sum_{i=0}^{n} \binom{2n+1}{i} (-1)^i (\alpha^{2n+1-i}\beta^i - \beta^{2n+1-i}\alpha^i)$$

$$= \frac{1}{\alpha - \beta} \sum_{i=0}^{n} \binom{2n+1}{i} \alpha^{2n+1-i}(-\beta)^i + \frac{1}{\alpha - \beta} \sum_{i=0}^{n} \binom{2n+1}{2n+1-i} \alpha^i (-\beta)^{2n+1-i}$$

$$= \frac{(\alpha - \beta)^{2n+1}}{\alpha - \beta} = (x^2 + 4s)^n.$$

The desired result is $(k^2 + 4)^n$, obtained for $x = k$ and $s = 1$.

Problem 4 (Problem 32, ME-1997) *Let $a_0 = 1996$ and $a_{n+1} = \frac{a_n^2}{a_n+1}$ for $n = 0, 1, 2, \ldots$. Prove that $\lfloor a_n \rfloor = 1996 - n$ for $n = 0, 1, 2, \ldots, 999$, where $\lfloor x \rfloor$ is the greatest integer less than or equal to x.*

Solution Note that $a_n > 0$ implies $a_{n+1} > 0$ and

$$a_n - a_{n+1} = 1 - \frac{1}{a_n + 1} > 0,$$

hence the sequence is decreasing, i.e., $a_0 > a_1 > \cdots$. We can also write

$$\begin{aligned} a_n &= a_0 + (a_1 - a_0) + \cdots + (a_n - a_{n-1}) \\ &= 1996 - n + \frac{1}{a_0 + 1} + \cdots + \frac{1}{a_{n-1} + 1} \\ &> 1996 - n. \end{aligned}$$

For $1 \le n \le 999$, this gives

$$\frac{1}{a_0+1}+\cdots+\frac{1}{a_{n-1}-1}<\frac{n}{a_{n-1}+1}<\frac{999}{a_{998}+1}<\frac{999}{1996-998+1}=1.$$

This proves that $[a_n] = 1996 - n$.

Problem 5 (Problem 21, ME-1997) *Show that if a polynomial P satisfies* $P(2x^2-1)=\frac{P(x)^2}{2}$, *then it must be constant.*

Solution Define the sequence $u_1 = 1$, $u_2 = -1$ and $u_n = \sqrt{\frac{u_{n-1}+1}{2}}$ for $n \geq 3$. We have $u_n < u_{n+1} < 1$ for $n \geq 2$ and $P(u_n) = P(u_{n+1})^2/2 - 1$ for $n \geq 1$. Note that $P(u_n) \neq 0$ for $n \geq 1$, as otherwise $P(u_n) = 0$ would imply $P(u_{n-1}), P(u_{n-2}), \ldots, P(u_1)$ are rational. One can check that $P(1) = 1 \pm \sqrt{3}$. Differentiating the equation for P, we get $4xP'(2x^2-1) = P(x)P'(x)$. Since $P(1) \neq 4$, we get $P'(u_1) = P'(1) = 0$. This implies $0 = P'(u_2) = P'(u_3) = \cdots$. Therefore, $P'(x)$ is the zero polynomial, hence it is constant.

Problem 6 (Titu Andreescu, J55, MR-2007) *Let* $a_0 = 1$ *and* $a_{n+1} = a_0 \cdots a_n + 4$, $n \geq 0$. *Prove that* $a_n - \sqrt{a_{n+1}} = 2$, *for* $n \geq 1$.

First Solution Notice first that $a_n > 0$, for $n \geq 0$. Multiplying both sides of $a_n = a_0 \cdots a_{n-1} + 4$ by $a_0 \cdots a_{n-1}$ and adding 4, we get

$$\begin{aligned} a_0 \cdots a_{n-1}a_n + 4 &= (a_0 \cdots a_{n-1})^2 + 4a_0 \cdots a_{n-1} + 4 \\ &= (a_0 \cdots a_{n-1} + 2)^2. \end{aligned}$$

This leads to $(a_n - 4 + 2)^2 = a_{n+1}$, or equivalently, $a_n - 2 = \sqrt{a_{n+1}}$.

Second Solution One can show by induction that $a_{n+1} = (a_n - 2)^2$. Indeed, this is true for $n = 1$. Assuming it is also true for $k = 1, \ldots, n$, we obtain

$$\begin{aligned} a_{n+2} &= a_0 \cdots a_n a_{n+1} + 4 \\ &= (a_{n+1} - 4)a_{n+1} + 4 \\ &= a_{n+1}^2 - 4a_{n+1} + 4 = (a_{n+1} - 2)^2. \end{aligned}$$

The conclusion easily follows.

Third Solution Notice that the sequence $(a_n)_{n\geq 0}$ is strictly increasing. From the relation $a_{n+1} = a_0 \cdots a_n + 4$, one obtains

$$(a_{n+1} - 4)a_{n+1} = (a_0 \cdots a_n)a_{n-1},$$

to which adding 4 and collecting terms, we obtain

$$(a_{n+1} - 2)^2 = a_{n+2},$$

which ends the proof.

Problem 7 (APMO, 2014) *A sequence of real numbers $(a_n)_{n\geq 0}$ is said to be good if the following three conditions hold:*

1. *The value of a_0 is a positive integer.*
2. *For each nonnegative integer i we have $a_{i+1} = 2a_i + 1$ or $a_{i+1} = \frac{a_i}{a_i+2}$.*
3. *There exists a positive integer k such that $a_k = 2014$. Find the smallest positive integer n such that there exists a good sequence $(a_n)_{n\geq 0}$ of real numbers with the property that $a_n = 2014$.*

First Solution Note that

$$a_{i+1} + 1 = 2(a_i + 1) \text{ or } a_{i+1} + 1 = \frac{a_i + a_i + 2}{a_i + 2} = \frac{2(a_i + 1)}{a_i + 2}.$$

Hence

$$\frac{1}{a_{i+1} + 1} = \frac{1}{2} \cdot \frac{1}{a_i + 1} \text{ or } \frac{1}{a_{i+1} + 1} = \frac{a_i + 2}{a_i + a_i + 2} = \frac{1}{2} \cdot \frac{1}{a_i + 1} + \frac{1}{2}.$$

Therefore

$$\frac{1}{a_k + 1} = \frac{1}{2^k} \cdot \frac{1}{a_0 + 1} + \sum_{i=1}^{k} \frac{\varepsilon_i}{2^{k-i+1}}, \tag{8.17}$$

where $\varepsilon_i = 0$ or 1. Multiply both sides by $2^k(a_k + 1)$ and put $a_k = 2014$ to obtain

$$2^k = \frac{2015}{a_0 + 1} + 2015 \left(\sum_{i=1}^{k} \varepsilon_i 2^{i-1} \right),$$

where $\varepsilon_i = 0$ or 1. As $\gcd(2, 2015) = 1$, we have $a_0 + 1 = 2015$ and $a_0 = 2014$. Therefore, we have

$$2^k - 1 = 2015 \left(\sum_{i=1}^{k} \varepsilon_i 2^{i-1} \right),$$

where $\varepsilon_i = 0$ or 1. We now need to find the smallest k such that 2015 divides $2^k - 1$. Since $2015 = 5 \cdot 13 \cdot 31$, from Fermat's little theorem we obtain $5 \mid 2^4 - 1$, $13 \mid 2^{12} - 1$, and $31 \mid 2^{30} - 1$. We also have $\text{lcm}(4, 12, 30) = 60$, hence $5 \mid 2^{60} - 1$, $13 \mid 2^{60} - 1$, and $31 \mid 2^{60} - 1$, which gives $2015 \mid 2^{60} - 1$.

Notice also that $5 \nmid 2^{30} - 1$, hence $k = 60$ is the smallest positive integer such that $2015 \mid 2^k - 1$. To conclude, the smallest positive integer k such that $a_k = 2014$ is $k = 60$.

Second Solution Clearly, all sequence terms are positive real numbers. For each positive integer i, we have $a_i = \frac{a_{i+1}-1}{2}$ or $a_i = \frac{2a_{i+1}}{1-a_{i+1}}$. Since $a_i > 0$ we deduce that

$$a_i = \begin{cases} \frac{a_{i+1}-1}{2} & \text{if } a_{i+1} > 1, \\ \frac{2a_{i+1}}{1-a_{i+1}} & \text{if } a_{i+1} < 1. \end{cases}$$

Thus a_i is uniquely defined from a_{i+1}. Starting from $a_k = 2014$, we can run the sequence backwards until we reach a positive integer, as shown below:

$$\frac{2014}{1}, \frac{2013}{2}, \frac{2011}{4}, \frac{2007}{8}, \frac{1999}{16}, \frac{1983}{32}, \frac{1951}{64}, \frac{1887}{128}, \frac{1759}{256}, \frac{1503}{512}, \frac{991}{1024}, \frac{1982}{33},$$
$$\frac{1949}{66}, \frac{1883}{132}, \frac{1751}{264}, \frac{1487}{528}, \frac{959}{1056}, \frac{1918}{97}, \frac{1821}{194}, \frac{1627}{388}, \frac{1239}{776}, \frac{463}{1552}, \frac{926}{1089}, \frac{1852}{163},$$
$$\frac{1689}{326}, \frac{1363}{652}, \frac{711}{1304}, \frac{1422}{593}, \frac{829}{1186}, \frac{1658}{357}, \frac{1301}{714}, \frac{587}{1428}, \frac{1174}{841}, \frac{333}{1682}, \frac{666}{1349}, \frac{1332}{683},$$
$$\frac{649}{1366}, \frac{1298}{717}, \frac{581}{1434}, \frac{1162}{853}, \frac{309}{1706}, \frac{618}{1397}, \frac{1236}{779}, \frac{457}{1558}, \frac{914}{1404}, \frac{1828}{187}, \frac{1641}{374}, \frac{1267}{748},$$
$$\frac{519}{1496}, \frac{1038}{977}, \frac{61}{1954}, \frac{122}{1893}. \frac{244}{1774}, \frac{488}{1527}, \frac{976}{1039}, \frac{1952}{63}, \frac{1889}{126}, \frac{1763}{252}, \frac{1511}{504}, \frac{1007}{1008}.$$

The next iteration produces $\frac{2014}{1}$, so the answer is $k = 60$ steps.

Problem 8 (Dorin Andrica) *Define $x_n = 2^{2^{n-1}} + 1$ for all $n \geq 1$. Prove that*

1. $x_n = x_1 x_2 \cdots x_{n-1} + 2$, $n \geq 1$;
2. $\gcd(x_k, x_l) = 1$, *for distinct* $k, l \in \mathbb{N}$;
3. x_n *ends in* 7 *for all* $n \geq 3$.

Solution

1. We have

$$x_k = 2^{2^{k-1}} + 1 = 2^{2^{k-2} \cdot 2} + 1 = (x_{k-1} - 1)^2 + 1 = x_{k-1}^2 - 2x_{k-1} + 2,$$

hence

$$x_k - 2 = x_{k-1}(x_{k-1} - 2).$$

Multiplying these relations for $k = 2, \ldots, n$, one obtains

$$x_n - 2 = x_{n-1} \cdots x_2 x_1 (x_1 - 2).$$

Since $x_1 = 3$, it follows that

$$x_n = x_1 x_2 \cdots x_{n-1} + 2. \tag{8.18}$$

An alternative proof can be obtained by the identity

$$\frac{x^{2^{k-1}} - 1}{x - 1} = \prod_{k=1}^{n-1} \left(x^{2^{k-1}} + 1 \right).$$

2. Since the terms $x_n, n \geq 1$ are all odd, from (8.18) one obtains

$$\gcd(x_n, x_1) = \gcd(x_n, x_2) = \cdots = \gcd(x_n, x_{n-1}), \quad n \geq 2,$$

hence $\gcd(x_k, x_l) = 1$, for distinct $k, l \in \mathbb{N}$.
3. Since $x_2 = 5$ and $x_1 x_2 \cdots x_{n-1}$ is odd, by the relation (8.18) it follows that the last digit of x_n is 7 for all integers $n \geq 3$.

Appendix A
Complex Geometry and Number Theory

In this section we expose some basic notions of complex plane geometry and number theory that have been used throughout this book. Basic elements of complex geometry include the triangle inequality, as well as various types of star polygons and directed graphs [6], [54, Chapter 2], and [110, Chapter 1].

Some definitions of number theory concepts are then presented, including Euler's totient function, least common multiple, and greatest common divisor [5, 24]. Basic properties of lcm and gcd are then discussed in the pairwise context, together with results concerning the number of pairs having the same lcm or gcd. A link between lcm and gcd for integer tuples obtained by Vălcan and Bagdasar is given [160]. These are useful for formulating results regarding periodic complex recurrences of second order (Chapter 5) and of arbitrary order (Chapter 6).

We conclude with enumeration theorems related to Stirling numbers [40], and density results established by Weyl [165], Hardy [74], Gologan [69], or Andrica and Buzeţeanu [19], used in the proofs of density results for Horadam sequences and their generalizations (Chapters 5–7).

A.1 Complex Geometry

A.1.1 The Triangle Inequality

Any two complex numbers u and v satisfy the inequalities [6], [168, p. 18]

$$||u| - |v|| \leq |u + v| \leq |u| + |v|.$$

This result allows us to establish inner and outer boundaries for the periodic and stable Horadam orbits. These are plotted in many illustrations.

In general, for $m \geq 3$ and the complex numbers $x_1, \ldots, x_m$ then

D. Andrica, O. Bagdasar, *Recurrent Sequences*, Problem Books in Mathematics,
https://doi.org/10.1007/978-3-030-51502-7

$$|x_1 + \cdots + x_m| \le |x_1| + \cdots + |x_m|.$$

Lower boundaries that only involve $|x_1|, \cdots, |x_m|$ are presented in the monograph of Dragomir [61, Chapter 3], under some assumptions on $|x_1|, \cdots, |x_m|$. However, it may happen that $x_1 + \cdots + x_m = 0$. The graphs illustrating periodic orbits of generalized Horadam sequences in Chapter 4 only display the outer boundary.

A.1.2 Regular Star Polygons and Multipartite Graphs

Star polygons and multipartite graphs with geometric symmetries can be recovered from periodic orbits of complex recurrent sequences.

Definition A.1 (star polygons) For integers k and p the regular star polygon denoted by the Schläfli symbol $\{k/p\}$ can be considered as being constructed by connecting every pth point out of k points regularly spaced in a circular placement (see [54, Chapter 2] and [55, Chapter 6]).

Definition A.2 (multipartite graph) For k a natural number, a k-partite graph W is a graph whose vertex set V is partitioned into k parts, with edges between vertices of different parts only (a 2-partite graph is simply called bipartite), e.g., $G = (V_0, \ldots, V_{k-1}, E)$ with $E \subset \{uv : u \in V_i, v \in V_j, i \neq j\}$. The vertices of $V_i, i = 1, \ldots, k-1$ are called the ith level of G [110, p.4].

A.2 Key Elements of Number Theory

A.2.1 The lcm *and* gcd *of Integer Pairs*

The least common multiple of two natural numbers a and b is often denoted by $\text{lcm}(a, b)$ or $[a, b]$, and is the smallest number divisible by both a and b [74, § 5.1, p. 48]. The dual notion is the greatest common divisor, denoted by $\gcd(a, b)$ or (a, b), which is the largest number that divides both a and b.

Assume that $a = p_1^{a_1} p_2^{a_2} \cdots p_k^{a_k}$, $b = p_1^{b_1} p_2^{b_2} \cdots p_k^{b_k}$, where $p_1 < p_2 < \ldots < p_k$ are primes and a_i, b_i are nonnegative integers. The following identities hold:

$$\gcd(a, b) = p_1^{\min(a_1, b_1)} p_2^{\min(a_2, b_2)} \cdots p_k^{\min(a_k, b_k)}$$
$$\text{lcm}(a, b) = p_1^{\max(a_1, b_1)} p_2^{\max(a_2, b_2)} \cdots p_k^{\max(a_k, b_k)}.$$

This ensures that

$$a \cdot b = \text{lcm}(a, b) \cdot \gcd(a, b) = p_1^{a_1+b_1} p_2^{a_2+b_2} \cdots p_k^{a_k+b_k}. \tag{A.1}$$

Numerous properties and results regarding these notions can be found in the literature.

For $n = p_1^{n_1} p_2^{n_2} \cdots p_r^{n_r}$, then the lcm and gcd can be expressed by number of ordered pairs (a, b) having the same lcm n is

$$|\{(a, b) :\ \text{lcm}\,(a, b) = n\}| = (2n_1 + 1)(2n_2 + 1) \cdots (2n_r + 1), \tag{A.2}$$

where $|\ |$ represents the cardinality of a set [8]. If n is square-free we have $3^{\omega(n)}$ [57], where $\omega(n)$ denotes the number of prime divisors of n.

The integer sequence whose nth term is (A.2) is indexed in the Online Encyclopedia of Integer Sequences (OEIS) [157] as A048691.

The number of relatively prime ordered pairs (a, b) with same lcm n is (see [153])

$$|\{(a, b) :\ \text{gcd}\,(a, b) = 1,\ \text{lcm}\,(a, b) = n\}| = 2^{\omega(n)}.$$

Many related notions are linked to various number sequences indexed in the OEIS.

A.2.2 *The* lcm *and* gcd *of Integer Tuples*

The lcm and gcd can be defined for k-tuple of integers $a_1, \ldots, a_k > 0$, where $k \geq 2$. Later on we extend results like (A.2) for general tuples of integers.

The number of k-tuples of positive integers having the same lcm n is

$$\text{LCM}(n; k) = |\{(a_1, \ldots, a_k) :\ \text{lcm}\,(a_1, \ldots, a_k) = n\}|.$$

Some identities and inequalities involving the above arithmetic function are presented. A number of sequences indexed in the OEIS are obtained from particular instances of this function. A detailed list is given in Section 5.3.

The number of ordered k-tuples with gcd d and lcm n is defined by

$$\text{GL}(d, n; k) = |\{(a_1, \ldots, a_k) :\ \text{gcd}\,(a_1, \ldots, a_k) = d,\ \text{lcm}\,(a_1, \ldots, a_k) = n\}|,$$

and certain properties of this function are analyzed. Other enumeration functions for increasing and strictly increasing integer tuples will be considered in Chapter 5

$$\text{LCM}^{\leq}(k, n) = |\{(a_1, \ldots, a_k) :\ [a_1, \ldots, a_k] = n,\ 1 \leq a_1 \leq \cdots \leq a_k \leq n\}|;$$

$$\text{LCM}^{<}(k, n) = |\{(a_1, \ldots, a_k) :\ [a_1, \ldots, a_k] = n,\ 1 \leq a_1 < \cdots < a_k \leq n\}|.$$

A.2.3 Links Between the lcm and gcd of Integer Tuples

A link between lcm and gcd of k-tuples of integers was proved by Vălcan and Bagdasar [160]. An inclusion–exclusion derivation is suggested in [68].

Theorem A.1 *Let $k \geq 2$ and $a_1, \ldots, a_k > 0$ be integers. We have*

$$\operatorname{lcm}(a_1, a_2, \ldots, a_k) = \frac{\prod\limits_{1 \leq i_1 < \cdots < i_u \leq k} \gcd\left(a_{i_1}, \ldots, a_{i_u}\right)}{\prod\limits_{1 \leq i_1 < \cdots < i_v \leq k} \gcd\left(a_{i_1}, \ldots, a_{i_v}\right)}, \tag{A.3}$$

where u is odd and v is even.

The dual of this theorem can be written as follows.

Theorem A.2 *Let $k \geq 2$ and $a_1, \ldots, a_k > 0$ be integers. We have*

$$\gcd(a_1, a_2, \ldots, a_k) = \frac{\prod\limits_{1 \leq i_1 < \cdots < i_u \leq k} \operatorname{lcm}\left(a_{i_1}, \ldots, a_{i_u}\right)}{\prod\limits_{1 \leq i_1 < \cdots < i_v \leq k} \operatorname{lcm}\left(a_{i_1}, \ldots, a_{i_v}\right)}, \tag{A.4}$$

where u is odd and v is even.

Proof If a prime p has multiplicities $m_1, \ldots, m_k$ in $a_1, \ldots, a_k$, then the relation (A.3) reduces to

$$\begin{aligned}\max(m_1, \ldots, m_k) = & \sum_{1 \leq i_1 < \ldots < i_u \leq n} \min(m_{i_1}, \ldots, m_{i_u}) \\ & - \sum_{1 \leq i_1 < \ldots < i_v \leq n} \min(m_{i_1}, \ldots, m_{i_v}),\end{aligned}$$

where u is odd and v is even. To this end one just need to count the terms in the two sides. This argument can also be checked using an inclusion–exclusion principle. □

In particular, for $k = 3$ the formulae (A.3) and (A.4) produce the identities

$$\operatorname{lcm}(a_1, a_2, a_3) = \frac{a_1 \cdot a_2 \cdot a_3 \cdot \gcd(a_1, a_2, a_3)}{\gcd(a_1, a_2) \cdot \gcd(a_1, a_3) \cdot \gcd(a_2, a_3)},$$

and the dual relation

$$\gcd(a_1, a_2, a_3) = \frac{a_1 \cdot a_2 \cdot a_3 \cdot \operatorname{lcm}(a_1, a_2, a_3)}{\operatorname{lcm}(a_1, a_2) \cdot \operatorname{lcm}(a_1, a_3) \cdot \operatorname{lcm}(a_2, a_3)}.$$

A.2.4 Euler's Totient Function

Euler's totient is one of the arithmetic functions which feature in number theoretic results and algorithms. For an integer $n \in \mathbb{N}$, $\varphi(n)$ represents the number of integers $1 \leq k \leq n$ relatively prime with n.

If is known that if p is prime and $\gcd(m, n) = 1$ and $k > 1$, then

$$\begin{aligned} \varphi(p) &= p - 1 \\ \varphi(p^k) &= p^{k-1}(p-1) \\ \varphi(mn) &= \varphi(m)\varphi(n). \end{aligned} \tag{A.5}$$

If the factorization of n is $n = p_1^{a_1} p_2^{a_2} \cdots p_k^{a_k}$, the following identity holds:

$$\varphi(n) = n\left(1 - \frac{1}{p_1}\right)\left(1 - \frac{1}{p_2}\right) \cdots \left(1 - \frac{1}{p_k}\right).$$

The following valid identity is useful in what follows.

Proposition A.1 *For any positive integers $a, b \in \mathbb{N}$ one has*

$$\varphi(gcd(a,b)) \cdot \varphi(lcm(a,b)) = \varphi(a) \cdot \varphi(b). \tag{A.6}$$

Proof Denote $d = \gcd(a, b)$ and the numbers a', b' such that $a = da'$ and $b = db'$. By (A.1), one has $\operatorname{lcm}(a, b) = da'b'$. Notice that any two of the numbers d, a', b' are relatively prime. By the multiplicity of φ (A.5), (A.6) becomes

$$\varphi(d) \cdot \varphi(da'b') = \varphi(d)\varphi(a') \cdot \varphi(d)\varphi(b') = \varphi(a)\varphi(b).$$

□

A.2.5 The "Stars and Bars" Argument

The stars and bars argument is used to prove certain combinatorial identities. There are two basic versions of this argument. In the first one, the question is to find the number of k-tuples of positive integers whose sum is n, for given numbers n and k. In the second, the number of k-tuples of nonnegative integers with the same property. It is not too difficult to prove the following results.

Let n and k be nonnegative integers. The following results hold.

Problem 1 *The number of k-tuples of positive integers with sum n is*

$$S_+^*(n,k) = \binom{n-1}{k-1}.$$

Diagram (A.7) *shows a feasible configuration for* $n = 9$ *and* $k = 3$.

$$* * * * * * * * * \qquad\qquad * \mid * * \mid * * * \mid * * * \tag{A.7}$$

To count the number of configurations, one needs to place $(k-1)$ *bars in the* $(n-1)$ *gaps between the n stars. This is equal to the number of* $(k-1)$*-element subsets of a set with* $n-1$ *elements. For the example above, the formula gives* $\binom{9-1}{3-1} = 36$.

Problem 2 *The number of k-tuples of nonnegative integers with sum n is*

$$S^+(n,k) = \binom{n+k-1}{k-1}.$$

Diagram (A.8) *shows a feasible configuration for* $n = 9$ *and* $k = 4$.

$$* * * * * * * * * \qquad\qquad * \mid * * \mid \mid * * * * * * \mid \tag{A.8}$$

To count the number of configurations, one has a total of $n + k - 1$ *objects (n stars and* $k-1$ *bars). Any choice of* $k-1$ *spaces for the bars determines the configuration. For the example above, the formula gives* $\binom{8+3-1}{3-1} = 55$.

For more details and examples one may consult [1, 5, 8, 64], or [68].

A.2.6 Partitions of Numbers and Stirling Numbers

The Stirling numbers of the second kind are usually denoted by $S(n,k)$ (as discussed by Knuth in [99], the notation was first used by Richard Stanley) and count the number of ways to partition a set of n labeled objects into k nonempty unlabeled subsets. Clearly, the following identities hold $S(n,n) = S(n,1) = 1$.

Recurrence Stirling numbers of the second kind obey the recurrence

$$S(n+1,k) = kS(n,k) + S(n,k-1).$$

To prove this note that a given partition of the $n+1$ objects into k nonempty subsets, may or may not contain the $(n+1)$th object as a singleton.

This element is a singleton for $S(n,k-1)$ configurations, as the remaining n objects are partitioned into the available $k-1$ subsets.

Otherwise, the $(n+1)$th object belongs to a subset containing other objects. This can happen in $kS(n,k)$ ways, as all objects other than the $(n+1)$th are partitioned into k subsets, and then there are k choices for inserting object $n+1$. The desired results is obtained by summation.

Explicit formula Stirling numbers of second kind are given by [151]

$$S(n,k) = \frac{1}{k!}\sum_{j=0}^{k}(-1)^{k-j}\binom{k}{j}j^n.$$

Identities

1. A simple identity is $S(n, n-1) = \binom{n}{2}$.
   To prove this result, note that n elements can be divided into $n-1$ sets only if one set is of size 2, while the rest $n-2$ sets are of size 1. Each of these configurations is fully determined by the choice of these two elements.
2. Another identity is $S(n,2) = 2^{n-1} - 1$.
   There are 2^n ordered pairs of complementary subsets A and B. We can discard the two cases when one is empty and get $2^n - 2$ ordered pairs of subsets. To obtain unordered pairs, the last number was divided by 2.
   Another explicit expansion of the recurrence relation is known.

$$S(n,2) = \frac{\frac{1}{1}(2^{n-1} - 1^{n-1})}{0!}$$

$$S(n,3) = \frac{\frac{1}{1}(3^{n-1} - 2^{n-1}) - \frac{1}{2}(3^{n-1} - 1^{n-1})}{1!}$$

$$S(n,4) = \frac{\frac{1}{1}(4^{n-1} - 3^{n-1}) - \frac{2}{2}(4^{n-1} - 2^{n-1}) + \frac{1}{3}(4^{n-1} - 1^{n-1})}{2!}$$

$$S(n,5) = \frac{\frac{1}{1}(5^{n-1} - 4^{n-1}) - \frac{3}{2}(5^{n-1} - 3^{n-1}) + \frac{3}{3}(5^{n-1} - 2^{n-1}) - \frac{1}{4}(5^{n-1} - 1^{n-1})}{3!}.$$

Similar relations hold for $S(n,k)$ when $k \geq 6$.

A.2.7 Linear (In)dependence and Density Results

Here we present some useful linear independence and density results. The notations $\lfloor x \rfloor = \max\{m \in \mathbb{Z} : m \leq x\}$ and $\{x\} = x - \lfloor x \rfloor$ are used for the (resp.) floor and fractional part of x. Clearly, the latter is periodic and $\{x+1\} = \{x\}$ for $x \in \mathbb{R}$.

Definition A.3 The numbers and $x_1, \ldots, x_k \in \mathbb{R}$, $k \geq 1$ are called linearly dependent over $\mathbb{Q}$ (or $\mathbb{Z}$) if there are coefficients $p_1, \ldots, p_k \in \mathbb{Q}$, such that

$$a_1x_1 + a_2x_2 + \cdots + a_kx_k = 0, \quad \text{and} \quad (a_1, \ldots, a_k) \neq (0, \ldots, 0). \tag{A.9}$$

If (A.9) only holds when $(a_1, \ldots, a_k) = (0, \ldots, 0)$, then the numbers $x_1, \ldots, x_k$ are called linearly independent.

It is a simple exercise to show that if numbers $x_1, \ldots, x_k$ are linearly dependent over rationals, they are also linearly dependent over integers.

We provide the following illustrative example of linear independence.

Proposition A.2 *For any prime number p, the numbers $1, \sqrt[3]{p}, \sqrt[3]{p^2}$ are linearly independent over $\mathbb{Z}$.*

Proof Assuming that the above triplet is linearly dependent over $\mathbb{Q}$, by (A.9) we can find the coefficients a_0, a_1, a_2 with the property

$$a_0 + a_1\sqrt[3]{p} + a_2\sqrt[3]{p^2} = 0. \tag{A.10}$$

We may assume without loss of generality that a_0, a_1, a_2 are relatively prime

$$\gcd(a_0, a_1, a_2) = 1. \tag{A.11}$$

Simple computations show that

$$(x + y + z)^3 = x^3 + y^3 + z^3 + 3(x + y + z)(xy + yz + zx) - 3xyz,$$

which ensures that whenever $x + y + z = 0$, the following relation holds:

$$x^3 + y^3 + z^3 = 3xyz.$$

Applying the latter identity to (A.10), one obtains

$$a_0^3 + a_1^3\, p + a_2^3\, p^2 = 3a_0a_1a_2\, p. \tag{A.12}$$

As p is prime, p is also a divisor of a_0^3, hence of a_0 itself. Dividing (A.12) by p, we can show that p also divides a_1. Using the same argument once more, one can prove that p divides a_2, in contradiction with (A.11). We conclude that numbers $1, \sqrt[3]{p}, \sqrt[3]{p^2}$ are linearly independent over $\mathbb{Z}$. □

For $k = 1$, the linear independence of 1 and x_1 implies that $x_1 \in \mathbb{R} \setminus \mathbb{Q}$, and one obtains the well-known lemma of Kronecker [74, Theorem 339], [69].

Theorem A.3 *If x is irrational, then $\{nx\}$ is dense in the interval $[0, 1]$.*

The following stronger result is a consequence of a property discovered by Weyl [165], [74, Theorem 445] (apparently also by Sierpínski and Bohl at about the same time) holds

Theorem A.4 *If x is irrational, then $\{nx\}$ are uniformly distributed in* $[0, 1]$.

The result of Weyl is given in a more general form in [19, Lemma 3.2]

Theorem A.5 *Let $P(X) = a_p X^p + \cdots + a_1 X + a_0 \in \mathbb{R}[X]$ be a polynomial such that at least one of the coefficients $a_p, \ldots, a_1$ is irrational. Then*

$$\lim_{N\to\infty} \frac{1}{N} \sum_{0}^{N} e^{2\pi i\, P(n)} = 0.$$

This theorem has numerous applications in approximation theory and probabilities. The density and uniformity results are used in what follows to prove the density of sequence terms and the uniform distribution of the argument of certain Horadam sequences.

Theorem A.3 can be found under the following reformulation:

Theorem A.6 *If x is irrational, the set*

$$A = \{n + mx : m \in \mathbb{N}, \quad n \in \mathbb{Z}\}$$

is dense everywhere in $\mathbb{R}$.

A generalization was proposed by Andrica and Buzeţeanu [19, Lemma 3.2].

Theorem A.7 *Let $s > 0$, $a \geq 0$, $b \geq 0$ be integer numbers and x an irrational number. Then the set $A = \{n - mx : m \in \mathbb{N}, \quad n \in \mathbb{Z}, n = a \pmod s), m = b \pmod s))\}$ is dense everywhere in $\mathbb{R}$.*

The equivalence between the above density results is based on the periodicity of $\{x\}$. In [19], the authors formulated several types of density results, involving T-relatively periodic functions defined as

Definition A.4 Let $P \in \mathbb{R}[X]$ and $T \in \mathbb{R} \setminus \{0\}$. A function $f : \mathbb{R} \to \mathbb{R}$ is T-relatively periodical with respect to P if $f(P(n) + mT) = f(P(n))$.

Although this class of functions seems to be larger than the class of periodical functions of period T, the authors prove that the two classes coincide for $P \in \mathbb{Q}[X]$, T irrational, and f continuous.

A multi-dimensional version of Kronecker's lemma will also be required.

Theorem A.8 ([74, Theorem 442]) *If $1, x_1, x_2, \ldots, x_k$ are linearly independent (over $\mathbb{N}$), $\alpha_1, \alpha_2, \ldots, \alpha_k$, and N and ε are positive, then there are integers $n > N$, $p_1, \ldots, p_k$ such that*

$$|nx_m - p_m - \alpha_m| < \varepsilon, \quad m = 1, \ldots, k.$$

Another form of this theorem is the following.

Theorem A.9 ([74, Theorem 443]) *If* $1, x_1, x_2, \ldots, x_k$ *are linearly independent (over* $\mathbb{N}$*), then the set of points*

$$(\{nx_1\},\ \{nx_2\}, \ldots,\ \{nx_k\}),$$

is dense in the unit k-dimensional hypercube.

The following result illustrates properties of linearly dependent triples, which will be used in the classification of Horadam stable orbits.

Proposition A.3 *Let* $x_1, x_2 \in \mathbb{R}$. *If* $(1, x_1, x_2)$ *are linearly independent over* $\mathbb{Q}$ *(or* $\mathbb{Z}$*), then sequence* $(\{nx_1\}, \{nx_2\})$ *is dense within* $[0, 1] \times [0, 1]$. *Otherwise,* $(1, x_1, x_2)$ *are linearly dependent over* $\mathbb{Q}$ *(or* $\mathbb{Z}$*), hence one can find* a_0, a_1, a_2 *with the property*

$$a_0 + a_1x_1 + a_2x_2 = 0.$$

The following cases are possible:

1. $x_1, x_2 \in \mathbb{Q}$ $(a_0 = -a_1x_1 - a_2x_2)$.
   In this case, the sequence $(\{nx_1\}, \{nx_2\})$ *is periodic.*
2. $x_1 = p/k \in \mathbb{Q}$ *(irreducible),* $x_2 \in \mathbb{R} \setminus \mathbb{Q}$ $(a_1 = 0,\ a_0 = -a_2x_2)$.
   Here the sequence $(\{nx_1\}, \{nx_2\})$ *is dense within* $\{0, \frac{1}{k}, \ldots, \frac{k-1}{k}\} \times [0, 1]$.
3. $x_1, x_2 \in \mathbb{R} \setminus \mathbb{Q}$ $(x_2 = -\frac{a_1}{a_2}x_1 - \frac{a_0}{a_2} = b_1x_1 + b_0)$.
   In this case the sequence $(\{nx_1\}, \{nx_2\})$ *is dense within the graph of the function* $f : [0, 1] \to \mathbb{R}^2$ *defined by* $f(x) = (x, b_1x + b_0)$.
   Important instances are $b_1 = 0$ *(i.e.,* $x_2/x_1 \in \mathbb{Q}$*) or* $b_1 = 1$ *(i.e.,* $x_2 - x_1 \in \mathbb{Q}$*).*

Remark A.1 The distinct cases above will result in distinct types of Horadam orbits, which will be finite (dimension zero), dense within distinct circles or other closed curves (dimension one), or dense within certain annuli in the complex plane (dimension two).

A.3 Numerical Implementation of LRS General Terms

The methods presented in Chapter 6 for generating the first terms of the generalized Horadam sequence can be refined to allow the direct computation of the sequence terms with a given index set $I = \{i_1, \ldots, i_N\}$. We detail below the computer algorithm, using matrix operations implemented in Matlab®.

A.3.1 Distinct Roots

In this case matrix $V_{N,m}(z_1,\ldots,z_m)$ in (6.72) can be replaced by the matrix $V_{I,m}(z_1,\ldots,z_m)$ defined for each set $I=\{i_1,\ldots,i_N\}$ as

$$V_{I,m}(z_1,\ldots,z_m)=\begin{pmatrix} z_1^{i_1} & z_2^{i_1} & \cdots & z_m^{i_1}\\ \vdots & \vdots & & \vdots\\ z_1^{i_N} & z_2^{i_N} & \cdots & z_m^{i_N}\end{pmatrix}. \tag{A.13}$$

In Matlab® syntax, matrix (A.13) can be implemented as

$$V_{I,m}(z_1,\ldots,z_m)=[\text{ones}(N,1)*(z_1,\ldots,z_m)].\wedge[(i_1,\ldots,i_N)'*\text{ones}(1,m)],$$

where $\mathbf{z}'$ denotes the transpose of vector $\mathbf{z}$.

A.3.2 Equal Roots

In this case matrix $\mathcal{V}_{n,m}(z)$ in (6.82) can be replaced by the matrix $\mathcal{V}_{I,m}(z_1,\ldots,z_m)$ defined for each set $I=\{i_1,\ldots,i_N\}$ as

$$\mathcal{V}_{I,m}(z)=\begin{pmatrix} z^{i_1} & i_1z^{i_1} & \cdots & i_1^{m-1}z^{i_1}\\ z^{i_2} & i_2z^{i_2} & \cdots & i_2^{m-1}z^{i_2}\\ \vdots & \vdots & & \vdots\\ z^{i_N} & i_Nz^{i_N} & \cdots & i_N^{m-1}z^{i_N}\end{pmatrix}. \tag{A.14}$$

The above matrix can be written as

$$\mathcal{V}_{I,m}(z)=\begin{pmatrix} 1 & i_1 & \cdots & i_1^{m-1}\\ 1 & i_2 & \cdots & i_2^{m-1}\\ \vdots & \vdots & & \vdots\\ 1 & i_N & \cdots & i_N^{m-1}\end{pmatrix}.*\begin{pmatrix} z^{i_1} & z^{i_1} & \cdots & z^{i_1}\\ z^{i_2} & z^{i_2} & \cdots & z^{i_2}\\ \vdots & \vdots & & \vdots\\ z^{i_N} & z^{i_N} & \cdots & z^{i_N}\end{pmatrix}, \tag{A.15}$$

where $.*$ is the element-by-element matrix product in Matlab®.

In this case the two matrices in (A.15) can be implemented as

$$\mathcal{V}_{I,m}^1(z)=[(i_1,\ldots,i_N)'*\text{ones}(1,m)].\wedge[\text{ones}(N,1)*(0,1,\ldots,m-1)],$$

$$\mathcal{V}_{I,m}^2(z)=\left[(\text{ones}(N,1)*z).\wedge(i_1,\ldots,i_N)'\right]*\text{ones}(1,m)],$$

where $.\wedge$ denotes the element-by-element power function implemented in Matlab®.

A.3.3 Distinct Roots $z_1, \ldots, z_m$ of Higher Multiplicities $d_1, \ldots, d_m$

In general, for an ordered set of indices $I = \{i_1, \ldots, i_N\}$ one can directly obtain the terms x_n, $n \in I$ of the sequence by considering the matrix

$$\mathcal{W}_{I,d_1,\ldots,d_m}(z_1, \ldots, z_m) = \left(\mathcal{V}_{I,d_1}(z_1)\Big|\cdots\Big|\mathcal{V}_{I,d_m}(z_m)\right),$$

where the matrix components $\mathcal{V}_{I,d_i}(z_i)$, $i = 1, \ldots, m$, are defined in (A.14).

References

1. Albertson, M.O., Hutchinson, J.P.: Discrete Mathematics with Algorithms. Wiley, New York (1988)
2. Alter, R., Kubota, K.K.: Multiplicities of second order linear recurrences. Trans. Am. Math. Soc. **178**, 271–284 (1973)
3. André-Jeannin, R.: Summation of reciprocals in certain second-order recurring sequences. Fibonacci Quart. **35**(1), 68–74 (1997)
4. Andreescu, T., Andrica, D.: 360 Problems for Mathematical Contests. GIL, Zalău (2003)
5. Andreescu, T., Andrica, D.: Number Theory. Structures, Examples, and Problems. Birkhäuser Verlag, Boston/Berlin/Basel (2009)
6. Andreescu, T., Andrica, D.: Complex Numbers from A to ... Z, 2nd edn. Birkhäuser, Boston (2014)
7. Andreescu, T., Andrica, D.: Quadratic Diophantine Equations. Developments in Mathematics. Springer, Cham (2015)
8. Andreescu, T., Zuming, F.: A Path to Combinatorics for Undergraduates: Counting Strategies. Birkhäuser, Basel (2004)
9. Andrejic, V.: On Fibonacci powers. Univ. Beograd. Publ. Elektrotehn. Fak. Ser. Mat. **17**, 38–44 (2006)
10. Andrica, D.: On a combinatorial sum. Gazeta Mat. **5**, 158 (1989)
11. Andrica, D., Bagdasar, O.: On cyclotomic polynomial coefficients. Malays. J. Math. Sci. In: Proceedings of "Groups, Group Rings, and Related Topics - 2017" (GGRRT 2017), 19–22 November 2017. Khorfakan, UAE (to appear)
12. Andrica, D., Bagdasar, O.: The Cauchy integral formula with applications to polynomials, partitions and sequences. In: Proceedings of the XVth International Conference on Mathematics and its Applications. Timişoara, November 1–3, 2018, pp. 12–25. Editura Politehnica, Timişoara (2019)
13. Andrica, D., Bagdasar, O.: On some results concerning the polygonal polynomials. Carpathian J. Math. **35**, 1–12 (2019)
14. Andrica, D., Bagdasar, O.: A new formula for the coefficients of Gaussian polynomials. An. Şt. Univ. Ovidius Constanţa **27**(3), 25–36 (2019)
15. Andrica, D., Bagdasar, O.: Remarks on a family of complex polynomials. Appl. Anal. Discr. Math. **13**, 605–618 (2019)
16. Andrica, D., Bagdasar, O.: On some new arithmetic properties of the generalized Lucas sequences. Mediterr. J. Math. Med. J. Math., to appear (2021)

D. Andrica, O. Bagdasar, *Recurrent Sequences*, Problem Books in Mathematics,
https://doi.org/10.1007/978-3-030-51502-7

17. Andrica, D., Buzeţeanu, Ş.: The reduction of a second-order linear recurrence and some consequences. G.M. Perfecţionare metodică şi metodologică în matematică şi informatică, **3–4**, 148–152 (1982, in Romanian)
18. Andrica, D., Buzeţeanu, Ş.: On the reduction of the linear recurrence of order r. Fibonacci Quart. **21**(1), 81–84 (1985)
19. Andrica, D., Buzeţeanu, Ş.: Relatively dense universal sequences for the class of continuous periodical functions of period T. Mathematica-L'Analyse Numérique et la Théorie de l'Approximation. **16**, 1–9 (1987)
20. Andrica, D., Marinescu, D.-Ş.: Sequences interpolating some geometric inequalities. Creat. Math. Inform. **28**(1), 9–18 (2019)
21. Andrica, D., Toader, Gh.: On systems of linear recurrences. Itinerant Seminar on Functional Equations, Approximation and Convexity, "Babeş-Bolyai" University, Cluj-Napoca, vol. 7, pp. 5–12 (1986)
22. Andrica, D., Toader, Gh.: On homographic recurrences. Seminar on Mathematical Analysis, "Babeş-Bolyai" University, Cluj-Napoca, vol. 4, pp. 55–60 (1986)
23. Andrica, D., Crişan, V., Al-Thukair, F.: On Fibonacci and Lucas sequences modulo a prime and primality testing. Arab J. Math. Sci. **24**(1), 9–15 (2018)
24. Bagdasar, O.: Concise Computer Mathematics: Tutorials on Theory and Problems. Springer Briefs in Computer Science. Springer, Cham (2013)
25. Bagdasar, O.: On some functions involving the LCM and GCD of integer tuples. Appl. Maths. Inform. Mech. **6**(2), 91–100 (2014)
26. Bagdasar, O., Chen, M.: A Horadam-based pseudo-random number generator. In: Proceedings of 16th UKSim, Cambridge, pp. 226–230 (2014)
27. Bagdasar, O., Larcombe, P.J.: On the characterization of periodic complex Horadam sequences. Fibonacci Quart. **51**(1), 28–37 (2013)
28. Bagdasar, O., Larcombe, P.J.: On the number of complex periodic complex Horadam sequences. Fibonacci Quart. **51**(4), 339–347 (2013)
29. Bagdasar, O., Larcombe, P.J.: On the masked periodicity of Horadam sequences: a generator-based approach. Fibonacci Quart. **55**(4), 332–339 (2013)
30. Bagdasar, O., Larcombe, P.J.: On the characterization of periodic generalized Horadam sequences. J. Differ. Equ. Appl. **20**(7), 1069–1090 (2014)
31. Bagdasar, O., Popa I.-L.: On the geometry of certain periodic non-homogeneous Horadam sequences. Electron. Notes Discrete Math. **56** (2016), 7–13. Proceedings of the 1st IMA TCDM 2016
32. Bagdasar, O., Larcombe, P.J., Anjum, A.: Particular orbits of periodic Horadam sequences. Octogon. Math. Mag. **21**(1), 87–98 (2013)
33. Bagdasar, O., Larcombe, P.J., Anjum, A.: On the structure of periodic complex Horadam sequences. Carpathian J. Math. **32**(1), 29–36 (2016)
34. Bagdasar, O., Hedderwick, E., Popa I.-L.: On the ratios and geometric boundaries of complex Horadam sequences. Electron. Notes Discrete Math. **68** (2018), 63–70. Proceedings of TREPAM 2017
35. Bastida, J.R., DeLeon, M.J.: A quadratic property of certain linearly recurrent sequences. Fibonacci Quart. **19**(2), 144–146 (1981)
36. Bellman, R.: Introduction to Matrix Analysis. McGraw-Hill Book Company, New York/Toronto/London (1960)
37. Berstel, J., Mignotte, M.: Deux propriétés décidables des suites récurrentes linéaires. Bull. Soc. Math. France **104**, 175–184 (1976)
38. Bibak, Kh., Shirdareh, H.: Some trigonometric identities involving Fibonacci and Lucas Numbers. J. Int. Seq. **12** (2009). Article 09.8.4
39. Boole, G.: Calculus of Finite Differences. 5th edn. (1860). Chelsea Publishing, Chelsea (1970)
40. Branson, D.: Stirling numbers and Bell numbers: their role in combinatorics and probability. Math. Sci. **25**, 1–31 (2000)
41. Brânzei, D.: Recurrent Sequences in College. GIL, Zalău (1996, in Romanian)

42. Bruckman, P.S.: On the infinitude of Lucas pseudoprimes. Fibonacci Quart. **32**(2), 153–154 (1994)
43. Buschman, R.G.: Fibonacci numbers, Chebyshev polynomials generalizations and difference equations. Fibonacci Quart. **1**(4), 1–7 (1963)
44. Cahill, N.D., D'Enrico, J.D., Spencer, J.P.: Complex factorizations of the Fibonacci and Lucas numbers. Fibonacci Quart. **41**(1), 13–19 (2003)
45. Carlitz, L.: Generating functions for powers of certain sequences of numbers. Duke Math. J. **29**, 521–537 (1962)
46. Carlitz, L.: Some determinants containing powers of Fibonacci numbers. Fibonacci Quart. **4**(2), 129–134 (1966)
47. Carson, T.R.: Periodic recurrence relations and continued fractions. Fibonacci Quart. **45**(4), 357–361 (2007)
48. Cartan, H.: Elementary Theory of Analytical Functions in One or More Complex Variables. Hermann, Paris (1961, in French)
49. Cerlienco, L., Piras., F.: Powers of a matrix. Boll. Un. Mat. Ital. **6**(2B), 681–690 (1983)
50. Cerlienco, L., Mignotte, M., Piras., F.: Linear Recurrent Sequences: (Algebraic and Arithmetic Properties). Publ. Inst. Rech. Math. Avancée (1984, in French)
51. Challacombe, M., Schwegler, E., Almlöf, J.: Recurrence relations for calculation of the Cartesian multipole tensor. Chem. Phys. Lett. **241**, 67–72 (1995)
52. Cheney, W., Kinkaid, D.: Numerical Mathematics and Computing, 7th edn. Brooks/Cole Cengage Learning, Pacific Grove (2013)
53. Cobzaş, Ş.: Mathematical Analysis (Differential Calculus). Cluj University Press, Cluj Napoca (1997, in Romanian)
54. Coxeter, H.S.M.: Introduction to Geometry, 2nd edn. Wiley, New York (1969)
55. Coxeter, H.S.M.: Regular Polytopes. Courier Corporation, North Chelmsford (1973)
56. Crandall, R., Dilcher, K., Pomerance, C.: A search for Wieferich and Wilson primes. Math. Comp. **66**(5), 433–449 (1997)
57. Crandall, R., Pomerance, C.: Prime Numbers: A Computational Perspective, 2nd edn. Springer, New York (2005)
58. Cuculescu, I.: A simple proof of a formula of Perron. An. Univ. "C. I. Parhon" Bucureşti, Mat. Fiz. **25**, 7–8 (1960, in Romanian)
59. Davis, J.P.: Circulant Matrices. Wiley, New York/Chichester/Brisbane (1979)
60. Deza, M.: On minimal number of terms in representation of natural numbers as a sum of Fibonacci numbers. Fibonacci Quart. **15**(4), 237–238 (1977)
61. Dragomir, S.S.: Advances in Inequalities of the Schwarz, Triangle and Heisenberg Type in Inner Product Spaces. Nova Science Publishers, New York (2007)
62. Everest, G., van der Poorten, A., Shparlinski, I., Ward, T.: Recurrence Sequences. Mathematical Surveys and Monographs, vol. 104. American Mathematical Society, Providence (2003)
63. Fairgrieve, S., Gould, H.W.: Product difference Fibonacci identities of Simson, Gelin-Ces'aro, Tagiuri and generalizations. Fibonacci Quart. **43**(2), 137–141 (2005)
64. Feller, W.: An Introduction to Probability Theory and Its Applications, vol. 1, 3rd edn. Wiley, New York (1968)
65. Finch, S.R.: Mathematical Constants. Cambridge University Press, Cambridge (2003)
66. Fredman, M.L., Tarjan, R.E.: Fibonacci heaps and their uses in improved network optimization algorithms. J. Assoc. Comput. Mach. **34**(3), 596–615 (1987)
67. Garnier, N., Ramaré, O.: Fibonacci numbers and trigonometric identities. Fibonacci Quart. **46**, 1–7 (2008)
68. Gelca, R., Andreescu, T.: Putnam and Beyond. Springer, Cham (2007)
69. Gologan, R.: Aplicaţii ale teoriei ergodice. Editura Tehnică (1989, in Romanian)
70. Graham, I., Kohr, G.: Geometric Function Theory in One and Higher Dimensions. CRC Press, Boca Raton (2003)
71. Guiaşu, S.: Applications of Information Theory: Dynamical Systems, Cybernetical Systems. Ed. Acad. R. S. Romania, Bucharest (1968, in Romanian)

72. Halava, V., Harju, T., Hirvensalo, M.: Positivity of second order linear recurrent sequences. T.U.C.S. Tech. Rep. No. 685, Turku Centre for Computer Science, University of Turku, Finland (2005)
73. Halton, J.: Some properties associated with square Fibonacci numbers. Fibonacci Quart. **5**(4), 347–354 (1967)
74. Hardy, G.H., Wright, E.M.: An Introduction to the Theory of Numbers, 5th edn. Oxford University Press, Oxford (1979)
75. Haukkanen, P.: A note on Horadam's sequence. Fibonacci Quart. **40**, 358–361 (2002)
76. Hellekalek, P.: Good random number generators are (not so) easy to find. Math. Comput. Simul. **46**, 485–505 (1998)
77. Hilton, A.J.W.: On the partition of Horadam's generalized sequences into generalized Fibonacci and generalized Lucas sequences. Fibonacci Quart. **12**(4), 339–345 (1974)
78. Horadam, A.F.: A generalized Fibonacci sequence. Am. Math. Monthly **68**, 455–459 (1961)
79. Horadam, A.F.: Basic properties of a certain generalized sequence of numbers. Fibonacci Quart. **3**(3), 161–176 (1965)
80. Horadam, A.F.: Generating functions for powers of a certain generalised sequence of numbers. Duke Math. J. **32**, 437–446 (1965)
81. Horadam, A.F.: Special properties of the sequence $w_n(a, b; p, q)$. Fibonacci Quart. **5**(5), 424–434 (1967)
82. Horadam, A.F.: Tschebyscheff and other functions associated with the sequence $\{w_n(a, b; p, q)\}$. Fibonacci Quart. **7**(1), 14–22 (1969)
83. Horadam, A.F.: Associated sequences of general order. Fibonacci Quart. **31**(2), 166–172 (1993)
84. Horadam, A.F., Shannon, A.G.: Generalization of identities of Catalan and others. Port. Math. **44**, 137–148 (1987)
85. Hu, H., Sun, Z.-W., Liu, J.-X.: Reciprocal sums of second-order recurrent sequences. Fibonacci Quart. **39**(3), 214–220 (2001)
86. Ivanov, N.V.: Linear Recurrences (2008). Preprint. http://www.mth.msu.edu/~ivanov/Recurrence.pdf
87. Jacobsen, L.: Composition of linear fractional transformations in terms of tail sequences. P. Am. Mat. Soc. **97**(1), 97–104 (1986)
88. Jeffery, T., Pereira, R.: Divisibility Properties of the Fibonacci, Lucas, and Related Sequences, 5 pp. ISRN Algebra, Hindawi (2014). Article 750325
89. Jeske, J.A.: Linear recurrence relations - part I. Fibonacci Quart. **1**(2), 69–74 (1963)
90. Just, E.: Problem E 2367. Am. Math. Monthly **7**, 772 (1972)
91. Kiefer, J.: Sequential minimax search for a maximum. P. Am. Mat. Soc. **4**, 502–506 (1953)
92. Kiliç, E.: The Binet formula, sums and representations of generalized Fibonacci p-numbers. Eur. J. Combin. **29**(3), 701–711 (2008)
93. Kiliç, E., Tan, E.: More general identities involving the terms of $\{W_n(a, b; p, q)\}$. Ars Comb. **93**, 459–461 (2009)
94. Kiliç, E., Tan, E.: On binomial sums for the general second order linear recurrence. Integers Elec. J. Comb. Num. Theory **10**, 801–806 (2010)
95. Kiliç, E., Ulutaş, Y.T., Ömür, N.: A formula for the generating functions of powers of Horadam's sequence with two additional parameters. J. Int. Seq. **14**, 8 pp. (2011). Article 11.5.6
96. Kiss, P., Phong, B.M., Lieuwens, E.: On Lucas pseudoprimes which are products of s primes. In: Philippou, A.N., Bergum, G.E., Horadam, A.F. (eds.) Fibonacci Numbers and Their Applications, vol. 1, pp. 131–139. Reidel, Dordrecht (1986)
97. Knopfmacher, A., Tichy, R.F., Wagner, S., Ziegler, V.: Graphs, partitions and Fibonacci numbers. Discrete Appl. Math. **155**, 1175–1187 (2007)
98. Knuth, D.E.: The Art of Computer Programming, vol. 1 Addison Wesley, Boston (1975)
99. Knuth, D.E.: Two notes on notation. Am. Math. Monthly **99**(5), 403–422 (1992)
100. Knuth, D.E.: The Art of Computer Programming, vol. 3, 2nd edn. Addison Wesley, Boston (2003)

101. Koshy, T.: Fibonacci and Lucas Numbers with Applications. Wiley, Hoboken (2001)
102. Koshy, T.: Pell and Pell-Lucas Numbers with Applications. Springer, New York (2014)
103. Lando, S.K.: Lectures on Generating Functions. Student Mathematical Library, vol. 23. AMS, Providence (2003)
104. Laohakosol, V., Kuhapatanakul, K.: Reciprocal sums of generalized second order recurrence sequences. Fibonacci Quart. **46/47**(4), 316–325 (2009)
105. Larcombe, P.J., Fennessey, E.J.: On Horadam sequence periodicity: a new approach. Bull. Inst. Combin. Appl. **73**, 98–120 (2015)
106. Larcombe, P.J., Fennessey, E.J.: On the phenomenon of masked periodic Horadam sequences. Utilitas Math. **96**, 111–123 (2015)
107. Larcombe, P.J., Bagdasar, O., Fennessey, E.J.: Horadam sequences: a survey. Bull. Inst. Combin. Appl. **67**, 49–72 (2013)
108. Larcombe, P.J., Bagdasar, O., Fennessey, E.J.: On a result of bunder involving Horadam sequences: a proof and generalization. Fibonacci Quart. **51**(2), 174–176 (2013)
109. Larcombe, P.J., Bagdasar, O., Fennessey, E.J.: On a result of bunder involving Horadam sequences: a new proof. Fibonacci Quart. **52**(2), 175–177 (2014)
110. Latapy, M., Phan, T.H.D., Crespelle, C., Nguyen, T.Q.: Termination of multipartite graph series arising from complex network modelling. In: Combinatorial Optimization and Applications. Lecture Notes in Computer Science, vol. 6508, pp. 1–10. Springer, Cham (2010)
111. Lee, J.Z, Lee, J.S.: Some properties of the sequence $\{W_n(a, b; p, q)\}$. Fibonacci Quart. **25**(3), 268–278, 283 (1987)
112. Lehmer, E.: On the infinitude of Fibonacci pseudoprimes. Fibonacci Quart. **2**(3), 229–230 (1964)
113. Mansour, T.: A formula for the generating functions of powers of Horadam's sequence. Aust. J. Comb. **30**, 207–212 (2004)
114. Markouchevitch, A.: Four Courses in Mathematics (Recurrent Sequences). Mir, Moscou (1973, in French)
115. Martin, G.E.: Counting: The Art of Enumerative Combinatorics. Springer, New York (2001)
116. May, R.M.: Simple mathematical models with very complicated dynamics. Nature **261**(5560), 459–467 (1976)
117. McLaughlin, R.: Sequences—some properties by matrix methods. Math. Gaz. **64**, 281–282 (1980)
118. Melham, R.S.: Summation of reciprocals which involve products of terms from generalized Fibonacci sequences. Fibonacci Quart. **38**(4), 294–298 (2000)
119. Melham, R.S.: Summation of reciprocals which involve products of terms from generalized Fibonacci sequences—part II. Fibonacci Quart. **39**(3), 264–267 (2001)
120. Melham, R.S.: A Fibonacci identity in the spirit of Simson and Gelin-Cesàro. Fibonacci Quart. **41**(2), 142–143 (2003)
121. Melham, R.S., Shannon, A.G.: Some congruence properties of generalized second-order integer sequences. Fibonacci Quart. **32**(5), 424–428 (1994)
122. Melham, R.S., Shannon, A.G.: A generalization of the Catalan identity and some consequences. Fibonacci Quart. **33**(1), 82–84 (1995)
123. Mező, I.: Several generating functions for second-order recurrence sequences. J. Int. Seq. **12**, 16 pp. (2009). Article 09.3.7
124. Milne-Thompson, L.M.: The Calculus of Finite Differences. Macmillan and Company, New York (1933)
125. Mitrinovic, D.S., Sándor, J., Crstici, B.: Handbook of Number Theory. Kluwer, Dordrecht (1995)
126. Morgado, J.: Note on some results of A.F. Horadam and A.G. Shannon concerning a Catalan's identity on Fibonacci numbers. Port. Math. **44**, 243–252 (1987)
127. Muir, T.: The Theory of Determinants in the Historical Order of Development, vol. 1. Dover, New York (1960)
128. Muntean, I., Popa, D.: The Method of Recurrent Sequences. GIL, Zalău (1995, in Romanian)
129. Mureşan, M.: A Concrete Approach to Classical Analysis. Springer, Berlin (2008)

130. Newell, A.C., Pennybacker, M.: Fibonacci patterns: common or rare? Procedia IUTAM **9**, 86–109 (2013)
131. Noonea, C.J., Torrilhonb, M., Mitsosa, A.: Heliostat field optimization: a new computationally efficient model and biomimetic layout. Solar Energy **86**(2), 792–803 (2012)
132. Oohama, Y.: Performance analysis of the internal algorithm for random number generation based on number systems. IEEE Trans. Inform. Theory **57**(3), 1177–1185 (2011)
133. Ouaknine, J., Worrell, J.: Ultimate Positivity is decidable for simple linear recurrence sequences (2013). CoRR: abs/1309.1914
134. Ouaknine, J., Worrell, J.: On the positivity problem for simple linear recurrence sequences. Proc. ICALP'14. CoRR: abs/1309.1550
135. Ouaknine, J., Worrell, J.: Positivity problems for low-order linear recurrence sequences. In: Proc. SODA'14. ACM-SIAM (2014)
136. Panneton, F., L'Ecuyer, P., Matsumoto, M.: Improved long-period generators based on linear recurrences modulo 2. ACM Trans. Math. Software **32**, 1–16 (2006)
137. Pawar, A.: Mandelbrot set and Julia set. MATLAB Central File Exchange. Retrieved February 26, 2020. https://www.mathworks.com/matlabcentral/fileexchange/24740-mandelbrot-set-and-julia-set
138. Raab, J.A.: A generalization of the connection between the Fibonacci sequence and Pascal's triangle. Fibonacci Quart. **1**(3), 21–31 (1963)
139. Rabinowitz, S.: Algorithmic manipulation of second-order linear recurrences. Fibonacci Quart. **37**(2), 162–177 (1999)
140. Reiter, C.A.: Exact Horadam numbers with a Chebyshevish accent. Vector **16**, 122–131 (1999)
141. Robinson, D.W.: The rank and period of a linear recurrent sequence over a ring. Fibonacci Quart. **14**(3), 210–214 (1976)
142. Rotkiewicz, A.: Lucas and Frobenius pseudoprimes. Ann. Math. Sil. **17**, 17–39 (2003)
143. Roy, S.: What's the next Fibonacci number? Math. Gaz. **64**(425), 189–190 (1980)
144. Santana, S.F., Diaz-Barrero, J.L.: Some properties of sums involving Pell numbers. Miss. J. Math. Sci. **18**(1), 33–40 (2006)
145. Sasu, B., Sasu, S.L.: Discrete Dynamical Systems. Editura Politehnică, Timişoara (2013, in Romanian)
146. Serway, R.A., Jewett, J.W.: Physics for Scientists and Engineers, 6th edn. Thomson Brooks/Cole, Pacific Grove (2004)
147. Shannon, A.G.: Generalized Fibonacci numbers as elements of ideals. Fibonacci Quart. **17**(4), 347–349 (1979)
148. Shannon, A.G.: A generalization of Hilton's partition of Horadam's sequences. Fibonacci Quart. **17**(4), 349–357 (1979)
149. Shannon, A.G., Horadam, A.F.: Some properties of third-order recurrence relations. Fibonacci Quart. **10**(2), 135–145 (1972)
150. Shannon, A.G., Horadam, A.F.: Special recurrence relations associated with the sequence $\{w_n(a, b; p, q)\}$. Fibonacci Quart. **17**(4), 294–299 (1979)
151. Sharp, H.: Cardinality of finite topologies. J. Combin. Theory **5**, 82–86 (1968)
152. Silvester, J.R.: Fibonacci properties by matrix methods. Math. Gaz. **63**(425), 188–191 (1979)
153. Sitaramachandra, R.R., Suryanabayana, D.: The number of pairs of integers with L.C.M. $\leq x$. Arch. Math. **21**, 490–497 (1970)
154. Stănică, P.: Generating functions, weighted and non-weighted sums for powers of second-order recurrence sequences. Fibonacci Quart. **41**(4), 321–333 (2003)
155. Sury, B.: On grand-aunts and Fibonacci. Math. Gaz. **92**, 63–64 (2008)
156. Sury, B.: Trigonometric expressions for Fibonacci and Lucas numbers. Acta Math. Univ. Comeniae (N.S.) **79**, 199–208 (2010)
157. The On-Line Encyclopedia of Integer Sequences. OEIS Foundation Inc. (2011). https://oeis.org
158. Toader, Gh.: Generalized double sequences. Rev. Anal. Numér. Théor. Approx. **16**, 81–85 (1987a)

159. Udrea, G.: A note on the sequence $(W_n)_{n\geq 0}$ of A.F. Horadam. Port. Math. **53**, 143–155 (1996)
160. Vălcan, D., Bagdasar, O.: Generalizations of some divisibility relations in $\mathbb{N}$. Creat. Math. Inform. **18**(1), 92–99 (2009)
161. Verhulst, P.-R.: Notice sur la loi que la population suit dans son accroissement. Correspondance mathématique et physique, 113–121 (1838)
162. Vince, A.: Period of a linear recurrence. Acta Arith. **39**, 303–311 (1981)
163. Vogel, H.: A better way to construct the sunflower head. Math. Biosci. **44**, 179–189 (1979)
164. Vorobiev, N.N.: Fibonacci Numbers. Birkhäuser Verlag, Basel/Boston (2002)
165. Weyl, H.: Über die gleichverteilung von zahlen mod. eins. Math. Ann. **77**(3), 313–352 (1916)
166. Wilf, H.: Generating Functionology. Academic, New York (1994)
167. Wu, H.: Complex factorizations of the Lucas sequences via matrix methods. J. Appl. Math. 6 pp. (2014). Article ID 387675. http://dx.doi.org/10.1155/2014/387675
168. Wunsch, D.A.: Complex Variables with Applications, 3rd edn. Pearson, London (2004)
169. Yazlik, Y., Taskara, N.: A note on generalized k-Horadam sequence. Comp. Math. Appl. **63**, 36–41 (2012)
170. Zeitlin, D.: Generating functions for products of recursive sequences. Trans. Am. Math. Soc. **116**, 300–315 (1965)
171. Zeitlin, D.: Power identities for sequences defined by $W_{n+2} = dW_{n+1} - cW_n$. Fibonacci Quart. **3**(4), 241–256 (1965)
172. Zeitlin, D.: On determinants whose elements are products of recursive sequences. Fibonacci Quart. **8**(4), 350–359 (1970)
173. Zeitlin, D.: General identities for recurrent sequences of order two. Fibonacci Quart. **9**(4), 357–388 (1971)
174. Zenkevich, I.G.: Recurrent relations for the approximation of the physicochemical constants of homologues. Russ. J. Phys. Chem. A **82**(5), 695–703 (2008)
175. Zhang, W.: Some identities involving the Fibonacci numbers. Fibonacci Quart. **35**(3), 225–229 (1997)
176. Zhang, Z.: Some identities involving generalized second-order integer sequences. Fibonacci Quart. **35**(3), 265–268 (1997)
177. Zhang, Z., Liu, M.: Generalizations of some identities involving generalized second-order integer sequences. Fibonacci Quart. **36**(4), 327–328 (1998)

Index

D. Andrica, O. Bagdasar, *Recurrent Sequences*, Problem Books in Mathematics,
https://doi.org/10.1007/978-3-030-51502-7